Sustainable Cosmeceuticals

Sadia Javed • Shazia Abrar • Muhammad Arshad
Editors

Sustainable Cosmeceuticals

Integrating Natural and Biotechnological Innovations

Springer

Editors
Sadia Javed
Department of Biochemistry
Government College University, Faisalabad
Faisalabad, Pakistan

Shazia Abrar
Department of Applied Chemistry
Government College University, Faisalabad
Faisalabad, Pakistan

Muhammad Arshad
Biochemistry Section, Jhang-campus
University of Veterinary and Animal
Sciences Lahore
Lahore, Pakistan

ISBN 978-3-031-86086-7 ISBN 978-3-031-86087-4 (eBook)
https://doi.org/10.1007/978-3-031-86087-4

This Springer imprint is published by the registered company Springer Nature Switzerland AG
The registered company address is: Gewerbestrasse 11, 6330 Cham, Switzerland

Preface

Cosmeceuticals have become a part of modern life, but their use can pose health concerns. However, "eco-cosmeceuticals" offer a potential solution to this issue. This book highlights the value of natural products derived from plants, marine organisms, and microorganisms, providing a unique source of prospective cosmeceutical active ingredients for inclusion into new formulations.

This book aims to provide comprehensive texts on eco-cosmeceuticals, with a strong emphasis on sustainability and the impact of climate change. The major focus is on ensuring the sustainable availability of eco-cosmeceutical products. In particular, it seeks to shed light on counterintuitive assessments, conclusions, and new perspectives on various products derived from different natural sources and their applications.

This book is divided into three parts: The first part, comprising three chapters, focuses on introducing cosmeceuticals and advancements in cosmetic science within the context of SDGs. It also provides an overview of the skin's structure, function, and the delivery of active agents derived from herbal extracts.

Eco-cosmeceuticals formulated with natural colorants from plants, as well as zooceuticals derived from microbial, marine, and algal sources are included in the second part of this book. This part also includes chapters on mushrooms and their role in skin health, highlighting the advanced techniques being employed in this industry.

In the final section of this book, future trends, legislative laws, and life cycle assessment of cosmeceuticals are discussed. Overall, this book covers a wide range of the scientific and technical aspects of eco-cosmeceuticals in the context of climate change. This book will be of interest to students, researchers, academicians, and industrialists working in the field of cosmeceuticals.

Faisalabad, Pakistan — Sadia Javed
Faisalabad, Pakistan — Shazia Abrar
Lahore, Pakistan — Muhammad Arshad

Acknowledgments

The successful completion of this book involved the assistance and expertise of numerous researchers. Special thanks to all our colleagues who provided their best knowledge in the form of book chapters. Several of our colleagues served as reviewers of the book chapters and aided immeasurably in providing useful suggestions during revisions. This helped in improving the scientific writing of this book. We are grateful to all the contributors for their insightful and thought-provoking aids. Finally, thanks to the publisher for their commitment to disseminating this vital knowledge.

Sadia Javed
Shazia Abrar
Muhammad Arshad

Contents

Chapter 1
Advancements in the Cosmetic Science and SDGs: An Appraisal

Shazia Abrar, Krishma Munir, Sarosh Iqbal, Shumaila Kiran, and Muhammad Arshad

Contents

Abstract The cosmetic industry is more concerned about the environmental, economic, and social impacts of the manufacturing and applications of cosmetic products. In the formulation of cosmetics, sustainability is addressed via sourcing and

S. Abrar (✉) · K. Munir · S. Iqbal · S. Kiran
Department of Applied Chemistry, Government College University Faisalabad, Faisalabad, Pakistan
e-mail: shaziaabrar@gcuf.edu.pk

M. Arshad
Biochemistry Section, Jhang-campus, University of Veterinary and Animal Sciences Lahore, Lahore, Pakistan

S. Javed et al. (eds.), *Sustainable Cosmeceuticals*,
https://doi.org/10.1007/978-3-031-86087-4_1

selection of the raw materials, which is achieved by replacing the synthetic chemical ingredients with sustainable natural alternatives. Therefore, herbal and natural cosmetics have gained huge recognition among the consumers who are more concerned about their health with beauty. A range of cosmetic products are available, and among them, skin cleanser is important as basic tool for self-care. Facial cleansers can be categorized into cream-based, gel-based, fluid-based, and powered-based. Skin-lightening and anti-aging agents like alpha-arbutin, ligninolytic enzymes (laccase, manganese peroxidase, lignin peroxidase), cytidine inhibitors, and all polyphenols extracted from natural resources are used in fairness creams, moisturizers, and sunscreens. Sunscreens are developed as inorganic blockers and organic absorbers to deal with skin premature aging due to UV radiations. All depigmentation agents like hydroquinine, kojic acid, and retinoic acid have wide applications in skin-lightening products. The natural active agents from different sustainable natural resources have been added in the cosmeceuticals to achieve the desired health benefits in accordance with sustainable development goals. Cosmetic formulations with natural active ingredients such as polyphenols, proteins, amino acids, natural colorant extracts, Gaur gur (GG)-based nanomaterial and nanoemulsion of arbutin loaded with coumaric acid have become widespread in cosmetic industry. Natural colorants gained attention in the cosmetic industry. These are extracted from plants leaves, flowers, roots, etc. In this chapter, we will discuss in detail the advancements in cosmetic science accordance with sustainable development goals via replacement of synthetic chemicals with natural, eco-friendly, and biodegradable alternatives.

Keywords Cosmeceuticals · Skincare · Sustainability · Advancements in cosmetics · Active agents from plants

1.1 History of Cosmetics

The art of beautifying the body has begun since antiquity and different civilizations in the world have contributed to the development of modern cosmetic science. In the ancient ages, people used cosmetics made from natural ingredients such as plant extracts, pastes, minerals, and animal-based components. These ingredients were considered safe, non-toxic, biodegradable, and eco-friendly. To enhance the natural appearance of the face, natural pigments were used for coloring the lips and face. Both men and women were used to painting their bodies with dried plant parts to make the body more attractive (Blanco-Dávila 2000). The first people to utilize "cosmetics" to draw attention to a particular area of the body were the Egyptians. The color of henna was used for painting the facial and body skin along with hairs. Many colors of orange were created on nails and lips with henna. In Egyptian culture, the color red was considered mystical, and red makeup was forbidden. The Egyptians made instruments for applying makeup out of wood, ivory, stones, and shells (Nayak and Ligade 2021). Eye paint was used by the Egyptians as early as

3000 B.C., according to available data. Crushed ant eggs or powdered kohl were the ingredients in the paint. Additional components utilized to create eye paint were crushed plant stems, galena, and malachite. After being reduced to a powder on a slab of stone, the ingredients were added to a container. The eye paint's copper content protected the eyes against sun glare-induced suppuration. Additionally, it avoided a few eye conditions. Because of the many medicinal benefits of eye makeup, both male and female adults were urged to wear it. Because eye paint was so vital, the builders were forced to halt construction on the pyramids until eye paint arrived to help shade and protect the workers' eyes from the sun's strong beams. Women were used to taking baths with milk and rose water for cleansing and beautifying purposes. Later, they used soaps, and hydrolysates of the animal fat by using wood ash for cleaning and bathing (Arshad et al. 2014; McMullen and Dell'Acqua 2023). In Western culture, bathing was not that much popular until in the eighteenth and nineteenth centuries (Smith 2008).

Natural flowers and plant parts were used to overcome the bad odors of the body. The first people to establish the craft of perfumery were the Arabs, who were also considered to be the early chemists. The first perfume was a fragrant flower, and it remains so now. Aristotle's contemporaries most likely employed the dried lavender sprig we use in the clothes press for the same reason (Blanco-Dávila 2000; Latif 2021). It should come as no surprise to those acquainted with the culture of the Greeks, one of the most highly civilized people in antiquity, that they were known for a wide variety of fragrant substances and oils perfumed with them, or perfumes in the modern sense of the word. The Greeks loved the perfume of violets; in addition, they employed the scents of various mints, thyme, marjoram, and other fragrant herbs (Blanco-Dávila 2000). Since 1875, the ancient compositions based on myths and mysticism were replaced by more technical cosmetic products. Cosmetic chemistry was first introduced after the publication of the Handbook of Cosmetic Chemistry in 1920. After the World War II, the Society of Cosmetic Chemists was founded in the United States. The idea behind this was to expand the cosmetic industry by exchange of scientific knowledge and expertise. Later on, other scientific societies were established in different countries and joined into International Federation of Societies of Cosmetic Chemists (Arshad et al. 2022; Millikan 2001).

1.2 Cosmetics and Types of Cosmetics

Cosmetics constitute a group of specialized products, formulated to keep the external body parts clean, attractive, and beautiful, and preserve or protect them from external environmental damage. In 1938, the revised Food and Drug Act regulated the definition of cosmetic products as an article or substance that is applied, rubbed, sprayed, or sprinkled on the human body parts for cleaning, beautifying, increasing attractiveness, altering the appearance, and preserving the integrity. According to the above definition, a clear difference exists between cosmetics and drugs (Wallack 2019). However, certain skincare products, i.e., anti-aging formulations, act as

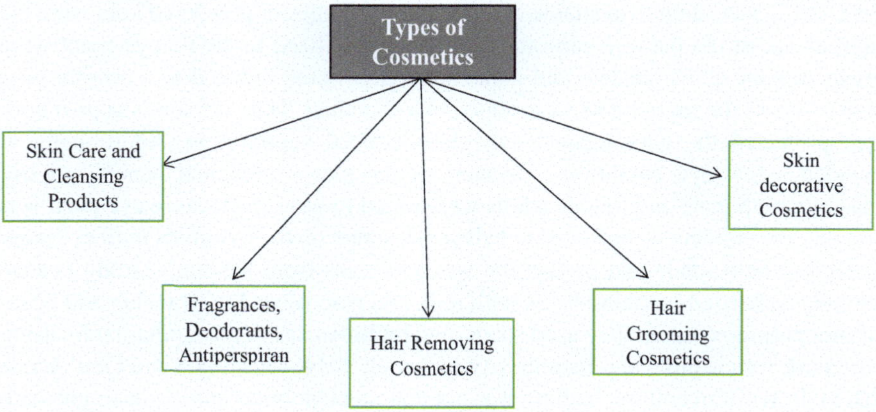

Fig. 1.1 Types of cosmetics

cosmetics due to their beautifying effects and as drugs as they change the skin structure. The term cosmeceutical has been used for such a type of product that has a dual function. In recent years, FDA has concluded that certain skincare products can be recognized as over-the-counter (OTC) drugs and are recognized and considered as cosmetics, for example, anti-acne, anti-perspirants, anti-aging, and sunscreening products. These cosmetics are available to the public without a prescription. The cosmetic industry includes manufacturers who manufacture these products, distributors, suppliers who supply them directly to the consumers, and suppliers of raw and packaging materials (Cheng et al. 2021). Different groups of cosmetics have been categorized based on their functions and regardless of their form as follows:

(a) Skincare and moisturizing products (moisturizers, emollients, toners and sunscreens, anti-aging and anti-acne cosmetics)
(b) Cleansing cosmetics (soaps, facial cleansers, bath gels, and shampoos)
(c) Fragrances, deodorants, and anti-perspirants
(d) Hair-removing cosmetics (shaving and depilatory formulations)
(e) Hair-grooming cosmetics (hair-conditioning, waving, straightening, styling, and coloring products)
(f) Decorative cosmetics (skin-coloring products, lipsticks, foundation creams, eye-coloring, and face-beautifying products)
(g) Cosmetics used for delivering active ingredients for improving skin health, i.e., cosmeceuticals (vitamin-containing serums, anti-aging cosmetics, skin fairness cosmetics, etc.) (Fig. 1.1)

All the above categories of cosmetics contain a variety of chemical components each of which imparts a specific function to the overall performance of the product. These cosmetic products are applied to improve the skin appearance and quality or alleviate the minor skin problems. All the cosmetic preparations are designed and formulated for external use and mainly contain the following basic components: (1) the main active ingredient, (2) the delivery vehicle, and (3) the additives/auxiliary components. The active ingredients are incorporated in a suitable form that makes them available for absorption and action in the skin for a specific purpose (Baki 2022).

1.3 Sustainability and Skincare Cosmetics

Sustainability is "meeting the needs of the present without compromising the ability of future generations to meet their own need." In the Oxford Dictionary, sustainability is "the ability to be maintained at a certain level or rate." Thus, sustainability is attaining a dynamic equilibrium between the growth, development, and resources required to maintain these (Waseem and Kota 2017). The depletion in natural resources due to an overgrowing population, and a change in the climate are posing challenges to support sustainability. In 2015, the UN adopted the 2030 Sustainable Development Agenda which contains 17 Sustainable Development Goals (SDGs) for the prosperity of the people and planet (Arshad 2024; Hák et al. 2016). Cosmetic companies form an essential part of the sustainability changes, and sustainable business models are increasing worldwide. Their main objective is to decrease the environmental, economic, and social impacts of their formulations, consume natural resources without their depletion, and avoid the harmful ecological effects along with a positive contribution to the economy. Consumers are more concerned about the health, environmental, and social impact of the cosmetic products they buy. This leads to the exponential growth of the sustainable cosmetic market. According to a published report, 60% of the consumers were willing to change their choice to reduce the environmental impact of their product and the majority claimed the importance of sustainability for them (Yildirim 2022). Consumers are willing to pay more to buy environmentally friendly and sustainable cosmetics.

The important step to ensure sustainability in the formulation of skincare cosmetics is selection of the raw material. The main points that must be addressed for raw materials selection are: (a) the biodegradability of the ingredients, (b) sources of the raw materials (animal, plant or microbial origin), and (c) how are the raw materials synthesized or extracted? Thus, different sources of raw materials can be classified as natural that are obtained from plants, animals, microbes, or mined from ores. Second are the naturally derived ingredients that are obtained from chemical processing of ingredients found in natural resources. The third one is the naturally identical components synthesized in the laboratories. The classification of cosmetic ingredients is challenging (Barros and Barros 2020). The USA Natural Cosmetic Act 2019 defines cosmetics containing ingredients that are at least 70% from natural sources (Barros and Barros 2020). It is also important to note that not all natural ingredients are 100% sustainable and not all synthetic chemicals are 100% non-sustainable. Sustainable cosmetic ingredients can also be obtained from green chemistry and upcycling. Ingredients in cosmetics offer several important functions such as efficacy, enhancing beauty, safety, and physiochemical stability. The important ingredients in cosmetics are: water, surfactants, emulsifiers, colorants, polymers, UV filters, antioxidants, colorants, preservatives, etc.

1.4 Functions of Skincare Cosmetics

Recently, a global tendency for eco-friendly cosmetics has motivated the personal care cosmetic industry to focus on the development of formulations containing environmental-friendly and sustainable ingredients from natural resources (Cui et al. 2023). Skincare cosmetics include a range of products that are formulated for the specific purpose of skincare and maintenance. This includes the preparations for softening and moisturizing, protecting from UV radiations, and repairing the skin. The flexibility and elasticity of the skin depend upon the concentration of the water. To preserve the skin's elasticity and keep it healthy, it is necessary to decrease the loss of water contents or increase the moisture level of the skin. The purpose of skin moisturizing formulations is to keep the balance of skin moisture contents by keeping the skin hydrated, reducing the loss of skin water contents, and preserving its elasticity (Zhang et al. 2020). For this purpose, cosmetic moisturizers containing oil and water components are prepared into emulsion systems (creams and lotions) that serve to keep the skin moist and fresh. The components of the emulsions that attract water from the surrounding environment and keep a balanced skin moisture level are called humectants. The list of such components is provided in Table 1.1. Some non-volatile hydrophobic chemical ingredients of the moisturizing formulation like petrolatum make an occlusive layer on the surface of the skin and decrease the transepidermal water loss (TEWL) of the skin. This traps the water in the stratum corneum layer of the skin causing the hydration of the skin cell. These emollients make the skin smooth and reduce the dryness of the skin by providing enough moisture and lipids for the skin. Some components of the emulsions penetrate the skin like lipids, emulsifiers, and non-volatile components and remain there (Salvioni et al. 2021). The emollients based on petrolatum and mineral oil have sustainability problems. It is difficult to find sustainable alternatives to this petroleum-based ingredients; however, research has shown that various vegetable oils have comparable occlusive properties as that of petrolatum. Another important sustainable emollient is derived from natural resources is squalane derived from vegetables (olive oil) or from microbial fermentation (Ahmadi-Ashtiani et al. 2020). An important component of the skin moisturizer product is the fatty alcohols such as cetyl alcohol, octyldodecanol, and isostearyl alcohol that act as emulsion stabilizer along with its emolliency and moisturizing ability. Fatty alcohol can be obtained from vegetable oils or by hydrogenation of fats from animal source (Kelm and Wickett 2017). Several exogenous and endogenous factors cause to decrease the skin elasticity and an increase the skin aging with the appearance of wrinkles in the skin. Also, exposure to UV radiation increases skin pigmentation with the appearance of dark spots on the surface of the skin (Coelho et al. 2009). Some skincare formulations claim the presence of active agents that reduce skin pigmentation and some aging effects.

Table 1.1 Active Ingredients in skincare

Name	Introduction	Benefits	Sources	References
Humectants	A hygroscopic substance, a component of cream and water emulsions	Attract water from surroundings and keep a balance skin moisture level	Hyaluronic acid Glycerin Panthenol Polyglutamic acid Honey	Fowler (2012)
Petrolatum	Non-volatile hydrophobic chemical ingredient	Make an occlusive layer on surface of skin and decrease the transepidermal water loss (TEWL)	Natural petroleum resources	Morrison (2005)
Emollients	Moisturizing treatments	They cover the skin with a protective film to trap in moisture by filling in crevices in the skin	Rosehip oil Argan oil Linoleic acid All plant-derived oils	Alander (2012)
Ceramides	Intracellular natural lipids	Protect the skin and prevent it from dryness and infection	Olive oil Jojoba oil Safflower oil Grapeseed oil	Coderch et al. (2003)
Alpha hydroxy acids	Group of acids found naturally in fruits	These include exfoliation Improved skin texture, acne reduction, reducing dark spots and scars	Fruits Vegetables Milk	Tang and Yang (2018)
Titanium dioxide	Titanium dioxide or titania, an inorganic compound derived from titanium	It helps to reflect and scatter UVA and UVB rays Also help to prevent sunburn and sun damage to skin	Minerals like: Ilmenite Rutile Anatase	Dreno et al. (2019)
Zinc oxide	A largely inert white compound	Treat sunburn and protect from sun's ultraviolet rays Treats skin irritation like cuts, burns, or diapers	Beef Oysters Cashew nuts Pumpkin seeds Eggs Lentils	Lu et al. (2015)

(continued)

Table 1.1 (continued)

Name	Introduction	Benefits	Sources	References
Retinoid	A class of chemical compounds; natural derivatives of vitamin A	Reduce skin wrinkles and acne and prevent acne scars formation	Liver Carrots Rosehip seed oil Spinach Sweet potato	Mambwe et al. (2024)
Collagen	A main structural protein provides structural support to extracellular space of connective tissues	It improves aging effects, improves hydration, and decreases transepidermal water loss (TEWL)	Bone broth Citrus fruit Eggs Berry Fish Garlic Meat	Avila Rodríguez et al. (2018)

These active components, i.e., alpha hydroxyl acids, trans-retinoic acids, hyaluronic acids, and collagen, have been known to improve skin wrinkles and aging effects (Ahmed et al. 2020; Arshad et al. 2023). The exposure to UV radiation can be decreased by using anti-UV ingredients in creams and lotions that protect the skin from the absorption of UV radiation. Both UVA and UVB radiations can cause skin burning and photogenic reactions in the skin resulting in wrinkling, keratosis, and also in extreme carcinogenic reactions in the skin. The purpose of sun-blocking agents is to physically block UV radiation by reflecting them and protecting the skin. Titanium oxide and zinc oxides can be used as sun-blocking agents having that can reflect this radiation without causing an undesirable effect (Shaath 1990).

1.5 Skin Cleansing Formulations

The purpose of skin cleaners is to remove the excess oil, dirt, and exogenous particles from the outer surface of the skin. Such unwanted materials are removed by the use of cleansing agents mainly based on surfactants and abrasive agents. Surfactants are molecules that have hydrophilic groups and hydrophobic long chains that tend to remove oil and dirt by making them soluble in the water and carrying them away from the surface of the skin (Chelazzi et al. 2020). Surfactants perform various important functions in the cosmetic formulations such as emulsifier, dispersant, humectant, and solubilizer. Mostly anionic surfactants are used in the cosmetics such as alkyl sulfates such as sodium lauryl sulfates, ammonium lauryl sulfates, and ethoxylated sulfates as sodium lauryl ether sulfates (Painter 1992). They have good foaming and cleansing abilities. Other anionic surfactants that have been used in the cosmetics are alkyl benzene sulfonates, acyl sarcocinates, and isethionates. Cationic surfactants are used in the hair-care products such as shampoo and conditioners to reduce the static charge of the hair. Some cationic surfactants are quaternary ammonium compounds (Gonçalves et al. 2023). All the above surfactants are synthetic, cause skin irritation, and are toxic for aquatic life and non-biodegradable. A good alternative to synthetic surfactants are biodegradable, renewable plant-based non-ionic surfactants such as alkyl polyglucosides (APGs) (Lourith and Kanlayavattanakul 2009). These compounds are derived from fatty alcohol (derived from palm and coconut oils) and glucose. These have low toxicity, non-irritant, and compatible for the skin. Another important function of the surfactants is making the emulsion stable. The conventional emulsifier such as polyethylene glycol has sustainability problems. Therefore, the natural alternatives to PEGs are polyglyceryl- and glucose-based emulsifiers. Some other biodegradable compounds as sucrose esters, glyceril, sorbitan, ethoxylated sorbitan, polyglyceril, and olive-oil-based esters are also used as emulsifiers (Sarathchandraprakash et al. 2013). Microbial-based surfactants and emulsifiers are also good alternative to conventional synthetic counterparts. They have low toxicity and perform good(Rosenberg and Mitchell 1985). This solution of these surfactants has good foaming qualities, water solubility, and pH in the range that is suitable for cleaning the facial and body skin (PH 5-6). Further, the addition of moisturizing agents in the cleansers can leave the skin soft and smooth after cleansing, decreasing the dryness effects (Mijaljica et al. 2022). To increase the emollient effect on the skin, some amphoteric and non-ionic compounds are added to the cleansers such as alkyl betaines, fatty acids, alkyl betaines, and some quaternary compounds. The acidic pH of such mild cleansers helps in controlling the growth of microbial species on the skin surface. Foam boasters and perfumes are added to the cleanser to further increase the aesthetic effects and make them attractive for use on the face and body. Creamy emulsion-type cleansers containing oils and waxes can also be used for removing the skin makeup, leaving skin soft and lubricant. The oil contents of the cream act as a solvent for removing the sebum, excess oil, and makeup from the skin surface. The additives in such cleansers include emulsifiers and perfumes (Draelos 2018a, b).

1.6 Skin-Coloring Cosmetics

Skin-coloring cosmetics are decorative products that are applied on the face, intended to enhance the beauty of the skin by adding color to the body or concealing some physical defects on the skin. One important function of skincare cosmetics is to sustain the skin health along with beauty. Therefore, the development of coloring cosmetics containing biodegradable, eco-friendly, and non-toxic ingredients is necessary to achieve the purpose of sustainable development goals. There are a variety of products included in this category, i.e., facial foundation creams, facial powders, compact bases, facial, coloring, eye, and lip-coloring products. The main coloring component in all coloring cosmetics is the pigment or dye that is stable at the condition of preparation and storage of the product (Lanzendörfer-Yu 2021). Strong color shades of dyes or pigments are used to highlight the color of the lips making them more attractive. These color shades are mainly obtained from organic dyes and lakes that are used in different concentrations to obtain the desired color shades of lipsticks. New color shades of lipsticks have been developed according to the demand and fashion of modern society. Lip-coloring cosmetics are available in a variety of shapes and textures according to the demands of the consumers (Engasser 2000). Most commonly used lipsticks are in the form of hard sticks, tinted lip gloss, and liquids. The long-lasting color shades of the lipsticks can be produced using some hydrophilic solvents. The more sophisticated and specialized lipsticks can be obtained by adding antioxidants, UV protectants, and non-toxic fragrances in the base of the lipstick (Heusèle et al. 2022). The formulation of the lipsticks contains components like waxes such as bees wax and carnauba wax, and oils such as castor oil and 2-octyldodecanol. The synthetic dyes and pigments used in lip- or eye-coloring cosmetics have various side effects, mainly causing skin irritation, redness, pigmentation, and allergic reactions in the skin (Arshad et al. 2023; Malabadi et al. 2022). These chemicals are hazardous not only for humans but also for environment and ecosystem and needs to be replaced with non-toxic, eco-friendly, and sustainable alternatives for their application in skincare cosmetics. Like lipsticks, the formulation used for coloring the areas surrounding the eyes has been used since antiquity. For this purpose, a range of color shades of eye shadows are available in the market. The use of mascara makes an eye more attractive and also lengthens the eyelashes. Eyeliners are used to highlight the areas of the eyelids and make the eye more prominent in appearance. Eye-coloring cosmetics are available in the form of compact powder, gloss, and liquid colors (Draelos 2001). Eyeliners are available in the form of emulsion that is applied with the help of a very fine brush. This emulsion leaves a waterproof lining around the eyes. Shiny eyeliners contain non-toxic pigments suspended in the polymers such as acrylic copolymers. The cosmetic industry must ensure the environmental sustainability of raw materials, along with minimum harmful effects on the ecosystem. To minimize the impact of chemical toxicity on the environment and aquatic life, the raw materials and end products should be biodegradable, non-toxic, and eco-friendly (Table 1.2).

1.7 Cosmeceuticals

The word cosmeceutical, first coined by Raymond Reed in 1961, represents a combination of two scientific terms "Cosmetics" and "Pharmaceutical." It represents a new category of formulation that is used to enhance the beauty as well as the health of the skin due to therapeutic actions (Arshad et al. 2014; Draelos 2008). It was not known to the world unless in the later 1970s. Dr. Albert M. Kligman prepared a formulation of retinoic acid for improving skin wrinkles. He also proposed that other therapeutic agents such as vitamins, antioxidants, peptides, and hormones can also be used in the skincare formulation (Kligman 2000). Skin being the largest body organ is more exposed to environmental pollutants and radiations from the sun, producing reactive oxygen species called free radicals in the cell. These excessive free radicals cause significant damage to the DNA and cell membranes. The oral intakes of antioxidants and vitamins cannot combat the harmful effects caused

Table 1.2 Skin-coloring cosmetics

Type of cosmetic	Ingredients	Working/benefits of ingredients	References
Eyeshadow	Dimethicone	Used to soften the skin and prevent itching/flaking	Song et al. (2019), Hunter et al. (2015)
	Ethylhexyl palmitate	Used to adjust the consistency of makeup	
	Zinc stearate	Used as a lubricant and a water repel	
Lipsticks	Bees wax	Seals in moisture and protects against dryness	Niederer et al. (2016), Niederer et al. (2016), Draelos (2018a, b), Mawazi et al. (2022), Mawazi et al. (2022)
	Mineral	Minerals gave staying power that resists smudging and feathering for a perfect pout	
	Natural plant oil	Provides hydration and nutrition to lips	
	Rosemary	Rosemary is an anti-aging secret of nature	
	Vitamin E	Vitamin E is antioxidant and protects lips from damage from free radicals by UV rays	
Lip tint	Shea butter	It has excellent effect for moisturizing and nourishment of lips	Tijare et al. (2022)
	Bees wax	It acts as a guardian angel for lips for creating a protective barrier to lips	
	Coconut oil	Coconut oil hydrates the lips and removes dead skin	
	Hyaluronic acid	It improves lips volume, hydration, and softness	
	Jojoba oil	It acts as balancer for lips	
	Castor oil	Castor oil is a gloss maker	

(continued)

Table 1.2 (continued)

Type of cosmetic	Ingredients	Working/benefits of ingredients	References
Eyeliner	Thickening agents	They stabilize the eyeliner and make it stay on the skin These are wax, gums, and clays	Draelos (2001), Ng et al. (2016)
	Natural preservatives	These include sugar, vitamin E, citric acid, and oils. They help to extend the shelf life of eyeliner	
	Argon oil and jojoba gel	They provide nourishment and protection to lashes	
Mascara	Mineral oil	The petrolatum derivatives are used in mascara to boost its viscosity	Ng et al. (2012), Giacomel et al. (2013), Draelos (2001)
	Lanolin	It is used as an emollient, derived from sheep wool	
	Floral waxes	Some mascaras contain blossom wax, cassia flower wax, rose wax, and blue lotus wax that coats and protects lashes	
	Dimethicone copolyol	It adheres to lashes and naturally repels water. It is a derivative of silica	
	Iron oxide	It is used as a coloring agent to create black color	
Primer	Benzoic acid	It regulates the pH of the skin and also pH of the product	Board et al. (1992), Solish et al. (2019)
	Vitamins A, B, C	These make skin smooth	
	Hyaluronic acid	It hydrates the skin and reduces wrinkles	
Concealer	Dimethicone	It gives long-lasting skin-smoothing benefits	Rukpracha et al. (2023)
	Talc	A common emollient that makes skin smooth	
	Rosehip seed oil	It acts as an antioxidant and repairs skin after sun damage	
	Benzyl alcohol	It retards the growth of bacteria and fungi in concealers	

(continued)

Table 1.2 (continued)

Type of cosmetic	Ingredients	Working/benefits of ingredients	References
Liquid foundation	Vitamin C	It is a powerful natural antioxidant and deals with dull skin	Mohiuddin (2019), Anwar and Rizkamiarty (2020)
	Hyaluronic acid	It works to combat dehydration, dullness, and signs of aging	
	Salicylic acid	It helps to unclog pores and keep the dermis clean	
	SPF	Liquid foundation with SPF deals with dull skin and damage by UV rays from sun	
Face powder	Talc	It absorbs moisture from the skin and prevents skin rashes	Draelos (2000), Benson et al. (2019)
	Natural starches	It is also a drying agent and relieve yeast infections like rash, pimple, etc	
	Calcium carbonate	It is known for translucency and skin-protecting properties. It has adhesion power for a long time	
	Kieselguhr (diatomite)	It helps to absorb excess sebum and cleanse the skin	
Natural hair dyes	Henna, indigo, coffee, beetroot, and other plant-derived ingredients	These natural colors give color to hair. They are free from harsh chemicals	Camargo Jr et al. (2022), Shahi et al. (2017)
Synthetic hair dyes	Keratin	It is known for strengthening effects; it helps to improve overall health and resilience of hair and prevent breakage	Da França et al. (2015); Guerra-Tapia and Gonzalez-Guerra (2014)
	Ammonia and alcohols	These are responsible for opening up hair cuticles so the color can penetrate hair but they can harm hair	
	Natural oils, shea butter, essential oils	They provide moisture and nourishment to hair, prevent dryness, and promote shine by reducing frizz	

(continued)

Table 1.2 (continued)

Type of cosmetic	Ingredients	Working/benefits of ingredients	References
Nail polish	Plasticizers	They work to make non-brittle films	Bryson (2022), Lanzendörfer-Yu (2021)
	Adhesive polymers	They provide intracellular adherence to the nail surface	
	Thickening agents	They maintain the sparkling particles in suspension	

by excessive production of free radicals in the skin cells (Makrantonaki and Zouboulis 2007). Therefore, the topical applications of additional therapeutic agents such as antioxidants and vitamins can help to neutralize the chain reactions initiated by free radicals (Allemann and Baumann 2008). Along with neutralizing the free radicals, the vitamins also aid in the production of elastin and collagen and improve skin keratinization, anti-inflammatory, and depigmentation effects. With the advancements in cosmetic science, the efficacy of vitamins and other active agents has been improved by the invention of a new delivery system to enhance the absorption and retention of these active agents in the skin (Vinardell and Mitjans 2015). New drug delivery systems have been introduced that have various advantages over the conventional ones (Ascenso et al. 2011). All the topical formulations containing the active agents have some disadvantages such as poor solubility and availability of the therapeutic agents to be available for the site of action. Some novel delivery systems such as liposomes, proniosomes, niosomes, and nanostructured lipids facilitate the delivery, stability, availability, and therapeutic response of the active agents in the skin (Ricci et al. 2024).

Cosmetic agents have been developed and used for the improvement of skin function and combating the effects of environmental pollutants and UV radiation. Scientific data has proved the most effective therapeutic agents are vitamins that are essential for the functioning of the body's systems (Kumar et al. 2024). Vitamins are organic compounds that are required as micronutrients in the food and also exhibit beneficial effects for the external application on the skin. The commonly known vitamins in the cosmetics are Vitamin A, B3, C, E, and K (Shapiro and Saliou 2001). These vitamins act as antioxidants that neutralize the free radicals by donating electrons, thus preventing the oxidative damage to the skin cells. The most important damage is photo-aging (George et al. 2022). Photo-aging is the condition in which skin loses its elasticity and develops wrinkles along with some pigmentation effects due to prolonged exposure to UV radiations. For preventing the UV-induced skin damage, sunscreen lotion or creams containing sunscreening agents are used. The following chemical compounds can be used as active agents in cosmeceuticals (Kokande et al. 2024) (Fig. 1.2).

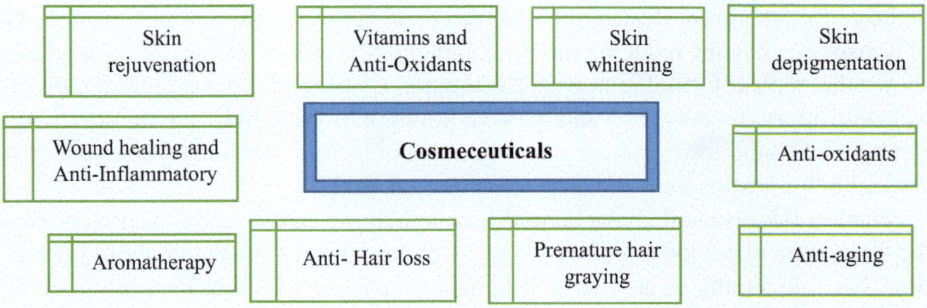

Fig. 1.2 Different types of cosmeceuticals

1.8 Sustainable Active Agents in Cosmeceuticals

1.8.1 Vitamins

The purpose of cosmeceutical is to ensure the skin health and beauty, thus ensuring the sustainable development goal of good health and well-being of the people. Vitamin A or retinoids is group of fat-soluble vitamins that can be applied in various forms such as tretinoin, tazarotene, and adapalene (Zasada and Budzisz 2019). Retinoids promote the proliferation of keratinocytes and inhibit the degradation of collagen and activity of enzyme metalloproteinase. This group of compounds is thus used in cosmetics for preventing the skin from aging and treatment of skin inflammation along with oxidative skin damage by UV radiations. Tretinoin inhibits the enzyme metalloproteinases and thus reduces the breakdown of protein collagen, inducing the production of precollagen in the skin. This helps thicken the skin's dermis layer and reduces the fine wrinkles, filled in with glycosaminoglycans (Kafi et al. 2007). Depending upon the concentration used in the formulation, some of the retinoids produce skin irritation and some dryness in the over-the-counter products. The retinoids have a complex mechanism of action in the body due to the low solubility in the body fluids and need special types of proteins for their intracellular transportation. In the cell cytoplasm, a special type of protein called cytosolic retinol-binding proteins (CRBP I and II) has an affinity for retinol and the concentration of the retinol in the cells depends on their binding efficacy to these proteins (Zasada and Budzisz 2019). Similarly, the concentration of retinoic acid in the cells is controlled by cytosolic retinoic acid-binding proteins (CRABP I and II) which have an affinity for retinoic acids. These proteins are responsible for activating the nuclear retinoid acid receptors (RAR) and retinoid X receptors (RXR) and thus mediate in the functioning of vitamins in the cells (Kong et al. 2016).

Vitamin E is the group of fat-soluble vitamin available in the form of alpha-tocopherol and four tocotrienols that have been used as antioxidant in topic formulation for reducing the skin wrinkles and hyperpigmentation along with acting as sunscreening agent (Halliday et al. 1998; Yarovaya and Khunkitti 2019). This vitamin functions by scavenging the free radicals and preventing the lipid peroxidation in the cell membranes. Research has shown that Vitamin E also inhibits the activity of tyrosinase enzyme, reducing the skin pigmentation effects. Research studies have shown that topical formulation with 2% Vitamin E can reduce the skin burn induced by sun to approximately 20% and with 5% strength of Vitamin E, and the skin wrinkling and UV-induced erythema can be well improved when the formulation is applied before exposure to sunlight (Younis et al. 2022).

Vitamin B3, also called niacinamide—a cofactor to NADP and NADPH, is used for the treatment of skin conditions such as photo-aging, acne, and hyperpigmentation (Na Takuathung et al. 2023). It helps in repairing the cell DNA by blocking ADP-ribose polymerase-1 (PARP) and transcriptional factors which are responsible for DNA damage due to their activation by UV radiations. It also helps in reducing skin pigmentation by blocking the transfer of tyrosinase protein 1, melanocyte protein, and tyrosinase protein 2, from the cells melanocytes to the keratinocytes (Wohlrab and Kreft 2014). Niacinamide acts as anti-inflammatory agent by inhibiting the anti-inflammatory mediators that caused overproduction of sebum and inflammation in the acne-prone skin. As a precursor of NAD and NADPH, niacinamide stimulates the production of epidermal proteins, keratin, collagen, filaggrin, and involucrin, and some extracellular matrices such as elastin, fibulin, and fibronectin. This prevents the wrinkles and fine lines by maintaining the skin moisture level and reduction of transepidermal water loss, thus relieving the signs of skin aging (Joshi et al. 2023; Wohlrab and Kreft 2014).

Vitamin C also called L-ascorbic acid is obtained from dietary sources. It is used in the form of its derivatives such as sodium ascorbate, magnesium ascorbyl phosphate, and ascorbyl-6-palmitate which is the skincare formulation. Due to its low bioavailability for the skin when taken orally, it is preferred to be used topically in dermatological treatments (Boo 2022; Farris 2014). Vitamin C acts as a powerful antioxidant that neutralizes the oxygen-free radicals produced by exposure to UV radiation, which helps in combating oxidative stress. This prevents the effect of photo-aging, sunburn, and erythema. Vitamin C also acts as a depigmenting agent by inhibiting the enzyme tyrosinase, which reduces melanin production in the skin (Zaid and Al Ramahi 2019). Vitamin C is also responsible for the synthesis of collagen protein by enhancing the production of prolysyl and lysyl hydroxylase which are responsible for stabilizing and crosslinking the collagen molecules. It also stimulates the collagen gene expression by increasing the production of malondialdehyde via lipid peroxidation (Lis and Baar 2019).

1.8.2 Caffeine

Caffeine is found naturally as an alkaloid in the leaves, seeds, and fruits. It is particularly extracted from the seeds of *Coffea Arabica* plant, natively found in Ethiopia. It can also be extracted from the leaves of *Camellia sinensis,* along with other polyphenols (Watts et al. 2024). Caffeine acts as stimulant of nervous system, mainly consumed in the form of coffee, tea, and other drinks on daily basis for keeping mind active and focused. Caffeine due to its important biological activities and health benefits has been used as a popular cosmeceutical agent (Watts et al. 2024). It showed several beneficial skin effects and has been used as an active agent in cosmeceutical due to its antioxidant, UV-protectant, anti-inflammatory, anti-carcinogenic, and anti-aging activities. Caffeine can penetrate the skin and get absorbed in the epidermis (Faudone et al. 2021). After 24 h of application, it was found that high concentrations (280ug/tissue) of caffeine were found in the epidermis. It has been shown in several studies that oral consumption as well as topical application of caffeine on the skin has lowered the incidence of UV-induced skin cancer (Visconti et al. 2020). Literature has shown that caffeine can help to reduce non-melanoma skin cancer, skin pigmentation, and aging effects due to its ability to scavenge the reactive oxygen species (ROS) (Blanco-Llamero et al. 2024). Due to its strong antioxidant activity, caffeine is very effective in topical formulation to prevent oxidative damage to the skin induced by ROS (Lu et al. 2002; Ma et al. 2015). Caffeine application is also associated with a reduction in skin inflammation dermatoses such as atopic dermatitis, rosacea, and psoriasis. Skin inflammation is caused by skin barrier dysfunction, and this condition is further worsened by changes in the environmental humidity (Ojeh et al. 2016). Topical application of caffeine along with other polyphenols can significantly reduce skin inflammations and ultimately skin redness. Caffeine has shown synergistic effect when used in combination with other cosmeceutical agents such as polyphenols (Herman and Herman 2012).

1.8.3 Curcumin as Cosmeceutical Agent

Curcumin, an active ingredient in the turmeric rhizomes (*Curcuma longa*), has several health benefits and has been used as an active agent in cosmeceuticals. It also represents one of the sustainable ingredients of cosmetics as it offers good health along with beauty, enhancing the well-being of the people. All the natural ingredients obtained from renewable resources are biodegradable and eco-friendly (Di Lorenzo et al. 2023; Rafiee et al. 2019). The other components of turmeric are essential oils, tumerin, and curcuminoids (Li et al. 2011). Although its concentration in turmeric is only 2–8%, curcumin is the most active component responsible for the biological activities of the turmeric rhizomes. It has been used in cosmeceuticals due to its anti-inflammatory activities. The hydrophobic curcumin molecule is poorly absorbed when taken orally; however, its absorption by skin epidermis is quite good, when applied topically on the skin (Algama et al. 2023). Curcumin-based creams can be used to treat skin inflammations due to their strong anti-inflammatory properties. Curcumin acts as an inhibitor of phosphorylase kinase (PhK), which is responsible for multiple signaling pathways, during skin inflammation. These reactions can lead to skin lesions and thinning, disturbed cell cycle, and increased tumor rate formation (Vollono et al. 2019). Therefore, the cosmetic formulation containing turmeric extracts, especially curcumin, has been used for treating skin damages caused by sun, photocarcinogenesis, skin wrinkling, and quick healing of surgical wound and burns (Thangapazham et al. 2013).

1.8.4 Green Tea

The most abundant polyphenol of the green tea is Epicatechin-3-gallate, EGCG (50–80%), responsible for its antioxidant, anti-inflammatory, chemo-protective, anti-aging, sunscreen, and antimicrobial activities. The application of green tea extracts in the cosmetic formulations also fulfills the sustainable development goals of good health and well-being (Liao et al. 2001). The other polyphenols present in the green tea are catechin, epicatechin, and epigallocatechin. EGCG has been used in skincare formulation due to multiple therapeutic activities (Gianeti et al. 2013). The breakdown of the skin elastin and collagen due to exposure to UVA and UVB radiation results in the skin photo-aging. Numerous investigations have shown the effectiveness of the EGCG in protecting the photo-induced skin damages. Therefore, the topical formulations containing the green tea extracts are effective for treating the skin inflammations, photo-aging, and sunburns. It has also been used effectively in treating other skin problems such as skin lesions, skin lesions, dermatitis, keloids, psoriasis, and acne condition.

1.8.5 Lipids

Skin dryness is common both in men and women, and it results from the impaired skin barrier function due to excessive loss of the skin water contents. This causes skin to lose elasticity and look rough and dull. Other metabolic and genetic factors also result in skin conditions such as eczema, dermatitis, and seborrhea. Lipids form the major part of the skin epidermis constituting a diversified composition such as fatty acids, triglycerides, waxes, fat-soluble vitamins, and phospholipids. Linoleic acid is one of the most abundant skin fatty acid constituents and plays a specific role in keeping the skin epidermal water barrier functional. When applied systemically and topically, linoleic acid is known to regenerate the skin lipid barrier and resume the skin moisture level (Mohiuddin 2019). The other important fat component of the stratum corneum is ceramide, and the most of the linoleic acid is bound to ceramide in the skin. The deficiency of the skin linoleic acid disturbs the lipid bilayer and causes the breakdown of the skin barrier functions. Metabolites of the linoleic acid such as 13-HODE (13-hydroxy-9,11-octadecadienoic acid) also function to reverse the hyperproliferation of the skin to the normal state, which supports the view that 13-HODE may act as a mediator to restore the symptoms of deficiency of essential fatty acids. Gamma linoleic acid (GLA) is a polyunsaturated fatty acid with 18 carbons and three double bonds, omega-6 (n-6), produced from linoleic acid by the action of the enzyme delta-6-desatrase (D6D). Vegetable oils rich in GLA have been evaluated for restoring the skin barrier function and maintaining the moisture contents in the epidermis. Therefore, topical applications of the formulations containing linoleic acids have been known to improve skin conditions by providing smoothness and mechanical strength by normalizing the transepidermal water loss (TEWL). The skin conditions that result from the loss of the skin lipids due to the effect of detergents or the result of lipid- a deficient diet can be reversed by the use of gamma linoleic acid-rich vegetable oils. These above findings suggest that the topical application of the lipids may help in improving skin health and beauty (Millikan 2001).

1.8.6 Hydroxy Acids

Hydroxy acids (α- or β) have gained popularity in cosmetics and in over-the-counter drugs as well since the past 20 years. α-Hydroxy acids (AHAs) include citric acid, glycolic acid, lactic acid, malic acid, and tartaric acid. Though commonly referred to as "fruit acids," these compounds can also originate from plant parts, such as the juice of *Saccharum officinalis* (sugar cane), which is the primary source of glycolic acid, and *Acer saccharum* (sugar maple) (Tang and Yang 2018). They can also be found in fruits such as apple, lemon, orange, and most other fruits in the citrus genus, bilberry, and blackcurrant. Salicylic acid is a common example of β-hydroxy acid (BHA) used in over-the-counter (OTC) medications and cosmetic products. The bark of willows (*Salix alba or nigra*) and wintergreen leaves (*Gaultheria procumbens*) are used to extract salicylic acid.

Brightening and smoothing the skin are two uses for AHAs and BHAs. They help in the removal of dead skin cells from the skin's surface, hydration, and even the regeneration of skin cells on the face, body, and scalp. Their ability to plump the dermis and epidermis is due to their stimulation of collagen and hyaluronic acid synthesis, even though they also encourage the renewal of epidermal cells.

1.8.7 Polyphenols

Plant components used topically to prevent skin against UVR-induced photodamage have drawn a lot of interest. These plants include both culinary and medicinal herbs. Particularly researched extracts include those from tropical ferns, blueberries, coffee berries, milk thistle, feverfew, pomegranate, and green tea. They include a diverse range of polyphenols called "flavonoids," which are further classified based on their chemical structure into flavonols, flavones, catechins, and isoflavonoids. To counteract oxidative stress-induced cellular damage, plants produce them in tandem with ascorbic acid, Vitamin E, GSH, and several antioxidant enzymes (Ahmed et al. 2020).

Polyphenols are powerful antioxidants with several uses. The polyphenol-derived radical's capacity to bind (i.e., chelate) transition metal ions like Fe(II), stops the hydroxyl radical formation, Polyphenols are thought to smother superoxide anion radicals, peroxyl radicals, singlet oxygen, and hydroxyl radicals in addition. Also beneficial for aging and damaged skin are polyphenolic chemicals' anti-inflammatory and other qualities (Farris 2014).

Condensed proanthocyanidins, quinic acid, ferulic acid, and chlorogenic acid are only a few of the several (poly)phenolic chemicals found in the coffee berry or unripe coffee bean. In comparison to a vehicle, a skincare routine containing 1% coffee berry extract significantly reduced the indications of skin aging in a clinical investigation (Hsu 2005). A human reconstituted skin model demonstrated significant antioxidant and anti-inflammatory capabilities and reduced UVB-mediated damage in response to pomegranate fruit extract containing the polyphenol ellagic acid. Feverfew, another natural extract, has the ability to scavenge free radicals over a broad spectrum of reactive oxygen species (ROS) by depleting parthenolide levels. Topical administration of a tropical fern extract decreased erythema and UVA caused rapid pigment darkening, and delayed tanning when administered prior to UV exposure, according to a human research examining phenolic plant extracts (Gianeti et al. 2013).

1.8.8 Protein and Amino Acids

Molecules with both an amino and a carboxylic group are known as amino acids. Except for glycine, there are 20 other types of naturally occurring amino acids having optical active structures at the α-position (L-amino acids). The markets for several amino acids are also expanding quickly in the health food and pharmaceutical industries. In recent decades, the characteristics of amino acids have gained practical significance for cosmetic applications. The fact that amino acids are sustainable and ecologically friendly resources and that consumers' worries about the health and environmental effects of conventional chemical compounds are contributing to the expand usage of amino acids in cosmetic preparations. Common examples are using arginine (Arg) as a neutralizer instead of triethanolamine (TEA) and glutamic acid (Glu) as a substitute acid for hydrochloric acid (Tyagi et al. 2023) (Table 1.3).

1.9 Conclusion

To achieve sustainable development goals, the cosmetic industry needs to align the process of cosmetic formulation starting from the selection of raw materials, product development, packaging, and finally distribution to the consumers. The formulation process must be environmentally friendly and meet the sustainable development goal of sustainable consumption and production. Therefore, the selection of raw

Table 1.3 Beneficial amino acids in Cosmetics

Amino acid	Introduction	Benefit for skin	Reference
Arginine	It participates as building blocks of collagen	They make skin smooth and deals with fine lines and wrinkles	Zhou et al. (2021)
Methionine	One of the nine body's essential amino acids, responsible for tissue repair and growth	It protects the skin from harmful substances	Kim et al. (2017)
Histidine	It is also an essential amino acid known for its antioxidant properties	It is used to treat inflammatory skin conditions like dermatitis	Lintner (2007)
Lysine	It is an essential amino acid necessary for human health	It stimulates collagen and reduces wrinkles	Lintner (2007)
Glycine	It serves as a structural protein found in connective tissues	It is added to moisturizers, body lotions, and anti-aging creams	Waqas et al. (2015)
Glutamine	It is an essential amino acid. It increases the ability to make antioxidants	It protects the skin from skin impurities	Solano (2020)

materials plays an important role in fulfilling the concepts of sustainable development goals of good health and well-being. The formulation of cosmeceuticals for skin and hair has been made sustainable and eco-friendly with the introduction of active agents from natural resources. For example, anti-aging, antimicrobial, antioxidant, anti-inflammatory, and anti-cancer properties of plant-based chemically active agents have been employed for the formulation of multifunctional skin and hair-care cosmetics. Synthetic chemical ingredients in cosmetics have been successfully replaced with new therapeutically active components of natural origin, making the cosmeceutical according to the sustainable development goals. There is a need to develop the technology, to upgrade and upscale the laboratory scale research and development to the commercial level at the industrial scale.

References

Ahmadi-Ashtiani H-R, Baldisserotto A, Cesa E, Manfredini S, Sedghi Zadeh H, Ghafori Gorab M, Khanahmadi M, Zakizadeh S, Buso P, Vertuani S (2020) Microbial biosurfactants as key multifunctional ingredients for sustainable cosmetics. Cosmetics 7(2):46

Ahmed IA, Mikail MA, Zamakshshari N, Abdullah A-SH (2020) Natural anti-aging skincare: role and potential. Biogerontology 21:293–310

Alander JT (2012) Chemical and physical properties of emollients. In: Treatment of dry skin syndrome: the art and science of moisturizers. Springer, pp 399–417

Algama CH, Siriwardane I, Mubarak AM (2023) Curcumin against skin aging: a multimechanistic solution to a multi-origin issue. J Cosmet Sci 74(6):404

Allemann IB, Baumann L (2008) Antioxidants used in skin care formulations. Skin Therapy Lett 13(7):5–9

Anwar E, Rizkamiarty S (2020) Formulation and evaluation of cosmetic foundation using epigallocatechin gallate as a sun protection. Int J Appl Pharm 12:130–134

Arshad M (2024) Sustainability of renewable energy options as compared to coal-fired power plants in Pakistan. Green Low Carbon Econ 2(3):162–173

Arshad M, Adil M, Sikandar A, Hussain T (2014) Exploitation of meat industry by-products for biodiesel production: Pakistan's perspective. Pakistan J Life Soc Sci 12:120–125

Arshad M, Ansari AR, Qadir R, Tahir MH, Nadeem A, Mehmood T, Alhumade H, Khan N (2022) Green electricity generation from biogas of cattle manure: an assessment of potential and feasibility in Pakistan. Front Energy Res 10:911485

Arshad M, Syed-Hassan SSA, Masood R, Ansari AR, Mumtaz A, Rahman A, Batool S, Atif FA, Muhammad A, Rehman AU (2023) Utilization of waste animal fat for sustainable biodiesel production. In: Climate changes mitigation and sustainable bioenergy harvest through animal waste: sustainable environmental implications of animal waste. Springer, pp 149–169

Ascenso A, Margarida Ribeiro H, Cabral Marques H, Simoes S (2011) Topical delivery of antioxidants. Curr Drug Deliv 8(6):640–660

Avila Rodríguez MI, Rodríguez Barroso LG, Sánchez ML (2018) Collagen: A review on its sources and potential cosmetic applications. J Cosmet Dermatol 17(1):20–26

Baki G (2022) Introduction to cosmetic formulation and technology. Wiley

Barros C, Barros RBG (2020) Natural and organic cosmetics: definition and concepts. J Cosmetol Trichol 6:2

Benson HA, Newton A-F, Leite-Silva VR, Leonardi GR (2019) Powders in cosmetic formulations. In: Cosmetic formulation. CRC Press, pp 209–220

Blanco-Dávila F (2000) Beauty and the body: the origins of cosmetics. Plast Reconstr Surg 105(3):1196–1204

Blanco-Llamero C, Macário HF, Guedes BN, Fathi F, Oliveira MBP, Souto EB (2024) Bioactives in nutricosmetics: a focus on caffeine from tea to coffee. Cosmetics 11(5):149

Board AA, Larsen WG, Jackson EM, Barker MO, Bednarz RM, Engasser PG, O'Donoghue MN, Strauss JS (1992) A primer on cosmetics. J Am Acad Dermatol 27(3):469–484

Boo YC (2022) Ascorbic acid (vitamin C) as a cosmeceutical to increase dermal collagen for skin antiaging purposes: emerging combination therapies. Antioxidants 11(9):1663

Bryson PH (2022) Colored nail cosmetics and hardeners. In: Cosmetic dermatology: products and procedures. Wiley, pp 280–288

Camargo FB Jr, Minami MM, Rossan MR, Magalhaes WV, Porto Ferreira VT, Maia Campos PMBG (2022) Prevention of chemically induced hair damage by means of treatment based on proteins and polysaccharides. J Cosmet Dermatol 21(2):827–835

Chelazzi D, Bordes R, Giorgi R, Holmberg K, Baglioni P (2020) The use of surfactants in the cleaning of works of art. Curr Opin Colloid Interface Sci 45:108–123

Cheng A-D, De La Garza H, Maymone MB, Johansen VM, Vashi NA (2021) Skin-lightening products: consumer preferences and costs. Cureus 13(8):e17245

Coderch L, López O, de la Maza A, Parra JL (2003) Ceramides and skin function. Am J Clin Dermatol 4:107–129

Coelho SG, Choi W, Brenner M, Miyamura Y, Yamaguchi Y, Wolber R, Smuda C, Batzer J, Kolbe L, Ito S (2009) Short-and long-term effects of UV radiation on the pigmentation of human skin. J Investig Dermatol Symp Proc 14:32

Cui H, Cai R, Hua Z, Tang Y (2023) Plant colorants for natural hair coloration: dyeing optimization and photostability assessment. Sustain Chem Pharm 36:101285

Da França SA, Dario MF, Esteves VB, Baby AR, Velasco MVR (2015) Types of hair dye and their mechanisms of action. Cosmetics 2(2):110–126

Di Lorenzo R, Forgione F, Bernardi A, Sacchi A, Laneri S, Greco G (2023) Curcumin as a topical agent in clinical studies. Skin Pharmacol Physiol 36:235–248

Draelos ZD (2000) Colored facial cosmetics. Dermatol Clin 18(4):621–631

Draelos ZD (2001) Special considerations in eye cosmetics. Clin Dermatol 19(4):424–430

Draelos ZD (2008) The cosmeceutical realm. Clin Dermatol 26(6):627–632

Draelos ZD (2018a) Lip care and cosmetics. Aesthet Dermatol Curr Perspect 6:50–54
Draelos ZD (2018b) The science behind skin care: cleansers. J Cosmet Dermatol 17(1):8–14
Dreno B, Alexis A, Chuberre B, Marinovich M (2019) Safety of titanium dioxide nanoparticles in cosmetics. J Eur Acad Dermatol Venereol 33:34–46
Engasser PG (2000) Lip cosmetics. Dermatol Clin 18(4):641–649
Farris PK (2014) Cosmeceutical vitamins: vitamin C. In: Cosmeceuticals e-book: procedures in cosmetic dermatology series. Elsevier, pp 37–44
Faudone G, Arifi S, Merk D (2021) The medicinal chemistry of caffeine. J Med Chem 64(11):7156–7178
Fowler J (2012) Understanding the role of natural moisturizing factor in skin hydration. Pract Dermatol 9:36–40
George J, Sneed K, Pathak Y (2022) The skin aging process and anti-aging strategies. Biomed J Sci Tech Res 42:33377–33386
Giacomel C, Dartora G, Dienfethaeler H, Haas S (2013) Investigation on the use of expired make-up and microbiological contamination of mascaras. Int J Cosmet Sci 35(4):375–380
Gianeti MD, Mercurio DG, Maia Campos PM (2013) The use of green tea extract in cosmetic formulations: not only an antioxidant active ingredient. Dermatol Ther 26(3):267–271
Gonçalves RA, Holmberg K, Lindman B (2023) Cationic surfactants: A review. J Mol Liq 375:121335
Guerra-Tapia A, Gonzalez-Guerra E (2014) Hair cosmetics: dyes. Actas Dermosifiliogr 105(9):833–839
Hák T, Janoušková S, Moldan B (2016) Sustainable development goals: A need for relevant indicators. Ecol Indic 60:565–573
Halliday GM, Yuen KS, Bestak R, Barnetson RSC (1998) Sunscreens and vitamin E provide some protection to the skin immune system from solar-simulated UV radiation. Australas J Dermatol 39(2):71–75
Herman A, Herman A (2012) Caffeine's mechanisms of action and its cosmetic use. Skin Pharmacol Physiol 26(1):8–14
Heusèle C, Cantin H, Bonté F (2022) Lips and lipsticks. In: Cosmetic dermatology: products and procedures. Blackwell, pp 252–258
Hsu S (2005) Green tea and the skin. J Am Acad Dermatol 52(6):1049–1059
Hunter M, Bhola R, Yappert MC, Borchman D, Gerlach D (2015) Pilot study of the influence of eyeliner cosmetics on the molecular structure of human meibum. Ophthalmic Res 53(3):131–135
Joshi M, Hiremath P, John J, Ranadive N, Nandakumar K, Mudgal J (2023) Modulatory role of vitamins A, B3, C, D, and E on skin health, immunity, microbiome, and diseases. Pharmacol Rep 75(5):1096–1114
Kafi R, Kwak HSR, Schumacher WE, Cho S, Hanft VN, Hamilton TA, King AL, Neal JD, Varani J, Fisher GJ (2007) Improvement of naturally aged skin with vitamin A (retinol). Arch Dermatol 143(5):606–612
Kelm GR, Wickett RR (2017) The role of fatty acids in cosmetic technology. In: Fatty acids. Elsevier, pp 385–404
Kim KT, Kim JS, Kim M-H, Park J-H, Lee J-Y, Lee W, Min KK, Song MG, Choi C-Y, Kim W-S (2017) Effect of enhancers on in vitro and in vivo skin permeation and deposition of S-methyl-L-methionine. Biomol Ther 25(4):434
Kligman D (2000) Cosmeceuticals. Dermatol Clin 18(4):609–615
Kokande AM, Surana KR, Ahire ED, Keservani RK, Mahajan SK (2024) Vitamins as nutraceuticals for skin care. In: Preventive and therapeutic role of vitamins as nutraceuticals. Apple Academic Press, pp 235–252
Kong R, Cui Y, Fisher GJ, Wang X, Chen Y, Schneider LM, Majmudar G (2016) A comparative study of the effects of retinol and retinoic acid on histological, molecular, and clinical properties of human skin. J Cosmet Dermatol 15(1):49–57
Kumar AV, Garg VK, Buttar HS, Choudhary S, Sharma S, Grover A (2024) The protective role of vitamins in skincare and cosmeceutical products: mechanisms involved to re-engineer the

skin towards a healthy state. In: Hydrophilic vitamins in health and disease. Springer, Cham, pp 105–124

Lanzendörfer-Yu G (2021) Colorants in cosmetic applications. Phys Sci Rev 6(12):835–877

Latif A (2021) Chapter 1: History of cosmetics. In: Herbal cosmeto-dermatology. Sankalp Publication, pp 1–4

Li S, Yuan W, Deng G, Wang P, Yang P, Aggarwal B (2011) Chemical composition and product quality control of turmeric (Curcuma longa L.). Pharm Crop 5(1):28–54

Liao S, Kao Y-H, Hiipakka RA (2001) Green tea: biochemical and biological basis for health benefits. Vitam Horm 62:1–94

Lintner K (2007) Peptides, amino acids and proteins in skin care. Cosm Toilet 122(10):26

Lis DM, Baar K (2019) Effects of different vitamin C–enriched collagen derivatives on collagen synthesis. Int J Sport Nutr Exerc Metab 29(5):526–531

Lourith N, Kanlayavattanakul M (2009) Natural surfactants used in cosmetics: glycolipids. Int J Cosmet Sci 31(4):255–261

Lu Y-P, Lou Y-R, Xie J-G, Peng Q-Y, Liao J, Yang CS, Huang M-T, Conney AH (2002) Topical applications of caffeine or (−)-epigallocatechin gallate (EGCG) inhibit carcinogenesis and selectively increase apoptosis in UVB-induced skin tumors in mice. Proc Natl Acad Sci 99(19):12455–12460

Lu P-J, Huang S-C, Chen Y-P, Chiueh L-C, Shih DY-C (2015) Analysis of titanium dioxide and zinc oxide nanoparticles in cosmetics. J Food Drug Anal 23(3):587–594

Ma H, Yu M, Lei M, Tan F, Li N (2015) A novel topical targeting system of caffeine microemulsion for inhibiting UVB-induced skin tumor: characterization, optimization, and evaluation. AAPS PharmSciTech 16:905–913

Makrantonaki E, Zouboulis C (2007) Molecular mechanisms of skin aging: state of the art. Ann N Y Acad Sci 1119(1):40–50

Malabadi RB, Kolkar KP, Chalannavar R (2022) Plant natural pigment colorants-health benefits: toxicity of synthetic or artificial food colorants. Int J Innov Sci Res Rev 4:3418–3429

Mambwe B, Mellody KT, Kiss O, O'Connor C, Bell M, Watson RE, Langton AK (2024) Cosmetic retinoid use in photoaged skin: a review of the compounds, their use and mechanisms of action. Int J Cosmet Sci 47:45

Mawazi SM, Azreen Redzal NAB, Othman N, Alolayan SO (2022) Lipsticks history, formulations, and production: a narrative review. Cosmetics 9(1):25

McMullen RL, Dell'Acqua G (2023) History of natural ingredients in cosmetics. Cosmetics 10(3):71

Mijaljica D, Spada F, Harrison IP (2022) Skin cleansing without or with compromise: soaps and syndets. Molecules 27(6):2010

Millikan LE (2001) Cosmetology, cosmetics, cosmeceuticals: definitions and regulations. Clin Dermatol 19(4):371–374

Mohiuddin AK (2019) Skin care creams: formulation and use. Dermatol Clin Res 5(1):238–271

Morrison DS (2005) Petrolatum. In: Dry skin and moisturizers. CRC Press, pp 305–314

Na Takuathung M, Klinjan P, Sakuludomkan W, Dukaew N, Inpan R, Kongta R, Chaiyana W, Teekachunhatean S, Koonrungsesomboon N (2023) Efficacy and safety of the genistein nutraceutical product containing vitamin E, vitamin B3, and ceramide on skin health in postmenopausal women: A randomized, double-blind, placebo-controlled clinical trial. J Clin Med 12(4):1326

Nayak M, Ligade VS (2021) History of cosmetic in Egypt, India, and China. J Cosmet Sci 72(4):432

Ng A, Evans K, North R, Purslow C (2012) Eye cosmetic usage and associated ocular comfort. Ophthalmic Physiol Opt 32(6):501–507

Ng A, Evans K, North RV, Jones L, Purslow C (2016) Impact of eye cosmetics on the eye, adnexa, and ocular surface. Eye Contact Lens 42(4):211–220

Niederer M, Stebler T, Grob K (2016) Mineral oil and synthetic hydrocarbons in cosmetic lip products. Int J Cosmet Sci 38(2):194–200

Ojeh N, Stojadinovic O, Pastar I, Sawaya A, Yin N, Tomic-Canic M (2016) The effects of caffeine on wound healing. Int Wound J 13(5):605–613

Painter H (1992) Anionic surfactants. In: Detergents. Springer, pp 1–88

Rafiee Z, Nejatian M, Daeihamed M, Jafari SM (2019) Application of curcumin-loaded nanocarriers for food, drug and cosmetic purposes. Trends Food Sci Technol 88:445–458

Ricci A, Stefanuto L, Gasperi T, Bruni F, Tofani D (2024) Lipid Nanovesicles for antioxidant delivery in skin: liposomes, ufasomes, ethosomes, and niosomes. Antioxidants 13(12):1516

Rosenberg E, Mitchell R (1985) Microbial surfactants. Crit Rev Biotechnol 3(2):109–132

Rukpracha W, Sripalakit P, Saraphanchotiwitthaya A (2023) Formulation of herbal concealer containing fennel seed oil and lavender oil with anti-Propionibacterium acnes activity. *Asia-Pacific*. J Sci Technol 29(01):APST–29

Salvioni L, Morelli L, Ochoa E, Labra M, Fiandra L, Palugan L, Prosperi D, Colombo M (2021) The emerging role of nanotechnology in skincare. Adv Colloid Interf Sci 293:102437

Sarathchandraprakash N, Mahendra C, Prashanth S, Manral K, Babu U, Gowda D (2013) Emulsions and emulsifiers. Asian J Exp Chem 8:30–45

Shaath NA (1990) The chemistry of sunscreens. In: Sunscreens: development, evaluation and regulatory aspects, vol 10. Marcel Dekker, New York, pp 211–233

Shahi Z, Mehrizi MK, Hadizadeh M (2017) A review of the natural resources used to hair color and hair care products. J Pharm Sci Res 9(7):1026

Shapiro SS, Saliou C (2001) Role of vitamins in skin care. Nutrition 17(10):839–844

Smith VS (2008) Clean: A history of personal hygiene and purity. Oxford University Press

Solano F (2020) Metabolism and functions of amino acids in the skin. In: Wu G (ed) Amino acids in nutrition and health. Advances in experimental medicine and biology, vol 1265. Springer, Cham, pp 187–199

Solish N, Bertucci V, Percec I, Wagner T, Nogueira A, Mashburn J (2019) Dynamics of hyaluronic acid fillers formulated to maintain natural facial expression. J Cosmet Dermatol 18(3):738–746

Song S-m, Yu S-j, Kim J-H (2019) Sensory evaluation of semi-permanent eyebrow cosmetics depending on different pH levels. Asian J Beauty Cosmetol 17(2):179–185

Tang S-C, Yang J-H (2018) Dual effects of alpha-hydroxy acids on the skin. Molecules 23(4):863

Thangapazham RL, Sharad S, Maheshwari RK (2013) Skin regenerative potentials of curcumin. Biofactors 39(1):141–149

Tijare RB, Wargantiwar SS, George S, Shastri DR, Watkar MR, Bodhankar MM, Bobad DK (2022) Formulation of tinted lip gloss from lotus flower and date seed oil. Magna Sci Adv Biol Pharm 6:1–10

Tyagi N, Tyagi A, Rastogi R, Singh B, Nagarajan K (2023) Unlocking radiance: the dynamic duo of flaxseeds and aloe vera in hair mask for bioactive brilliance. Proteins 22:29–22

Vinardell MP, Mitjans M (2015) Nanocarriers for delivery of antioxidants on the skin. Cosmetics 2(4):342–354

Visconti MJ, Haidari W, Feldman SR (2020) Therapeutic use of caffeine in dermatology: a literature review. J Dermatol Dermatol Surg 24(1):18–24

Vollono L, Falconi M, Gaziano R, Iacovelli F, Dika E, Terracciano C, Bianchi L, Campione E (2019) Potential of curcumin in skin disorders. Nutrients 11(9):2169

Wallack G (2019) Rethinking FDA'S regulation of cosmetics. Harv J Legis 56:311

Waqas MK, Akhtar N, Mustafa R, Jamshaid M, Khan H, Murtaza G (2015) Dermatological and cosmeceutical benefits of Glycine max (soybean) and its active components. Acta Pol Pharm 72(1):3–11

Waseem, N., & Kota, S. (2017). *Sustainability definitions—an analysis*. In Research into design for communities, volume 2: Proceedings of ICoRD 2017,

Watts SS, Pal K, Asthana N, Bhattu M, Verma M (2024) Green synthetic approaches of caffeine extraction for cosmeceutical application: a review. J Mol Struct 1305:137733

Wohlrab J, Kreft D (2014) Niacinamide-mechanisms of action and its topical use in dermatology. Skin Pharmacol Physiol 27(6):311–315

Yarovaya L, Khunkitti W (2019) Effect of grape seed extract as a sunscreening booster. Songklanakarin J Sci Technol 41:708–715
Yildirim S (2022) The consumer role for sustainable development: how consumers contribute sustainable development goals. In: Research anthology on measuring and achieving sustainable development goals. IGI Global, pp 872–888
Younis MM, Ayoub IM, Mostafa NM, El Hassab MA, Eldehna WM, Al-Rashood ST, Eldahshan OA (2022) GC/MS profiling, anti-collagenase, anti-elastase, anti-tyrosinase and anti-hyaluronidase activities of a Stenocarpus sinuatus leaves extract. Plan Theory 11(7):918
Zaid AN, Al Ramahi R (2019) Depigmentation and anti-aging treatment by natural molecules. Curr Pharm Des 25(20):2292–2312
Zasada M, Budzisz E (2019) Retinoids: active molecules influencing skin structure formation in cosmetic and dermatological treatments. Postepy Dermatol Alergol 36(4):392–397
Zhang L, Adique A, Sarkar P, Shenai V, Sampath M, Lai R, Qi J, Wang M, Farage MA (2020) The impact of routine skin care on the quality of life. Cosmetics 7(3):59
Zhou Y, Liu G, Huang H, Wu J (2021) Advances and impact of arginine-based materials in wound healing. J Mater Chem B 9(34):6738–6750

Chapter 2
Skin: Structure, Function, and Permeability Studies for Delivery of Active Agents

Ch. Moazzam Hassan, Maryam Habiba, Shazia Abrar, Khayala Mammadova, and Azqa Amjad

Contents

C. M. Hassan
Department of Applied Chemistry, Government College University Faisalabad, Faisalabad, Pakistan

Punjab Forensic Science Agency, Lahore, Punjab, Pakistan

M. Habiba
Department of Applied Chemistry, Government College University Faisalabad, Faisalabad, Pakistan

School of Chemistry, Minhaj University Lahore, Lahore, Punjab, Pakistan

S. Abrar (✉) · A. Amjad
Department of Applied Chemistry, Government College University Faisalabad, Faisalabad, Pakistan
e-mail: shaziaabrar@gcuf.edu.pk

K. Mammadova
Medical and Biological Physics Department, Azerbaijan Medical University, Baku, Azerbaijan

S. Javed et al. (eds.), *Sustainable Cosmeceuticals*,
https://doi.org/10.1007/978-3-031-86087-4_2

Abstract Skin is the most visible and viable organ of the human body that constitutes more than 6% of the body by weight. Keratinized cells are the basic constituent of the skin which organize themselves in multiple layers, namely epidermis, dermis, and hypodermis. All of the three layers are well established to perform their specific function. Melanin present in the upper layer of the skin acts as a protective pigment and contributes to skin appearance as well. The few millimeter thick skin while acting as a protective barrier of the human body suffers unwanted alterations. These alterations may be augmented with solar radiations, oxidative stress, hormones, environment, diet, and other factors leading to serious skin concerns. When absorbed through the epidermis layer of the skin, various synthetic chemicals and toxic contaminants in substandard cosmetic products can affect the skin barrier functioning. Due to its high sensitivity to toxic substances, skin can suffer from allergic reaction, irritation, inflammation, and in extreme cases skin cancer. Therefore, it is important to understand the consequences of beauty products on the skin and ecosystem. The phenomenon of beautification must relate to Sustainable Development Goal 3 (SDG3) to ensure that beauty products for skin should not affect the health of the consumers. The semipermeability of skin somehow favors the therapeutic treatment by cosmetic and pharmaceutical formulations via topical application. Active agents are transported through skin epidermis via various delivery vehicles such as emulsion, gels, vesicles, and liposomes. A variety of products have been developed to enhance the aesthetic appearance and the health of the human skin, thus increasing research in this field.

Keywords Human skin · Skin structure · Skin layers · Skin permeability

2.1 Introduction

Human skin has been factually categorized as the most enormous organ of the body (McGrath et al. 2004; Kabashima et al. 2019; Montagna 2012; Guzmán and Lucia 2021). Different viewpoints ranked skin as a 15–18% weight contributor to the human body, initially considered as 7% (Kanitakis 2002; Edwards and Marks 1995; Adeel et al. 2022). Moreover, this vast spread of skin layer 1.8–2.0 m^2 also accounts for approximately 3.6 kg of body weight in an adult human being (Carvalho et al. 2020; Kanitakis 2002). The skin coating while extending from the head to the toes flawlessly alters its thickness from 0.5 mm below the eyelids to 4.0 mm at volar areas of the hands and toes (Chopra et al. 2015; Ha et al. 2005).

Human skin remains the fundamental organ of body aesthetics (McGrath et al. 2004; Jablonski 2004). In literature, skin has been considered aesthetic, beautiful, glowing, and luminous since old times (Misery et al. 2017; Venus et al. 2010). It tends to be the most visible organ of the body owing to its ample proportion (Walters and Roberts 2002; Montagna 2012; Adeel et al. 2022). Thus, the enormity of skin demands an enhanced skin care requirement. Cosmetic chemistry has evolved to deal with the interactions of skin (Youn et al. 2005; Connor 2004; Adeel et al. 2022). This chapter summarizes the basics of human skin in coordination with cosmetics.

The main purpose of skin care cosmetics is to make the skin beautiful, along with maintaining its health. Synthetic chemicals in skin care products not only affect the human skin but also have a terrible impact on the environment. The continuous application of synthetic chemicals, toxic substances, and prohibited drugs can severely affect the skin's physicochemical properties and functions. Cosmetic toxicity refers to the abnormal physiochemical changes in the skin structure and function. The long-term and continuous application of harmful chemicals used in the cosmetic products can result in irritation, allergic reactions, dermatitis, and skin cancer. For protecting the skin from external harmful effects and toxicity of synthetic ingredients, it is the need of the hour to introduce the novel and sustainable cosmetic products that are based on the biodegradable, nontoxic, eco-friendly, and sustainable raw materials. Herbal skin care cosmetics are formulated from natural, green, and biodegradable plant extracts, which ensure the skin health along with beauty, thus meeting the Sustainable Development Goal (SDG) of good health and well-being. The use of natural raw materials also ensures the responsible production and consumption according to the SDG12. There, herbal ingredients also offer nourishment besides cure for various skin ailments. The bioactive ingredients from plants may include vitamins, minerals, antioxidants, essential oils, alkaloids, tannins, and dyes. The cosmetic products that are used on a daily basis are moisturizing agents, lip care products, facial beauty creams, shampoos, conditioners, and hair grooming products. Therefore, it is extremely important to maintain the skin health and well-being via a sustainable approach while selecting beauty products in accordance with the SDG3.

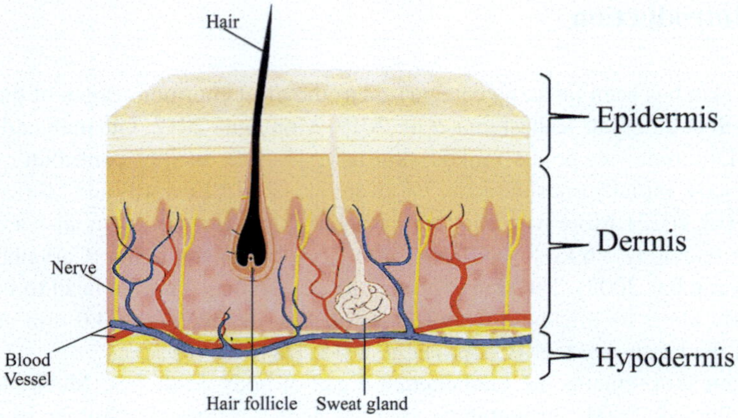

Fig. 2.1 Structure of human skin (Adobe Photoshop CC 2019-20.0.5)

2.2 Structure of Human Skin

The complex structure of skin originates from a nexus of three major layers, namely:

1. Epidermis
2. Dermis
3. Hypodermis

These layers persist uniformly throughout the human body, despite the varying thicknesses of skin at different locations (Ng and Lau 2015; Montagna 2012; Honari 2017) (Fig. 2.1).

2.2.1 Epidermis

The epidermis is the exposed and more precisely the foremost and outermost layer of the human skin and undoubtedly human body. It is a thin layer of 0.08-mm to 0.5-mm thickness, comprised of squamous epithelial cells. Four types of cells are found in the epidermis. Keratinocytes, the major building unit of the epidermis, account for 90–95% of epidermal cells. Melanocytes along with the Langerhans and Merkel cells comprise 5–10% of epidermal cells. The epidermis is well designed by the pragmatic combination of further 4–5 strata of cells varying upon the location of the skin (Montagna and Lobitz 2013; Boulais and Misery 2008; Landmann 1986; Baroni et al. 2012; Yousef et al. 2017; Whitear 1986).

2.2.1.1 Stratum Basale

This innermost or deepest layer of the epidermis is comprised of basal cells. The basal cells undergo mitotic cell division in the stratum basale or the basal cell layer. New cells are produced here; hence, it may also be called as the germination layer. Some of the basal cells are pushed upward to the stratum spinosum. Melanocytes are formed and retained in this layer. Melanocytes produce organelles called melanosomes which cater for the production of melanin in the skin. The presence of Merkel cells has also been discovered in stratum basale where they are closely associated with nerve endings. Merkel cells are very few and localized to the epidermis of volar areas and genitals (McGovern et al. 2013; Castellano-Pellicena et al. 2021; Murphrey et al. 2018; Xie et al. 2019; Yousef et al. 2017; Cordero and Casadevall 2020).

2.2.1.2 Stratum Spinosum

Stratum spinosum is also termed the squamous cell layer. It is the thickest layer of the epidermis. In this layer, the basal cells received from the basal layer are held together by proteins called desmosomes. The assembly of cells in this layer may appear as a spine or prickles. These mature cells with intercellular junctions are now keratinocytes or squamous cells. The mature keratinocytes produce keratin in this layer and are gradually forwarded to the stratum granulosum.

The Langerhans cells are also found in stratum granulosum. Langerhans cells tend to bind with the antigen and boost up and activate skin repair (Khavkin and Ellis 2011; Wolff and Stingl 1983; Baroni et al. 2012; Yousef et al. 2017; Xie et al. 2019).

2.2.1.3 Stratum Granulosum

This thin granular layer of the epidermis acquires the keratinocytes from the stratum spinosum. Keratinocytes may acquire a bigger size and in return become dehydrated. Thus, cells began to lose lipids and organelles in this layer and keratin-rich cells migrated outward. A water-resistant barrier of phospholipids (ceramides), fats, and cholesterol is formed which regulates the permeability of skin (Yoshida et al. 2013; Matsui 2023; Haftek et al. 2011).

2.2.1.4 Stratum Corneum

The superficial and the outermost layer of the epidermis is the stratum corneum (SC). It is often referred to as the horny layer owing to its hard and toughened appearance. Despite the loss of cell organelles, keratin production is still witnessed in this layer. Corneocytes are the dead keratinocytes that lack a nucleus and are arranged at the topmost layers of the stratum corneum. Corneocytes add to the hardness of the stratum corneum. Cells shed off the body from the stratum corneum and get replaced by cells from the inner layers (Menon et al. 2012; Harding 2004; Elias 2005; Egelrud 2000; Marks 2004).

2.2.1.5 Stratum Lucidum

An additional fifth layer of stratum lucidum has also been reported in thick skin regions such as palms, toes, and digits. This transparent layer may often be indistinguishable. It exists right below the stratum corneum in alliance with the stratum granulosum. In stratum lucidum, nuclei are distorted vigorously and a high keratin production occurs.

The epidermis is joined with the dermis via a narrow dermoepidermal junction. It cohesively binds the two layers together via rete ridges also termed as dermal papilla. Owing to capillary loops in the dermal papilla, the exchange of oxygen and nutrients occurs between the epidermis and dermis. Since the epidermis lacks any blood supply, it relies on the dermis for nourishment and supplementary support (Yousef et al. 2017; Montagna and Lobitz 2013; Honari 2017).

2.2.2 *Dermis*

The dermis is the middle layer of skin. The dermis varies from 1 to 5 mm contributing to 90% of skin thickness. It plays a crucial role in the transportation of nutrients, oxygen, waste, and water to the epidermis. The major constituents of the dermis are collagen and elastin besides blood vessels, nerve endings, lymph vessels, sweat glands, sebaceous glands, hair follicles, and nerve endings.

Collagen is a tough and insoluble protein that adds to the durability of the epidermis, whereas the elastin protein extends the stretchability and flexibility of skin. Blood vessels carry out the transportation of nutrients and other essentials to and from the epidermis (Landmann 1986; Brown and Krishnamurthy 2018; Lapiere 1990; Prost-Squarcioni et al. 2008; Krieg and Aumailley 2011).

The dermis is further subdivided into two layers.

2.2.2.1 Papillary Layer

A thin layer of collagen fiber is responsible for the supply of nutrients to epidermal layers. It also maintains blood flow and regulates the temperature (Brown and Krishnamurthy 2018; Woodley 2017; Rippa et al. 2019; Mizukoshi et al. 2016; Lynch et al. 2022; Wang et al. 2015).

2.2.2.2 Reticular Layer

A thick layer of collagen fiber is relatively denser than the papillary layer. It supports the appendages such as hair follicles, sweat glands, and sebaceous glands. The reticular layer of the dermis manages the strength of skin while adding to its structure and elasticity (Ogawa 2017; Brown and Krishnamurthy 2018; Rogovaya et al. 2021; Korosec et al. 2019).

2.2.3 Hypodermis

Hypodermis is the innermost layer of skin often termed as the subcutis or subcutaneous layer. It is a few mm thick; however, thickness alters from person to person and the location of the body. Largely composed of fat followed by collagen cells with interlaced blood vessels and nerve endings, the hypodermis is present beneath the dermis. Primarily, the hypodermis provides insulation and shock absorption to the skin (Monfort et al. 2013; Kim et al. 2019; Zimoch et al. 2021; Yousef et al. 2017; Khanal and Knoche 2014).

2.3 Functions of Skin

The skin of human beings tends to be a comparatively complex organ when analyzed with other mammals. It has been well designed to contain and protect the rest of the body while adding to aesthetics and appearance (Arshad 2024; Montagna 2012; Walters and Roberts 2002). Diverse functions of skin are witnessed in the human body which are summarized as follows.

2.3.1 Barrier Against Physical Stresses

Nerve endings are responsible for skin sensations. The feeling of physical stresses such as heat, cold, and pain is effectively transferred to the brain. Signals from the brain thereby protect the body from additional damage (Roosterman et al. 2006; Bangert et al. 2011; Kennedy 2004; Provitera et al. 2007). Thus, skin plays an important role in protecting the body from external harmful effect of chemicals and xenobiotics, while helping to achieve the goal of good health and well-being.

2.3.2 Pathogenic Defense

Lymph vessels play the role of lymph transportation to counter infections and microorganisms crossing the barrier of the skin, thus boosting the self-defense of the body against external pathogens (Monteiro-Riviere 2010; Archer 2004; Skobe and Detmar 2000; Lund et al. 2016). This pathogenic defense system of the skin keeps the skin healthy and fresh by fighting the infectious microorganism, meeting the standards of sustainable development goals of good health and well-being, thereby minimizing the use of dreadful external synthetic chemicals.

2.3.3 Body Hydration

Sweat glands produce sweat even up to 2 L per hour. Production of sweat maintains the amount of hydration and concentration of salts in the body while regulating the temperature. Sweat production is higher in hot and humid environments, whereas it is minimized in cold and low temperatures (Chuong et al. 2002; Montagna 2012; Shibasaki and Crandall 2010; Baker 2019). Sweat production is therefore directly proportional to the consumption of water. Skin flexibility, elasticity, and permeability functions depend on the transepidermal water level of the skin. Maintaining the skin moisture content by providing clean and fresh water free of any toxic materials is extremely important as already identified important in SDG6.

2.3.4 Texture

Sebaceous glands produce sebum mainly composed of oils from the skin. Sebum enhances the smoothness of skin, especially the volar areas. The appearance of a person in terms of texture and glow is thereby maintained and regulated by sebum (Zouboulis et al. 2008; Picardo et al. 2009; Thiboutot 2004; Shi et al. 2015; Zouboulis 2004, 2009). The skin's proper functioning and texture must be maintained by the application of skin care products containing lipids, antioxidants, and essential minerals.

2.3.5 Hair

Hair is formed as appendages of skin from keratinocytes in the epidermis. Hair follicles are contained in the dermis of the skin where they are provided with nourishment. Increased health of hair follicles results in adding to the volume and appearance of hair enhancing the visual aspects of a person. Moreover, hair also provides a resistance to external shocks from damaging the inner layers of the skin (Lange-Asschenfeldt et al. 2011; Lavker et al. 1993; Lai-Cheong and McGrath 2013; Motter Catarino et al. 2023; Yu et al. 2006).

2.3.6 Skin Tone

Skin contains melanin which is a colored pigment and is directly associated with the shade of skin. The color tone of skin has always been a considerable feature of beauty and well-being. Moreover, the skin color also acts as a distinguishing feature among races of human beings (Adeel et al. 2022; Brenner and Hearing 2008; Alaluf et al. 2002; Karsten and Smit 2012; Matts et al. 2007; Cichorek et al. 2013; Gasque and Jaffar-Bandjee 2015).

2.3.7 Vitamin D Synthesis and Bone Health

The vitamin D synthesis cycle has been reported to initiate in response to the exposure of skin to sunlight via the biochemical process. Vitamin D is a vital contributor to bone health. It enhances the strength of the body and inhibits the onset of various bone diseases such as arthritis (Bikle 2010, 2011, 2012; Piotrowska et al. 2016; Bikle and Christakos 2020; Kechichian and Ezzedine 2018; Mostafa and Hegazy 2015).

2.3.8 Selective Permeability

Fats and proteins in the body regulate the body temperature and protect internal organs from damage while acting as shock absorbents. Furthermore, the lipid barrier of epidermis controls the to-and-fro movement of water and other biomaterials establishing a characteristic of skin permeation (McCusker and Grant-Kels 2010; Caton et al. 2017; Michalak et al. 2021; Tobin 2006; Sanders 2024; Niyonsaba et al. 2017; Schröder and Harder 2006; Mori et al. 2008; Harder and Schröder 2005).

Skin acts as a permeable membrane, and around 70% of products topically applied to human skin last in the upper layers of skin. To augment the benefits of products applied to human skin, understanding the concept of skin permeation is necessary. Skin permeation refers to the ability of the product to travel through the stratum corneum. Products may range from the cosmetic, pharmaceutical, or cosmeceutical family. Furthermore, the permeation across the skin is regulated not only by the skin itself but also by the physicochemical characteristics of products such as size, shape, surface charge, surface area, and hydrophobicity/hydrophilicity. Therefore, the calculation of skin permeation is a basic requirement of product development targeted for the cosmetic industry (Chen et al. 2018; Afornali and Lorencini 2016; Schaefer et al. 2013).

Skin represents the first line of the defense system of the body, protecting it from external perturbations such as chemical/mechanical stress, UV radiations, and microbial infections. Therefore, necessary nutrients are compulsory through either diet or topical formulations for maintaining the skin health and well-being. An unhealthy diet and contaminated cosmetic products may alter the skin structure and its biological function, resulting in poor skin barrier functioning.

Nutrients are added as active agents in cosmetics called cosmeceuticals for treating various skin ailments along with improving skin health. Both macro- and micronutrients are vital for the skin to combat the exogenous and endogenous harmful effects. Important nutrients include vitamins (A, B, C, and E), minerals, and different essential elements. Skin wrinkles and pigmentation caused by exposure to UV radiations can be treated with the application of antioxidant cosmetic formulations. The skin collagen protein makes a major component of the dermis layer and maintains the skin texture smooth. A decrease in collagen contents due to UV radiations and reactive oxygen species can be treated via the application of cosmeceuticals containing amino acids and antioxidants (Chedik et al. 2024).

2.4 Mechanism of Skin Permeability and Absorption

The skin on humans has a protective function that makes it hard for foreign substances to get through. This keeps the body safe from external risks like germs, toxins, and dangerous chemicals. The stratum corneum (SC) is the biggest obstacle for molecules to move through, and it can even stop some nanometer-sized artificial objects (Bäsler et al. 2016). It is interesting to note that tests conducted on live organisms have shown that over 70% of skin care products do not pass the SC. On the other hand, nanoparticles with a diameter of around 40 nm or less can readily penetrate the skin. Additionally, different kinds of nanoparticles can pass through the skin simultaneously. When compared to ellipsoidal particles, spherical particles seem to have a greater possibility of passing through the skin. Based on the findings of a research that analyzed the two kinds, it was discovered that ellipsoidal particles were mostly discovered in the layers of the superficial layer (SC) and the epidermis, while spherical particles were discovered deeper into the epidermis and the dermis (Lundborg et al. 2018) (Fig. 2.2).

There are two ways that active chemicals may enter the skin: either they can diffuse across the whole epidermis or they can travel via areas of the skin such as hair follicles and sweat glands, which together account for approximately 0.1% of the total area of the human skin and serve as shunt routes for molecules. Intercellular communication, transcellular communication, and follicular communication are the three primary pathways that may be used to penetrate the intact skin barrier. The term "intercellular lipid route" refers to the phenomenon in which substances pass across corneocytes by way of interlamellar regions of the cell membrane. In this

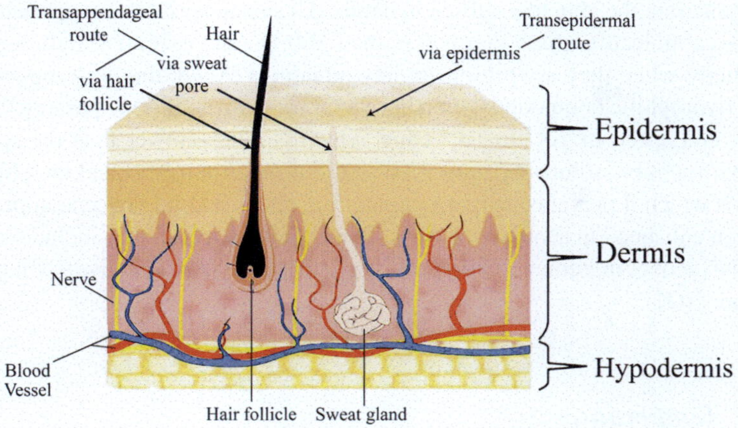

Fig. 2.2 Pathways of skin permeability (Adobe Photoshop CC 2019-20.0.5)

particular instance, fluid lipids indeed play a significant part in the transepidermal diffusion of molecules that are both lipidic and amphiphilic. The majority of hydrophilic molecules, on the other hand, diffuse "laterally" over the surfaces of water-filled gaps or via the empty spaces that exist between the lamella and the corneocyte outer membranes (Ruela et al. 2016). Transcellular paths involve molecules going across corneocytes, which have a keratin-rich macromolecular scaffold and lipids in between. Although it is believed that this route is shorter than intercellular pathways, the fact that chemicals have a more difficult time passing via transcellular diffusion means that it cannot be used for the transportation of illicit substances. Follicular pathways are often disregarded, despite the fact that they are of great significance. This is due to the fact that hair cells only constitute a tiny portion of the face. However, studies have been conducted to investigate the efficacy of follicular pathways once again. Follicular entrance is a sophisticated process that may have a lot of potential for absorbing items below the surface of the skin. This is because of the way that hair follicles are constructed (Kumar et al. 2015).

2.5 Factor Influencing Skin Permeability

There are three possible ways for a material to get into living skin: through the continuous stratum corneum, the hair shafts and the oil glands that are connected to them, or the sweat ducts. Transepidermal penetration is affected by both biological and physical factors. The health of the skin, age of the skin, blood flow, and metabolism of the skin are all examples of biological variables. Skin moisture, temperature, pH, diffusion coefficient, drug content, partition coefficient, molecular size, and shape are all examples of characteristics that are considered to be physical parameters (Dąbrowska et al. 2018). To be of any value, the majority of active substances do not penetrate the skin to a sufficient depth. All substances that are applied to the skin have a molecular mass that is less than 500 Da and exhibit a high degree of lipophilicity when they are present in tiny quantities. According to Azagury et al. (2014), hydrophilic compounds often have a far more difficult time passing through the skin at a speed of 10^{-7} to 10^{-8} cm/s, which is much slower than the speed at which hydrophobic solutes operate. As a result of the skin barrier and the effectiveness with which it prevents items from entering, chemical and physical approaches have been enhanced in a variety of ways. The chemistry techniques include ways to make the active ingredient with very high thermodynamic activity (Supe and Takudage 2021).

2.5.1 Temperature

The warmth of the skin can change how permeable it is. Conditions like being in the sun's infrared rays, having a fever, or having more blood vessels (more blood flow or more capillaries) raise the skin's temperature. Skin transparency and a rise in skin warmth may be linked to the fluidization of intercorneocyte lipid tails, which makes the dermis clear out more. It was found that heating the skin to about 40 °C increased the release of both testosterone and fentanyl through the skin. Controlled heat application could greatly improve nicotine uptake and blood flow to the skin from patches. However, it has been said that raising the skin's temperature may make it less likely for drugs to be absorbed because a gaseous penetrant evaporates more quickly and there are fewer of the compounds that are actually working on the skin's surface (Neupane et al. 2020).

2.5.2 Topical Delivery System

A topical delivery device, also called a "vehicle," is the thing that brings a drug in touch with and through the skin. Getting drugs through the skin barrier is hard for external drug release. The amount of the active drug that gets into the body can vary depending on the transport system. This is mostly because of the physiochemical features of the parts that make up that system. The right drug transport method will be chosen based on the patient's wishes and the active spot of the disease (Salim et al. 2016).

2.6 Types of Topical Delivery System in Cosmetics

2.6.1 *Creams*

These are dosage forms that are about half solid and include one or more medications that have been dissolved or spread out in a base that is acceptable. In general, this term has been used to refer to semisolids that are comprised of oil-in-water (O/W) or water-in-oil (W/O) emulsions that are reasonably soft and spreadable. On the other hand, as of late, the phrase has been restricted to solely referring to products that are composed of oil-in-water emulsions or watery microcrystalline dispersions of long-chain fatty acids or alcohols that can be washed off with water and have a more appealing appearance (Kim et al. 2020).

(O/W) creams, which are composed of tiny oil droplets that are spread out in a continuous phase, and water-in-oil (W/O) creams, which are composed of small water droplets that are spread out in a continuous oily phase, are the two primary varieties of creams. The emulsifying agent and the quantities of the two liquid phases are the primary factors that decide which of the two liquid phases, the oil phase or the water phase, will become the predominant dispersed phase. As a result of this, an oil-in-water (O/W) emulsion is one in which the oil is dispersed throughout the water phase in the form of droplets. When water is the solute and oil is the carrier, the kind of emulsion that is being discussed is referred to as a water-in-oil (W/O) type. In general, oil-in-water creams are more comfortable and seem more attractive to the majority of individuals. This is because they are less sticky and simpler to remove with water. The use of water-in-oil creams is more difficult than working with oil-in-water creams; nevertheless, many medications that are combined into creams are slippery and will be easier to remove from a water-in-oil cream than they would be from an oil-in-water cream. Additionally, water-in-oil creams are able to maintain a higher level of moisture in your skin because they create a sticky barrier on top of the stratum corneum, which is the top layer of the skin. It is essential for pharmacists to be aware of the kind of emulsion they are working with or have created, since this may have a significant impact on the way the emulsion functions and the characteristics it has. Since there are a few distinct procedures that have the potential to provide you with inaccurate findings, you should always utilize a second approach to demonstrate the kind of emulsion that was discovered by the first method (Costa and Santos 2017).

2.6.2 *Gels*

The pharmaceutical sciences have come up with a lot of new and improved ways to give medicines to treat a wide range of illnesses. The cases are designed to hold or encase the drug so that it can be delivered to the right place. Drugs are put into this container material with three-dimensional structures that take biocompatibility, stability, and flexibility into account. Gels are the name for the things in this group (Bhuyan et al. 2021). In the late 1800s, the word "gel" was created to give names to some semisolid substances based on how they behave physiologically. Gels have a lot of liquid in them, but they don't move when they're at rest. They're made up of a two-part semisolid system that is cross-linked and very weak. Gels are a popular way to deliver drugs because they keep the drug's chemical stability, have a good network structure, and are compatible with living things. The United States Pharmacopeia (USP) says that a gel is a semisolid system made up of a distribution of either large organic molecules or small artificial molecules that are connected and surrounded by liquid (Plamen et al. 2015). A house-of-cards shape is made by the solid particle. Gels are made up of two-phase systems. In these systems, solid particles are not dissolved; instead, they are spread out over the continuous phase and twist around randomly in the flexible chains. An increasing number of creams are being used in medicines and beauty products. A gel is mostly liquid and colloid, and its surface tension keeps it in place. The gelling agent then arranges the molecules into a large network of fibers (Choudhury et al. 2017).

Gel is a material that may be created from natural or man-made polymers that adhere to one another in three dimensions on top of a liquid that is particularly fond of water or a medium for the dissemination of substances. It is the gelling agent that makes gel stiff because it connects the particles and makes a network. What makes up the network and what the gel's properties are depend on the shape and maturity of the particles that form the links. As soon as the gel is put on the skin, the liquid disappears, leaving behind a thin film of gel-forming matrix that covers the skin (Antunes et al. 2017). The gel should be easy to use and put together; should not do anything, be safe, and work with other ingredients; shouldn't change how the drug works biologically; should stay steady while being stored; and should have qualities like being thixotropic, not coloring, moisturizing, and oil-free. The gelling agents in the mixture should give it a solid-like consistency while it's being stored, which can be easily broken by squeezing the tube, creating shear force by shaking the bottle, or applying it topically. It shouldn't stick to anything. Ideally, it should have the right antimicrobial agent to stop microbes from attacking (Coviello et al. 2015).

2.6.3 Patches

Cosmetic patches are one-of-a-kind ways to give vitamins, alpha-hydroxy acids, and other ingredients to the skin. They also make it possible to apply active ingredients directly to the skin. These common patches have several layers that are stacked on top of each other. The first layer is a backing layer that keeps the patch safe from the outside world (Lee and Prausnitz 2018). It surrounds the adhesive layer that goes on the inside and touches the skin during treatment. The adhesive layer is attached to the support layer and usually has one or more active chemicals in it. Pressure-sensitive adhesives (PSAs) are substances that stick to a surface when light force is applied and don't leave any residue when they are removed. Transdermal drug delivery systems (TDDS) also need pressure-sensitive glue to make sure that the drug-releasing part of the TDDS is in close touch with the skin. This is important for controlling the release of the drug (Sabbagh and Kim 2022). Pressure-sensitive glue, which has at least one substance that works on the skin's appearance, is also an important part of cosmetic patches. There is a release cover on the top layer. The release lining keeps the patch safe, but it is taken off before it is used. Keratolytic drugs, like lactic acid and glycolic acid, are now being used in cosmetics and skin care products. However, the ways that these chemicals work have not been well studied (Park et al. 2015).

2.7 Strategies for Enhancing Skin Penetration

The usage of transdermal routes dates back to the 1800s and has been used for quite some time. As a result of the significant increase in market value, transdermal drug delivery has emerged as one of the most rapidly expanding sectors within the pharmaceutical industry. A significant number of individuals are enthusiastic about the idea of applying medications to the skin, which may have effects not just locally but also throughout the body. Pharmaceuticals have a difficulty penetrating the skin, which is why strengthening procedures are often required (Yu et al. 2021). Topical application also avoids the body's way of turning drugs off and reduces the risk of stomach problems and other damaging effects (Champeau et al. 2019). There are a variety of technologies that have been developed in order to circumvent or alter the protective function of the skin in order to make it easier for medications to enter the dermal microcirculation. There are two categories that may be used to classify these technologies: physical and chemical (Dragicevic and Maibach 2017) (Fig. 2.3).

2.7.1 Physical Methods

Physical methods utilize physical approaches to enhance the penetration of active ingredients through the layers of skin. Most common methods include electric probes, heat, microneedling but not limited to these. Iontophoresis is a technique that involves applying a therapeutic material to the skin and then using an electric current to assist the chemical in penetrating deeper into the body. The drug is placed on an electrode that has the same charge as the drug, and a counter electrode that does not have any charge is placed on a separate area of the body on the other side of the body. Through the use of the active electrode, the active material is pushed into the skin and prevented from migrating from there. There are a number of different ways that iontophoresis might cause medications to pass deeper into the body. The first theory proposes that the medicine is transported over the skin by the straightforward force of charges that are, in some way, comparable to one another. It is possible for anionic medications to pass through the skin with the assistance of

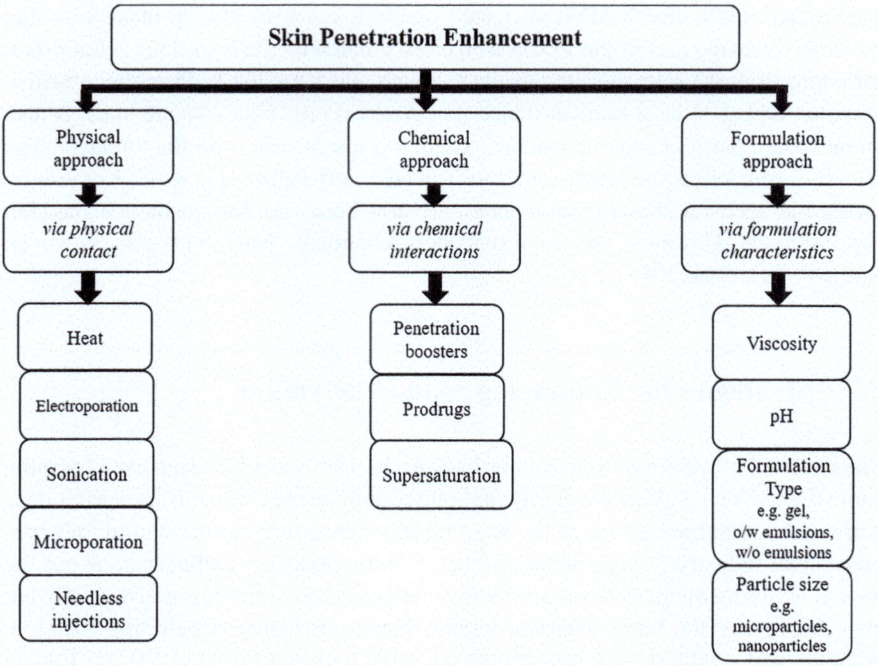

Fig. 2.3 Flowchart showing various strategies for skin penetration enhancement (Yu et al. 2021)

a functioning electrode that is negatively charged. It is also possible for cationic medications to flow through the skin when used in conjunction with a positively charged electrode. In the second idea, it is hypothesized that the electric current makes the penetration more severe by preventing the skin from defending itself against the penetration. The third one asserts that iontophoresis is a highly effective method for enhancing the entrance of water into the stratum corneum (SC) via the process of electro-osmosis. Iontophoresis through the skin would be a great way to deliver biotechnology-made hydrophilic drugs like peptides and oligonucleotides. The use of iontophoretic transport of drugs could help treat skin conditions like skin cancer, eczema, rashes, and hypertrophic marks. A lot of people have used it to fix problems with their eyes, ears, nose, teeth, and mouth. To get analytes (like glucose) out of the body, it has been used (Zheng et al. 2023).

2.7.1.1 Electroporation

The process of electroporation is yet another method for enhancing electrical impulses. Through the application of high voltage bursts that are just a few microseconds or milliseconds in duration, it is possible to achieve this effect. An assortment of mechanisms, including diffusion and electrophoresis, may be used to facilitate the movement of large molecules from the outside of the cell to the inside of the cell. Permeation is brought about by electric shocks, which result in the formation of temporary holes. The delivery of larger molecules, such as vaccines, oligonucleotides, microparticles, and insulin, has also been accomplished by the mechanism of electroporation. For the purpose of determining whether or not electroporation may enhance the skin absorption of certain model molecules, such as calcein and LHRH medications, research has also been conducted (Weaver and Chizmadzhev 2018).

2.7.1.2 Microporation

In the process of microporation, very small points are applied to the skin in such a way that they only penetrate the stratum corneum. Because of this, the skin becomes more porous. A microneedle is a needle that measures the length from 10 to 200 μm and the width from 10 to 50 μm. Due to the fact that microneedles do not exert pressure on nerves, the patient does not experience any discomfort or agony. The majority of the time, they consist of solid silicon projections that have been coated with medications or empty metal pins that have been loaded with pharmaceuticals (Mun et al. 2016).

2.7.1.3 Heat

The increase in bodily fluid flow, blood vessel wall permeability, rate-limiting membrane permeability, and drug solubility that occurs as a result of heat makes it simpler for medications to pass through the uppermost layer of the skin. Because of this, it is possible for medications to enter the systemic circulation with greater ease. When heat is added to the cell membrane, it is known to speed up the pace at which drug molecules, proteins, lipids, and carbohydrates move across the membrane. Another possibility is that an increase in temperature will make the medicine more soluble in the skin and in the patch. In a laboratory environment, we evaluated the effect of temperature on the cutaneous fentanyl flow for both 32 and 37 °C. During this five-degree range, the drug flow almost quadrupled again. According to the findings of further study, changes in temperature of around 5 °C are required in order to bring about changes that may be quantified in the permeability of cell membranes. Additionally, heat has the ability to alter the physiochemical properties of patches, cause you to sweat, and increase the amount of moisture in your skin, all of which make it simpler for medications to penetrate your skin (Dou et al. 2017).

2.7.1.4 Needleless Injection

It is possible to administer medications to the skin in a method that is painless by using a needleless shot. Particles, whether they are liquid or solid, are accelerated through the stratum corneum at rates that are greater than the speed of sound with this technique. It is necessary to push pressurized gases such as helium or nitrogen through the aperture in order for this to function properly (Arshad et al. 2018). It is believed that the drug particles are flying at a speed that allows them to enter the skin because they get entangled in the jet flow. This approach has a number of drawbacks, including the fact that it is costly to manufacture both the device and the dosage form and that it is not possible to program or regulate the release of the medicine to take into account variations in the degree to which different people's skin is permeable (Wu and Wang 2024).

2.7.2 Chemical Methods

Using things that improve permeation, for instance, the addition of penetration boosters changes the protective feature of the stratum corneum, which makes it convenient for drugs to be absorbed. An ideal penetration booster would possibly not influence the factors such as the mode of action of the active ingredient, irritation or allergic response, smell, taste, or color yet being safe for most drugs and excipients, cost-effective, and easily formulated. Over the last two decades, many new types of penetration enhancers have been created (Zsikó et al. 2019). Examples of these include surfactants (such as Tween, Span, and SLS), fatty acids (such as oleic acid), amines and amides (such as azone and *N*-methylpyrrolidone), terpenes (such as limonene), sulfoxides (such as dimethyl sulfoxide), and esters (such as isopropyl myristate). There are a number of different ways in which permeation enhancers can make the skin more permeable. Some of these ways include interaction with intercellular lipids in order to break up their structure and make them more fluid, removing lipids from the stratum corneum, moving bound water, loosening horny cells, delaminating the stratum corneum, increasing solubility and partitioning into the stratum corneum, interacting with intercellular protein, and denaturing keratin (Dragicevic and Maibach 2017).

2.7.2.1 Prodrug Approach

Prodrugs are chemical versions of drugs that work therapeutically. The medically active drug is made from a prodrug through a process called metabolism. The molecular features of a prodrug are different from those of its parent drug. It is more lipophilic. Different prodrugs for estradiol were made, and "Transdermal Bioactive Hormone Delivery" devices were made based on the findings. When transdermal bioactive hormones are delivered, the rate at which estradiol is released depends on how long the chain is at the 17th point of the ester group. Ketorolac alkyl ester prodrugs that are very good at sticking to fat could make transdermal release of ketorolac better. Also, using a prodrug is a very practical way to make protein/peptide drugs go deeper into the skin (Lobo and Yan 2018; Arshad 2024).

2.7.2.2 Supersaturation

The purpose of this technique is to improve skin penetration without causing any changes to the structure of the stratum corneum to occur. The increased thermodynamic activity of the medication is the basis for its mode of action. The occurrence of this event results in an increase in the concentration gradient (Co-Ci) in accordance with Fick's law: $\{(J = KD/h(\text{Co-Ci}))\}$, which in turn motivates the active principle to leave the stratum corneum. To produce supersaturated systems, one may use a wide range of methods, including the following: in order to improve the entry of hydrophilic and lipophilic components, one of the most important strategies is to optimize the hydration of the stratum corneum. It is thus possible that the partitioning of the permeant vehicle into the membrane might be affected by potential variations in the solubility of a permeant in the stratum corneum that are caused by the presence of free water inside the tissue (Cilurzo et al. 2015).

2.7.2.3 Advanced Formulation Approaches

The use of colloidal carriers is primarily responsible for the enhanced penetration that may be accomplished via certain formulation techniques. Nanoparticles with a size of less than 1 μm are intended to transport active substances that have been trapped in the skin. Liposomes, transferosomes, ethosomes, niosomes, nanoemulsions, and solid-lipid nanoparticles are many examples of the types of transporters that fall under this category. As a supplementary approach to enhance the delivery of pharmaceuticals via the skin, liposomes have garnered a lot of widespread support. Typically, phospholipids and cholesterol are the components that make up these vesicles, which are very small bilayer structures. They include components that are hydrophilic and lipophilic, and they have the potential to act as carriers for both polar and nonpolar medicinal preparations (Osborne and Musakhanian 2018). In order to release their medicament, the liposomes get entangled in the superficial layer of stem cells and interact with the lipids that are found in the skin. Liposomes that have been changed provide transferosomes with an expanded capacity to permeate the skin in an effective manner. The components that make up their makeup include phospholipids, cholesterol, and other surfactant molecules such as sodium cholate. It is possible for the transferosomes to endure compression via pores in the stratum corneum that are smaller than one tenth of their diameter because the surfactant molecules perform the job of edge activators. This provides the transferosomes with outstanding deformability. Transferosomes have been used as a carrier for a wide variety of chemicals, including proteins, immunomodulators, corticosteroids, nonsteroidal anti-inflammatory drugs (NSAIDs), anticancer medicines, and more. Ethosomes are structures that are mostly composed of phospholipids, concentrated alcohols, and glycols, and they are primarily composed of water. Their ability to enhance the penetration into deep tissues and the circulation throughout the body is one of their core capabilities (Arshad 2018). It is hypothesized that alcohol causes the ethosomal lipids and intercellular lipids of the stratum corneum to become fluid, which in turn makes it possible for the ethosome, which is soft and flexible, to penetrate the stratum corneum. It has been shown that they are able to transport acyclovir, minoxidil, propranolol, and testosterone in an effective manner. Additionally, they enhance the transdermal permeability of the skin for a number of different substances. The lipid microvesicles known as niosomes, which are composed of nonionic surfactants, have been evaluated for their potential usage as carriers in a variety of cosmetic and medicinal applications (Yu et al. 2021).

The creation of microemulsions is yet another type of formulation that aims to increase the skin's ability to absorb the substance. The formation of a transparent and thermodynamically stable liquid is the result of the formation of microemulsions, which are polymeric mixes of water, oil, and surfactant. Microemulsions are distinguished by a number of key features, including their ability to dissolve hydrophobic and hydrophilic ingredients, their thermodynamic stability, and their optical clarity. The increase in drug concentration is the key factor that leads to penetration enhancement in microemulsions. This rise in drug concentration causes a substantial concentration gradient to occur between the vehicle and the skin. Researchers have recently started looking at the possibility of using solid-lipid nanoparticles as transporters for the purpose of improving the delivery of sunscreens, vitamins A and E, triptolide, and glucocorticoids to the skin (Shukla et al. 2018).

2.7.3 Impact of Formulation Characteristics on Delivery Efficacy

The formulation properties of cosmetic products have a substantial influence on the delivery efficiency by determining the effectiveness with which active components reach and permeate the skin. Key considerations include the stability and compatibility of components, the formulation type (emulsions, gels, or serums), and the particle size of the active components. Enhanced penetration and stability may be achieved via delivery methods such as liposomes or nanoparticles, whereas the absorption rate is influenced by pH and viscosity. Humectants and occlusive agents aid in the retention of moisture and facilitate the delivery of ingredients, while penetration enhancers may further increase this effectiveness. Collectively, these parameters dictate the efficiency with which the active components are transported to and assimilated by the skin, therefore influencing the total efficacy of the product (Abdulbaqi et al. 2016).

2.8 Cosmetic Industry and Sustainable Development Goals

To make the cosmetic industry environmentally responsible, economically accessible, and socially equitable, sustainable development goals must be achieved. The purpose of sustainable development goal was to mark up the objective of making future generations have the same standards and quality of life as today. The main objective was to prepare an agenda planned for 2030. This agenda is a combined effort of countries to attain the goals of more bright future, clean environment, climate change, and clean energy along with promoting well-being and prosperity for future generations. The cosmetic industry can put an effort into achieving the sustainable development goals. This can be achieved by making the industrial process environmentally friendly, using nontoxic, biodegradable, sustainable raw materials from natural resources, cleaner energy production and consumption, biodegradable and environmentally friendly packaging, and making cosmetics that promote the physical health of the people along with beautification.

With the increase in consumption rate and increase in production, the cosmetic industry also has socio-environmental consequences. One important concern is the generation of microplastics and plastics in the environment. These plastics can be present as components of the formulation as exfoliating agents and secondary plastics come in the form of cosmetic product packaging. Since plastics are resistant to biodegradation, they can cause concern as persistent organic pollutants, ultimately affecting the aquatic environment and marine life.

Another important thing is using raw materials in cosmetics that are safe, nontoxic, biodegradable, and eco-friendly. This can be achieved by replacing synthetic, toxic, and nonbiodegradable raw materials, with safer and nontoxic and biodegradable raw materials, according to the SDG3 and SDG12 of responsible consumption and production.

2.9 Challenges

Innovation is crucial for maintaining competitiveness in the cosmetics sector. Nevertheless, the process of generating and studying novel product concepts may be arduous, particularly in light of progressive safety regulations and changing customer expectations. In order to effectively generate novel concepts for cosmetic products, producers must achieve a harmonious equilibrium between creativity and adherence to regulations. Undertaking comprehensive research and development is crucial to guarantee that new formulations adhere to safety standards and comply with legislation. By engaging in collaboration with safety assessors and ensuring transparent communication with consumers, businesses may successfully introduce groundbreaking items to the market while upholding regulatory required standards. Manufacturing a cosmetic product that is both safe and compliant is a challenging endeavor, since each product has several components such as raw ingredients, physical properties such as pH and viscosity, accurate labeling, adherence to good manufacturing procedures (GMP), packaging conditions, and shelf-life testing. The constituent components of cosmetic goods are interdependent. Prior to commencing product assembly, it is essential to ensure that your source of raw materials and packaging materials is well organized. Swift adaptability to changes is essential. These modifications may vary due to new client demands, legislation revisions, or suppliers unable to provide a certain raw material. Utilizing a simplified process is the only method to manage it. Your ability to get a comprehensive overview of the statuses of all your items is essential. Utilizing a central dashboard will provide you with a comprehensive insight. Furthermore, this capability enables you to identify the influence of the various modifications on your whole investment portfolio (Franca and Ueno 2020).

An inherent difficulty in procuring high-quality ingredients is the need to guarantee punctual delivery while simultaneously managing the trade-off between quality and expenses. To ensure uninterrupted production of cosmetic items, it is crucial to maintain a sufficient inventory of raw materials. In a dynamic and evolving global landscape, supply chains will inevitably be affected, e.g., the significant shipping disruptions during the COVID pandemic, the catastrophic Evergrande incident, or the repercussions of the conflict in Ukraine on sunflower oil demand. You may really need to completely alter your formulas. Such modifications may also affect the safety of these items, even if the modification is only applicable to a single batch of products. It is important that customers may consistently assume the safety of the items they use. Any minor modifications that affect the safety of the product must be reviewed by a safety assessor and documented in the Product Information File. In order to avoid an unnecessary rise in the administrative workload, it is advisable for you and your safety assessor to use a procedure or technological solution that may effectively minimize this strain (Oliveira et al. 2022).

2.10 Conclusion

Human skin, being the most captivating feature of the human body, requires a pragmatic approach for its nourishment and beautification. Skin care must be carefully dealt with. Skin care science must be keenly examined at each step from raw materials, formulations, by-products, excipients, and waste associated with each step. This approach is much needed to ensure compliance with the guidelines of sustainable development goals while coping up with the innovations of the modern era.

References

Abdulbaqi IM, Darwis Y, Khan NAK, Assi RA, Khan AA (2016) Ethosomal nanocarriers: the impact of constituents and formulation techniques on ethosomal properties, in vivo studies, and clinical trials. Int J Nanomedicine 11:2279–2304

Adeel S, Habiba M, Kiran S, Iqbal S, Abrar S, Hassan CM (2022) Utilization of colored extracts for the formulation of ecological friendly plant-based green products. Sustainability 14(18):11758

Afornali A, Lorencini M (2016) Nanoemulsions to prevent photoaging. In: Nanoscience in dermatology. Academic, New York, pp 237–246

Alaluf S, Atkins D, Barrett K, Blount M, Carter N, Heath A (2002) Ethnic variation in melanin content and composition in photoexposed and photoprotected human skin. Pigment Cell Res 15(2):112–118

Antunes AF, Pereira P, Reis C, Rijo P, Reis C (2017) Nanosystems for skin delivery: from drugs to cosmetics. Curr Drug Metab 18(5):412–425

Archer CB (2004) Functions of the skin. In: Rook's textbook of dermatology, vol 1. Wiley, New York, p 4.1

Arshad M (2018) Cogeneration policy to contest climate changes and the future of sustainable energy. IOP Conf Ser Earth Environ Sci 200:012050

Arshad M (2024) Sustainability of renewable energy options as compared to coal-fired power plants in Pakistan. Green Low Carbon Econ 2(3):162–173

Arshad M, Bano I, Younus M, Khan A, Rahman A (2018) Health concerns associated with biofuel production. In: Perspectives on water usage for biofuels production: aquatic contamination and climate change. Springer, Berlin, pp 97–105

Azagury A, Khoury L, Enden G, Kost J (2014) Ultrasound mediated transdermal drug delivery. Adv Drug Deliv Rev 72:127–143

Baker LB (2019) Physiology of sweat gland function: the roles of sweating and sweat composition in human health. Temperature (Austin) 6(3):211–259

Bangert C, Brunner PM, Stingl G (2011) Immune functions of the skin. Clin Dermatol 29(4):360–376

Baroni A, Buommino E, De Gregorio V, Ruocco E, Ruocco V, Wolf R (2012) Structure and function of the epidermis related to barrier properties. Clin Dermatol 30(3):257–262

Bäsler K, Bergmann S, Heisig M, Naegel A, Zorn-Kruppa M, Brandner JM (2016) The role of tight junctions in skin barrier function and dermal absorption. J Control Release 242:105–118

Bhuyan C, Saha D, Rabha B (2021) A brief review on topical gels as drug delivery system. J Pharm Res Int 33:344–357

Bikle DD (2010) Vitamin D and the skin. J Bone Miner Metab 28:117–130

Bikle DD (2011) Vitamin D metabolism and function in the skin. Mol Cell Endocrinol 347(1–2):80–89

Bikle DD (2012) Vitamin D and the skin: physiology and pathophysiology. Rev Endocr Metab Disorders 13:3–19

Bikle D, Christakos S (2020) New aspects of vitamin D metabolism and action—addressing the skin as source and target. Nat Rev Endocrinol 16(4):234–252

Boulais N, Misery L (2008) The epidermis: a sensory tissue. Eur J Dermatol 18(2):119–127

Brenner M, Hearing VJ (2008) The protective role of melanin against UV damage in human skin. Photochem Photobiol 84(3):539–549

Brown TM, Krishnamurthy K (2018) Histology, dermis. In: StatPearls. StatPearls Publishing, Treasure Island

Carvalho CP, Costa-Júnior JFS, Ferreira da Silva FF, D Agostinho JG, Krüger MAV, Pereira WCDA (2020) Using high-resolution ultrasound imaging to characterize dermal striae in human skin. Skin Res Technol 26(4):465–475

Castellano-Pellicena I, Morrison CG, Bell M, O'Connor C, Tobin DJ (2021) Melanin distribution in human skin: influence of cytoskeletal, polarity, and centrosome-related machinery of stratum basale keratinocytes. Int J Mol Sci 22(6):3143

Caton PW, Evans EA, Philpott MP, Hannen RF (2017) Can the skin make you fat? A role for the skin in regulating adipose tissue function and whole-body glucose and lipid homeostasis. Curr Opin Pharmacol 37:59–64

Champeau M, Vignoud S, Mortier L, Mordon S (2019) Photodynamic therapy for skin cancer: how to enhance drug penetration? J Photochem Photobiol B Biol 197:111544

Chedik L, Baybekov S, Cosnier F, Marcou G, Varnek A, Champmartin C (2024) An update of skin permeability data based on a systematic review of recent research. Sci Data 11(1):224

Chen CP, Chen CC, Huang CW, Chang YC (2018) Evaluating molecular properties involved in transport of small molecules in stratum corneum: a quantitative structure-activity relationship for skin permeability. Molecules 23(4):911

Chopra K, Calva D, Sosin M, Tadisina KK, Banda A, De La Cruz C et al (2015) A comprehensive examination of topographic thickness of skin in the human face. Aesthet Surg J 35(8):1007–1013

Choudhury H, Gorain B, Pandey M, Chatterjee LA, Sengupta P, Das A et al (2017) Recent update on nanoemulgel as topical drug delivery system. J Pharm Sci 106(7):1736–1751

Chuong CM, Nickoloff BJ, Elias PM, Goldsmith LA, Macher E, Maderson PA et al (2002) What is the 'true' function of skin? Exp Dermatol 11(2):159–187

Cichorek M, Wachulska M, Stasiewicz A, Tymińska A (2013) Skin melanocytes: biology and development. Adv Dermatol Allergol 30(1):30–41

Cilurzo F, Casiraghi A, Selmin F, Minghetti P (2015) Supersaturation as a tool for skin penetration enhancement. Curr Pharm Des 21(20):2733–2744

Connor S (2004) The book of skin. Cornell University Press, Ithaca

Cordero RJ, Casadevall A (2020) Melanin. Curr Biol 30(4):R142–R143

Costa R, Santos L (2017) Delivery systems for cosmetics—from manufacturing to the skin of natural antioxidants. Powder Technol 322:402–416

Coviello T, Trotta AM, Marianecci C, Carafa M, Di Marzio L, Rinaldi F et al (2015) Gel-embedded niosomes: preparation, characterization and release studies of a new system for topical drug delivery. Colloids Surf B: Biointerfaces 125:291–299

Dąbrowska AK, Spano F, Derler S, Adlhart C, Spencer ND, Rossi RM (2018) The relationship between skin function, barrier properties, and body-dependent factors. Skin Res Technol 24(2):165–174

Dou Y, Hynynen K, Allen C (2017) To heat or not to heat: challenges with clinical translation of thermosensitive liposomes. J Control Release 249:63–73

Dragicevic N, Maibach HI (eds) (2017) Percutaneous penetration enhancers physical methods in penetration enhancement (no. 25247). Springer, Berlin

Edwards C, Marks R (1995) Evaluation of biomechanical properties of human skin. Clin Dermatol 13(4):375–380

Egelrud T (2000) Desquamation in the stratum corneum. Acta Derm Venereol 80:44

Elias PM (2005) Stratum corneum defensive functions: an integrated view. J Invest Dermatol 125(2):183–200

Franca CCV, Ueno HM (2020) Green cosmetics: perspectives and challenges in the context of green chemistry. Desenvolvimento e Meio Ambiente 53:133–150

Gasque P, Jaffar-Bandjee MC (2015) The immunology and inflammatory responses of human melanocytes in infectious diseases. J Infect 71(4):413–421

Guzmán E, Lucia A (2021) Essential oils and their individual components in cosmetic products. Cosmetics 8(4):114

Ha RY, Nojima K, Adams WP Jr, Brown SA (2005) Analysis of facial skin thickness: defining the relative thickness index. Plast Reconstr Surg 115(6):1769–1773

Haftek M, Callejon S, Sandjeu Y, Padois K, Falson F, Pirot F et al (2011) Compartmentalization of the human stratum corneum by persistent tight junction-like structures. Exp Dermatol 20(8):617–621

Harder J, Schröder JM (2005) Antimicrobial peptides in human skin. In: Mechanisms of epithelial defense, vol 86. S.Karger AG, Basel, pp 22–41

Harding CR (2004) The stratum corneum: structure and function in health and disease. Dermatol Ther 17:6–15

Honari G (2017) Skin structure and function. In: Sensitive skin syndrome. CRC Press, Boca Raton, pp 16–22

Jablonski NG (2004) The evolution of human skin and skin color. Annu Rev Anthropol 33(1):585–623

Kabashima K, Honda T, Ginhoux F, Egawa G (2019) The immunological anatomy of the skin. Nat Rev Immunol 19(1):19–30

Kanitakis J (2002) Anatomy, histology and immunohistochemistry of normal human skin. Eur J Dermatol 12(4):390–401

Karsten AE, Smit JE (2012) Modeling and verification of melanin concentration on human skin type. Photochem Photobiol 88(2):469–474

Kechichian E, Ezzedine K (2018) Vitamin D and the skin: an update for dermatologists. Am J Clin Dermatol 19:223–235

Kennedy WR (2004) Opportunities afforded by the study of unmyelinated nerves in skin and other organs. Muscle Nerve 29(6):756–767

Khanal BP, Knoche M (2014) Mechanical properties of apple skin are determined by epidermis and hypodermis. J Am Soc Hortic Sci 139(2):139–147

Khavkin J, Ellis DA (2011) Aging skin: histology, physiology, and pathology. Facial Plast Surg Clin 19(2):229–234

Kim BS, Gao G, Kim JY, Cho DW (2019) 3D cell printing of perfusable vascularized human skin equivalent composed of epidermis, dermis, and hypodermis for better structural recapitulation of native skin. Adv Healthc Mater 8(7):1801019

Kim B, Cho HE, Moon SH, Ahn HJ, Bae S, Cho HD, An S (2020) Transdermal delivery systems in cosmetics. Biomed Dermatol 4:1–12

Korosec A, Frech S, Gesslbauer B, Vierhapper M, Radtke C, Petzelbauer P, Lichtenberger BM (2019) Lineage identity and location within the dermis determine the function of papillary and reticular fibroblasts in human skin. J Invest Dermatol 139(2):342–351

Krieg T, Aumailley M (2011) The extracellular matrix of the dermis: flexible structures with dynamic functions. Exp Dermatol 20(8):689–695

Kumar S, Zakrewsky M, Chen M, Menegatti S, Muraski JA, Mitragotri S (2015) Peptides as skin penetration enhancers: mechanisms of action. J Control Release 199:168–178

Lai-Cheong JE, McGrath JA (2013) Structure and function of skin, hair and nails. Medicine 41(6):317–320

Landmann L (1986) Epidermis and dermis. In: Biology of the integument: 2 vertebrates. Springer, Berlin, pp 150–187

Lange-Asschenfeldt B, Marenbach D, Lang C, Patzelt A, Ulrich M, Maltusch A et al (2011) Distribution of bacteria in the epidermal layers and hair follicles of the human skin. Skin Pharmacol Physiol 24(6):305–311

Lapiere CM (1990) The ageing dermis: the main cause for the appearance of 'old' skin. Br J Dermatol 122(s35):5–11

Lavker RM, Miller S, Wilson C, Cotsarelis G, Wei ZG, Yang JS, Sun TT (1993) Hair follicle stem cells: their location, role in hair cycle, and involvement in skin tumor formation. J Invest Dermatol 101(1):S16–S26

Lee JW, Prausnitz MR (2018) Drug delivery using microneedle patches: not just for skin. Expert Opin Drug Deliv 15(6):541–543

Lobo S, Yan G (2018) Improving the direct penetration into tissues underneath the skin with iontophoresis delivery of a ketoprofen cationic prodrug. Int J Pharm 535(1–2):228–236

Lund AW, Medler TR, Leachman SA, Coussens LM (2016) Lymphatic vessels, inflammation, and immunity in skin cancer. Cancer Discov 6(1):22–35

Lundborg M, Wennberg CL, Narangifard A, Lindahl E, Norlén L (2018) Predicting drug permeability through skin using molecular dynamics simulation. J Control Release 283:269–279

Lynch B, Pageon H, Le Blay H, Brizion S, Bastien P, Bornschlögl T, Domanov Y (2022) A mechanistic view on the aging human skin through ex vivo layer-by-layer analysis of mechanics and microstructure of facial and mammary dermis. Sci Rep 12(1):849

Marks R (2004) The stratum corneum barrier: the final frontier. J Nutr 134(8):2017S–2021S

Matsui T (2023) Epidermal barrier development via corneoptosis: a unique form of cell death in stratum granulosum cells. J Dev Biol 11(4):43

Matts PJ, Dykes PJ, Marks R (2007) The distribution of melanin in skin determined in vivo. Br J Dermatol 156(4):620–628

McCusker MM, Grant-Kels JM (2010) Healing fats of the skin: the structural and immunologic roles of the ω-6 and ω-3 fatty acids. Clin Dermatol 28(4):440–451

McGovern JA, Heinemann JR, Burke LJ, Dawson R, Parker TJ, Upton Z et al (2013) Stratum basale keratinocyte expression of the cell-surface glycoprotein CDCP1 during epidermogenesis and its role in keratinocyte migration. Br J Dermatol 168(3):496–503

McGrath JA, Eady RAJ, Pope FM (2004) Anatomy and organization of human skin. In: Rook's textbook of dermatology, vol 1. Wiley, New York, pp 3–2

Menon GK, Cleary GW, Lane ME (2012) The structure and function of the stratum corneum. Int J Pharm 435(1):3–9

Michalak M, Pierzak M, Kręcisz B, Suliga E (2021) Bioactive compounds for skin health: a review. Nutrients 13(1):203

Misery L, Stander S, Szepietowski JC, Reich A, Wallengren J, Evers AW et al (2017) Definition of sensitive skin: an expert position paper from the special interest group on sensitive skin of the international forum for the study of itch. Acta Derm Venereol 97(1):4–6

Mizukoshi K, Nakamura T, Oba A (2016) The relationship between dermal papillary structure and skin surface properties, color, and elasticity. Skin Res Technol 22(3):295–304

Monfort A, Soriano-Navarro M, García-Verdugo JM, Izeta A (2013) Production of human tissue-engineered skin trilayer on a plasma-based hypodermis. J Tissue Eng Regen Med 7(6):479–490

Montagna W (2012) The structure and function of skin. Elsevier, Amsterdam

Montagna W, Lobitz WC (eds) (2013) The epidermis. Elsevier, Amsterdam

Monteiro-Riviere NA (2010) Structure and function of skin. In: Toxicology of the skin. CRC Press, Boca Raton, pp 15–32

Mori S, Yoshizuka N, Takizawa M, Takema Y, Murase T, Tokimitsu I, Saito M (2008) Expression of uncoupling proteins in human skin and skin-derived cells. J Invest Dermatol 128(8):1894–1900

Mostafa WZ, Hegazy RA (2015) Vitamin D and the skin: focus on a complex relationship: a review. J Adv Res 6(6):793–804

Motter Catarino C, Cigaran Schuck D, Dechiario L, Karande P (2023) Incorporation of hair follicles in 3D bioprinted models of human skin. Sci Adv 9(41):eadg0297

Mun JY, Shin KK, Kwon O, Lim YT, Oh DB (2016) Minicircle microporation-based non-viral gene delivery improved the targeting of mesenchymal stem cells to an injury site. Biomaterials 101:310–320

Murphrey MB, Miao JH, Zito PM (2018) Histology, stratum corneum. In: StatPearls. StatPearls Publishing, Treasure Island

Neupane R, Boddu SH, Renukuntla J, Babu RJ, Tiwari AK (2020) Alternatives to biological skin in permeation studies: current trends and possibilities. Pharmaceutics 12(2):152

Ng KW, Lau WM (2015) Skin deep: the basics of human skin structure and drug penetration. In: Percutaneous penetration enhancers chemical methods in penetration enhancement: drug manipulation strategies and vehicle effects. Springer, Berlin, pp 3–11

Niyonsaba F, Kiatsurayanon C, Chieosilapatham P, Ogawa H (2017) Friends or foes? Host defense (antimicrobial) peptides and proteins in human skin diseases. Exp Dermatol 26(11):989–998

Ogawa R (2017) Keloid and hypertrophic scars are the result of chronic inflammation in the reticular dermis. Int J Mol Sci 18(3):606

Oliveira C, Coelho C, Teixeira JA, Ferreira-Santos P, Botelho CM (2022) Nanocarriers as active ingredients enhancers in the cosmetic industry—the European and North America regulation challenges. Molecules 27(5):1669

Osborne DW, Musakhanian J (2018) Skin penetration and permeation properties of Transcutol®—neat or diluted mixtures. AAPS PharmSciTech 19:3512–3533
Park Y, Park J, Chu GS, Kim KS, Sung JH, Kim B (2015) Transdermal delivery of cosmetic ingredients using dissolving polymer microneedle arrays. Biotechnol Bioprocess Eng 20:543–549
Picardo M, Ottaviani M, Camera E, Mastrofrancesco A (2009) Sebaceous gland lipids. Dermatoendocrinology 1(2):68–71
Piotrowska A, Wierzbicka J, Żmijewski M (2016) Vitamin D in the skin physiology and pathology. Acta Biochim Pol 63(1):17–29
Plamen KIRILOV, Le CAK, Halima R, Rum S, Carla V, Haftek M, Pirot F (2015) Organogels for cosmetic and dermo-cosmetic applications. Househ Pers Care Today 10:16–20
Prost-Squarcioni C, Fraitag S, Heller M, Boehm N (2008) Functional histology of dermis. Annales de dermatologie et de venereologie 135(1 Pt 2):1S5–1S120
Provitera V, Nolano M, Pagano A, Caporaso G, Stancanelli A, Santoro L (2007) Myelinated nerve endings in human skin. Muscle Nerve 35(6):767–775
Rippa AL, Kalabusheva EP, Vorotelyak EA (2019) Regeneration of dermis: scarring and cells involved. Cells 8(6):607
Rogovaya OS, Zupnik AO, Izmailova LS, Vorotelyak EA (2021) Morphofunctional characteristics of fibroblasts of the papillary and reticular layers of human skin dermis. Mosc Univ Biol Sci Bull 76(4):225–231
Roosterman D, Goerge T, Schneider SW, Bunnett NW, Steinhoff M (2006) Neuronal control of skin function: the skin as a neuroimmunoendocrine organ. Physiol Rev 86(4):1309–1379
Ruela ALM, Perissinato AG, Lino MEDS, Mudrik PS, Pereira GR (2016) Evaluation of skin absorption of drugs from topical and transdermal formulations. Braz J Pharm Sci 52(03):527–544
Sabbagh F, Kim BS (2022) Recent advances in polymeric transdermal drug delivery systems. J Control Release 341:132–146
Salim N, Ahmad N, Musa SH, Hashim R, Tadros TF, Basri M (2016) Nanoemulsion as a topical delivery system of antipsoriatic drugs. RSC Adv 6(8):6234–6250
Sanders TA (2024) Introduction: the role of fats in human diet. In: Functional dietary lipids. Elsevier, Amsterdam, pp 1–28
Schaefer H, Schalla W, Zesch A, Stüttgen G (2013) Skin permeability. Springer Science & Business Media, Berlin
Schröder JM, Harder J (2006) Antimicrobial skin peptides and proteins. Cell Mol Life Sci 63:469–486
Shi VY, Leo M, Hassoun L, Chahal DS, Maibach HI, Sivamani RK (2015) Role of sebaceous glands in inflammatory dermatoses. J Am Acad Dermatol 73(5):856–863
Shibasaki M, Crandall CG (2010) Mechanisms and controllers of eccrine sweating in humans. Front Biosci (Schol Ed) 2:685
Shukla T, Upmanyu N, Agrawal M, Saraf S, Saraf S, Alexander A (2018) Biomedical applications of microemulsion through dermal and transdermal route. Biomed Pharmacother 108:1477–1494
Skobe M, Detmar M (2000) Structure, function, and molecular control of the skin lymphatic system. J Investig Dermatol Symp Proc 5(1):14–19
Supe S, Takudage P (2021) Methods for evaluating penetration of drug into the skin: a review. Skin Res Technol 27(3):299–308
Thiboutot D (2004) Regulation of human sebaceous glands. J Invest Dermatol 123(1):1–12
Tobin DJ (2006) Biochemistry of human skin—our brain on the outside. Chem Soc Rev 35(1):52–67
Venus M, Waterman J, McNab I (2010) Basic physiology of the skin. Surgery (Oxford) 28(10):469–472
Walters KA, Roberts MS (2002) The structure and function of skin. In: Dermatological and transdermal formulations. CRC Press, Boca Raton, pp 19–58
Wang Y, Xu R, He W, Yao Z, Li H, Zhou J et al (2015) Three-dimensional histological structures of the human dermis. Tissue Eng Part C Methods 21(9):932–944
Weaver JC, Chizmadzhev Y (2018) Electroporation. In: Biological and medical aspects of electromagnetic fields, 4th edn. CRC Press, Boca Raton, pp 243–284

Whitear M (1986) Epidermis. In: Biology of the integument: 2 vertebrates. Springer, Berlin, pp 8–38
Wolff K, Stingl G (1983) The Langerhans cell. J Invest Dermatol 80:S17
Woodley DT (2017) Distinct fibroblasts in the papillary and reticular dermis: implications for wound healing. Dermatol Clin 35(1):95–100
Wu Y, Wang F (2024) A novel percutaneous delivery system—needle-free injection and its clinical applications in dermatology. Int J Dermatol Venereol 7:E54–E87
Xie J, Hao T, Li C, Wang X, Yu X, Liu L (2019) Automatic evaluation of stratum basale and dermal papillae using ultrahigh resolution optical coherence tomography. Biomed Signal Process Control 53:101527
Yoshida K, Yokouchi M, Nagao K, Ishii K, Amagai M, Kubo A (2013) Functional tight junction barrier localizes in the second layer of the stratum granulosum of human epidermis. J Dermatol Sci 71(2):89–99
Youn SW, Na JI, Choi SY, Huh CH, Park KC (2005) Regional and seasonal variations in facial sebum secretions: a proposal for the definition of combination skin type. Skin Res Technol 11(3):189–195
Yousef H, Alhajj M, Sharma S (2017) Anatomy, skin (integument), epidermis. In: StatPearls. StatPearls Publishing, Treasure Island
Yu M, Finner A, Shapiro J, Lo B, Barekatain A, McElwee KJ (2006) Hair follicles and their role in skin health. Expert Rev Dermatol 1(6):855–871
Yu YQ, Yang X, Wu XF, Fan YB (2021) Enhancing permeation of drug molecules across the skin via delivery in nanocarriers: novel strategies for effective transdermal applications. Front Bioeng Biotechnol 9:646554
Zheng H, Pu Z, Wu H, Li C, Zhang X, Li D (2023) Reverse iontophoresis with the development of flexible electronics: a review. Biosens Bioelectron 223:115036
Zimoch J, Zielinska D, Michalak-Micka K, Rütsche D, Böni R, Biedermann T, Klar AS (2021) Bio-engineering a prevascularized human tri-layered skin substitute containing a hypodermis. Acta Biomater 134:215–227
Zouboulis CC (2004) Acne and sebaceous gland function. Clin Dermatol 22(5):360–366
Zouboulis CC (2009) Sebaceous gland receptors. Dermatoendocrinology 1(2):77–80
Zouboulis CC, Baron JM, Böhm M, Kippenberger S, Kurzen H, Reichrath J, Thielitz A (2008) Frontiers in sebaceous gland biology and pathology. Exp Dermatol 17(6):542–551
Zsikó S, Csányi E, Kovács A, Budai-Szűcs M, Gácsi A, Berkó S (2019) Methods to evaluate skin penetration in vitro. Sci Pharm 87(3):19

Chapter 3
Herbal Extracts in Cosmeceuticals: Perspective Role to Achieve SDGs

Tahir Mehmood, Iqra Aman, Fareeha Nadeem, Maha Gul Zafar, and Muhammad Arshad

Contents

T. Mehmood (✉)
Institute of Microbiology and Molecular Genetics, University of the Punjab, Quaid-e-Azam Campus, Lahore, Pakistan
e-mail: tahir.mmg@pu.edu.pk

I. Aman
Centre for Applied Molecular Biology (CAMB), University of the Punjab, Lahore, Punjab, Pakistan

F. Nadeem
Institute of Biochemistry and Biotechnology, University of Veterinary and Animal Sciences, Lahore, Pakistan

M. G. Zafar · M. Arshad
Biochemistry Section, Jhang-campus, University of Veterinary and Animal Sciences Lahore, Lahore, Pakistan

S. Javed et al. (eds.), *Sustainable Cosmeceuticals*,
https://doi.org/10.1007/978-3-031-86087-4_3

Abstract The idea of beauty and beauty care products is essentially as old as humankind and civilization, with individuals, especially women, striving for a youthful and attractive appearance. This quest has led it to herbal cosmetics that use components of plants, such as leaves, bark, roots, seeds, and flowers known for their healing, nourishing, and protective properties. Herbal cosmetics are widely favored for their minimal side effects, and these underscore various skin issues such as aging, acne, and dryness. Commonly used herbal products are facewash, shampoo, conditioners, and soaps that offer effective solutions for skin care. A hybrid of cosmetics and pharmaceuticals is termed "cosmeceuticals." Many medicinal benefits are achieved from cosmeceuticals, and depending upon their functional components, cosmeceuticals affect the structure and biological functioning of skin, overall promoting skin health to enhance both the appearance and feel of the skin. The use of green chemistry concepts in herbal cosmetics promotes sustainability, lessens the impact on the environment by using renewable resources, reduces waste, and protects from harmful chemicals. This is in line with the Sustainable Development Goals (SDGs) of the UN. In addition to satisfying customer needs for safer, more environmentally friendly beauty products, the sector supports worldwide efforts in sustainable development by emphasizing eco-friendly formulas and ethical component sourcing. This chapter explores the development of herbal cosmetics, as well as their advantages, functional ingredients, potential for various treatments, and alignment with SDG and sustainability concepts.

Keywords Cosmetics · Cosmeceuticals · Herbs · Skincare · Herbal extracts

3.1 Introduction

Cosmetic products that contain active ingredients with potential therapeutic benefits for the skin are called cosmeceuticals (Pandey et al. 2019). Cosmeceuticals are basically skincare products that have both cosmetic and pharmaceutical properties. They contain active ingredients that can provide therapeutic benefits to the skin, going beyond just enhancing its appearance. These products are formulated to address specific skin concerns and can help improve the overall health and condition of your skin. They're like your secret weapon for a healthy and radiant complexion (Manela-Azulay and Bagatin 2009). They improve the appearance and health of the skin, targeting specific concerns such as aging, acne, or hyperpigmentation (Martin and Glaser 2011). Cosmeceuticals are like the ultimate skincare multitaskers. Cosmeceuticals are compounds and mixtures that are applied topically to treat a range of conditions, such as hyperpigmentation, dry skin, wrinkles, photoaging, dark spots, and hair damage. The skin and other human body components respond very well to these therapies. Growing consumer desire for customized and inventive products based on more in-depth scientific understanding has been a defining feature of the cosmetics industry's explosive growth (Salvioni et al. 2021).

3.2 Herbal Cosmetics

Effective herbal cosmetics are made using herbal Cosmetics, often known as products made from diverse natural herbal elements. Herbal remedies are not designed to provide an immediate response (Chatterjee et al. 2022). They give a means of realigning the body with nature because they are gentle on the skin and have no negative side effects, and herbal medications are becoming more and more popular (Sharma et al. 2022). In 1961, Raymond Reed, a member of the United States Society of Cosmetics, was a chemist who first used the term "cosmeceuticals." He employed the term to define cutting-edge and potent cosmetics. In 1984, Dr. Albert Kligman kept using the same term to characterize medications with both therapeutic and cosmetic uses. Cosmetic-pharmaceutical hybrids, referred to as cosmetic-pharmaceuticals and chemical compounds, influence the biological structure and functionality of the skin to enhance both its health and aesthetic attributes (Ferreira et al. 2021).

3.3 Advantages of Herbal Cosmetics over Synthetic Cosmetics

3.3.1 Easily Accessible

Throughout the whole product lifecycle, including the phases of harvesting, manufacturing, packing, distribution, consumer use, and post-use, sustainability techniques are implemented. Harvesting sustainably is essential for preserving biodiversity and lowering carbon emissions. Sustainable production, manufacturing, distribution, and consumer-related practices align with sustainable development goals (SDGs) (Kunatsa and Katerere 2021).

Natural ingredients used to make herbal cosmetics are free of all potentially dangerous synthetic compounds (Jadhav et al. 2017). Although made from organically occurring plant components and extracts, herbal cosmetics may be just as effective as synthetic products such as coconut oil and aloe vera gel (Joshi and Pawar 2015). They also include organic ingredients that beautify and nourish the skin, such as vitamin C and vitamin E. For example, aloe vera is a readily available naturally occurring herbal plant species in the Liliaceae family.

3.3.2 *Safe Use*

Scientists are investigating a variety of botanical resources in response to the increasing demand for sustainable and natural ingredients in cosmetic compositions (Gwanya et al. 2024). Herbal cosmetics are safe to use as comparatively synthetic items. They are less poisonous and less allergic, and dermatologists have examined and approved their safety for use because they are created with natural ingredients (Arshad et al. 2022; Jain et al. 2022).

3.3.3 *Fewer Adverse Effects*

Synthetic beauty products might itch your skin, create skin rash and occasionally obstruct skin pores, and make oily and dry skin (Shivanand et al. 2010). Herbal cosmetics don't contain parabens, the most common preservative in cosmetics that can invade the skin as well as being suspected of disrupting the endocrine system (Jain et al. 2022).

3.3.4 *Affordable to Use*

Socioeconomic characteristics such as gender, education, age, and marital status have a significant impact on the frequency of herbal product purchases and brand preference. This emphasizes the necessity for marketers to customize their products to target customers (Muthuswamy and Manoharan 2024). Natural cosmetics are generally more cost-effective, and in certain instances, they exhibit a lower price point compared to their synthetic counterparts (Katiyar et al. 2012). Traditional herbal remedies are accessible, affordable, and relatively safe and are presently encouraged and recommended by the World Health Organization in the context of natural healthcare programs (Jain et al. 2022).

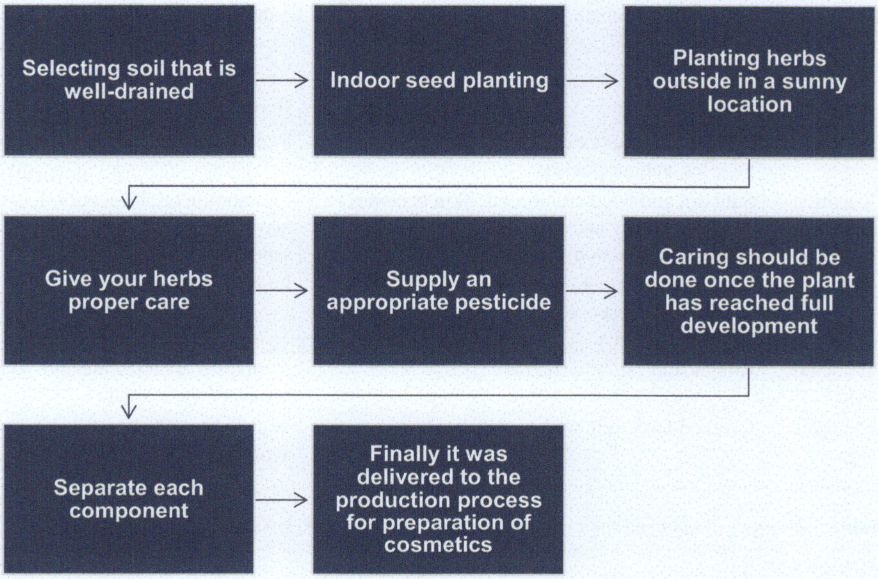

Fig. 3.1 Schematic diagram for herbing a plant (Saudagar and Sisodiya 2018)

3.3.5 How to Grow Herbs for Cosmetics

Figure 3.1 provides an overview of the process of growing herbal cosmetics.

3.4 Skin Cosmetic-Containing Herbals

Herbals containing cosmetics are restorative things that have mending or prescription-like advantages that can affect the natural functions of the pores and skin as per their substances (Simoes et al. 2015). All people have different skin types and have different textures of skin. Table 3.1 shows which herbal remedy should be applied according to the skin type (D'Orazio et al. 2013).

3.4.1 Types of Skin and Its Care

Types of skin always vary from person to person and understanding which skin type we have is very crucial (Table 3.2). Before using herbal cosmetics on your skin, it is important first to perform a patch test to know that rather it is having any adverse or bad effect on your skin or not. Usually, herbal cosmetics do not carry any side effects (Arshad et al. 2023; Celleno and Tamburi 2009).

Table 3.1 Types of skin problems and their herbal remedies (Ashawat et al. 2009; Bhowmik et al. 2010)

Problem of skin	Texture of skin	Herbal remedies
Acne	Sack of pollutants exhibiting red wounds, pimples, and air pockets	Red sandalwood oil
Sensitive skin	Sunburn and wind are normal. Skin becomes delicate and dry and is leaned to a negatively defenseless reaction	Chamomile oil, lavender oil, neroli oil, rose oil and sandalwood oil
Sallow skin	Without shading or concealing look, dull skin without the presence of essentialness, warmth, and cold sensitive	Intake of vitamin B
Withered skin	Rough skin full of wrinkles	A mixture of egg white and honey, carrot extract
Chapped skin	Cracked flaky skin with a rough texture	Warm olive oil massage and avocado

Table 3.2 Types of skin and their care (Aburjai and Natsheh 2003; Tabassum and Hamdani 2014)

Category of skin	Skin characteristics	Herbal-type skincare	Essential oils skincare
Normal skin	Flawless, oily flakes free, smooth textured skin with uniform tone	Pomegranate leaves extract	Sandalwood oil, fennel oil, chamomile oil, lemon oil, rose oil, lavender oil, geranium oil
Dry skin	Skin having low sebum level and inclined to sensitivity, has a dry look and tight texture. Drying and breakings are indications of dehydrated and very dry skin	Extracts of Aloe vera and Calendula comfrey	Almond oil, sandalwood oil, fennel oil, chamomile oil, lemon oil, rose oil, lavender oil, geranium oil
Oily skin	Shimmering, dull concealed, and thick. Smooth skin has pimples, coarse pores, and different blemishes. Leaned to obstructed pores	Aloe vera extract, extract of burdock root, horsetail, oat straw, lavender, lemon grass, chamomile thyme	Lemon oil, sage oil, geranium oil, evening primrose oil, bergamot oil, lavender oil, juniper oil
Combination skin	Dried out and scabrous skin, while the T zone of the face is oily	Aloe vera extract, extract of mint leaves or menthol, witch hazel, sweet flag, turmeric, sweet flag, wheat germ	Jasmine oil, citrus oils, and sandalwood oil

3.5 Anti-aging Herbal Extracts

3.5.1 Rhodiola rosea

Rhodiola rosea, often referred to as simply Rhodiola, is an herb with a long history

Fig. 3.2 *Rhodiola rosea*

of use in traditional medicine, particularly in regions like Siberia and the Arctic (Amsterdam and Panossian 2016) as shown in Fig. 3.2. While it's primarily known for its potential to combat stress, increase energy, and improve mental clarity, some proponents also suggest that it may have anti-aging properties (Anghelescu et al. 2018). As an adaptogen, Rhodiola may help the body adapt to a variety of stressors, both physical and mental. Chronic stress is considered a factor in premature aging, and adaptogens like Rhodiola may help mitigate these effects (Hassen et al. 2022). Rhodiola contains compounds with antioxidant properties, such as salidroside and rosavin. Chemicals called free radicals can damage cells and hasten aging. Antioxidants help to counteract them. It's important to note that while *Rhodiola rosea* has promising properties that could potentially support a healthy aging process, the scientific evidence regarding its specific anti-aging effects is still limited (Adaptogen 2001).

3.5.1.1 Bioactivity of *Rhodiola rosea* Constituents

Its pharmacology is very complex and contains numerous active substances, including flavonoids, monoterpenes, phenylpropanoids, triterpenes, phenolic acids, and derivatives of phenyl ethanol. Antioxidant, anticarcinogenic, cardioprotective, and neuroendocrine characteristics are only a few of its many impacts (Adaptogen 2001). Its psychopharmacological effects could include serotonin and dopamine agonist characteristics (perhaps as a result of monoamine oxidase inhibition), as well as an impact on opioid peptides (such as beta-endorphins). Rhodiola has been demonstrated to have certain antidepressant and psychostimulant effects.

Fig. 3.3 Carrot

3.5.2 Carrot

Carrots and carrot-derived ingredients are often included in natural and organic cosmeceutical products that focus on harnessing the benefits of plant-based ingredients for skincare (Tiwari and Cummins 2013) as shown in Fig. 3.3. When considering cosmeceutical products containing carrot extracts or carrot seed oil, if you have certain skincare difficulties or sensitivities, it's imperative that you go over the ingredient list of the product and speak with a dermatologist or skincare specialist. Carrot seed oil is derived from the seeds of the carrot plant (*Daucus carota*). It is rich in antioxidants, particularly beta-carotene and vitamin E. Carrot seed oil is used in cosmeceutical products for its antioxidant properties, which help protect the skin from free radical damage and premature aging (Dawid et al. 2015). It is often found in anti-aging skincare products. These are a natural source of vitamin A and its precursor, beta-carotene. Vitamin A is a key ingredient in many cosmeceutical formulations, as it promotes skin cell turnover, helps reduce the appearance of fine lines and wrinkles, and supports overall skin health (Carrot 2011).

3.5.2.1 Bioactivity of Carrot Constituents

The natural pigments in carrots can help to improve skin tone and reduce the appearance of dark spots and abnormalities in pigmentation. Carrot extracts are used in cosmeceuticals designed to brighten and rejuvenate the skin (Prasad et al. 2017). Carrots contain natural sugars and water, making them hydrating and beneficial for maintaining skin moisture (Singh et al. 2022). This property is used in cosmeceutical products to provide hydration and prevent dryness (Moelants 2013). The enzymes in carrots can help gently exfoliate the skin, promoting a smoother complexion. Some cosmeceutical formulations may include carrot-derived ingredients for their exfoliating effects. Antioxidants, which are abundant in carrots, help shield

Fig. 3.4 Gingko

the skin from pollution, UV rays, and environmental stress. Antioxidants can help maintain skin health and combat signs of aging (Carrot 2011).

3.5.3 Gingko

Ginkgo biloba, commonly referred to as Ginkgo, is an ancient tree species known for its fan-shaped leaves and therapeutic properties (Shahrajabian et al. 2022) as shown in Fig. 3.4. While it is primarily recognized for its potential health benefits when taken orally as a dietary supplement, Ginkgo is also used in cosmeceutical products for its skincare advantages (Lin et al. 2022). Ginkgo-based cosmeceuticals are available in various forms, such as serums, creams, masks, and under-eye treatments. When considering cosmeceutical products containing Ginkgo, it's important to review the ingredient list, consider your skin type, should get advice from a dermatologist or other skincare specialist if you have any particular sensitivities or concerns about your skin. Additionally, ginkgo-based products should be used in conjunction with a well-rounded skincare routine that includes sunscreen for comprehensive protection against premature aging and skin damage (Diamond et al. 2000).

3.5.3.1 Bioactivity of Gingko Constituents

Ginkgo extract is supposed to enhance both general blood circulation and skin microcirculation. Improved blood flow can promote a healthier complexion and may assist in reducing the appearance of dark circles under the eyes (Baek et al. 2015). Ginkgo has anti-inflammatory properties that can help soothe and calm irritated or sensitive skin. It may be included in cosmeceutical formulations designed for individuals with redness or skin conditions related to inflammation (Ude et al. 2013). Some cosmeceuticals incorporate Ginkgo for its potential to improve skin elasticity and firmness. These properties can be beneficial in anti-aging products, as they may help reduce the appearance of wrinkles and fine lines (Dziwenka and Coppock 2021). Ginkgo extracts can provide moisture to the skin, helping to keep it hydrated and plump. This hydration can be advantageous in skincare products aimed at maintaining skin suppleness and reducing dryness. Ginkgo contains enzymes that can help exfoliate the skin gently. Some cosmeceuticals include Ginkgo-derived ingredients for their exfoliating effects, promoting a smoother complexion (Chatterjee et al. 2003).

Ginkgo may assist in wound healing and scar reduction due to its potential to enhance blood circulation and promote tissue regeneration. While not a substitute for sunscreen, Ginkgo's antioxidant properties may offer some protection against UV damage when used in skincare products (Diamond et al. 2000).

3.6 Skin Brightening Herbal Extracts

3.6.1 Licorice Root Extract

Licorice root extract is a popular ingredient in skincare products, especially those designed to promote skin brightening and address hyperpigmentation issues (Fukuchi et al. 2016) as shown in Fig. 3.5. Licorice root extract is commonly used in various skincare products, including serums, creams, masks, and spot treatments,

Fig. 3.5 Licorice root extract

intended to address issues related to uneven skin tone and pigmentation (Kwon et al. 2020). Licorice root extract contains compounds, particularly glabridin and liquiritin that have demonstrated skin-brightening properties. These compounds can help fade dark spots, even out skin tone, and promote a more radiant complexion (Pan et al. 2000).

3.6.1.1 Bioactivity of Licorice Root Extract

Licorice root is a native plant of Europe and Asia. Its extract is frequently utilized in herbal treatments and skincare products (Račková et al. 2007). Due to its numerous health advantages, including its possible anti-inflammatory, antioxidant, and skin-brightening characteristics, it has been utilized for generations. Glabridin, which helps to prevent melanin formation and encourage a more uniform skin tone, is the primary active ingredient in licorice root (Ohno et al. 2014). It is renowned for its calming effects on the skin as well. Overall, licorice root is a well-liked component of skincare goods and can be a wonderful addition to your regimen (Fukuchi et al. 2016).

3.6.2 Mulberry Extract

Your first experience with elongated blackberries may come from mulberry. Its botanical name is Morus, and it is a member of the family Moraceae of flowering plants as shown in Fig. 3.6. They thrive in temperate to tropical settings and are quite simple to grow (Kadam et al. 2019). Three varieties of mulberries are typically farmed worldwide. The white mulberry, which is native to China, is grown in roughly 48 states in the United States (Killedar et al. 2017). Fruit from black mulberries is grown in

Fig. 3.6 Mulberry

Western Asia and Europe. The Eastern United States is where red mulberries were first domesticated. These fruits are very well-liked in China, Europe, and the United States (Wen et al. 2019). Mulberries typically appear in colors of white, red, and deep purple. Trees that bear mulberries are deciduous (Bendre et al. 2023).

3.6.2.1 Bioactivity of Mulberry Extracts

Mulberry extracts are another organic component that has several advantages for the skin. They come from the mulberry plant and are renowned for their ability to brighten the skin (Park et al. 2013). Arbutin and resveratrol, two substances found in mulberry extracts, may aid in suppressing the synthesis of melanin and lessen the appearance of dark patches and hyperpigmentation. Aside from that, mulberry extracts offer antioxidant qualities that help shield the skin from environmental harm and support a young appearance (Hemeg et al. 2020; Shao et al. 2013). For skin that is more radiant and healthy-looking, think about including products with mulberry extracts into your skincare routine. The chemical ingredients of the tree suppress the development of melanin and tyrosinase. These mulberry leaf extract chemical compounds were found utilizing a mushroom test, which is affordable and effective in screening for inhibitors of melanin formation (Bendre et al. 2023).

3.7 Acne-Fighting Herbal Extracts

3.7.1 Tea Tree Oil

Tea tree oil is a well-known herbal extract used in skincare (Fig. 3.7), particularly for its potent acne-fighting properties (Shao et al. 2013). Tea tree oil can be a valuable addition to your skincare routine for managing acne. Still, it should be used

Fig. 3.7 Tea tree oil

with care and in conjunction with other acne-fighting products and a proper skincare regimen.

3.7.1.1 Anti-inflammatory and Anti-bacterial Potential of Tea Tree Oil

Compounds like terpinen-4-ol, which is present in tea tree oil, have strong antibacterial properties. It can lessen irritation and stop the development of new pimples by aiding in the destruction of the acne-causing bacteria *Propionibacterium acnes* (*P. acnes*). Tea tree oil has anti-inflammatory properties that can help soothe redness and swelling associated with acne. It can also alleviate discomfort caused by inflamed pimples (Puvača et al. 2018). Tea tree oil has the ability to penetrate the skin and dissolve the oil and debris that clog pores. This helps in preventing the formation of whiteheads and blackheads. Tea tree oil can help regulate oil production in the skin, which is essential for preventing acne (Luo et al. 2021).

3.7.2 Witch Hazel

Witch hazel is a herbal extract as shown in Fig. 3.8 that is commonly used in skincare products, including those designed to combat acne (Abbas et al. 2020). It offers several properties that make it effective for addressing acne. Witch hazel contains tannins, which are natural compounds with astringent properties (Rasooly et al. 2020). When applied to the skin, witch hazel can help tighten and constrict tissues. This makes it beneficial for reducing the size of enlarged pores, which can become clogged and contribute to acne (Warade and Bhosale 2023).

Fig. 3.8 Hazel witch

3.7.2.1 Anti-acne Potential of Hazel Witch

Witch hazel's anti-inflammatory properties can help soothe and reduce the redness and swelling associated with acne lesions. It can provide relief to individuals with inflamed or irritated skin and has mild antimicrobial properties, which means it can help control the growth of acne-causing bacteria on the skin's surface (Ünlü 2022). This property can aid in preventing new acne breakouts. Witch hazel can help exfoliate the skin gently by removing dead skin cells and excess oil. This can contribute to a smoother complexion and reduce the likelihood of clogged pores. Witch hazel has a cooling and soothing effect when applied to the skin (Baumann 2007).

3.7.3 Neem Extract

Neem extract, derived from the neem tree (*Azadirachta indica*), is a well-known herbal remedy that is often used in skincare for its potential acne-fighting properties (Yasmin et al. 2020) as shown in Fig. 3.9. Neem extract is generally considered safe for most skin types, but individual responses may vary. Incorporating it into your skincare regimen, along with other acne-fighting ingredients, can contribute to clearer and healthier skin (Boelsma et al. 2001).

3.7.3.1 Bioactivity of Neem Extract

Neem extract contains compounds such as nimbin and nimbidin, which have strong antibacterial properties (Giri et al. 2019). These compounds help combat the proliferation of acne-causing bacteria, particularly *Propionibacterium acnes* (*P. acnes*),

Fig. 3.9 Neem

which contribute to the formation of pimples (Kalaskar et al. 2021). Neem has anti-inflammatory properties that can help reduce redness, swelling, and discomfort associated with acne lesions. It can provide relief to individuals with inflamed or irritated skin (Dua and Sharma 2023). Neem extract is rich in antioxidants, such as quercetin and catechin, which help protect the skin from free radical damage. Antioxidants can contribute to skin health and slow down the aging process (Pallai et al. 2021). Neem extract can help balance sebum (skin oil) production. By reducing excess oil on the skin's surface, it minimizes the risk of clogged pores and the formation of new pimples (Bansal et al. 2019). It has organic exfoliating ingredients that aid in getting rid of dead skin cells and keep them from clogging pores. This promotes a smoother complexion and reduces the likelihood of acne breakouts. It may aid in minimizing the appearance of acne scars and blemishes. Its anti-inflammatory and wound-healing properties can support the skin's recovery process and reduce scarring (Majeed et al. 2020).

3.8 Soothing and Healing Herbal Extract

3.8.1 Chamomile

Chamomile is a well-known herbal extract widely used in skincare for its soothing and healing properties. It is derived from the dried flowers of the chamomile plant, and it offers several benefits for skin health (Al-Dabbagh et al. 2019) as shown in Fig. 3.10. Chamomile is generally considered safe for most skin types, including

Fig. 3.10 Chamomile

sensitive skin. It is a gentle and versatile herbal extract that can benefit various aspects of skin health, from soothing irritation to promoting overall skin wellness (Poswal et al. 2019).

3.8.1.1 Bioactivity of Chamomile Constituents

Chamomile contains compounds like chamazulene and alpha-bisabolol, which have potent anti-inflammatory properties. These properties help reduce redness, swelling, and skin irritation, making chamomile ideal for calming sensitive or inflamed skin (Sayed et al. 2022). It has a natural calming effect on the skin. It can alleviate itching, burning sensations, and discomfort associated with various skin conditions, including sunburn, insect bites, and rashes. It can help lock in moisture and prevent dryness, making it suitable for dry or dehydrated skin. It provides a protective barrier that helps retain skin hydration (Holleran et al. 2020; Samuelson et al. 2020). Chamomile supports the skin's natural healing process. It can aid in the recovery of minor wounds, cuts, and scratches by promoting tissue repair and reducing the risk of infection and rich in antioxidants, such as flavonoids, which help combat free radicals. Antioxidants contribute to overall skin health and may help slow down the aging process (Silvestre et al. 2019).

3.9 Anti-oxidants and Anti-inflammatory Herbal Extracts

3.9.1 Vitamin C

It is a powerful antioxidant and has anti-inflammatory properties that make it highly beneficial for skincare. It is a water-soluble vitamin found in various fruits and vegetables, and when used topically in skincare products. Vitamin C is suitable for various skin types and concerns, and it is often included in anti-aging, brightening, and overall skin-enhancing skincare products (Enescu et al. 2022).

3.9.1.1 Photoprotective and Antiaging Potential of Vitamin C

Strong antioxidants like vitamin C aid in defending the skin from free radical damage brought on by pollution, UV rays, and other environmental stresses. This protection can slow down the aging process and contribute to overall skin health. A brighter complexion can be attained by reducing the visibility of black patches, hyperpigmentation, and uneven skin tone. It prevents the overproduction of melanin, the pigment that imparts skin its color. Vitamin C is essential for collagen synthesis, a protein that helps maintain skin elasticity and reduce the appearance of fine lines and wrinkles. Regular use of vitamin C can support firmer and more youthful-looking skin. Because of its anti-inflammatory qualities, it can help reduce swelling and redness that come with rosacea and other skin irritations. It can provide relief to individuals with sensitive or inflamed skin (Boo 2022; Krutmann et al. 2021).

3.9.2 Carotenoids

Carotenoids are natural pigments found in various fruits and vegetables, and they are known for their antioxidant properties. Carotenoids are often found in skincare products that focus on antioxidant protection, anti-aging, and overall skin enhancement. These products can include serums, moisturizers, and creams formulated with carotenoid-rich botanical extracts, such as carrot seed oil, marigold (calendula) extract, and red palm oil. Carotenoids offer several skincare benefits due to their antioxidant and anti-inflammatory qualities (Bin-Jumah et al. 2021).

3.9.2.1 Potential Role of Carotenoids in Skincare

Carotenoids, particularly beta-carotene, and strong antioxidants like zeaxanthin and lutein aid in shielding the skin from oxidative stress brought on by free radicals. This protection can delay the signs of aging and maintain overall skin health. Some carotenoids, like beta-carotene and astaxanthin, provide additional protection against UV radiation. While they are not a substitute for sunscreen, they can complement sun protection measures and reduce the harmful effects of UV exposure. They support skin health by promoting an even skin tone and reducing the appearance of dark spots and hyperpigmentation. They contribute to a healthier complexion and a more youthful appearance. Carotenoids can stimulate collagen production in the skin, helping to maintain elasticity and reduce the appearance of fine lines and wrinkles (Wathoni et al. 2018).

3.9.3 Caffeine

Caffeine is a natural compound found in coffee, tea, and various plants, and it is known for its antioxidant and anti-inflammatory properties. While it is not a traditional herbal extract, caffeine offers several benefits for skincare due to its unique characteristics. Caffeine is commonly found in skincare products designed to address various concerns, such as anti-aging, reducing puffiness, and enhancing overall skin health. These products may include eye creams, serums, moisturizers, and body lotions formulated with caffeine (Kole et al. 2005).

3.9.3.1 Benefits

Strong antioxidants like caffeine can shield the skin from harm caused by free radicals. It can neutralize harmful free radicals caused by UV radiation, pollution, and environmental stressors, thereby reducing the signs of premature aging (Ren and Chen 2020). Caffeine has anti-inflammatory properties that can help reduce redness, swelling, and skin irritation. It can provide relief to individuals with sensitive or inflamed skin, making it suitable for soothing conditions like rosacea. It can temporarily tighten and firm the skin by improving blood flow and stimulating collagen production (Bordeaux and Lieberman 2022; Kim and Priefer 2021). This effect can lead to a smoother and more youthful appearance.

3.9.4 Beta-Carotene

Beta-carotene is a natural pigment found in various fruits and vegetables, particularly those that are orange, red, and green in color. It is a type of carotenoid, which is a group of antioxidant compounds that have numerous health benefits (Larsson et al. 2007). Beta-carotene is found in various foods, including carrots, sweet potatoes, spinach, kale, and apricots. Consuming a diet rich in beta-carotene can contribute to healthy skin from the inside out. Additionally, beta-carotene is often included in topical skincare products, such as serums, moisturizers, and creams, to enhance skin health and appearance (Boelsma et al. 2001).

3.9.4.1 Benefits

Beta-carotene is a powerful antioxidant that helps protect the skin from free radical damage caused by UV radiation, pollution, and other environmental stressors. This protection can slow down the signs of aging and contribute to overall skin health (Wang et al. 2022). It promotes healthy skin by reducing the risk of skin disorders, such as acne, eczema, and psoriasis. It can also contribute to a more youthful complexion by improving skin texture and reducing the appearance of fine lines and wrinkles.

3.9.5 Retinol (Diet A)

Retinol, also known as vitamin A1, is a fat-soluble vitamin that functions as an antioxidant and anti-inflammatory compound in the body (Boonsong et al. 2011). Retinol-based skincare products, such as serums, creams, and lotions, are widely available and can be beneficial for addressing various skin concerns, including signs of aging, acne, and uneven skin tone (Colvard et al. 2006).

3.9.5.1 Benefits

Retinol acts as a powerful antioxidant, helping to protect the skin from the harmful effects of free radicals (Kim and Priefer 2021). Free radicals are unstable molecules that can damage skin cells, collagen, and DNA, contributing to premature aging and various skin concerns. Retinol promotes collagen production, a protein that provides structural support to the skin. Collagen is vital for maintaining skin elasticity, firmness, and a youthful appearance (Cohen and Portugal-Cohen 2020). By stimulating collagen synthesis, retinol helps reduce the appearance of fine lines and wrinkles. Retinol supports the skin's natural repair mechanisms, aiding in the healing of minor skin injuries, reducing the risk of scarring, and promoting an even skin tone.

3.10 Marketed Formulation

There are many products that are being used in markets and are referred to as herbal products for cosmetics. Some of them are shown in Table 3.3.

Table 3.3 Marketed formulations containing herbal additives (Bedi and Shenefelt 2002; Cohen et al. 2000; Kumari and Khurana 2013; Mattia et al. 2021)

Ingredient	Action	Source	Marketed formulation
Aloe vera	Softens the skin	Aloe vera	Natures gel, aloe vera gel
B-carotene	Lessens peroxidation activity as an antioxidant	Carrot and tomato	Evion cream
Calendula	Soothes and softens skin and promotes cell formation	*Calendula officinalis*	Calendula cream
Cucumber	Antioxidant, refreshes and tights pores	*Cucumis sativus*	Everyuth cream
Green tea extract	Antioxidant	Green tea	Alba botanica moisturizer
Licorice extract	Act as a bleaching agent or skin brightening	*Glycyrrhiza glabra*	Liquorice balm
Neem oil	Antimicrobial properties	*Azadirachta indica*	Himalaya neem face wash
Rosemary extract	Antioxidant and antimicrobial action	*Rosemarinus officinalis*	Loreal body cream
Turmeric	Antioxidant and antimicrobial action	*Curcuma longa*	Vicco turmeric cream
Vitamin A	Antioxidant	Vitamins A, C, E	Everyuth peel
Witch hazel extract	Tones skin	*Hamamelis virginiana*	Witch hazel cream

3.11 Environmental Impact of Cosmetic Waste and SDGs

The plant matter that is not utilized for food, feed, or fiber is left to naturally decompose over extended periods of time, producing fossil fuels (Arshad 2024). The clean and environmentally friendly energy is the need of the day (Arshad 2018). The cosmetic industry is rapidly growing, which has led to increased waste and chemical effluents, including surfactants, parabens, dyes, and titanium dioxide, which pose risks to humans as well as the environment, which highlights the need for advanced treatment techniques and sustainability practices. The packaging of cosmetic products is also very important to ensure sustainable practice. The use of biopolymers offers a more sustainable alternative to petroleum-based plastics in cosmetic packaging, addressing SDG rules 12, 13, 14, and 15. SDGs rule 12 targets responsible production and consumption, along with proper waste disposal and management, while 14 and 15 mainly target the conservation of marine and terrestrial ecosystems by combating pollution and environmental impact imposed by chemicals, heavy metals, and microplastics. So, the implementation of SDGs in cosmetic practice not only ensures the product's sustainability and efficacy but also ensures the sustainability of the ecosystem and climate.

3.12 Conclusion

In conclusion, herbal extracts play a beneficial role in cosmeceuticals. They offer a natural approach to skincare in anti-aging, skin brightening, and skin acne. These herbal products are available in markets as they are safe to use and their source is a natural plant or a species. Many herbal extracts also have anti-inflammatory and anti-oxidant properties, which protect the skin from oxidative stress and reduce redness or itching. These cosmetic products are assumed to be safe for longer periods of time. The demand for herbal cosmetics is rapidly increasing day by day because these herbal products are of low cost, side effects free, and are environment friendly, and they are having great future ahead as compared to synthetic cosmetics.

References

Abbas TF, Abbas MF, Lafta AJ (2020) Antibacterial activity and medical properties of witch hazel Hamamelis virginiana. Ann Trop Med Public Health 23(11):46

Aburjai T, Natsheh FM (2003) Plants used in cosmetics. Phytother Res 17(9):987–1000

Adaptogen APP (2001) Rhodiola rosea: a possible plant adaptogen. Altern Med Rev 6(3):293–302

Al-Dabbagh B, Elhaty IA, Elhaw M, Murali C, Al Mansoori A, Awad B, Amin A (2019) Antioxidant and anticancer activities of chamomile (Matricaria recutita L.). BMC Res Notes 12(1):1–8

Amsterdam JD, Panossian AG (2016) Rhodiola rosea L. as a putative botanical antidepressant. Phytomedicine 23(7):770–783

Anghelescu I-G, Edwards D, Seifritz E, Kasper S (2018) Stress management and the role of Rhodiola rosea: a review. Int J Psychiatry Clin Pract 22(4):242–252
Arshad M (2018) Cogeneration policy to contest climate changes and the future of sustainable energy. IOP Conf Ser Earth Environ Sci 200:012050
Arshad M (2024) Sustainability of renewable energy options as compared to coal-fired power plants in Pakistan. Green Low Carbon Econ 2(3):162–173
Arshad M, Ansari AR, Qadir R, Tahir MH, Nadeem A, Mehmood T, Alhumade H, Khan N (2022) Green electricity generation from biogas of cattle manure: an assessment of potential and feasibility in Pakistan. Front Energy Res 10:911485
Arshad M, Syed-Hassan SSA, Masood R, Ansari AR, Mumtaz A, Rahman A, Batool S, Atif FA, Muhammad A, Rehman AU (2023) Utilization of waste animal fat for sustainable biodiesel production. In: Climate changes mitigation and sustainable bioenergy harvest through animal waste: sustainable environmental implications of animal waste. Springer, Berlin, pp 149–169
Ashawat M, Banchhor M, Saraf S, Saraf S (2009) Herbal cosmetics: "trends in skin care formulation". Pharmacogn Rev 3(5):82
Baek K-H, Sharma A, Bajpai VK (2015) Antibacterial mode of action of Ginkgo biloba leaf essential oil: effect on morphology and membrane permeability. Bangladesh J Pharmacol 10(2):337–350
Bansal V, Gupta M, Bhaduri T, Shaikh SA, Sayed FR, Bansal V, Agrawal A (2019) Assessment of antimicrobial effectiveness of neem and clove extract against Streptococcus mutans and Candida albicans: an in vitro study. Nigerian Med J 60(6):285
Baumann LS (2007) Less-known botanical cosmeceuticals. Dermatol Ther 20(5):330–342
Bedi MK, Shenefelt PD (2002) Herbal therapy in dermatology. Arch Dermatol 138(2):232–242
Bendre S, Karodi R, Dange S, Rajad S (2023) Formulation and evaluation of skin alleviating suntan SLN lotion of herbal extracts. J Coast Life Med 11:1691–1712
Bhowmik D, Chiranjib YJ, Tripathi K, Kumar K (2010) Herbal remedies of Azadirachta indica and its medicinal application. J Chem Pharm Res 2(1):62–72
Bin-Jumah M, Alwakeel SS, Moga M, Buvnariu L, Bigiu N, Zia-Ul-Haq M (2021) Application of carotenoids in cosmetics. In: Carotenoids: structure and function in the human body. Springer, Berlin, pp 747–756
Boelsma E, Hendriks HF, Roza L (2001) Nutritional skin care: health effects of micronutrients and fatty acids. Am J Clin Nutr 73(5):853–864
Boo YC (2022) Ascorbic acid (vitamin C) as a cosmeceutical to increase dermal collagen for skin antiaging purposes: emerging combination therapies. Antioxidants 11(9):1663
Boonsong P, Laohakunjit N, Kerdchoechuen O, Matta FB (2011) Detection of pigments and natural colorants from Thai herbal plants for possible use as coloring dyes. HortScience 46(2):265–272
Bordeaux B, Lieberman HR (2022) Benefits and risks of caffeine and caffeinated beverages. UpToDate
Carrot W (2011) Carrot: history and iconography. Chronica 51(2):13
Celleno L, Tamburi F (2009) Structure and function of the skin. In: Nutritional cosmetics. Elsevier, Amsterdam, pp 3–45
Chatterjee S, Kondratskaya E, Krishtal O (2003) Structure-activity studies with Ginkgo biloba extract constituents as receptor-gated chloride channel blockers and modulators. Pharmacopsychiatry 36(Suppl 1):68–77
Chatterjee S, Kataruka A, Mukherjee S, Adhikari A, Bhattacharjee A (2022) A review on medicinal benefits and applicability of herbal ingredients in cosmeceuticals. Int J Pharm Sci Invent 11(3):1–9
Cohen D, Portugal-Cohen M (2020) Safe retinol-like skin biological effect by a new complex, enriched with retinol precursors. J Cosmet Dermatol Sci Appl 10(2):59
Cohen SM, Rousseau ME, Robinson EH (2000) Therapeutic use of selected herbs. Holist Nurs Pract 14(3):59–68

Colvard MD, Cordell GA, Villalobos R, Sancho G, Soejarto DD, Pestle W, Echeverri TL, Perkowitz KM, Michel J (2006) Survey of medical ethnobotanicals for dental and oral medicine conditions and pathologies. J Ethnopharmacol 107(1):134–142

D'Orazio J, Jarrett S, Amaro-Ortiz A, Scott T (2013) UV radiation and the skin. Int J Mol Sci 14(6):12222–12248

Dawid C, Dunemann F, Schwab W, Nothnagel T, Hofmann T (2015) Bioactive C17-polyacetylenes in carrots (Daucus carota L.): current knowledge and future perspectives. J Agric Food Chem 63(42):9211–9222

Diamond BJ, Shiflett SC, Feiwel N, Matheis RJ, Noskin O, Richards JA, Schoenberger NE (2000) Ginkgo biloba extract: mechanisms and clinical indications. Arch Phys Med Rehabil 81(5):668–678

Dua S, Sharma M (2023) Preparation of herbal face pack cum scrub using papaya and neem extracts

Dziwenka M, Coppock RW (2021) Ginkgo biloba. In: Nutraceuticals. Elsevier, Amsterdam, pp 835–852

Enescu CD, Bedford LM, Potts G, Fahs F (2022) A review of topical vitamin C derivatives and their efficacy. J Cosmet Dermatol 21(6):2349–2359

Ferreira MS, Magalhães MC, Oliveira R, Sousa-Lobo JM, Almeida IF (2021) Trends in the use of botanicals in anti-aging cosmetics. Molecules 26(12):3584

Fukuchi K, Okudaira N, Adachi K, Odai-Ide R, Watanabe S, Ohno H, Yamamoto M, Kanamoto T, Terakubo S, Nakashima H (2016) Antiviral and antitumor activity of licorice root extracts. In Vivo 30(6):777–785

Giri RP, Gangawane AK, Giri SG (2019) Neem the wonder herb: a short review. Int J Trend Sci Res Dev 3(3):962–966

Gwanya HY, Mbuyiswa A-AN, Phokwe OJ, Gxaba N, Manganyi MC (2024) Harnessing the potential of Helinus integrifolius in cosmeceutical research: toward sustainable natural cosmetics. Cosmetics 11(4):126

Hassen G, Belete G, Carrera KG, Iriowen RO, Araya H, Alemu T, Solomon N, Bam DS, Nicola SM, Araya ME (2022) Clinical implications of herbal supplements in conventional medical practice: a US perspective. Cureus 14(7):e26893

Hemeg HA, Moussa IM, Ibrahim S, Dawoud TM, Alhaji JH, Mubarak AS, Kabli SA, Alsubki RA, Tawfik AM, Marouf SA (2020) Antimicrobial effect of different herbal plant extracts against different microbial population. Saudi J Biol Sci 27(12):3221–3227

Holleran G, Scaldaferri F, Gasbarrini A, Currò D (2020) Herbal medicinal products for inflammatory bowel disease: a focus on those assessed in double-blind randomised controlled trials. Phytother Res 34(1):77–93

Jadhav V, Dhande S, Kadam V (2017) Cosmetics side effects. World J Pharm Pharm Sci 6(1):327

Jain S, Trivedi M, Raikwar M, Lodhi M, Ali MY, Purohit A, Nema P (2022) A review on herbal cosmetics and cosmeceuticals. Asian J Dent Health Sci 2(4):9–16

Joshi LS, Pawar HA (2015) Herbal cosmetics and cosmeceuticals: an overview. Nat Prod Chem Res 3(2):170

Kadam RA, Dhumal ND, Khyade VB (2019) The mulberry, Morus alba (L.): the medicinal herbal source for human health. Int J Curr Microbiol Appl Sci 8(4):2941–2964

Kalaskar AR, Bhowate RR, Kalaskar RR, Ghonmode S (2021) Novel neem leaves extract mouthwash therapy for oral lichen planus. J Herb Med 26:100408

Katiyar SK, Singh T, Prasad R, Sun Q, Vaid M (2012) Epigenetic alterations in ultraviolet radiation-induced skin carcinogenesis: interaction of bioactive dietary components on epigenetic targets. Photochem Photobiol 88(5):1066–1074

Killedar SG, Pawar AV, Suresh Killedar C (2017) Preparation of herbal tea from mulberry leaves. J Med Plants Stud 5(2):325–328

Kim N, Priefer R (2021) Retinol binding protein 4 antagonists and protein synthesis inhibitors: potential for therapeutic development. Eur J Med Chem 226:113856

Kole PL, Jadhav HR, Thakurdesai P, Nagappa AN (2005) Cosmetic potential of herbal extracts. Indian J Nat Prod Resour 4(4):315–321
Krutmann J, Schalka S, Watson REB, Wei L, Morita A (2021) Daily photoprotection to prevent photoaging. Photodermatol Photoimmunol Photomed 37(6):482–489
Kumari S, Khurana S (2013) Cosmeceuticals: current trends and market preparations. IOSR J Pharm Biol Sci 8:45–48
Kunatsa Y, Katerere DR (2021) Checklist of African soapy saponin—rich plants for possible use in communities' response to global pandemics. Plants 10(5):842
Kwon Y-J, Son D-H, Chung T-H, Lee Y-J (2020) A review of the pharmacological efficacy and safety of licorice root from corroborative clinical trial findings. J Med Food 23(1):12–20
Larsson SC, Bergkvist L, Näslund I, Rutegård J, Wolk A (2007) Vitamin A, retinol, and carotenoids and the risk of gastric cancer: a prospective cohort study. Am J Clin Nutr 85(2):497–503
Lin HY, Li WH, Lin CF, Wu HR, Zhao YP (2022) International biological flora: Ginkgo biloba. J Ecol 110(4):951–982
Luo B, Kastrat E, Morcol T, Cheng H, Kennelly E, Long C (2021) Gaultheria longibracteolata, an alternative source of wintergreen oil. Food Chem 342:128244
Majeed M, Majeed S, Nagabhushanam K, Mundkur L, Neupane P, Shah K (2020) Clinical study to evaluate the efficacy and safety of a hair serum product in healthy adult male and female volunteers with hair fall. Clin Cosmet Investig Dermatol 13:691–700
Manela-Azulay M, Bagatin E (2009) Cosmeceuticals vitamins. Clin Dermatol 27(5):469–474
Martin KI, Glaser DA (2011) Cosmeceuticals: the new medicine of beauty. Mo Med 108(1):60
Mattia M, Biehler-Gomez L, Palamenghi A, Nichetti D, Caccia G, Sguazza E, De Angelis D, Galimberti PM, Fedeli AM, Slavazzi F (2021) "Man is what he eats". Plant residues from dental calculus in the ancient population of Milano from Roman times to modern age. J Archaeol Sci Rep 39:103180
Moelants K (2013) Effect of structural characteristics on the rheology and nutritional quality of plant-tissue-based food suspensions. A study on carrot- and tomato-derived products
Muthuswamy VV, Manoharan K (2024) SDG: consumer's perception towards herbal/organic products with respect to sustainable development goals. J Lifestyle SDGs Rev 4(4):e01994
Ohno H, Miyoshi S, Araho D, Kanamoto T, Terakubo S, Nakashima H, Tsuda T, Sunaga K, Amano S, Ohkoshi E (2014) Efficient utilization of licorice root by alkaline extraction. In Vivo 28(5):785–794
Pallai DR, Chandrashekar S, Nayak D, Banu A, Kumar P, Pai V, Ram AH, Raj R (2021) Development and evaluation of herbal hand wash with neem alcoholic extract. Res J Pharm Technol 14(1):308–310
Pan X, Liu H, Jia G, Shu YY (2000) Microwave-assisted extraction of glycyrrhizic acid from licorice root. Biochem Eng J 5(3):173–177
Pandey A, Jatana GK, Sonthalia S (2019) Cosmeceuticals. In: StatPearls. StatPearls Publishing, Treasure Island
Park E, Lee S-M, eun Lee J, Kim J-H (2013) Anti-inflammatory activity of mulberry leaf extract through inhibition of NF-κB. J Funct Foods 5(1):178–186
Poswal FS, Russell G, Mackonochie M, MacLennan E, Adukwu EC, Rolfe V (2019) Herbal teas and their health benefits: a scoping review. Plant Foods Hum Nutr 74:266–276
Prasad K, Haq R-U, Bansal V, Siddiqui MW, Ilahy R (2017) Carrot: secondary metabolites and their prospective health benefits. In: Plant secondary metabolites, vol 2. Apple Academic Press, Palm Bay, pp 133–220
Puvača N, Čabarkapa I, Bursić V, Petrović A, Aćimović M (2018) Antimicrobial, antioxidant and acaricidal properties of tea tree (Melaleuca alternifolia). J Agron Technol Eng Manag 1(1):29–38
Račková L, Jančinová V, Petríková M, Drábiková K, Nosáľ R, Štefek M, Košťálová D, Prónayová N, Kováčová M (2007) Mechanism of anti-inflammatory action of liquorice extract and glycyrrhizin. Nat Prod Res 21(14):1234–1241

Rasooly R, Molnar A, Do P, Morroni G, Brescini L, Cirioni O, Giacometti A, Apostolidis E (2020) Witch hazel significantly improves the efficacy of commercially available teat dips. Pathogens 9(2):92

Ren X, Chen J-F (2020) Caffeine and Parkinson's disease: multiple benefits and emerging mechanisms. Front Neurosci 14:602697

Salvioni L, Morelli L, Ochoa E, Labra M, Fiandra L, Palugan L, Prosperi D, Colombo M (2021) The emerging role of nanotechnology in skincare. Adv Colloid Interf Sci 293:102437

Samuelson R, Lobl M, Higgins S, Clarey D, Wysong A (2020) The effects of lavender essential oil on wound healing: a review of the current evidence. J Altern Complement Med 26(8):680–690

Saudagar R, Sisodiya M (2018) Review on herbal cosmetics. World J Pharm Res 7(7):573–591

Sayed R, Hussein OE, Omran AA (2022) Method optimization and validation for the determination of mancozeb in chamomile by modified QuEChERS and liquid chromatography–tandem mass spectrometry. J Food Compos Anal 111:104646

Shahrajabian MH, Sun W, Cheng Q (2022) Ginkgo biloba: a famous living fossil tree and an ancient herbal traditional Chinese medicine. Curr Nutr Food Sci 18(3):259–264

Shao X, Cheng S, Wang H, Yu D, Mungai C (2013) The possible mechanism of antifungal action of tea tree oil on Botrytis cinerea. J Appl Microbiol 114(6):1642–1649

Sharma RR, Deep A, Abdullah ST (2022) Herbal products as skincare therapeutic agents against ultraviolet radiation-induced skin disorders. J Ayurveda Integr Med 13(1):100500

Shivanand P, Nilam M, Viral D (2010) Herbs play an important role in the field of cosmetics. Int J PharmTech Res 2(1):632–639

Silvestre WP, Medeiros FR, Agostini F, Toss D, Pauletti GF (2019) Fractionation of rosemary (Rosmarinus officinalis L.) essential oil using vacuum fractional distillation. J Food Sci Technol 56:5422–5434

Simoes MF, Sousa JS, Pais AC (2015) Skin cancer and new treatment perspectives: a review. Cancer Lett 357(1):8–42

Singh B, Karmakar P, Singh P, Singh J (2022) Exploring the phytochemical potential of purple carrot (Daucus carota L. subsp. sativus var. atrorubens Alef.) developed at ICAR-IIVR, Varanasi, UP. Veg Sci 49(2):133–144

Tabassum N, Hamdani M (2014) Plants used to treat skin diseases. Pharmacogn Rev 8(15):52

Tiwari U, Cummins E (2013) Factors influencing levels of phytochemicals in selected fruit and vegetables during pre-and post-harvest food processing operations. Food Res Int 50(2):497–506

Ude C, Schubert-Zsilavecz M, Wurglics M (2013) Ginkgo biloba extracts: a review of the pharmacokinetics of the active ingredients. Clin Pharmacokinet 52:727–749

Ünlü B (2022) Herbal medicine and common dermatologic diseases. In: Herbal medicines. Elsevier, Amsterdam, pp 329–345

Wang J, Hu X, Chen J, Wang T, Huang X, Chen G (2022) The extraction of β-carotene from microalgae for testing their health benefits. Food Secur 11(4):502

Warade DD, Bhosale MP (2023) A review on herbal cosmetics used in skin and hair care. Int Res J Modern Eng Technol Sci 5(4)

Wathoni N, Haerani A, Yuniarsih N, Haryanti R (2018) A review on herbal cosmetics in Indonesia. Int J Appl Pharm 10(5):13–16

Wen P, Hu T-G, Linhardt RJ, Liao S-T, Wu H, Zou Y-X (2019) Mulberry: a review of bioactive compounds and advanced processing technology. Trends Food Sci Technol 83:138–158

Yasmin A, Chia S, Looi Q, Omar A, Noordin M, Ideris A (2020) Herbal extracts as antiviral agents. In: Feed additives. Elsevier, Amsterdam, pp 115–132

Chapter 4
Sustainable Cosmeceuticals Based on Natural Colorants from Plants

Shazia Abrar, Anbreen Anjum, Amna Younas, Sidra Abbas, Shumaila Kiran, Sarosh Iqbal, and Hafiz Qasim Naveed

Contents

Abstract Plants being rich sources of phytochemicals behave as therapeutic agents and also provide a variety of natural dyes for applications in the textile, cosmetic, and pharmaceutical industries. In recent decades, plant-based natural colorants have gained increasing importance due to their non-toxic, eco-friendly, therapeutic, and sustainable characteristics. Thus, plant-based natural colorants meet the sustainable development goal which is related to the good health and well-being. Medicinal plants have been investigated for their therapeutic properties to find their applications in skin- and hair-care cosmetics. Natural hair colors have been used to dye gray hairs to replace synthetic hair dyes. Recently, formulations have been developed to apply natural colorants in multipurpose lip-care cosmetics. Due to their

S. Abrar (✉) · A. Anjum · A. Younas · S. Kiran · S. Iqbal · H. Q. Naveed
Department of Applied Chemistry, Government College University Faisalabad, Faisalabad, Pakistan
e-mail: shaziaabrar@gcuf.edu.pk

S. Abbas
Department of Zoology, University of Jhang, Jhang, Pakistan

S. Javed et al. (eds.), *Sustainable Cosmeceuticals*,
https://doi.org/10.1007/978-3-031-86087-4_4

different chemical structures and compositions, natural colorants require various extraction methodologies to obtain greater yield and maintain their chemical integrity. For skincare applications, the study of therapeutic properties and cytotoxicity of extracts of natural dyes is essential. Antimicrobial, antioxidant, anti-inflammatory, anti-cancer, cytotoxicity, and anti-aging potential of natural colorants provide additional benefits for producing multifunctional skincare cosmetics. Natural colorants can be extracted from plant wastes such as fruit and vegetable peels; therefore, they meet the sustainable development goals of good health and well-being and also ensure the sustainable consumption and production of cosmetic products employing natural plant waste products. In this chapter, we will provide a detailed overview of plant-based natural colorants, chemistry, extraction methods, and therapeutic properties for their application in cosmetic industry.

Keywords Cosmeceuticals · Natural colorants · Colored cosmetics · Sustainable development goals · Therapeutic agents in cosmetics

4.1 Introduction

Natural colorants represent a class of biodegradable, non-toxic, sustainable, and multifunctional therapeutic agents from natural resources. Due to their unique multifunctional activities, natural colorants have gained popularity for their applications in food, cosmetics, and pharmaceuticals (Luzardo-Ocampo et al. 2021). The harmful effects of synthetic chemicals and dyes in the cosmetics have made people aware about the beneficial properties of natural colorants from plants. An important application of natural colors is cosmetic industry, where consumers are more concerned about the effects of active chemical constituents on skin health. This chapter will inform the readers about the classification of natural colorants, their therapeutic properties, and their application in the development of sustainable cosmeceuticals. On the basis of their sources, chemistry, and applications, plant-based natural colorants have been grouped into classes. Anthocyanin belongs to a class of flavonoids that gives red to blue shades to fruits and vegetables. More than 700 different types of anthocyanin have been identified to date, including 6 common aglycones and various forms of glycosylations and acylations. Anthocyanins making a major part of dietary consumption have strong potential antioxidant activity due to which they also exhibit multifunctional therapeutic properties, such as anti-inflammatory, anti-carcinogenic, and antimicrobial activities. Different fruits and vegetables such as red cabbage, carrots, sweet potato, cherries, and berries contain anthocyanin as their active coloring constituent (He and Giusti 2010). Anthocyanins are less stable during food processing due to their high reactivity and ease of degradation. Betalains are also the other important class of natural colorants classified into betacyanins and betaxanthins, which have betalamic acid functional group in their structure. These colorants give bright color shades to fruits and vegetables such as orange, yellow, purple, and red colors. Another important class of organic compounds is carotenoids

Table 4.1 Colorant shades and their sources according to the nature

Sr. #	Colorants	Colors	Sources	References
1	Anthocyanins	Dark blue to red	Plum, various berries like blackberry, strawberry, blueberry, cranberry, red cabbage, and purple grapes	Mortensen (2006), Sigurdson et al. (2017), Simon et al. (2017), Yusuf et al. (2017)
2	Betalains	Yellow to violet	Beetroot, dragon fruit, etc.	
3	Chlorophylls	Blue-green,yellow-green	All green photosynthesizing plants	
4	Carotenoids	Yellow, orange, and red	Green vegetables, red and orange fruits, and vegetables	

which are lipophilic and give plants the orange and yellow colors (Mortensen 2006). They are present in the *cis* and *trans* forms that are categorized as hydrocarbon carotenoids and xanthophylls (Riaz et al. 2021). Carotenoids are major coloring agents in vegetables such as carrots, spinach, and paprika. Due to the conversion of carotenoid to vitamin A, an essential micronutrient, it is important for the maintenance of growth. Chlorophyll is a green color pigment, amphiphilic, and considered insoluble in polar solvents due to its long carbon chain and is classified as chlorophyll *a* and *b* (Sigurdson et al. 2017). It is permitted and well-established green food colorant with standard number E140 (Table 4.1 and Fig. 4.1).

4.2 Extraction Techniques

Various procedures have been used for the extraction of natural colorants. The cheap and easy way of obtaining natural colorants is to extract them using conventional methods such as solvent extraction using Soxhlet apparatus, maceration, and hydrodistillation. However, one has to compromise on the extraction efficiency in terms of time and yield of extraction (Boo et al. 2012; Prado et al. 2020). Some more efficient and advanced separation methods have been evolved that give higher yields, i.e., microwave-assisted extraction, supercritical fluid extraction, ultrasonic-assisted extraction, pressurized liquid extraction, enzyme-assisted extraction, pulse-electric field-assisted, and extraction using ionic liquids (Alvarez-Rivera et al. 2020; Cadoni et al. 1999; Destandau et al. 2013; Nadar et al. 2018; Sheikh et al. 2016). The use of greener solvents for extraction such as ethanol, water, ionic liquids, glycerol, and esters of fatty acids is very effective and less toxic alternative to organic solvents. The advancement in extraction technology has resulted in increased extraction yield, use of minimum solvent, less time consumption, and minimum degradation of natural colorants (Derrien et al. 2018) (Fig. 4.2 and Table 4.2).

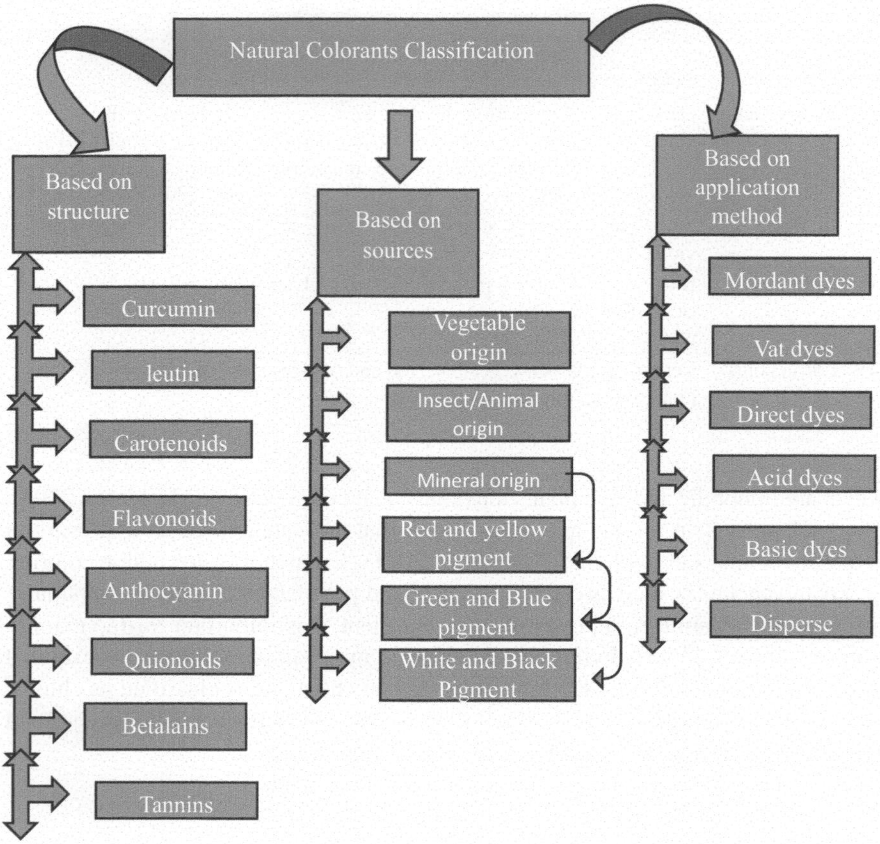

Fig. 4.1 Classification of natural pigments

4.3 Therapeutic Properties of Natural Colorants

In addition to their coloring properties, natural plant extracts exhibit various medicinal effects that have been utilized for various health benefits since ancient times. Natural colorants showed multifunctional effects that have been studied extensively in recent decades. These properties of natural colorants meet the sustainable development goal of good health and well-being by providing cosmeceuticals with no side effects. Natural colorants showed various biological and therapeutic properties such as antioxidant, anti-inflammatory, antimicrobial, and anti-cancer. (Smeriglio et al. 2016). It has been shown that betalain-containing supplements used by athletes to enhance their ability to run for longer times (Van Hoorebeke et al. 2016). As an essential micronutrient, it is involved in maintaining normal development and reproduction. The deficiency of vitamin A is a major health issue in developing countries (Rao and Rao 2007). Carotenoids exhibit anti-obesity and anti-diabetic properties. It has been seen as beneficial after being used by people of different age

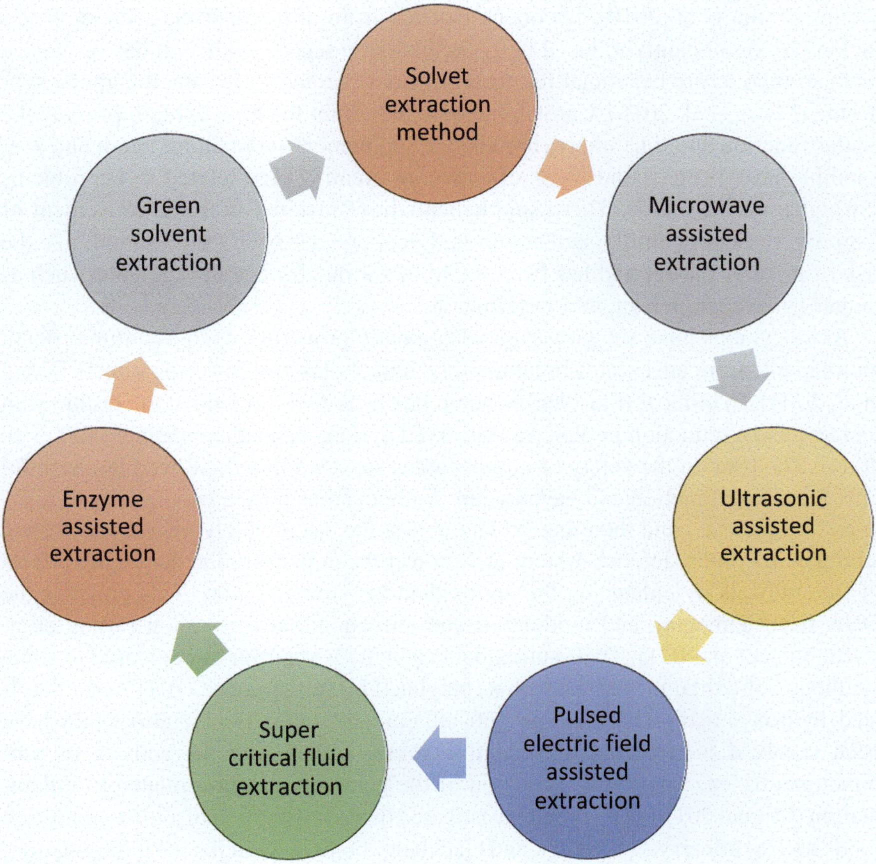

Fig. 4.2 Extraction methods for natural pigments

Table 4.2 Some color-yielding plants

Sr. #	Species name	Colors	References
1	Annato	Fugitive orange and red	Preston and Rickard (1980)
2	Brazil wood	Bright red color	de Oliveira et al. (2002)
3	Cudbear	Lilac color	Gardner (1913)
4	Cutch	Rich brown color	Win and Ni (2019)
5	Indigo	Blue color	Kabish et al. (2023)
6	Lac	Bright red color	Srivastava et al. (2013)
7	Maadar	Red color	Blackburn (2017)
8	Quercitron	Brown to yellow colors	Hansen (2011)
9	Log wood	Purple on wool, blue and black on silk	Kahr et al. (1998)

groups (Yahia et al. 2010). Lycopene isolated from different fruits such as watermelon has significantly reduced body weight. In research studies, it has been seen that carotenoids are beneficial for curing various cardiac issues and metabolic syndrome (Maria et al. 2015). Carotenoids also help filter the blue light, improving the ocular function and thus improving vision. Supplements containing lutein and zeaxanthin have been found very effective in treating age-related eye problems (Stringham et al. 2017). These supplements have resulted in the improvement of memory, reasoning ability, and attention. Lycopene has been used to minimize the risk of prostate cancer and has been found in various fruits and vegetables such as tomato, guava, grapefruit, and watermelon.

Anthocyanins have shown various therapeutic properties such as antimicrobial, antioxidant, anti-cancer, anti-inflammatory, and cholesterol lowering effects (Kong et al. 2003). In clinical trials, anthocyanin has been found effective in chronic pain therapy, and a reduction in pain was observed in dose-dependent manner (Tall et al. 2004). The trial on the safety of anthocyanins revealed that there were no harmful effects on liver function and metabolism. Anthocyanins have a good ability to scavenge free radicals, and their antioxidant ability has been widely investigated. They stimulate the body defense system and protect the cells from the damaging effects of free radicals by enhancing the antioxidant enzyme activities. This prevents the DNA from damaging and modulating the mitochondrial respiration (Wang et al. 2010; Yang et al. 2021). Their antioxidant capacity is superior as compared to some synthetic antioxidants such as Trolox, butylated hydroxytoluene (BHT), and butylated hydroxyanisole (BHA). The anti-inflammatory effects of anthocyanin have been involved in relieving of oxidative stress, suppressing the activity of pro-inflammatory enzymes, and regulation of the gene transcription related to inflammation (Miguel 2011). It has been investigated that consumption of anthocyanin-rich foods such as bilberry, strawberry, and blackberry helps in reducing the inflammation-causing bacteria in the gut and thus reducing the chances of type 2 diabetes and obesity. It also improves the insulin sensitivity and relieving the patients with type 2 diabetes and insulin resistance (Ghosh and Konishi 2007). Anthocyanin stimulates the energy consumption in body by regulating the lipid metabolism and suppressing the absorption of lipids. Due to its good anti-cancer activity, dietary intake of anthocyanin reduces the occurrence of various types of cancers such as blood, liver, breast, gastrointestinal, and cervical cancers (Chen et al. 2022). The mechanism of anti-cancer activity of anthocyanin is based on their ability to suppress the cell metastasis and boost the body immune system (Fig. 4.3).

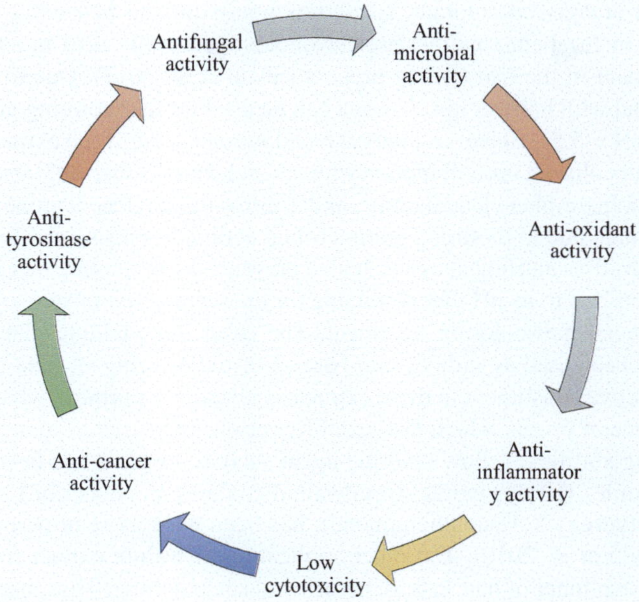

Fig. 4.3 Therapeutic properties of natural colorants

4.4 Applications of Natural Colorants as Cosmeceutical in Accordance with SDGs

4.4.1 Application of Natural Colorants in Hair Cosmeceuticals

Strongly alkaline compounds such as sodium or ammonium hydroxide, ammonium thioglycolate, sodium or lithium hydroxide, and guanidine have high pH. They can produce cracks in hair, making it weak. The choice of hair-care cosmetics can reduce the effects of harsh chemical treatments. One important hair grooming treatment is covering gray hair by dyeing it with suitable colorants. Synthetic dyes work as oxidative hair treatment and are used in combination with peroxide- and ammonia-based chemicals. This type of hair grooming treatment decreases the hair strength and affects its integrity (Palaniappan et al. 2023). Plant-based natural colorants provide an alternative to the above harmful chemicals of synthetic dyes. Natural hair dyes are more sustainable, multifunctional, and sustainable coloring agents (Cui et al. 2022). Natural hair dyes can be extracted from various plants such as henna, beetroot, walnut, and turmeric. Their extracts have multifunctional effects for hair health, i.e., hair conditioning, promoting hair growth, along with hair coloring. The phytochemicals in the dye extracts are polyphenols, tannins, catechins, flavonoids, and vitamins. They can provide various benefits for hair such as anti-dandruff, reduction of hair graying, and inhibiting the microbial action on scalp. A natural hair colorant *Lawsonia inermis* L. (henna) containing high concentration of

Lawsone, an orange-red colorant, is used as non-oxidizing hair color. The other components in the henna extract are flavonoids, and gallic acid is acting as an organic mordant in the hair dyeing process (Shahi et al. 2017). Due to its natural affinity to bind with hair keratin, *Lawson* can be used for hair coloring under acidic conditions (pH = 5.5). However, allergic reactions can occur if the extract of henna is used for hair coloring in combination with other chemicals such as 3-aminophenol, 4-aminophenolpara-phenylenediamine, and 2-nitro-4-phenylenediamine. The color extract of henna, due to its strong antimicrobial action, has been used for the treatment of dandruff as antifungal agent. It also prevents the premature hair falling and graying effects due to its pH that is suitable for the scalp, as compared to the highly alkaline artificial hair-coloring treatment. The other dye-yielding plants such as madder have been used as anthocyanin-based hair red coloring agents (Yadav et al. 2014). The active constituent in these extracts is alizarin or purpurin which depends on their source of origin, which also exhibits various bioactivities along with their good coloring abilities. Yellow coloring agent such as apigenin has been extracted from chamomile. The dye extract from walnut contains an isomer of Lawson, i.e., Juglone (2-Hydroxy-1,4-naphthoquinone), has been used as semi-permanent hair colorant (Beiki et al. 2018). The other application of walnut extract for hair-care includes the treatment of hair loss, dandruff, and scalp itching. Thus, hair colorants based on natural sources provide safe, non-toxic, and eco-friendly alternative for hair grooming cosmetics. Natural hair colorants fulfill the sustainable development goal of providing good health and well-being and also ensure sustainable consumption and production. Therefore, applications of natural coloring agents for the formulation of cosmeceuticals have been found beneficial, economical, and eco-friendly. In a recent research study, it has been shown that the extract of the skin of black current fruit can be used for dyeing the gray colors as natural source of anthocyanin, showing high yield and property to adsorb the hair fiber with stability to multiple hair washes (Fig. 4.4).

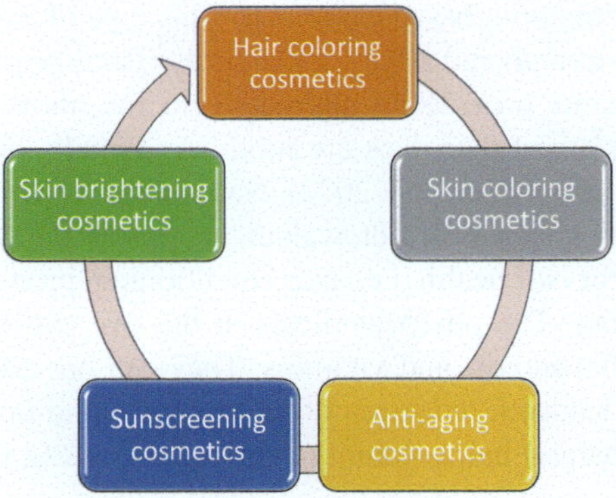

Fig. 4.4 Application of natural colorants in cosmeceuticals

4.4.2 Natural Colorants-Based Skin Cosmeceuticals

Skin, being the largest living organ, is involved in the protection of the internal body systems from the external environment. The stratum corneum that makes the outermost skin covering acts as selectively permeable membrane that keeps the skin moist and protects from the exogenous harmful substances. Stratum corneum maintains the skin health by keeping the balance of moisture and lipid contents. Any change in its structure due to external and internal factors can ultimately damage the skin barrier functioning, increasing the transepidermal water loss (TEWL). It is very important to keep the skin healthy via maintaining the functioning and integrity of stratum corneum. The function of cosmeceuticals is the beautification along with keeping the fresh and clean. Cosmeceuticals based on natural ingredients from plant extracts have been used since ancient times for treating the skin problems and improving skin health. The healthy skin is essential for a beautiful appearance. The plant-based natural colorants provide the ingredients as supplements that nourish the skin and provide the therapeutic agents to combat the external and internal reactive oxygen species (Brudzyńska et al. 2022; Cheng et al. 2008). Natural cosmetic industry is expanding rapidly due to non-toxic, eco-friendly, and sustainable effects of natural ingredients. The natural vitamins, minerals, antioxidants, proteins, and colorants in plant extracts provide skin with multifunctional benefits. These vitamins and antioxidants showed therapeutic and medicinal properties such as antioxidants, antimicrobial, UV-protecting, anti-cancer, anti-inflammatory, and non-cytotoxic effects (Chengaiah et al. 2010). A variety of multifunctional formulations containing natural colorants can be produced for various health benefits.

4.4.3 Anti-aging and UV-Blocking Cosmeceuticals

Skin aging is a systematic and natural process that leads to the loss of skin integrity, barrier functioning, along with the appearance of wrinkles. The main cause of skin aging is an increase in reactive oxygen species (ROS) due to external environmental factors or internal systematic biological process (Makrantonaki and Zouboulis 2007). The ROS causes skin damage by interacting with stratum corneum at cellular level, by reacting with lipids, protein, and DNA. Both the intercellular and extracellular oxidative reactions initiated by reactive oxygen species lead to skin aging effects, which are characterized by wrinkles, pigmentation, and loss of elasticity. Antioxidants are used to combat the effects of ROS and prevent symptoms of photo-induced skin aging (Masaki 2010). Natural colorants have been found to act as strong antioxidants and can be applied in the formulation of anti-aging cosmeceuticals. This property has been employed to combat UV-induced skin problems, hyperpigmentation, and photo-aging (Masaki 2010). Excessive exposure to the sun may cause an increase in the risk of damage, accelerate the aging agents, induce skin pigmentation, and impair the skin barrier functioning (Lowe 2006). Research studies have proved that application of natural colorants such as anthocyanin in very low concentration from plants has UV-blocking effects and can reduce the UV-induced skin problems. Anthocyanin acts via absorbing UV-radiations, blocking the radiation to be reached and absorbed in the skin epidermis; therefore, it is used as sun protection cosmetic (Karunarathne et al. 2021). The addition of other phytochemicals can enhance the sun blocking effect such as caffeic acid, coumaric acid, and ferulic acid (Girsang et al. 2020). Cosmetics containing pomegranate extracts showed the protection of protein and DNA by significantly reducing the protein oxidation (Aqil et al. 2012). It has been reported that anthocyanin protects the UV-induced damage to cultured keratinocytes and in vivo to naked skin mice. Blueberry extracts have also been shown to protect the skin from photo-damaging. Both in vivo and in vitro studies have been reported, which confirm the UV-protecting and reduced photo-aging of skin effects of anthocyanin.

The natural dye extract from *Alkanna tinctoria* roots contains *Alkannin* that has been used as reddish coloring agent in food and cosmetics (Jaradat et al. 2018). Its color gets changed with pH, and it showed deep red color at low pH and blue in high pH. Biological activities of Alkannin have been reported in the literature such as antioxidant, antibiotic, anti-inflammatory, antimicrobial, and anti-cancer activities. Its antioxidant properties can be incorporated in the cosmetic formulations to treat the skin aging effect that can used to treat the skin wrinkles by restoring the skin elasticity and slowing down the aging process (Sabih Ozer et al. 2010). The extracts can also reduce the UV-induced skin pigmentation. Similarly, carotenoid colorants from natural resources have been used as active cosmeceutical agent in the cosmetic formulation. These formulations protect the skin from exogenous and endogenous detrimental effects. Their antioxidant activities help to prevent the oxidative skin damage by combating the reactive oxygen species (Palace et al. 1999). These pharmacological actions prevent the cell proliferation and necrosis of skin and force apoptosis. Thus, it helps in preventing the skin dryness, photo-aging, and wrinkling by their action as UV filters. Carotenoids have ability to absorb UV-radiations and prevent the skin freckles, pigmentation, and discoloration (Anunciato and da Rocha Filho 2012). Formulations of moisturizing creams, anti-aging, and sunscreens can be made with carotenoid-rich extracts.

4.4.4 Skin-Coloring Cosmeceuticals

Coloring cosmetic products include a range of cosmetics used for enhancing facial beauty by highlighting the features. The most commonly used coloring cosmetics are lipsticks, eye shadows, foundation, blushes, mascaras, highlighters, and several other colored products. Colored products are very crucial in women life for improving their appearance and beauty. Making tattoos on the body skin has become very popular among the young generation who use synthetic or natural colorants for painting the body. Various side effects are associated with the use of the synthetic colorant in these tattoos (Ramesh and Muthuraman 2018). Moreover, the coloring agents used in the powdered or paste-type coloring cosmetics have mineral origin and may contain high levels of toxic heavy metals present as impurity. These heavy metals have been studied for their toxicity when absorbed by the body system (Al-Dayel et al. 2011). There is a great need to replace the synthetic and mineral coloring agents with those that have multifunctional health benefits. The extraction and application of natural colorants in the development of sustainable cosmetics have been studied for exploring their benefits. The application of anthocyanins in topical skin-coloring cosmetics has been enhanced because of the attractive and bright color shades. Various colorants such as anthocyanin have been applied as colorants in the formulation of lip-care cosmetics (Westfall 2015). The sources of anthocyanin that have been employed are carrots, pomegranate, and red cabbage. The formulated cosmetics also exhibit additional properties along with coloring effects for skin such as UV-protecting and anti-aging activities. These products also showed long shelf life due to their good antimicrobial effects. Being rich source of reddish pink colorant such as betalains, the extract of beetroot has been employed for the formulation of lip-coloring cosmetics (Rani et al. 2019). Since lipsticks are used to increase the beauty of the lips, it is important to use coloring agent that can maintain the skin health of this area. Most of the artificial lip colorants make lips discolor, dry, and inflamed, effecting the appearance and beauty. Therefore, the cosmeceuticals based on natural extracts of colorants can be used to overcome such problems to enhance the natural beauty of the lips. Natural colorant from beetroot represents a good source of vitamins and minerals that provide essential nutrients along with their coloring ability (Kaličanin and Velimirović 2016). Another important colorant for skin is obtained from the leaves of *Lawsonia inermis* (Henna) (Adeel et al. 2018). The active coloring agent in Henna is Lawsone that is known for its bright orange shade. It has been used for coloring hair, skin along with other biological activities. Both Lawsone and Juglone have protein binding capacity and showed strong antimicrobial, antiviral, and anti-inflammatory activities. These colorants have ability to bind with skin and get absorbed in hair cuticle. Lawson also acts as cooling agent and astringent for treating the skin infections (Pradhan et al. 2012). Juglone has been used for treating skin fungal infections and can be formulated into cosmeceutical products.

4.4.5 Cosmeceuticals for Treating Skin Pigmentation

One of the skin problems that have to be solved for enhancing the beauty of the person is skin discoloration and hyperpigmentation. Hyperpigmentation is recognized by excessive production and irregular distribution of melanin in the skin owing to exposure to UV radiation (Ortonne and Bissett 2008). UV-induced pigmentation mainly involves three distinct processes, i.e., darkening or immediate pigmentation, persistent or lasting pigmentation, and delaying darkening or pigmentation (neo-pigmentation). Pigmentation caused by UVB radiations is a delayed process that can lead to sunburn. Pigmentation caused by UVA is more lasting than by UVB. Continued exposure to UV radiation ultimately leads to skin aging effects. It is therefore very necessary to apply those cosmeceuticals that can inhibit the production of melanin in the skin and results in skin brightening and lightening. The active agents in cosmeceuticals reduce the activity of tyrosinase enzyme that catalyzes the production of melanin. These active agents target the melanin synthesis pathway and inhibit the key step in the process. The most commonly used active agent in treating pigmentation is hydroquinone. However, due to its increasing safety concerns such as causing DNA damage and carcinogenic effects, hydroquinone needs to be replaced with more natural, sustainable, safe, and eco-friendly active agents for treating hyperpigmentation. The more safe alternatives of active agents for hyperpigmentation are vitamin C, vitamin E, kojic acid, arbutin, and natural colorant extracts. The colorants from various plants, i.e., anthocyanin, have therapeutic properties and can inhibit the action of tyrosinase that is involved in the path of melanin synthesis. The tyrosinase inhibition of extracts from *Phaseolus vulgaris* L. has been investigated (Parvez et al. 2007). It has also been found that the pigment extracts from the seed coat are involved in the inhibition of tyrosinase. This property of natural extracts has been employed in skin-lightening cosmeceuticals. Cyanidin 3-*O*-*β*-D-glucoside and cyanidin were found to be the best tyrosinase inhibitors.

4.4.5.1 Natural Colorants Application and SDGs

In recent decades, a radical change has been seen in the cosmetic industry with the formulation of cosmetic products containing natural, non-toxic, organic, and all healthy chemical components. One of the important natural chemical constituents of natural origin is organic colorant extracted from various parts of plants and animals. These colorants exhibit wonderful multifunctional therapeutic properties that can be incorporated into eco-friendly cosmeceutical formulations. Cosmetic industry has thus provided an excellent opportunity to new business models that can introduce new products with non-toxic ingredients, thus fulfilling the sustainable development goals of good health and well-being, thus accomplishing the sustainable development goal 12 for responsible consumption and production by efficiently using the natural resources. This also helps in reducing poverty in accordance with sustainable development goal 1. Another important step is the implementation of 3R (reducing, reusing, and recycling) in the sustainable production of cosmetic products. It can be achieved by reducing the use of raw materials that are obtained from fossil fuels and synthesized artificially, thus reducing the environmental pollution, reusing the waste from plants and animals in efficiently, along with reducing the emission of carbon dioxide. The sustainability goals should also be implemented by using the environmental-friendly packaging materials, reducing the use of plastics and synthetic polymers. The research and development in the cosmetics have led the cosmetic companies to make their products sustainable by using natural raw materials, i.e., natural ingredients, reducing the environmental pollution, i.e., decrease the use of synthetic chemicals. The use of biodegradable packaging and green industrial process such as green extraction methods will also help to achieve the sustainable development goals.

4.5 Conclusion

This chapter focused on the natural colorants from plants, their composition, chemistry, extraction, therapeutic properties, and applications as active agent in the formulation of eco-friendly cosmeceuticals. The main focus was to provide insight into the sustainable development goals for formulation of natural colorant-based cosmetics. Consumers are more interested for buying the cosmetic products that are green, environmental-friendly, and organic. Thus, cosmetic industry needs to be aware of the environmental impact of their product and modify their manufacturing policies accordingly. The sustainable formulations start with the selection of ingredients that are natural and obtained from the sustainable sources as synthetic chemicals are posing threats to the environment. Therefore, the future of sustainability in cosmetic industry is to ensure the implementation of sustainable development goals throughout the life of the cosmetic products from selection of raw materials to the delivery of the cosmetics to the consumers. Natural colorants are rich sources of therapeutic active agents that provide multifunctional effects in the skincare, haircare, and coloring cosmetics. Natural colorants are used for the formulation of skincare cosmetics, such as creams, lotions, sunscreens, lip, and facial coloring cosmetics. The formulation of anti-aging and skin brightening cosmetics has been obtained through the use of natural colorants extracts such as anthocyanins, carotenoids, and betalains. Natural colorants have been successfully used for the formulation of hair dyeing cosmetics along with hair grooming cosmetics. We can conclude that the natural colorants represent an excellent source of multifunctional, sustainable, non-toxic coloring agents for the skin- and hair-care cosmetics, according to the sustainable development goal of good health and well-being.

References

Adeel S, Abrar S, Kiran S, Farooq T, Gulzar T, Jamal M (2018) Sustainable application of natural dyes in cosmetic industry. In: Handbook of renewable materials for coloration and finishing. Wiley, New York, pp 189–211

Al-Dayel O, Hefne J, Al-Ajyan T (2011) Human exposure to heavy metals from cosmetics. Orient J Chem 27(1):1

Alvarez-Rivera G, Bueno M, Ballesteros-Vivas D, Mendiola JA, Ibanez E (2020) Pressurized liquid extraction. In: Liquid-phase extraction. Elsevier, Amsterdam, pp 375–398

Anunciato TP, da Rocha Filho PA (2012) Carotenoids and polyphenols in nutricosmetics, nutraceuticals, and cosmeceuticals. J Cosmet Dermatol 11(1):51–54

Aqil F, Munagala R, Vadhanam MV, Kausar H, Jeyabalan J, Schultz DJ, Gupta RC (2012) Antiproliferative activity and protection against oxidative DNA damage by punicalagin isolated from pomegranate husk. Food Res Int 49(1):345–353

Beiki T, Najafpour GD, Hosseini M (2018) Evaluation of antimicrobial and dyeing properties of walnut (Juglans regia L.) green husk extract for cosmetics. Color Technol 134(1):71–81

Blackburn RS (2017) Natural dyes in madder (Rubia spp.) and their extraction and analysis in historical textiles. Color Technol 133(6):449–462

Boo H-O, Hwang S-J, Bae C-S, Park S-H, Heo B-G, Gorinstein SJIC (2012) Extraction and characterization of some natural plant pigments. Ind Crop Prod 40:129–135

Brudzyńska P, Kurzawa M, Sionkowska A, Grisel M (2022) Antioxidant activity of plant-derived colorants for potential cosmetic application. Cosmetics 9(4):81

Cadoni E, De Giorgi MR, Medda E, Poma GJD (1999) Supercritical CO2 extraction of lycopene and β-carotene from ripe tomatoes. Dyes Pigments 44(1):27–32

Chen J, Xu B, Sun J, Jiang X, Bai WJCR (2022) Anthocyanin supplement as a dietary strategy in cancer prevention and management: a comprehensive review. Crit Rev Food Sci Nutr 62(26):7242–7254

Cheng Y, Dong Y-Y, Dong M-X, Wang C, Su N, Sun Y-T et al (2008) Protection effect of cosmetics on human skin under simulated rigorous environment. Skin Res Technol 14(1):45–52

Chengaiah B, Rao KM, Kumar KM, Alagusundaram M, Chetty CM (2010) Medicinal importance of natural dyes—a review. Int J PharmTech Res 2(1):144–154

Cui H, Xie W, Hua Z, Cao L, Xiong Z, Tang Y, Yuan Z (2022) Recent advancements in natural plant colorants used for hair dye applications: a review. Molecules 27(22):8062

de Oliveira LF, Edwards HG, Velozo ES, Nesbitt M (2002) Vibrational spectroscopic study of brazilin and brazilein, the main constituents of brazilwood from Brazil. Vib Spectrosc 28(2):243–249

Derrien M, Aghabararnejad M, Gosselin A, Desjardins Y, Angers P, Boumghar YJL (2018) Optimization of supercritical carbon dioxide extraction of lutein and chlorophyll from spinach by-products using response surface methodology. LWT Food Sci Technol 93:79–87

Destandau E, Michel T, Elfakir C (2013) Microwave-assisted extraction. In: Natural product extraction: principles applications, vol 21. Royal Society of Chemistry, London, p 113

Gardner A (1913) The red coloring principle of cudbear. J Am Pharm Assoc 2(1):51–53

Ghosh D, Konishi T (2007) Anthocyanins and anthocyanin-rich extracts: role in diabetes and eye function. Asia Pac J Clin Nutr 16(2):200–208

Girsang E, Lister INE, Ginting CN, Bethasari M, Amalia A, Widowati W (2020) Comparison of antiaging and antioxidant activities of protocatechuic and ferulic acids. Mol Cell Biomed Sci 4(2):68–75

Hansen H (2011) The quest for quercitron: revealing the story of a forgotten dye. University of Delaware, Newark

He J, Giusti MM (2010) Anthocyanins: natural colorants with health-promoting properties. Annu Rev Food Sci Technol 1(1):163–187

Jaradat NA, Zaid AN, Hussen F, Issa L, Altamimi M, Fuqaha B et al (2018) Phytoconstituents, antioxidant, sun protection and skin anti-wrinkle effects using four solvents fractions of the root bark of the traditional plant Alkanna tinctoria (L.). Eur J Integr Med 21:88–93

Kabish AK, Abate MT, Alemar ZA, Girmay S (2023) The importance of natural indigo dye and its revitalization and Ethiopian potential for indigo growing. Adv Mater Sci Eng 2023(1):2135014

Kahr B, Lovell S, Subramony JA (1998) The progress of logwood extract. Chirality 10(1–2):66–77

Kaličanin B, Velimirović D (2016) A study of the possible harmful effects of cosmetic beauty products on human health. Biol Trace Elem Res 170(2):476–484

Karunarathne WAHM, Molagoda IMN, Lee KT, Choi YH, Yu S-M, Kang C-H, Kim G-Y (2021) Protective effect of anthocyanin-enriched polyphenols from Hibiscus syriacus L. (Malvaceae) against ultraviolet B-induced damage. Antioxidants 10(4):584

Kong J-M, Chia L-S, Goh N-K, Chia T-F, Brouillard R (2003) Analysis and biological activities of anthocyanins. Phytochemistry 64(5):923–933

Lowe N (2006) An overview of ultraviolet radiation, sunscreens, and photo-induced dermatoses. Dermatol Clin 24(1):9–17

Luzardo-Ocampo I, Ramírez-Jiménez AK, Yañez J, Mojica L, Luna-Vital DA (2021) Technological applications of natural colorants in food systems: a review. Food Secur 10(3):634
Makrantonaki E, Zouboulis C (2007) Molecular mechanisms of skin aging: state of the art. Ann N Y Acad Sci 1119(1):40–50
Maria AG, Graziano R, Nicolantonio DO (2015) Carotenoids: potential allies of cardiovascular health? Food Nutr Res 59(1):26762
Masaki H (2010) Role of antioxidants in the skin: anti-aging effects. J Dermatol Sci 58(2):85–90
Miguel MG (2011) Anthocyanins: antioxidant and/or anti-inflammatory activities. J Appl Pharm Sci 1(6):7–15
Mortensen A (2006) Carotenoids and other pigments as natural colorants. Pure Appl Chem 78(8):1477–1491
Nadar SS, Rao P, Rathod VK (2018) Enzyme assisted extraction of biomolecules as an approach to novel extraction technology: a review. Food Res Int 108:309–330
Ortonne J-P, Bissett DL (2008) Latest insights into skin hyperpigmentation. J Investig Dermatol Symp Proc 13(1):10–14
Palace VP, Khaper N, Qin Q, Singal PK (1999) Antioxidant potentials of vitamin A and carotenoids and their relevance to heart disease. Free Radic Biol Med 26(5–6):746–761
Palaniappan V, Karthikeyan K, Anusuya S (2023) Dermatological adverse effects of hair dye use: a narrative review. Indian J Dermatol Venereol Leprol:1–17
Parvez S, Kang M, Chung HS, Bae H (2007) Naturally occurring tyrosinase inhibitors: mechanism and applications in skin health, cosmetics and agriculture industries. Phytother Res 21(9):805–816
Pradhan R, Dandawate P, Vyas A, Padhye S, Biersack B, Schobert R et al (2012) From body art to anticancer activities: perspectives on medicinal properties of henna. Curr Drug Targets 13(14):1777–1798
Prado JM, Veggi PC, Náthia-Neves G, Meireles MAA (2020) Extraction methods for obtaining natural blue colorants. Curr Anal Chem 16(5):504–532
Preston H, Rickard M (1980) Extraction and chemistry of annatto. Food Chem 5(1):47–56
Ramesh M, Muthuraman A (2018) Flavoring and coloring agents: health risks and potential problems. In: Natural and artificial flavoring agents and food dyes. Elsevier, Amsterdam, pp 1–28
Rani GS, Pooja G, Harshavardhan V, Madhav BV, Pallavi B (2019) Formulation and evaluation of herbal lipstick from Beetroot (Beta vulgaris) extract. Res J Pharmacogn Phytochem 11(3):197–201
Rao AV, Rao LG (2007) Carotenoids and human health. Pharmacol Res 55(3):207–216
Riaz M, Zia-Ul-Haq M, Dou D (2021) Chemistry of carotenoids. In: Carotenoids: structure and function in the human body. Springer, Berlin, pp 43–76
Sabih Ozer M, Sarikurkcu C, Tepe B, Can S (2010) Essential oil composition and antioxidant activities of alkanet (Alkanna tinctoria subsp. tinctoria). Food Sci Biotechnol 19:1177–1183
Shahi Z, Mehrizi MK, Hadizadeh M (2017) A review of the natural resources used to hair color and hair care products. J Pharm Sci Res 9(7):1026
Sheikh J, Jagtap PS, Teli M (2016) Ultrasound assisted extraction of natural dyes and natural mordants vis a vis dyeing. Fibers Polymers 17(5):738–743
Sigurdson GT, Tang P, Giusti MM (2017) Natural colorants: food colorants from natural sources. Annu Rev Food Sci Technol 8(1):261–280. https://doi.org/10.1146/annurev-food-030216-025923
Simon JE, Decker EA, Ferruzzi MG, Giusti MM, Mejia CD, Goldschmidt M, Talcott ST (2017) Establishing standards on colors from natural sources. J Food Sci 82(11):2539–2553
Smeriglio A, Barreca D, Bellocco E, Trombetta D (2016) Chemistry, pharmacology and health benefits of anthocyanins. Phytother Res 30(8):1265–1286
Srivastava S, Ray D, Pandey S, Prasad K, Prasad M, Baboo B (2013) Pure lac dye: a potential natural food additive. Int J Emerg Technol Adv Eng 3(7):589–594
Stringham JM, Stringham NT, O'Brien K (2017) Macular carotenoid supplementation improves visual performance, sleep quality, and adverse physical symptoms in those with high screen time exposure. Food Secur 6(7):47

Tall JM, Seeram NP, Zhao C, Nair MG, Meyer RA, Raja SN (2004) Tart cherry anthocyanins suppress inflammation-induced pain behavior in rat. Behav Brain Res 153(1):181–188

Van Hoorebeke JS, Trias CO, Davis BA, Lozada CF, Casazza GA (2016) Betalain-rich concentrate supplementation improves exercise performance in competitive runners. Sports 4(3):40

Wang BC, He R, Li ZM (2010) The stability and antioxidant activity of anthocyanins from blueberry. Food Technol Biotechnol 48(1):42–49

Westfall A (2015) Evaluation of the efficacy of anthocyanins as biologically active ingredients in lipstick formulations. The Ohio State University, Columbus

Win S, Ni KT (2019) Studies on the fastness properties of cotton fabrics dyed with cutch (sha). Yadanabon Univ Res J 10(1)

Yadav N, Yadav R, Kharya MD (2014) Preparation and evaluation of natural hair colorant. World J Pharm 3(4):1314–1319

Yahia EM, Ornelas-Paz JJ (2010) Chemistry, stability, and biological actions of carotenoids. In: Fruit and vegetable phytochemicals: chemistry, nutritional value and stability. Wiley-Blackwell, Ames, pp 177–222

Yang S, Wang C, Li X, Wu C, Liu C, Xue Z, Kou X (2021) Investigation on the biological activity of anthocyanins and polyphenols in blueberry. J Food Sci 86(2):614–627

Yusuf M, Shabbir M, Mohammad F (2017) Natural colorants: historical, processing and sustainable prospects. Nat Prod Bioprospect 7:123–145

Chapter 5
Sustainability and Zooceuticals: Cosmetic Ingredients Derived from Animals

Muhammad Umar Ijaz, Muhammad Faisal Hayat, Asma Ashraf, Saima Shahzad Mirza, Sadia Javed, and Muhammad Arshad

Contents

M. U. Ijaz · M. F. Hayat
Department of Zoology, Wildlife and Fisheries, University of Agriculture, Faisalabad, Faisalabad, Pakistan

A. Ashraf (✉)
Department of Zoology, Government College, University, Faisalabad, Faisalabad, Pakistan
e-mail: asmaashraf@gcuf.edu.pk

S. S. Mirza
Microbial Bioengineering Laboratory, Department of Zoology, University of Education, Bank Road Campus, Lahore, Pakistan

S. Javed
Department of Biochemistry, Government College University Faisalabad, Faisalabad, Pakistan

M. Arshad
Biochemistry Section, Jhang-campus, University of Veterinary and Animal Sciences Lahore, Lahore, Pakistan

S. Javed et al. (eds.), *Sustainable Cosmeceuticals*,
https://doi.org/10.1007/978-3-031-86087-4_5

Abstract For decades, the use of cosmetic products has been increasing steadily over time and it is regarded as a part of "human beauty rituals". However, advancements in cosmetic science have created a fascinating link between beauty and animals. Animals possess exceptional bioactive compounds, which make them intrigue for the cosmetic industry. Various types of bioactive compounds derived from animals (zooceuticals) have emerged as "treasure of gems" in cosmetic industry which revolutionized the cosmetic products, i.e., skin care and personal care. These bioactive compounds include caviar extract, conchiolin, nacre, keratin, elastin, hyaluronic acid, collagen as well as allantoin. Other zooceutical compounds have showed remarkable results in terms of skin texture, rejuvenation, and aging. Various world-class cosmetic research institutes are continuously exploring the efficacy, Cutting-edge research and innovative formulations of zooceuticals to offer luxury and groundbreaking formulations with exceptional results. However, with a focus on ethical considerations as well as sustainability, the use of zooceuticals opens enormous pathways to enhance the efficiency of beauty products. However, the excessive capturing or killing the animals for cosmeceutical purposes can lead to the disruption of natural environment balance as well as directly impact the biodiversity. In conclusion, the rise of zooceuticals in the cosmetic industry shows the integration of beauty and nature. These hidden treasures push the boundaries of cosmetic industries and transform them into the way they approach self-care.

Keywords Zooceuticals · Cosmetics · Treasures of gems · Natural beauty · Beauty products

5.1 Introduction

A cosmetic product is defined as any compound that is used on external surface of body such as lips, teeth, hairs, epidermis, and genital organs for different purposes such as cleaning, perfuming, maintaining as well as protecting (César et al. 2017; França et al. 2017). The cosmetic industry is the most vibrant as well as ever-evolving business of the current era. A survey executed by Cosmetic Europe—The Personal Care Association revealed that Europeans spent about 450 million per day on different cosmetic products such as hair conditioner, shampoo, soap, skincare, makeup, shaving creams, and perfume (Paikray et al., 2024). Since the last two decades, the cosmetic industry has gained enormous innovation in a variety of products to protect and moisturize skin as well as against aging. Due to high demands and minimum side effects, cosmetic industry is turning toward the extraction of natural extracts from plants and animals. These active ingredients open a gateway for the recycling of old compounds, which were obtained from different natural resource products (Nunes et al. 2017; Rodrigues et al. 2016). These extractions totally align with the fundamental principles of green chemistry by using biodegraded resources as well as escalating sustainable recycling of these compounds.

Since the birth of a girl or boy, beauty plays a pivotal role in our society and acts as marker of physical attractiveness. For example, tattooed symbols on skin in indigenous Colombia while curvy skinny figures in Western nations are considered as landmarks of beauty (Franco García 2012; Salinas 2015). Nowadays, healthy and young skin is designated as beauty standard, which attracts the consumers towards smooth skin texture. Therefore, various research institutes are pursuing their research on skin-based products. Pharmaceutical companies have established efficient research and development institutions to find out new ingredients for the discovery and manufacturing of various medicines. However, these ingredients are not utilized in pharmaceuticals industry but also used in cosmetics, agriculture, and food (Abdullah et al. 2018).

Zooceuticals are zoologically derived compounds, which are extensively used in the pharmaceutical as well as cosmetical industry. These compounds are further divided into cosmeceuticals and biocosmetics. Zooceuticals include all the animal-based cosmetic products intended for human utilization. However, there are certain limitations regarding animal derived products, which include ethical as well as biodiversity conservations concerns (Fatima et al. 2020). The sustainable use of animals for cosmeceutical industries is pivotal as per the regulations of Sustainable Development Goals (SDGs). The 12th principle of SDGs known as "Responsible consumption and sustainable production" is considered as the primary focus regarding the sustainable use of animals in various fields (Nhani et al. 2024). Various types of raw materials are derived from animals, which are used in cosmetic products. The following are some key zoocueticals, which are extensively studied regarding their cosmetical aspects.

Anastas and Warner (2000) elucidated that green chemistry promotes the use of renewable and natural resources that ultimately reduce the needs of synthetic materials, thus preventing the onset of environmental pollution. The use of animals in cosmeceutical industry emphasizes the reduction of carbon footprint by using environmental friendly materials, which are alight with the fundamental goals of sustainability (Clark et al. 2012). Therefore, the use of natural compounds not only provides a strong alternative but also regulates the 12 SGGs (United Nations 2015). Moreover, the homology among sustainable use of zooceuticals signifies the value of natural substances to reduce the burden of environmental pollution (Sheldon 2014). Therefore, the use of zooceuticals provides a convergence stewardship, which leads to a sustainable development.

5.2 Zooceuticals and Their Use in Cosmetic Industry

5.2.1 *Beeswax*

Skin is the first line of defense against the adverse effects of external environment and thus referred as "skin barrier function." Therefore, any structural alternations may escalate the probability of getting infections from external pathogens (Kim et al. 2012). Lipid nanoparticles are considered as potential therapeutic agents in cosmetics for dermatological purposes. Owing to their tiny size, they get contacted with stratum corneum and ultimately enhance skin hydration. Beeswax is a natural fatty product, which is extensively used for the development of various drugs in pharmaceutical industry (Zamarioli et al. 2015; Gowda et al. 2011). Beeswax is highly hydrophobic, due to which it is considered as potential agent for the development of lipid nanoparticles for dermatological purposes. It is reported that beeswax is incorporated into various creams and other dermatological formulations and remains intact for a long period (Souza et al. 2017).

Beeswax is frequently used as a humectant as well as a thickener in the development of creams, ointment, lipsticks, and other cosmetic products (Attama and Müller-Goymann 2008; Kamairudin et al. 2015). It is derived from the honeycombs of *Apis mellifera,* which have diverse composition including monoester, diesters, triesters, hydrocarbons, long chain alcohols, saturated fatty acids, and myristic acid and are insoluble in water with 54 °C of melting point (Elder 1984; Rosiaux et al. 2015). The composition of beeswax depends upon season, geographical location, bees' diet as well as environmental effects. Apparently, beeswax is whitest in its natural form but turns yellow after being exposed to pollens and honey (Fratini et al. 2016). Various forms of beeswaxes are available due to their relatively low industrial cost. Moreover, in recent years the demand for beeswax has increased many times owing to its physical properties such as skin biocompatibility and biodegradability (Liu and Feng 2015; Mandawgade and Patravale 2008). An investigation was conducted to evaluate the effectiveness of beeswax-based nanoparticles on skin. It was found that nanoparticles of beeswax exhibited exceptional skin protecting properties and therefore can be used in various creams, lotions as well as other cosmetic-based products (Arshad et al. 2022; Souza et al. 2017) (Table 5.1 and Fig. 5.1).

Table 5.1 The cosmeceutical impacts of honey wax on skin

Cosmeceutical effects of honey wax	Skin barrier	Kim et al. (2012)
	Skin moisturizing	Souza et al. (2017)
	Acne prevention	Kamairudin et al. (2015)
	Wound healing	Liu and Feng (2015)
	Prevent skin infection	Kim et al. (2012)

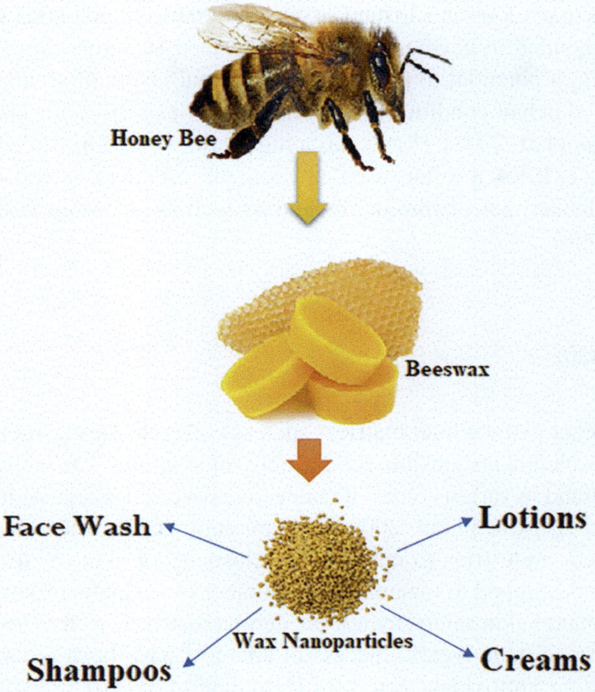

Fig. 5.1 Nanoparticles of beeswax and their use in cosmetic products

5.2.2 *Honey*

Since decades, honey is the topmost used natural substance due to its beneficial results on skin. From old times, it has been extensively used on skin as a traditional remedy. Archaeological records reported that various tribes of Tasian culture mixed fat, oil, spar, copper, malachite, and honey to make different kinds of cosmetic products for eyes (Lamprecht 1994). In ancient times, Egyptian women used honey with sodium bicarbonate for vaginal irrigation (Karou et al. 2007) owing to its antibacterial potential. They also chewed honey-made scented pills to sweeten their mouth breath. In ancient Rome, a paste of honey called honey-mint was used for oral health (Namias 2003). Due to its various potential as well as composition, it is used in lotions, creams as well as paste (Green 1979).

Currently, honey is included in the list of moisturizing products in "International Nomenclature of Cosmetic Ingredients" under the name of "Mel." Besides its multifaceted benefits, it is considered as an essential remedy for skin such as to avoid wrinkles and enhance the natural glamour (Rane 1980). The use of honey in various cosmetic products depends upon the type of cosmetic. For example, a very low quantity of honey is used in foaming products such as emulsions and creams, while a high proportion is used for anhydrous ointments (Krell 1996). It is reported that honey nurtures the skin and regulates the pH of the upper layer of skin. Furthermore, honey exhibits marvelous anti-irritant as well as demulcent potential; therefore, it is highly recommended in newborn babies due to their sensitive skin (Tilton 1991). Owing to its high potential to penetrate the external layers of skin and hairs, it is extensively used in hair conditioners to enhance the flexibility and elasticity of hairs (Umesh Hebbar et al. 2008). Honey has a high ratio of stickiness; therefore, only a little amount (1–10%) is often used in cosmetic products, which include tonic lotions, sun blocker, gels, hydrating creams as well as cleansing milks and shampoos (Krell 1996).

5.2.3 Elastin

Similarly, to other extracellular matrices such as collagen, the elastin is an extracellular protein, which is responsible for elasticity of skin, lungs as well as blood vessels and it is found in various types of connective tissues (Calleja-Agius et al. 2013; Momot et al. 2013; Chen et al. 2009; Sugitani et al. 2012). Therefore, it is widely used in cosmetic industries to enhance the elasticity of skin. A diverse range of techniques are developed to manufacture a number of cosmetic products from elastin fibers to enhance skin safety, hydration, dermal matrix as well as texture (Kramer et al. 2012). Since many years, successful attempts have been reported to isolate elastin containing antioxidant activities (Nadalian et al. 2015, 2019). A detailed investigation was conducted to evaluate the efficacy as well as stability of elastin in cosmetic products. It was revealed that the sample of elastin-based moisturizing cream remained stable at different temperatures (40 °C, 50 °C, and 25 °C) for a period of approximately 3 months. Furthermore, there were no significant alterations in the pH, color, and odor of the product. Additionally, the formulated cream showed marvelous results in terms of skin elasticity. Based on these findings, it is suggested that elastin is potent ingredient in cosmetic industry for various types of moisturizing formulations (Kamaruzaman and Yusop 2021).

5.2.4 Keratin

Keratin is a fibrous protein, which is abundant in epithelial cells of normal tissues. A plethora of literature highlighted the cosmetical as well as pharmaceutical effects of keratin on hair, skin, and other tissues. Keratin is derived from the wool of sheep, feathers of birds as well as hoof/horn of bovine (Chahal 2016). Since many years ago, human hairs were considered the best source of keratin, but their use is now limited due to ban on human-sourced materials for cosmetic industry. However, the purity and composition of keratin depend upon the source of extraction (Burnett et al. 2021). Hydrolyzates of keratin demonstrated various cosmetic applications such as its use on skin and hair. A long-term keratin treatment gives softness and shine to hair. On the other hand, keratin maintains the moisture level of skin and hairs (Barba et al. 2010). Owing to its thermal as well as mechanical abilities, it is highly recommended in various cosmetic products. It is reported that keratin peptides keep the skin supple and firm by reducing moisture loss from skin epidermis. Keratin hydrolyzate binds with water molecules from the inner layer of epidermis to the stratum corneum and thus acts as humectant. Numerous investigations revealed that keratin incorporates into nails and enhances the strength of nail plate (Lusiana et al. 2011).

5.2.5 Propolis

Propolis is a natural compound, which is derived from bees, and it is widely used in dermatology to fight against various types of fungal infections. Hidradenitis, thrush, purulent skin infections as well as intertrigo are commonly treated with propolis-based products (Olczyk et al. 2014). Propolis is reported to show efficient results against *Acne vulgaris*. A wide range of pharmaceutical as well as cosmeceuticals products are enriched with propolis for dermal use against *Acne vulgaris* (Ali et al. 2015). Propolis is considered as the key components of various acne cure creams and other products, which produce drugs to ameliorate the potential effects of fungal as well as bacterial infections (Bankova 2005). It is documented that propolis exhibits protective, repairing as well as regenerative effects against external factors. After application, propolis shows antibedsores properties and firmly interacts with dermal tissues, thus acting as a barrier against pathogen penetration (Sawicka and Borawska 2013). Propolis is widely used in sun-blocker products owing to the presence of ferulic acid, coumaric acid as well as caffeic acid, which directly absorb UV radiation (Sforcin and Bankova 2011).

Furthermore, the propolis is widely used to manufacture protective lipsticks. Various skin care products that constitute propolis showed tremendous potential against different skin infections owing to the presence of pinocembrin and pinobanksin (flavonoids) as well as terpenes (Sawicka and Borawska 2013). It is reported that isolating pinocembrin from propolis demonstrated inhibitory action against the growth of *Penicillium italicum* via damaging its cell membrane (Peng et al. 2012). Propolis-based shampoos are highly recommended by dermatologists for the treatment and prevention of dandruff. Moreover, it is documented that toothpaste with propolis decreased approximately 34.3% plaque formation in teeth. After 24 months of continuous use, the plaque formation further reduced to 12.4%. Moreover, a mouth wash with propolis solution (0.5%) maintains the health of the oral cavity. Interestingly, after 3 weeks usage of propolis-based mouth wash, the solution was able to decrease plaque formation (Kędzia 2011). Additionally, propolis showed marvelous results against wrinkle formation, thus ultimately demonstrating antiaging potential. The presence of flavonoids in propolis helps to keep skin fresh and hydrated (Sawicka and Borawska 2013).

5.2.6 Collagen

Collagen is an animal-derived structural protein, which is abundant in connective tissues such as skin, tendons, bones, and joints (Arshad and Abbas 2018; Arshad et al. 2018). Collagen primarily consists of amino acids, hydroxyproline, proline, and glycine. Due to its nontoxic nature, high biocompatibility, and biodegradability, collagen is used in cosmetics, food as well as pharmaceutical industries (León-López et al. 2019; Ricard-Blum 2011; Bhagwat and Dandge 2016). However, applications of collagen in cosmetic industry have gained much attention due to its film forming, regenerative as well as moisturizing properties (Rodrigues et al. 2016). Collagen demonstrates exceptional potential to bind with water, which make it an effective moisturizer of skin during the day. It is reported that collagen prevents hair and skin damage instigated by external factors. Numerous studies have elaborated that collage helps the tissue to regenerate as well as accelerate the healing process (Secchi 2008; Peng et al. 2004; Nagelschmidt and Struck 1974; Li et al. 2005; Jridi et al. 2015).

Subcutaneous injections are manufactured from soluble collagen, which are widely used for improving the density and quality of skin as well as mitigating dermatological defects (Pachence et al. 1987). Collage fillers are reported to restore the signs of aging as well as showing no allergic reaction. These fillers are generally characterized by high biodegradability, nontoxic nature and they are reproducible (Newman 2009; Requena et al. 2011; DeVore et al. 1996). Furthermore, collagen is the core component of various hydrogels, which are extensively used in the so-called beauty masks. These masks are used to restore the elasticity of skin as well as to eliminate the adverse effects of environmental factors (Mitura et al. 2020). Additionally, collagen is also used as thickening agent in various cosmeceutical products around the world (Sionkowska et al. 2017).

5.2.7 Hyaluronic Acid

Hyaluronic acid is an animal-derived compound, which is widely used in dermal fillers to maintain the texture of skin (Gold 2007; Matarasso et al. 2006). Naturally hyaluronic acid is present in the extracellular matrix of human tissues such as skin, vitreous fluid of eyes, synovial fluid of joints as well as cartilage (Laurent and Reed 1991; Almond 2007). An average human (70 kg) roughly contains approximately 15 g of hyaluronic acid in the body, which is continuously exposed to the process of degradation and synthesis every day (Laurent and Reed 1991; Hascell and Laurent 2008), while skin holds the largest concentration (7–8 g) per person. Furthermore, hyaluronic acid is extensively used in the manufacturing of dermal fillers, intradermal injections, hydrogels, creams, foams, scaffolds, and other dermatological products. It is reported that hyaluronic acid demonstrated immunomodulatory (Fiszer-Szafarz et al. 1988), tissue regenerator (Hussain et al. 2017), anti-inflammatory (Chen et al. 2018), anti-diabetic (Ievdokimova 2008), anti-proliferative (Safdar et al. 2018), and skin restoring agents (Narurkar et al. 2016), antiaging (Papakonstantinou et al. 2012) as well as various other cosmetical applications (Pavicic et al. 2011). The efficacy, safety, and tolerability of hyaluronic acid as a dermal filler or facial injection have been reported against various skin-related diseases. For instance, the use of hyaluronic acid along with lidocaine and other dermatological agents showed promising results in cosmeceutical as well as nutricosmetic effects such as skin tightness, antiaging, skin elasticity, skin rejuvenation, and tear trough rejuvenation (Bukhari et al. 2018).

Nowadays, hyaluronic acid is widely used in the cosmetical industry for reducing facial wrinkles. A wrinkle is a skin ridge or fold or a prominent crease in skin, which is considered as a common sign of aging (Danby 2010). There are multiple factors underlying the development of skin wrinkles such as sun damage, poor hydration, habitual facial expression, habitual sleeping positions as well as prolong immersion in water (Zimbler et al. 2001). However, numerous investigations have shown that hyaluronic acid has the potential to eradicate skin wrinkles. A detailed investigation was conducted on 76 female having periocular wrinkles. A cream formulation of 0.1% (w/w) with different quantities of hyaluronic acid was applied twice a day for 2 months. It was observed that hyaluronic acid enhanced skin elasticity, improved skin hydration as well as significantly reduced periocular wrinkles (Pavicic et al. 2011). In view of the above experimentation, researchers conducted different tests regarding nanoparticles of hyaluronic acid in different serums, creams, and lotions (Jegasothy et al. 2014).

5.2.8 Silk

Silk is a fibrous protein, which is derived from common silkworm (*Bombyx mori*) and different species of spiders such as *Araneus diadematus and Nephila clavipes.* These proteins are prepared exclusively in different specialized glands. Silk proteins are synthesized in epithelial cells and after attaining appropriate peptide length, these proteins are secreted into the lumen of glands. These glands act to store as well as spun these fibers into silk fibers (Joseph and Raj 2012). The silk derived from silkworms is widely used in various cosmetic products (Nikitakis and Breslawec 2013). Silk encompasses two essential proteins known as sericin and fibroin. Fibroin protein has been employed in the textile industry for manufacturing of different biomedical textiles. On the other hand, sericin along with other polymers has the potential to make film or scaffold, which is used in various pharmaceutical as well as cosmeceutical industries (Aramwit et al. 2012).

Skin tissues owing to various environmental factors leave lesions even after a complete treatment. Researchers are continuously investigating various types of materials, which can be used to make scarless skin. It is found that a scaffold prepared by using silk, hyaluronic acid, and natural nanofibers of silks showed promising results against skin healing (Wang et al. 2021). Furthermore, the combination of fibroin silk and glycidyl methacrylate hydrogels activated immense number of macrophages and inducted the population of hair follicles and stem cells, which result in a scarless skin (Liang et al. 2021). It is reported that fibroin silk escalated the rate of gases exchange as well as nutrient perfusion across the skin, which enhance and restore the thickness of skin (Liu et al. 2021). The chemical and physical properties of silk make it an attractive raw material for various types of cosmetic products. Silk proteins provide a solid support during skin and hair regeneration. Owing to these reasons, various beauty products such as lotions, creams, and hair products contain silk proteins (Dam et al. 2022). It is documented that silk proteins are used as active ingredients in gels, mousses, conditions, shampoos, bleaches, dyes, cleanings items, antiaging creams, and lotions (Aramwit et al. 2012; Natural Sourcing 2013; Voegeli et al. 1993).

Silk peptides are used in cosmetic formulations as hydrating, protective as well as antioxidant agents for the nourishment of skin (Papini 2010). The chemical structure of silk revealed that it contains carbohydrate moiety, hydrophilic amino acids and hydroxyl groups, which facilitate its binding with other hydrophilic proteins such as keratin. Owing to its water holding capacity, silk acts as an efficient moisturizing agent, which maintains soft texture of skin (Voegeli et al. 1993). Furthermore, biosilk is the top-notch treatment against hair damage. These formulations are weightless and they are used in various hair treatments such as repair, smoothness as well as replenishment (Yanqing 2004a). It is manifested that silk enhances the luster as well as elasticity of hair (Voegeli et al. 1993; Yanqing 2004b). A Japanese company named “Kanebo” was one of the first to use silk in various makeup products. Since silk has been used in various herbal medicines for skin care, Kanebo developed advanced technology to manufacture an aqueous solution of silk protein (fibroin). Owing to water resistant, silk was first used in silk powder cake (Fox 1994). Similarly, other cosmetic companies such as Shiseido and L’Oreal used the same patent to produce powders, suspensions, and emulsions (Otoi and Horikawa 1980) (Fig. 5.2).

In cosmetic industry, nails have been considered as important component of beauty and self-adornment for centuries (Barnett et al. 1991). Moreover, modern technology used silk wraps to improve the beauty and quality of nails (Haneke 2006). These silk containing products are used to split and mend fractured nails as well as they are used as nail extension. Broken nails are efficiently repaired by using different methods. The basic technique involves the thin coating of nail with the assistance of cyanoacrylate glue. A small piece of silk is used to employ over the nail by using additional coating of nail polish (Baran and Schoon 2004; Iorizzo et al. 2007). When silk is used as nail extension, it splits over the edges of nail plate. These nails are then highly saturated with various nail glue or polishes (Wong et al. 2014).

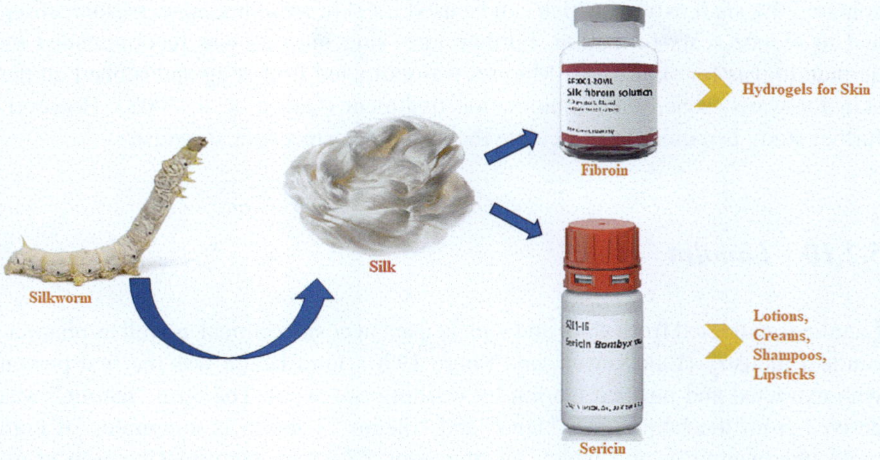

Fig. 5.2 Silk and its role in cosmeceutical industry

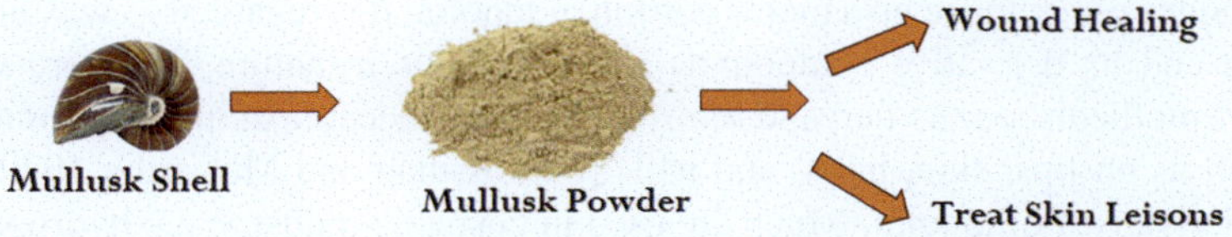

Fig. 5.3 Mollusk powder and its efficacy to treat wound healing and skin lesions

5.2.9 Mollusk-Based Cosmetics

Marine molluscans are a diverse group of Kingdom animalia, which constitute about 23% of living organisms under the water (Pati et al. 2015). These mollusks are economically significant owing to their usage in craft making, yarn, clothes, and dye cotton (Flores-Garza et al. 2012). Since decades, molluscans shells and their by-products are used to maintain the human skin. The shells of marine mollusks are used to cure skin wounds around the world. It is reported that mollusks shell stimulates the migration of fibroblasts, which fasten the healing process of wounds (Li et al. 2013). Furthermore, the hydrophilic matrices of the pearl promote the proliferation of collagen as well as fibroblast (Jian-Ping et al. 2010). However, the use of pearls for the maintenance of human skin is costlier owing to their limited numbers. Moreover, the nacreous layer of shell is used for regeneration of skin (Lopez et al. 2000) (Fig. 5.3).

Different species of snail and slugs are used sporadically for skin treatment since the times of Greeks. Various Hippocrates documented that paste of crushed snail is efficient to cure skin inflammation, while snail farmers reported that snail paste has potential to treat skin lesions without any scar. These investigations led to the formulation of "Elicina," a well-known snail slime-based product in marker (Ekin and Şeşen 2018). Since 2010, various cream manufacturers claimed that their creams contain 70% snail extract, which can be used for skin wrinkles, acne, pigmentation, and to reduce scars formation. Furthermore, slug mucous was recommended by dermatologists for skin warts. Mucous was extracted from slug and rubbed on the skin for warts, acne, and inflammation treatment (Quave et al. 2008). However, further study is required to elucidate their role in cosmeceutical industry.

5.2.10 Lanolin

Lanolin is extracted from wool and widely used in cosmeceutical as well as pharmaceutical industry (Baldwinson, and Shore 1990). Otto Braun was the first person who extracted and purified lanolin by washing raw wool. The term "lanolin" was derived from the latin words "lana" and "oleum." Lanolin is a complex of high molecular weight esters, which encompasses 87% raw lanolin (Chemtob et al. 1975). The remaining portion of raw lanolin consists of aliphatic alcohols, sterols, fatty acids, hydrocarbons, and various unidentified compounds (Chemtob et al. 1975). Since esters make up a major portion of lanolin, it is regarded as wax not fats. Lanolin and its derivative compounds are ubiquitous in nature. They are widely used as emollients, emulsifier, and moisturizer in cosmeceutical industry for dermal use such as on lips, skin, hairs, and nails (Schlossman and McCarthy 1979). The major derivatives of lanolin, which are used in cosmetic industry, are hydrogenated lanolin, acetylated hydrogenated lanolin, and hydroxylated lanolin. These ingredients are widely used in lip products, creams, and hair care products (Latheef and Wilkinson 2020).

5.2.11 Lecithin

The term lecithin was first coined to elaborate an orange colored, sticky, phosphorous, and nitrogen containing material, which is directly isolated from egg yolk, blood, brain, bile, and other organic materials (Wisniaka 2019). The word lecithin was derived from a Greek word called "lekythos," which means egg yolk. Later, it is found that the nitrogenous material in lecithin is choline, which is an organic base that naturally exists in bile (Andersen 2001). About 5% of lecithin is used in lotions, creams, shampoos, and various bar soaps (Haumann 1992; Wendel and Ghyczy 1990). Furthermore, lecithin is widely used as coating agent in various moisturizers to improve the adhesion of skin. Lecithin is directly applied to the skin to reduce skin irritation as well as it is used in cosmetic powders, which are used against various allergic reactions (Wendel and Ghyczy, 1990).

The expert panel for cosmetic ingredient safety issued a comprehensive report on the safety of lecithin in cosmetics, which elucidate that lecithin is completely a safer compound that can be used in rinse off products with a concentration of <15%

(Andersen 2001). Lecithin owing to its hydrophobic nature is extensively used to synthesize liposomes (Reva et al. 2015). Liposomes are spherical vesicles having an aqueous region surrounded by lamella (a bilayer membrane) and enclosed by various aquatic environments. These vehicles can vary in size owing to their function such as drug delivery and manufacturing of cosmetic products (Laouini et al. 2012; Reva et al. 2015). These liposomal vesicles act as cosmeceutical vehicles as well as active ingredients themselves. When skin layers are continuously affected owing to external damage or low moisture, the liposomes highly interact with lipids, carbohydrates, and proteins of skin to restore the stratum corneum from eczema (Nastruzzi et al. 1993). To penetrate the upper layer of skin, molecules should possess similar physiochemical properties such as solubility, low weight as well as low melting point (Dragicevic and Maibach 2017). To overcome these limitations, liposomes are exploited to act as carriers for dermatological as well as cosmeceutical ingredient, which assist in the penetration of drugs or cosmetic products (Nastruzzi et al. 1993).

5.2.12 Viper Serum

Cosmeceutical and pharmaceutical products play a pivotal role to fight against aging of skin particularly to prevent the appearance of skin wrinkles (Puizina-Ivić 2008). Botulinum toxin is one of the significant aesthetic medicines used for aesthetic purposes. It is documented that administration of toxin weakens the facial muscles, which ultimately relax the affected area of skin (Dressler and Adib Saberi 2005; Pellizzari et al. 1999). Skin aging or wrinkling is a dynamic process, which occurs owing to extrinsic as well as intrinsic factors (Bhardwaj et al. 2022; Campiche et al. 2021; Ruiz et al. 2007). Continuous exposure to extrinsic factors such as radiations, which include infrared and ultraviolet, causes significant loss of skin elasticity, which ultimately leads to skin aging as well as wrinkle formation (Fussell and Kelly 2020; Bielach-Bazyluk et al. 2021) (Fig. 5.4).

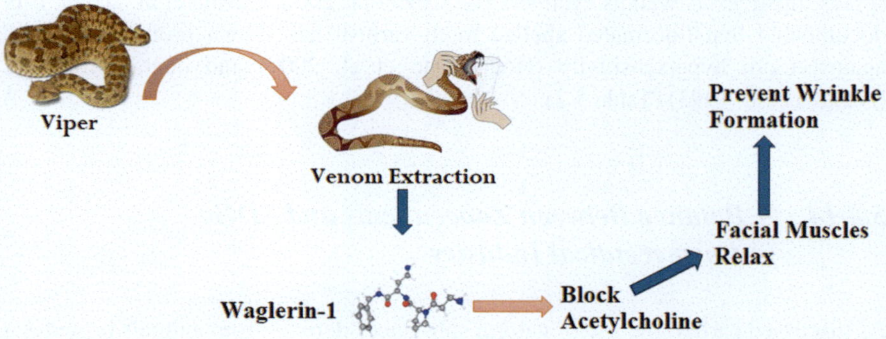

Fig. 5.4 Application of viper venom to prevent wrinkle formation

Dipeptide diaminobutyroyl benzylamide diacetate or Syn-Ake is a cosmeceutical agent, which is prepared synthetically on commercial scale to mimic the action of Waglerin-1 (a peptide derived from vipers). It is reported that Syn-Ake remarkably reduced the formation of wrinkles on facial skin (Zhang and Falla 2009). Syn-Ake stops the contractility of facial muscles by blocking the secretion of acetylcholine (neurotransmitters) from muscle receptors. Therefore, the muscles relax, which ultimately reduced the formation of wrinkles (Errante et al. 2020). An investigation executed on 37 volunteer females showed that application of Syn-Ake quickly reduced fine wrinkles, while the coarse wrinkles were decreased over time (Trookman et al. 2009). Furthermore, after applying 4% formulation of Sky-Ake on forehead for 4 weeks could reduce the size of wrinkle up to 52% (Errante et al. 2020). However, the safety and commercial efficacy of this product should need further research (Arshad 2024).

5.2.13 *Shellac*

Shellac is a purified form of a resin, which is derived from female insect Laccifer, found in India and Thailand (Le Coz et al. 2002; Hoang-Dao et al. 2009; Pascoe et al. 2010). These insects secret shellac from their pores as a by-product while consuming tree sap. After exposure to air, the shellac becomes hard and acts as a protective barrier for encasing cocoons, although this process ensures the success rate of progeny but ultimately culminates in the demise of adult insects. Therefore, Laccifer is known as "feast of death" (Hoang-Dao et al. 2009). Since earlier times, shellac has been widely used in medical industry as an emollient for the treatment of uncontrolled bleeding as well as to fill the gaps in farm animal's hooves (Hoang-Dao et al. 2009; Azouka et al. 1993). By the mid-1600s, it is widely used in high commodities for various purposes such as painting to protect their masterpieces (Hoang-Dao et al. 2009). Owing to its emollient film forming potential, shellac and its wax is used in the manufacturing of various cosmetic products such as hair dyes, sprays, lipsticks as well as eyeliners (Le Coz et al. 2002; Gowda et al. 2011). It is documented that fluorinated shellac tooth varnish has shown promising results against tooth hypersensitivity (Hoang-Dao et al. 2009) and dental prosthetic (Azouka et al. 1993) (Table 5.2).

5.2.14 *A Balance Between Zooceuticals and SDGs in Cosmeceutical Industry*

As discussed earlier, the use of natural substances derived from animals provides a decent solution against environmental challenges by decreasing our dependency on highly toxic substances, thereby sustaining the health of humans as well as their

Table 5.2 Various cosmetic products, their sources, and cosmeceutical uses

Source	Product	Cosmetical use	References
Honeybees	Beeswax	Skin hydration, protection	Liu and Feng (2015), Mandawgade and Patravale (2008)
Honeybees	Honey	Eyes protection, oral health, anti-wrinkle	Lamprecht (1994), Namias (2003), Rane (1980)
Connective tissues	Elastin	Skin elasticity, texture and hydration	Kramer et al. (2012)
Honeybees	Propolis	Anti-acne, skin infections	Ali et al. (2015), Bankova (2005)
Connective tissues	Collagen	Moisturizer, protect skin and hair against external factors	Secchi (2008), Peng et al. (2004), Nagelschmidt and Struck (1974)
Bombyx mori	Silk fibers	Anti-wrinkle	Healing (Wang et al. 2021)
Snail	Snail slime	Antiaging	Quave et al. (2008)
Viper (Snake)	Viper serum	Relax skin muscles, anti-wrinkle	Dressler and Adib Saberi (2005), Trookman et al. (2009)

environment. In the prospective of well-being and health, the use of animal-derived chemicals provides biocompatible, safer as well as cheaper alternatives against manmade chemicals. The marvelous therapeutic values including skin moisturizing, skin protection, antiaging, as well as anti-inflammatory further endorsed their use in cosmetic industry (Kim and Wijesekara 2010). These advancements significantly highlight the importance of zooceuticals in context of environmental sustainability. The use of animal by-products such as wool, membranes of eggshells, fish scales, honey, and silk promotes resource efficiency as well as diminishes the production of waste. These strategies reduce environmental footprint as well as optimize the cosmetic formulations (Arshad et al. 2023; Deng et al. 2022). Moreover, Anastas and Warner (2000) emphasized the importance of zooceuticals for the production of cosmetic products, which are aligned with the fundamental principles of green chemistry.

The techniques used to extract bioactive compounds from natural sources such as marine organisms are energy efficient and environmental friendly. For instance, marine chitosan derived from marine sources is biodegradable and has relatively low eco-impact to obtain, making it vital for green cosmetic formulation (Sheldon 2014). In addition, the practice of sustainable harvest of zooceuticals implies conservation of the biodiversity of the natural environment. The components of ethical sourcing that are used in sourcing raw materials are least intrusive to ecosystems, such as lanolin from sheep wool or snail mucin. These approaches clearly address the need to meet the growing quest for green cosmetics as they incorporate innovation and conservation (Clark et al. 2012). The integration of these practices helps to maintain ecosystem health as a path to promote the sustainable development of industry. Zooceuticals also fit within larger global goals like the carbon footprint and biodegradability of consumable products. Linkages of these innovations are closely associated with sustainability initiatives and decreased pollution and environmental impacts. The linkage of zooceuticals with the SDG also qualitatively

Table 5.3 Marine-derived bioactive compounds and their mode of action

Marine zooceuticals	Bioactive ingredient	Uses	Reference
Crustacean shells	Chitosan	Antiaging	Kim and Wijesekara (2010)
Sea cucumber	Saponins	Skin elasticity	Deng et al. (2022)
Marine sponges	Peptides	Skin nourishment	Deng et al. (2022)
Fish oils	Omega-3 fatty acids	Antiaging, anti-inflammatory	Sheldon (2014)
Star fish	Collagen	Skin rejuvenation, hydration, moisturizing, anti-acne	Kim and Wijesekara (2010)

underlines the cardinal role of zooceuticals as the potential solution provider in religious, health, environmental, and economical fronts (Table 5.3).

5.3 Conclusion

In the realm of cosmetic industry, various animal-derived compounds are undeniable gems owing to their exceptional properties. Among these, honey, beeswax, keratin, elastin, propolis, collagen, hyaluronic acid, silk amino acids, viper venom, and insects' shellac are prominent compounds, which play a pivotal role as humectant, emollient, moisturizers, hair dryers, and skin polishers. Owing to their abovementioned properties, these compounds are extensively used in the manufacturing of various cosmetic products such as soaps, creams, lotions, lipsticks, skin whiteners, perfume, body mist, eyeliners, antiaging, anti-acne, anti-wrinkle, hair shiners, hair shampoos as well as to cure various skin-related disorders. However, the sustainable use of these bioactive compounds supports environmental conservation and aligns with the principles of green chemistry outlined in the Sustainable Development Goals (SDGs).

References

Abdullah MS, Noordin MI, Ismail SI, Mustapha NM, Jasamai M, Danik MF, Ismail WA, Shamsuddin AF (2018) Recent advances in the use of animal-sourced gelatine as natural polymers for food, cosmetics and pharmaceutical applications. Sains Malays 47:323–336

Ali BM, Ghoname NF, Hodeib AA, Elbadawy MA (2015) Significance of topical propolis in the treatment of facial acne vulgaris. Egypt J Dermatol Venerol 35:29–36

Almond AH (2007) Hyaluronan. Cell Mol Life Sci 64:1591–1596

Anastas PT, Warner JC (2000) Green chemistry: theory and practice. Oxford University Press

Andersen FA (2001) Final report on the safety assessment of lecithin and hydrogenated lecithin. Int J Toxicol 20:21–45

Aramwit P, Siritientong T, Srichana T (2012) Potential applications of silk sericin, a natural protein from textile industry by-products. Waste Manag Res 30:217–224

Arshad M (2024) Sustainability of renewable energy options as compared to coal-fired power plants in Pakistan. Green Low-Carbon Econ 2(3):162–173

Arshad M, Abbas M (2018) Water sustainability issues in biofuel production. In: Perspectives on water usage for biofuels production: aquatic contamination and climate change, pp 55–76

Arshad M, Bano I, Younus M, Khan A, Rahman A (2018) Health concerns associated with biofuel production. In: Perspectives on water usage for biofuels production: aquatic contamination and climate change, pp 97–105

Arshad M, Ansari AR, Qadir R, Tahir MH, Nadeem A, Mehmood T, Alhumade H, Khan N (2022) Green electricity generation from biogas of cattle manure: an assessment of potential and feasibility in Pakistan. Front Energy Res 10:911485

Arshad M, Syed-Hassan SSA, Masood R, Ansari AR, Mumtaz A, Rahman A, Batool S, Atif FA, Muhammad A, Rehman AU (2023) Utilization of waste animal fat for sustainable biodiesel production. In: Climate changes mitigation and sustainable bioenergy harvest through animal waste: sustainable environmental implications of animal waste. Springer, pp 149–169

Attama AA, Müller-Goymann CC (2008) Effect of beeswax modification on the lipid matrix and solid lipid nanoparticle crystallinity. Colloids Surf A Physicochem Eng Asp 315:189–195

Azouka A, Huggett R, Harrison A (1993) The production of shellac and its general and dental uses: a review. J Oral Rehabil 20:393–400

Bankova V (2005) Recent trends and important developments in propolis research. Evid-Based Complement Altern Med 2:29–32

Baran R, Schoon D (2004) Nail fragility syndrome and its treatment. J Cosmet Dermatol 3:131–137

Barba, C., Martí, M., Roddick-Lanzilotta, A., Manich, A., Carilla, J., Parra, J., & Coderch, L. (2010). Effect of wool keratin proteins and peptides on hair water sorption kinetics. Journal of thermal analysis and calorimetry, 102(1), 43–48.

Barnett JM, Scher RK, Taylor SC (1991) Nail cosmetics. Dermatol Clin 9:9–17

Bhagwat PK, Dandge PB (2016) Isolation, characterization and valorizable applications of fish scale collagen in food and agriculture industries. Biocatal Agric Biotechnol 7:234–240

Bhardwaj V, Namkoong J, Tartar O, Diaz I, Mao J, Wu J (2022) In vitro and ex vivo mechanistic understanding and clinical evidence of a novel anti-wrinkle technology in single-arm, monocentric, open-label observational studies. Cosmetics 9(4):80

Bielach-Bazyluk A, Zbroch E, Mysliwiec H, Rydzewska-Rosolowska A, Kakareko K, Flisiak I, Hryszko T (2021) Sirtuin 1 and skin: implications in intrinsic and extrinsic aging—a systematic review. Cells 10:813

Bukhari SN, Roswandi NL, Waqas M, Habib H, Hussain F, Khan S, Sohail M, Ramli NA, Thu HE, Hussain Z (2018) Hyaluronic acid, a promising skin rejuvenating biomedicine: a review of recent updates and pre-clinical and clinical investigations on cosmetic and nutricosmetic effects. Int J Biol Macromol 120:1682–1695

Burnett CL, Bergfeld WF, Belsito DV, Hill RA, Klaassen CD, Liebler DC, Marks JG Jr, Shank RC, Slaga TJ, Snyder PW, Gill LJ (2021) Safety assessment of keratin and keratin-derived ingredients as used in cosmetics. Int J Toxicol 40:36–51

Calleja-Agius J, Brincat M, Borg M (2013) Skin connective tissue and ageing. Best Pract Res Clin Obstet Gynaecol 27:727–740

Campiche R, Pascucci F, Jiang L, Vergne T, Cherel M, Gougeon S, Préstat-Marquis E, François G, Laurent G, Gempeler M (2021) Facial expression wrinkles and their relaxation by a synthetic peptide. Int J Pept Res Ther 27:1009–1017

César FC, Carnevale F, Porto GS, Campos PM (2017) Patent analysis: a look at the innovative nature of plant-based cosmetics. Quím Nova 40:840–847

Chahal S (2016) Letter from Dr. Surinder Chahal concerning the CJR review of Keratin and Hydrolyzed Keratin. Unpublished data submitted by Personal Care Products Council

Chemtob C, Fawaz F, Puisieux F (1975) Analysis of ointments, oils and waxes. XX. The chemical composition of lanolin wax. 2. Alcohols. Ann Pharm Fr 33:243–251

Chen Z, Shin MH, Moon YJ, Lee SR, Kim YK, Seo JE, Kim JE, Kim KH, Chung JH (2009) Modulation of elastin exon 26A mRNA and protein expression in human skin in vivo. Exp Dermatol 18:378–386

Chen LH, Xue JF, Zheng ZY, Shuhaidi M, Thu HE, Hussain Z (2018) Hyaluronic acid, an efficient biomacromolecule for treatment of inflammatory skin and joint diseases: a review of recent developments and critical appraisal of preclinical and clinical investigations. Int J Biol Macromol 116:572–584

Clark JH, Luque R, Matharu AS (2012) Green chemistry, biofuels, and biorefinery. Annu Rev Chem Biomol Eng 3:183–207

Dam P, Altuntas S, Mondal R, Baudrit JR, Kati A, Ghorai S, Sadat A, Gangopadhyay D, Shaw S, Franco OL, Temur N (2022) Silk-based nano-biocomposite scaffolds for skin organogenesis. Mater Lett 327:133024

Danby FW (2010) Nutrition and aging skin: sugar and glycation. Clin Dermatol 2:409–411

Deng J, Zhu EQ, Xu GF, Naik N, Murugadoss V, Ma MG, Guo Z, Shi ZJ (2022) Overview of renewable polysaccharide-based composites for biodegradable food packaging applications. Green Chem 24:480–492

DeVore DP, Kelman CD, Fagien S, Casson P (1996) Autologen: autologous, injectable dermal collagen. Ophthalmic Plast Reconstr Surg 1:670–675

Dragicevic, N., & Maibach, H. I. (Eds.). (2017). Percutaneous penetration enhancers drug penetration into/through the skin: Methodology and general considerations. Springer.

Dressler D, Adib Saberi F (2005) Botulinum toxin: mechanisms of action. Eur Neurol 53:3–9

Ekin İ, Şeşen R (2018) Molluscs: their usage as nutrition, medicine, aphrodisiac, cosmetic, jewelry, cowry, pearl, accessory and so on from the history to today. Middle East J Sci 4:45–51

Elder RL (1984) Final report on the safety assessment of methylparaben, ethylparaben, propylparaben, and butylparaben. J Am Coll Toxicol 3:147–209

Errante F, Ledwoń P, Latajka R, Rovero P, Papini AM (2020) Cosmeceutical peptides in the framework of sustainable wellness economy. Front Chem 8:572923

Fatima N, Anwar S, Jaffar S, Hanif A, Hussain HM, Mukhtar MW, Khan D (2020) An insight into animal and plant halal ingredients used in cosmetics. Int J Innov Creat Chang 14:2020

Fiszer-Szafarz B, Rommain M, Brossard C, Smets P (1988) Hyaluronic acid-degrading enzymes in rat alveolar macrophages and in alveolar fluid: stimulation of enzyme activity after oral treatment with the immunomodulator RU 41740. Biol Cell 6:355–360

Flores-Garza R, Flores-Rodríguez P, Torreblanca-Ramírez C, Galeana-Rebolledo L, Valdés-González A, Suástegui-Zárate A, Violante-González J (2012) Commercially important marine mollusks for human consumption in Acapulco, México. Nat Resour 3:11

Fox C (1994) Advances in cosmetic science and technology. I: pigments and makeup. Cosmet Toiletries 109:39–51

França K, França AP, de França R (2017) Environmental psychodermatology: stress, environment and skin. In: Stress and skin disorders: basic clin aspects. Springer, Cham, pp 47–53

Franco García R (2012) Cariba malo: episodios de resistencia de un pueblo indígena aislado del Amazonas. Universidad Nacional de Colombia, Sede Amazonia

Fratini F, Cilia G, Turchi B, Felicioli A (2016) Beeswax: a minireview of its antimicrobial activity and its application in medicine. Asian Pac J Trop Med 9:839–843

Fussell JC, Kelly FJ (2020) Oxidative contribution of air pollution to extrinsic skin ageing. Free Radic Biol Med 151:111–122

Gold MH (2007) Use of hyaluronic acid fillers for the treatment of the aging face. Clin Interv Aging 2:369–376

Gowda DV, Gupta VK, Khan MS, Bathool A (2011) Encapsulation of clozapine into beeswax microspheres: preparation, characterization and release kinetics. Int J Pharm Tech Res 3:2199–2207

Green P (1979) Ars gratia cultus: Ovid as beautician. Am J Philol 100:381–392

Haneke E (2006) Onychocosmeceuticals. J Cosmet Dermatol 5:95–100

Hascell V, Laurent T (2008) Hyaluronan: structure and physical properties. Hyaluronan Today. (1997, Online)
Haumann BF (1992) Liposomes: their role in personal care products, cosmetics. Inform 3:1172–1178
Hoang-Dao BT, Hoang-Tu H, Tran-Thi NN, Koubi G, Camps J, About I (2009) Clinical efficiency of a natural resin fluoride varnish (Shellac F) in reducing dentin hypersensitivity. J Oral Rehabil 36:124–131
Hussain Z, Thu HE, Katas H, Bukhari SN (2017) Hyaluronic acid-based biomaterials: a versatile and smart approach to tissue regeneration and treating traumatic, surgical, and chronic wounds. Polym Rev 57:594–630
Ievdokimova NI (2008) Hyaluronic acid, receptor CD44, and their role in diabetic complications. Ukr Biokhim Zh(1999) 80:5–44
Iorizzo M, Piraccini BM, Tosti A (2007) Nail cosmetics in nail disorders. J Cosmet Dermatol 6:53–58
Jegasothy, S. M., Zabolotniaia, V., & Bielfeldt, S. (2014). Efficacy of a new topical nano-hyaluronic acid in humans. The Journal of clinical and aesthetic dermatology, 7(3), 27.
Jian-Ping D, Jun C, Yu-Fei B, Bang-Xing H, Shang-Bin G, Li-Li J (2010) Effects of pearl powder extract and its fractions on fibroblast function relevant to wound repair. Pharm Biol 48:122–127
Joseph B, Raj SJ (2012) Therapeutic applications and properties of silk proteins from Bombyx mori. Front Life Sci 6:55–60
Jridi M, Bardaa S, Moalla D, Rebaii T, Souissi N, Sahnoun Z, Nasri M (2015) Microstructure, rheological and wound healing properties of collagen-based gel from cuttlefish skin. Int J Biol Macromol 77:369–374
Kamairudin N, Abd Gani SS, Masoumi HR, Basri M, Hashim P, Mokhtar NM, Lane ME (2015) Modeling of a natural lipstick formulation using an artificial neural network. RSC Adv 5:68632–68638
Kamaruzaman N, Yusop SM (2021) Determination of stability of cosmetic formulations incorporated with water-soluble elastin isolated from poultry. J King Saud Univ Sci. 33:101519
Karou D, Nadembega WM, Ouattara L, Ilboudo DP, Canini A, Nikiéma JB, Simpore J, Colizzi V, Traore AS (2007) African ethnopharmacology and new drug discovery. Med Aromat Plant Sci Biotechnol 1:61–69
Kędzia B (2011) Propolis in the treatment of dental caries. Postępy Fitoterapii
Kim SK, Wijesekara I (2010) Development and biological activities of marine-derived bioactive peptides: a review. J Funct Foods 2:1–9
Kim DH, Park WR, Kim JH, Cho EC, An EJ, Kim JW, Oh SG (2012) Fabrication of pseudo-ceramide-based lipid microparticles for recovery of skin barrier function. Colloids Surf B: Biointerfaces 94:236–241
Kramer RM, Shende VR, Motl N, Pace CN, Scholtz JM (2012) Toward a molecular understanding of protein solubility: increased negative surface charge correlates with increased solubility. Biophys J 102:1907–1915
Krell R (1996) Value-added products from beekeeping. Food and Agriculture Organization
Lamprecht I (1994) bjt. Calorimetric investigations around a royal hieroglyph. Thermochim Acta 234:179–200
Laouini A, Jaafar-Maalej C, Limayem-Blouza I, Sfar S, Charcosset C, Fessi H (2012) Preparation, characterization and applications of liposomes: state of the art. J Colloid Sci Biotechnol 1:147–168
Latheef F, Wilkinson M (2020) Cosmetics and skin care products. Contact Derm 1–21
Laurent UB, Reed RK (1991) Turnover of hyaluronan in the tissues. Adv Drug Deliv Rev 7:237–256
Le Coz, C. J., Leclere, J. M., Arnoult, E., Raison-Peyron, N., Pons-Guiraud, A., Vigan, M., & Of Revidal-gerda, T. M. (2002). Allergic contact dermatitis from shellac in mascara. Contact Dermatitis, 46(3), 149–152.
León-López A, Morales-Peñaloza A, Martínez-Juárez VM, Vargas-Torres A, Zeugolis DI, Aguirre-Álvarez G (2019) Hydrolyzed collagen—sources and applications. Molecules 24:4031

Li GY, Fukunaga S, Takenouchi K, Nakamura F (2005) Comparative study of the physiological properties of collagen, gelatin and collagen hydrolysate as cosmetic materials. Int J Cosmet Sci 27:101–106

Li YC, Chen CR, Young TH (2013) Pearl extract enhances the migratory ability of fibroblasts in a wound healing model. Pharm Biol 51:289–297

Liang R, Shen X, Xie C, Gu Y, Li J, Wu H, Wen Y, Wu B, Zhao K, Wu Y, Yao X (2021) Silk gel recruits specific cell populations for scarless skin regeneration. Appl Mater Today. 23:101004

Liu Y, Feng N (2015) Nanocarriers for the delivery of active ingredients and fractions extracted from natural products used in traditional Chinese medicine (TCM). Adv Colloid Interf Sci 221:60–76

Liu J, Huang R, Li G, Kaplan DL, Zheng Z, Wang X (2021) Generation of nano-pores in silk fibroin films using silk nanoparticles for full-thickness wound healing. Biomacromolecules. 2:546–556

Lopez E, Le Faou A, Borzeix S, Berland S (2000) Stimulation of rat cutaneous fibroblasts and their synthetic activity by implants of powdered nacre (mother of pearl). Tissue Cell 32:95–101

Lusiana, B., van Noordwijk, M., Suyamto, D., Mulia, R., Joshi, L., & Cadisch, G. (2011). Users' perspectives on validity of a simulation model for natural resource management. International Journal of Agricultural Sustainability, 9(2), 364–378.

Mandawgade SD, Patravale VB (2008) Development of SLNs from natural lipids: application to topical delivery of tretinoin. Int J Pharm 363:132–138

Matarasso SL, Carruthers JD, Jewell ML, Restylane Consensus Group (2006) Consensus recommendations for soft-tissue augmentation with nonanimal stabilized hyaluronic acid (Restylane). Plast Reconstr Surg 11:3–4

Mitura S, Sionkowska A, Jaiswal A (2020) Biopolymers for hydrogels in cosmetics. J Mater Sci Mater Med 31:1–4

Momot KI, Powell SK, Mithieux SM, Weiss AS (2013) Biomechanics of synthetic elastin: insights from magnetic resonance microimaging. Adv Mater Res 699:457–463

Nadalian, M., Yusop, S. M., Babji, A. S., Mustapha, W. A. W., & Azman, M. A. (2015). Effects of Enzymatic Hydrolysis on the Antioxidant Activity of Water-Soluble Elastin Extracted from Broiler and Spent Hen Skin. 169090. https://pesquisa.bvsalud.org/portal/resource/pt/sea-169090

Nadalian M, Kamaruzaman N, Yusop MS, Babji AS, Yusop SM (2019) Isolation, purification and characterization of antioxidative bioactive elastin peptides from poultry skin. Food Sci Anim Resour 39:966

Nagelschmidt M, Struck H (1974) Collagen as cosmetic compound? Arch Dermatol 250:237–243

Namias N (2003) Honey in the management of infections. Surg Infect 4:219–226

Narurkar VA, Fabi SG, Bucay VW, Tedaldi R, Downie JB, Zeichner JA, Butterwick K, Taub A, Kadoya K, Makino ET, Mehta RC (2016) Rejuvenating hydrator: restoring epidermal hyaluronic acid homeostasis with instant benefits. J Drugs Dermatol 15:24–37

Nastruzzi C, Esposito E, Menegatti E, Walde P (1993) Use ano stability of liposomes in dermatological preparations. J Appl Cosmetol 11:77–91

Natural Sourcing (2013) Silk protein powders. www.naturalsourcing.com/downloads/NS_info_silk.pdf. Accessed 18 Jan 2013

Newman J (2009) Review of soft tissue augmentation in the face. Clin Cosmet Investig Dermatol:141–150

Nhani GB, Di Filippo LD, de Paula GA, Mantovanelli VR, da Fonseca PP, Tashiro FM, Monteiro DC, Fonseca-Santos B, Duarte JL, Chorilli M (2024) High-tech sustainable beauty: exploring nanotechnology for the development of cosmetics using plant and animal by-products. Cosmetics 11:112

Nikitakis J, Breslawec HP (eds) (2013) International cosmetic ingredient dictionary and handbook. PCPC

Nunes MA, Rodrigues F, Oliveira MB (2017) Grape processing by-products as active ingredients for cosmetic proposes. In: Handbook of grape processing by-products, pp 267–292

Olczyk P, Mencner Ł, Komosinska-Vassev K (2014) The role of the extracellular matrix components in cutaneous wound healing. Biomed Res Int 2014:747584

Otoi K, Horikawa Y, inventors; Kanebo Spun Silk Ltd, Kanebo Ltd, assignee (1980) Process for producing a fine powder of silk fibroin. United States patent 233:212

Pachence JM et al (1987) Collagen: its place in the medical device industry. Med Device Diagn Ind 9:49–55

Paikray, E., Bisoyi, D., Rout, A., Mishra, V., Paikray Sr, E., & ROUT, A. (2024). Analysis and reporting of adverse drug reactions of cosmetics at a tertiary care hospital. Cureus, 16(3).

Papakonstantinou E, Roth M, Karakiulakis G (2012) Hyaluronic acid: a key molecule in skin aging. Dermatoendocrinol 4:253–258

Papini AN (2010) Cosmetics toward peptide-based cosmeceutics. Chim Oggi-Chem Today 28:3–4

Pascoe D, Moreau L, Sasseville D (2010) Emergent and unusual allergens in cosmetics. DERM. 21:127–137

Pati P, Sahu BK, Panigrahy RC (2015) Marine molluscs as a potential drug cabinet: an overview. 961–970. http://nopr.niscpr.res.in/handle/123456789/34841

Pavicic T, Gauglitz GG, Lersch P, Schwach-Abdellaoui K, Malle B, Korting HC, Farwick M (2011) Efficacy of cream-based novel formulations of hyaluronic acid of different molecular weights in anti-wrinkle treatment. J Drugs Dermatol 10:990–1000

Pellizzari R, Rossetto O, Schiavo G, Montecucco C (1999) Tetanus and botulinum neurotoxins: mechanism of action and therapeutic uses. Philos Trans R Soc B: Biol Sci 354:259–268

Peng Y, Glattauer V, Werkmeister JA, Ramshaw JA (2004) Evaluation for collagen products for cosmetic application. Int J Cosmet Sci 26:313

Peng L, Yang S, Cheng YJ, Chen F, Pan S, Fan G (2012) Antifungal activity and action mode of pinocembrin from propolis against Penicillium italicum. Food Sci Biotechnol 21:1533–1539

Puizina-Ivić N (2008) Skin aging. Acta Dermatovenerol Alp Panonica Adriat 17:47–54

Quave CL, Pieroni A, Bennett BC (2008) Dermatological remedies in the traditional pharmacopoeia of Vulture-Alto Bradano, inland southern Italy. J Ethnobiol Ethnomed 4:5

Rane E (1980) A book of honey. Oxford University Press, Oxford

Requena L, Requena C, Christensen L, Zimmermann US, Kutzner H, Cerroni L (2011) Adverse reactions to injectable soft tissue fillers. J Am Acad Dermatol 64:1–34

Reva T, Vaseem A, Satyaprakash S, Khalid JM (2015) Liposomes: the novel approach in cosmaceuticals. World J Pharm Pharm Sci 4:1616–1640

Ricard-Blum S (2011) The collagen family. Cold Spring Harb Perspect Boil 3:004978

Rodrigues F, Gaspar C, Palmeira-de-Oliveira A, Sarmento B, Helena Amaral M, Oliveira PP, MB. (2016) Application of coffee silverskin in cosmetic formulations: physical/antioxidant stability studies and cytotoxicity effects. Drug Dev Ind Pharm 42:99–106

Rosiaux Y, Jannin V, Hughes S, Marchaud D (2015) Solid lipid excipients as matrix agents for sustained drug delivery. In: Excipient applications in formulation design and drug delivery, pp 237–271

Ruiz MA, Clares B, Morales ME, Cazalla S, Gallardo V (2007) Preparation and stability of cosmetic formulations with an anti-aging peptide. J Cosmet Sci 58:157–171

Safdar MH, Hussain Z, Abourehab MA, Hasan H, Afzal S, Thu HE (2018) New developments and clinical transition of hyaluronic acid-based nanotherapeutics for treatment of cancer: reversing multidrug resistance, tumour-specific targetability and improved anticancer efficacy. Artif Cells Nanomed Biotechnol 46:1967–1980

Salinas D (2015) Standards of beauty and culture in the manifestation of anorexia in young people from the township of Bellavista in Panama City. Zero Point:35–54. ISSN: 1815-0276

Sawicka D, Borawska MH (2013) The use of propolis in skin diseases. Derm Estet 1:13–17

Schlossman ML, McCarthy JP (1979) Lanolin and derivatives chemistry: relationship to allergic contact dermatitis. Contact Derm 5:65–72

Secchi G (2008) Role of protein in cosmetics. Clin Dermatol 26:321–325

Sforcin, J. M., & Bankova, V. (2011). Propolis: is there a potential for the development of new drugs?. Journal of ethnopharmacology, 133(2), 253–260.

Sheldon RA (2014) Green and sustainable manufacture of chemicals from biomass: state of the art. Green Chem 16:950–963

Shore, John. "Colourants and Auxiliaries." (1990): 251–251.

Sionkowska A, Skrzyński S, Śmiechowski K, Kołodziejczak A (2017) The review of versatile application of collagen. Polym Adv Technol 28:4–9

Souza C, de Freitas LA, Maia Campos PM (2017) Topical formulation containing beeswax-based nanoparticles improved in vivo skin barrier function. AAPS PharmSciTech 18:2505–2516

Sugitani H, Hirano E, Knutsen RH, Shifren A, Wagenseil JE, Ciliberto C, Kozel BA, Urban Z, Davis EC, Broekelmann TJ, Mecham RP (2012) Alternative splicing and tissue-specific elastin misassembly act as biological modifiers of human elastin gene frameshift mutations associated with dominant cutis laxa. J Biol Chem 287:22055–22067

Tilton H (1991) Global makeup. Chemical Marketing Reporter Special Report. p 310

Trookman NS, Rizer RL, Ford R, Ho E, Gotz V (2009) Immediate and long-term clinical benefits of a topical treatment for facial lines and wrinkles. J Clin Aesthet Dermatol 2:38

Umesh Hebbar H, Rastogi NK, Subramanian R (2008) Properties of dried and intermediate moisture honey products: a review. Int J Food Prop 11:804–819

United Nations (2015) Transforming our world: the 2030 agenda for sustainable development. United Nations, Department of Economic and Social Affairs, New York, pp 1–41

Voegeli R, Meier J, Blust R, Hofsteter R (1993) Sericin silk protein: unique structure and properties. Cosmet Toiletries. 108:101–108

Wang Q, Zhou S, Wang L, You R, Yan S, Zhang Q, Li M (2021) Bioactive silk fibroin scaffold with nanoarchitecture for wound healing. Compos Part B Eng 224:109165

Wendel A, Ghyczy M (1990) Liposomal cosmetics. Soap Cosmet Chem Spectrosc 6:33–37

Wisniaka J (2019) Theodore Nicolas Gobley: Animal physiology and plant principles. Revista CENIC. Ciencias Biológicas, 50(1), 1–16

Wong J, Chan HK, Chrzanowski W (2014) Silk for pharmaceutical and cosmeceutical applications. In: Silk biomaterials for tissue engineering and regenerative medicine, pp 519–545

Yanqing Y (2004a) A study of sericin protein on hair-care. J Text Res. 25:10–12

Yanqing Y (2004b) A study of the relationship between the molecular weight of sericin peptides and the effects of hair-care. J Text Res 25:14–15

Zamarioli CM, Martins RM, Carvalho EC, Freitas LA (2015) Nanoparticles containing curcuminoids (Curcuma longa): development of topical delivery formulation. Rev Bras 25:53–60

Zhang L, Falla TJ (2009) Host defense peptides for use as potential therapeutics. Curr Opin Invest Drugs 10:164–171

Zimbler MS, Kokoska MS, Thomas JR (2001) Anatomy and pathophysiology of facial aging. Facial Plast Surg Clin N Am 9:179–187

Chapter 6
Environment-Friendly Cosmetics from Microbial Sources

Tahir Mehmood, Shafaq Nawaz, Fareeha Nadeem, and Sibtain Ahmed

Contents

Abstract The development of eco-cosmeceutical products is an emerging trend in the cosmetic industry. Microbes, including bacteria, algae, and fungus, have been the subject of intense research as a potential source of several cosmeceuticals' rel-

T. Mehmood (✉)
Institute of Microbiology and
Molecular Genetics, University of the Punjab, Quaid-e-Azam Campus, Lahore, Pakistan
e-mail: tahir.mmg@pu.edu.pk

S. Nawaz
Department of Zoology, Government College Women University, Sialkot, Pakistan

F. Nadeem
Institute of Biochemistry and Biotechnology, University of Veterinary and Animal Sciences, Lahore, Pakistan

S. Ahmed
Department of Biochemistry, Bahauddin Zakariya University, Multan, Pakistan

S. Javed et al. (eds.), *Sustainable Cosmeceuticals*,
https://doi.org/10.1007/978-3-031-86087-4_6

evant composites, such as hyaluronic acid, enzymes, peptides, natural colors, surfactants, exopolysaccharides, and other secondary peptides, among other natural sources. Microbial proteolytic enzymes are being studied as efficient exfoliators and skin-brightening agents. Also, some antimicrobial peptides are used to treat acne caused by acne bacteria. The use of natural pigments from microbial sources is of great interest to cosmetic companies for their use in lipsticks and eye shadows due to their antioxidant and antiaging properties that produce beautiful colors, red, yellow, and green pigments, respectively. Apart from skincare, there is a huge scope for extracting microbial cosmeceutical products including shampoo, conditioners, and hair removal products. Bacteriocins and important vitamins from lactic acid bacteria and microalgae are good for hair follicle health. Eco-friendly cosmetics promote responsible consumption by reducing environmental impact, and fostering innovation in sustainable production processes therefore contributes to Sustainable Development Goals. Besides these findings, the development of such products needs extensive research to ensure the safety and dosage optimization before marketing such products.

Keywords Cosmetics · Bioactive compounds · Skin · Herbal · Synthetic

6.1 Introduction

The lotions, creams, emulsions, shampoos, conditioners for hair, sunscreens, color cosmetics, and face packs are examples of cosmetic formulations that are exclusively and extensively used to improve a person's appearance. People employed natural materials like minerals, herbal lotions, and oils to improve their look and health before the invention of cosmetics. Subsequently, several cosmetic products were created with the aim of improving the skin's appearance, texture, and structure by utilizing suitable active components (Fig. 6.1). In the market, cosmetic products typically are a blend of synthetic or natural compounds used for caring, cleaning, and enhancement of skin, hair, nails, and teeth. Cosmetic products containing synthetic ingredients can lead to complications such as skin irritation, allergies, acne, sensitivity, and hormonal disturbance (Bianchet et al. 2020). Many pigmented products, including eye shadow, lipstick, blush, and eyeliner, incorporate metals that are harmful to the human body and have the potential to disrupt human metabolism. Consequently, the cosmetic industry has witnessed a paradigm shift in recent years with an increasing demand for the utilization of bioactive and eco-friendly ingredients. The widespread dissemination of information has made the general population more aware of the complications and adverse consequences associated with the use of synthetic products. This has increased consumers favoring the use of natural, greener additives termed as cosmeceuticals. However, a very little fraction represents the natural cosmetic market in relation to conventional cosmetics due to limited guidelines regarding active ingredients (Gupta et al. 2019). In Pakistan, cosmetic industry is making revenue of amounts $5 billion in 2023 which is expected to grow

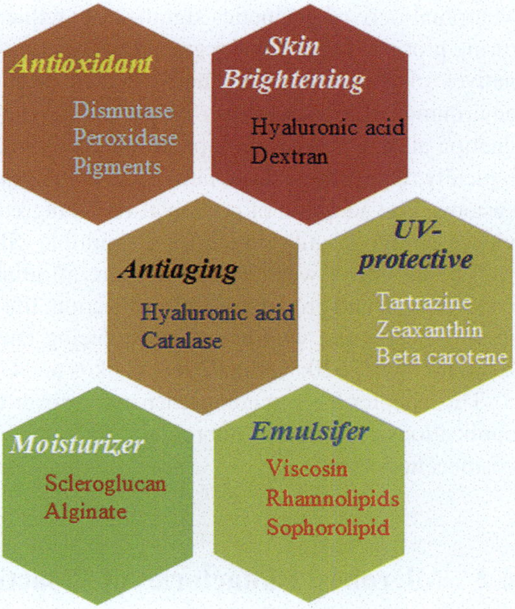

Fig. 6.1 Physiochemical properties of cosmetic ingredients

with a CAGR of 2.35% till 2028 as many strategies are being used by the companies to grow the market.

In the past, bioactive compounds were extracted from various sources, with plants being the predominant choice among others (Venil et al. 2020). The progress made in genetic modification of microbes has noticeably boosted the synthesis use of novel bioactive agents in cosmeceuticals (Arshad 2024; Choi and Lee 2023).

The Sustainable Development Goals (SDGs) are a universal framework of 17 interlinked goals adopted by the United Nations in 2015 as part of the Agenda for Sustainable Development with the aim to address global challenges, climate change, environmental degradation, etc. SDGs provide a holistic representation for achieving global sustainability to ensure a better future for generations to come. To implement, everyone including governments, businesses, civil society, and individuals has to play their roles in creating a more equitable and sustainable world. As this chapter discusses microbial-based cosmetics, their ingredients, processes, and products are eco-friendly and are aligned with many SDGs, increasing sustainability, reducing adverse impact on the environment, and improving health outcomes. These cosmetics, such as those using bacteria, fungi, and algae, provide safer, non-toxic alternatives to otherwise chemical-based formulations. Such products often feature natural, biobased ingredients that enhance skin health, reducing the exposure to harmful synthetic chemicals. Moreover, microbial production methods involve less water consumption, as microorganisms can be cultivated in water-efficient bioreactors, supporting cleaner production processes. Such products can be produced using less land, renewable energy, and fewer resources, thus encouraging more sustainable practices and promoting biodegradable ingredients.

Microbial-derived cosmetics significantly reduce the carbon footprint of manufacturing processes hence lower greenhouse gas emissions and adopt low-carbon alternatives to conventional methods. It also promotes aquaculture and reduces the environmental impact of overfishing and habitat destruction. Microbial-based production allows cosmetic companies to decrease their reliance on raw materials that typically require large-scale farming or extraction, helping preserve land-based ecosystems. The adoption of microbial-based ingredients in the cosmetic industry provides an opportunity to advance multiple SDGs, including good health and well-being, clean water and sanitation, affordable and clean energy, responsible consumption and production, climate action, life below water, and life on land, promoting health and well-being and reducing environmental degradation (Ahmadi-Ashtiani et al. 2020; Arshad 2018; Halkos and Gkampoura 2021).

This chapter aims to comprehensively cover the classification, significance, and applications of different microbial compounds as key ingredients in the field of cosmeceuticals.

6.2 Microbes: Biofactories of Bioactive Chemicals

Microbes being the exceptionally diverse sources of novel bioactive compounds are highly valuable for various innovative biotechnological processes. The microbial community is comprised of around 30 phyla and encompasses a diverse range of bacteria, fungi, algae, and actinomycetes (Kumar et al. 2022). A significant proportion of this community is still uncultured and unexplored. The tiny living organisms being remarkably adaptive exist over a range of habitats such as soil, oceans, ponds, and different extreme environments. Over 23,000 bioactive microbial metabolites have been identified (Shuikan et al. 2020). These bioactive compounds allow the organism to survive unfavorable conditions of light, temperature, ion concentration, etc. Such harsh conditions have an evolutionary impact in transforming normal microbes to extremophiles (Coker 2016; Duarte et al. 2019). Multiple studies have found these compounds to have a variety of applications in industries such as food, cosmetics, textiles, and pharmaceuticals. The cosmeceutical application of microbes is due to the production of different pigments, antioxidants, biosurfactants, exopolysaccharides (EPS), cyclodextrins, enzymes, and other secondary metabolites (Hashemi et al. 2020; Stoyneva-Gärtner et al. 2020) as shown in Fig. 6.2.

Pigments, like Riboflavin, β-carotene, astaxanthin, and carotenoids in cosmetics act as antioxidants, neutralize free radicals, protect skin from damage, and enhance color vibrancy. Biosurfactants like iturin, surfactin, and viscosin enhance cosmetics by stabilizing emulsions, promoting skin hydration, reducing microbial growth, and improving ingredient delivery and texture. Enzymes like keratinases degrade keratin for skin renewal, superoxide dismutase (SOD) neutralizes superoxide radicals, catalases break hydrogen peroxide, and glutathione peroxidase reduces lipid peroxides for antioxidant protection. Exopolysaccharides like dextran, alginate, and xanthan hydrate skin, enhance moisture retention, form protective barriers, and promote

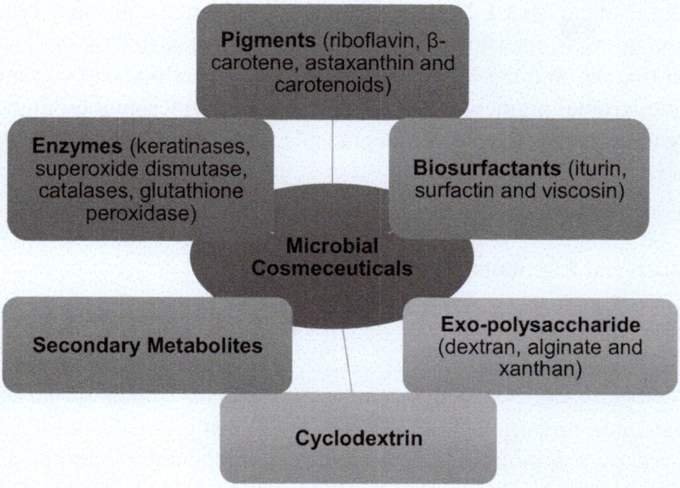

Fig. 6.2 Categorization of microbial cosmetic ingredients

healing by modulating cellular activity and antioxidant properties. Secondary metabolites in cosmetics act as antioxidants, antimicrobials, UV protectants, and anti-inflammatory agents, enhancing skin health, reducing oxidative stress, and improving texture and appearance (Dini et al. 2024; Kumar et al. 2023).

Historically, fermentation-based extraction of bioactive compounds from microbes was the common practice. However, significantly high operational cost, in recent decades, makes industries to shift toward synthetic chemicals. This transition poses high risk and adverse effects on human health. To offset this trend, advanced biotechnology tools are being leveraged. Therefore, cosmetic industry is eagerly investing on production of cost-effective microbial ingredients and thus holds a great capacity for the development of new cosmeceutical formulations, even though "natural" does not always mean healthy but potentially valued to human health.

6.2.1 Pigments

Pigments, due to impending applications not only in cosmetic industry but also in food, pharmaceutical, paper, and textile processes, are under extensive research (Venil et al. 2020). In general, synthetic dyes and pigments comprise toxic nonbiodegradable chemicals such as chromium, lead, mercury, copper, benzene, toluene, and many other carcinogenic substances which are threat to human health. Naturally, our skin contains antioxidants that inhibit reactive oxygen species (ROS) and prevent the cells from instability, autophagy, and carcinogenesis (Sajjad et al. 2020). However, the chronic UV-enriched light exposure may enhance ROS production that overcomes the level of antioxidants resulting in adverse effects. Therefore, sunscreens with added antioxidant pigments prevent the skin from UV light coming from the sun (Solano 2020). Moreover, the use of pigments in many colored cosmetic formulations needed to be less harmful or harmless. This initiated a wide research on toxicity and the large-scale production, a hallmark of commercialization of safe microbial pigments. The bioactive class of microbial products is being industrially important as they are biodegradable, safe, novel, and eco-friendly products (Verma et al. 2021) (Table 6.1).

6.2.1.1 Bacterial Pigments

In response to the considerable demands for better and greener cosmetic ingredients, companies are financing for large-scale, high-quality, and cost-effective production of colors. Besides skin enhancement, pigments have phenomenal antioxidant and UV protection properties. Research suggested that pigments produced by bacteria are protective biomolecules from harsh environmental conditions. Lycopene, astaxanthin, and β-carotene are among extensively studied photoprotectants owing to their potent antioxidant activities (Kiki 2023). β-Carotene, a precursor to provitamin A and a notable example of an antioxidant, acts as a scavenger of free radicals after accumulating in the skin. Apart from these, bacteriochlorophylls, phenazines, melanins, flavins, prodigiosin, and phycocyanin are being produced by different

Table 6.1 Cosmeceutically important pigments of microbial origin

Origin	Species	Bioactive compound	Cosmeceutical applications	References
Bacteria	*Paracoccus, Agrobacterium aurantiacum*	Astaxanthin	Skin enhancement, antioxidant	Gupta et al. (2019)
	Spirulina sp.	Phycocyanobilin	Photoprotective	Dufossé (2023)
	Chryseobacterium artocarpi	Tartrazine		Mandal and Majumdar (2023)
	Corynebacterium autotrophicum, Flavobacterium	Zeaxanthin	UV protectant, antioxidant, skin hydration	Muthusamy et al. (2022)
Fungi	*Thraustochytrids, Rhodotorula*	Astaxanthin	Skin enhancement, antioxidant	Fang et al. (2019), Li et al. (2020)
	Aspergillus variecolor, Periconia sp.	Quinones	Skin enhancement, antioxidant	Rather et al. (2022)
Algae	*Dunaliella salina, Chlorella* sp.	Beta-carotene	Photoprotective, antioxidant	Hashemi et al. (2020)
	Chlorella sp.	Sporopollenin	Antiwrinkle potential	Stoyneva-Gärtner et al. (2020)
	Scytonema sp., *Lyngbya aestuarii*	Scytonemin	UV protectant	Vega et al. (2020)

bacterial strains (Aman Mohammadi et al. 2022). Extensive research is done on the production of yellow and red pigments so far; researchers have widely worked on the production of red and yellow pigments compared to blue pigments which show very low yield. In previous studies, yellow pigments, tartrazine and zeaxanthin, are reported to be produced by *Chryseobacterium artocarpi*, *Flavobacterium*, *and Planococcus maritimus*, whereas red pigment compounds, prodigiosin, beta-carotene, lycopene, and astaxanthin, are produced in previous studies from canthaxanthin, *Brevibacterium* sp. *Gordonia* sp., and *Dietzia* sp. (Mandal and Majumdar 2023), *Hahella chejuensis* (Kim et al. 2008), *Pseudoalteromonas rubra* (Setiyono et al. 2020), *Streptomyces* sp. (Song et al. 2015), *Vibrio* sp. (Morgan et al. 2017), and *Serratia marcescens* sp. (Ibrahim et al. 2014), and from cyanobacterium *Synechococcus*, *Vibrio owensii*, *Flavicella marina*, *Gordonia terrae*, *Blastochloris tepida*, *Brevundimonas scallop, Corynebacterium glutamicum*, and *Brevundimonas* sp. (Nawaz et al. 2020; Vallejos-Almirall et al. 2022).

6.2.1.2 Fungal Pigments

A number of filamentous fungi like ascomycetous and basidiomycetous fungi and lichens yield natural pigments of a range of colors belonging to various classes such as quinones, azaphilones, carotenoids, oxopolyene, flavins, phenazine, and astaxanthin (Fang et al. 2019; Li et al. 2020). Tremendous valued biological activities including antioxidant, antimicrobial, anticancer, immunological suppressor, and anti-inflammatory properties make fungal pigments important in cosmetic industries (Lopes et al. 2013). There, the extensive use in the healthcare and cosmetics is due to absorbability for UV radiations and skin enhancement properties (Chen et al. 2019; Pailliè-Jiménez et al. 2020). Fungal pigments like ankaflavin and canthaxanthin from *Monascus* sp. and the Arpink red™ pigment (Natural red™) from *Penicillium oxalicum* are commercially available in market (Kaur et al. 2022; Mapari et al. 2009). Unlike laboratory-scale production of fungal pigments, commercial cultivation of fungi on a large industrial scale is impractical (Lin and Xu 2022). Studies are been done on advanced biotechnological process for upscaling the production of fungal secondary metabolites. However, halophilic fungi being capable of extracellular production of secondary metabolites with higher yields hold inordinate biotechnological significance (Kalra et al. 2020). This includes *Aspergillus variecolor* and *Periconia* sp. for the production of yellow (quinones, e.g., variecolorquinones) (Arshad and Abbas 2018; Rather et al. 2022) and blue pigments (Afroz Toma et al. 2023). The antioxidant, melanogenesis-inhibiting, and photoprotective properties of these pigments make them excellent for future cosmetics with health advantages (Kiki 2023).

6.2.1.3 Algal Pigments

Microalgae, a diverse group, thrive in a variety of environments including fresh water and seawater. Many of them including *Spirulina* and *Chlorella* produce pigments like carotene, sporopollenin, and scytonemin which have antioxidant properties. These attributes make microalgal metabolites promising ingredients of effective photoprotectant, antiaging, and sunscreen formulations. Microalgae being autotrophic can be cultivated in bioreactors, reducing the dependence on seasonal factors and thus making them favorable to use in the cosmetic industry. Many pigments like red-colored compounds lutein, canthaxanthin, and astaxanthin are produced by *Dichromanthus cinnabarinus* and *Haematococcus pluvialis*, which have been approved by the US FDA (Rather et al. 2022). Additional pigments include water-soluble chlorophyll, phycocyanins, and phycoerythrins from Rhodophyta, Cyanophyta, and *Cryptophyta* (Wu et al. 2021).

6.2.2 Exopolysaccharides

Cosmetic products including lotions, creams, emulsions, shampoos, hair conditioners, sunscreens, color cosmetics, and face packs are commonly used to enhance the appearance of a person. Such formulations are frequently supplemented with EPS. Suspensions, hydrogels, and encapsulations come under EPS used in cosmetics (Freitas et al. 2015). The most widely used EPS in cosmetic formulations based on emulsifiers, gels, and suspensions are xanthan, dextran, alginate, xanthan,

Table 6.2 Cosmeceutically important exo-polysaccharides of microbial origin

Origin	Species	Bioactive compound	Cosmeceutical applications	References
Bacteria	*Leuconostoc mesenteroides*, *Streptococcus mutans*	Dextran	Brightening and smoothing agent	Chen et al. (2021)
	Pseudomonas aeruginosa, *Azotobacter vinelandii*	Alginate	Moisturizer and gelling agent	Serrato (2024)
	Xanthomonas spp.	Xanthan	Thickening, gelling, and emulsification	Singhvi et al. (2019)
	Sinorhizobium meliloti, *Gluconacetobacter* sp.	Glucuronan	Gelling and thickening	Dubessay et al. (2021)
	Streptococcus thermophilus	Hyaluronic acid	Skin repair and antiaging	Zhong et al. (2023)
Fungi	*Aureobasidium pullulans*, *Pullularia pullulans*	Pullulan	Skin tightening	Muthusamy et al. (2022)
	Sclerotium glucanicum, *Sclerotium rolfsii*	Scleroglucan	Hydrating and moisturizing	Tan et al. (2019)
	Botryosphaeria rhodina	Botryosphaeran	Antioxidant	Keshavarz et al. (2022)

glucuronan, hyaluronic acid, and gellan (Balkrishna et al. 2018; Iurciuc et al. 2017). In addition, a few other EPS are also used as encapsulating agents and nanoparticles for carrying other bioactive compounds (Kaur and Dey 2022). Many exopolysaccharides have been studied from different microbes as potent biofactories of exopolysaccharides (Table 6.2).

Bacterial cellulose, because of its high hydration activity and healing effects, is used in skin regeneration, moisturizers, and skin repair products (Bianchet et al. 2020). Levan possesses cell proliferating, skin moisturizing, and skin whitening effects (da Silva et al. 2022). Xanthan gum also has various uses in cosmetics such as oral hygiene products, deodorants, baby products, skin conditioning agents, mucodermal products, hair colorants, and nail products (Freitas et al. 2015). These EPS make the formulation healthy as they are not only good brightening, smoothing, moisturizer, and gelling agents but also bioactive in skin repair and antiaging products (Chen et al. 2021; Serrato 2024).

6.2.2.1 Hyaluronic Acid

Out of all EPS currently being used in cosmetic industries, hyaluronic acid holds the most versatile applications. Due to the physiochemical aspects of hyaluronic acid including lubricity, viscosity, biocompatibility, and elasticity, it has the ability of controlling tissue permeation, hydration, and intercellular macromolecular transport and has been extensively exploited for its use in cosmetics making cosmetics containing HLA, one of the cost-intensive and economically valued products (Bukhari et al. 2018). Low molecular weight hyaluronic acids are effective antioxidants and stimulating agents for collagen development, cell division, rebuilding of soft tissue, and cytotoxicity, thus slowing down the aging process (Alipoor et al. 2022). This is also widely used as dermal filler in cosmetic surgical process. Additionally, many skin lotion and serums contain sodium hyaluronate as its active component to increase moisture retention, reduce skin wrinkles, and increase skin elasticity. Many microbes produce hyaluronic acid including bacterial species *Streptococcus equisub, Streptococcus zooepidemicus,* and *Streptococcus* sp. (Serra et al. 2023), and algal species *Chlorella zofingiensis, Scenedesmus* sp., *Nostoc* sp., *Chlorella vulgaris, Chlamydomonas reinhardtii, Rhodella violacea, Tetraselmis suecica*, and *Porphyridium cruentum* (Uhliariková et al. 2020; Zampieri et al. 2020).

6.2.2.2 Biosurfactants

Biosurfactants exhibit extraordinary potential like lower toxicity, skin compatibility, protection, and surface moisturizing, for effective skincare routine (Akbari et al. 2018; Lee et al. 2018). The use of chemical surfactants as ingredients of cosmetic formulations such as emulsifiers and foaming agents is extremely effective but is potentially harmful to the skin and its microbiome (Adu et al. 2020). Studies have suggested that such products alter microflora of skin, thus causing allergies and skin irritation as they interact with biomolecules present on epidermal cells. Besides, their extended use especially in high concentration may solubilize the epidermal and intracellular lipids, thus affecting biomolecular integrity of the skin (Arshad et al. 2018; Souto et al. 2022). Such outcomes shift consumers' demand toward less toxic, biodegradable, and biocompatible microbial surfactants (Pal et al. 2023). Bacteria including *Pseudomonas* sp. produce microbial biosurfactants viscosin and rhamnolipids that serve as a skin hydrating agent (Chauhan et al. 2023; Moldes et al. 2020), *Bacillus* sp. produce surfactin (Ganesan and Rangarajan 2021), *Botryosphaeria rhodina* produce botryosphaeran (antioxidant), and *Pseudozyma* sp., *Ustilago* sp., and *Candida antarctica* produce mannosylerythritol lipid and sophorolipid that serve as a detergent, foaming, emulsifying, and skin hydrating agent (Borah et al. 2021; Sharma 2021) (Table 6.3).

6.2.3 Enzymes

Along with many other microbial ingredients, proteins and peptides also have added huge value in the cosmetic industry, as conventionally used for improving the quality of the skin, hair, or nails. A variety of cosmeceutically important proteins, Superoxide dismutase (SOD) and peroxidase (catalase, glutathione peroxidases, lactoperoxidases), after a massive research done in the same prospect, have been identified synergistic to exfoliators (Gade et al. 2023). Incorporation of enzymes in cosmetic formulations makes them antioxidant, thus preventing the skin from adverse effects of prolong sun exposure. Research has been done on many extremophilic microbes, *Marinomonas* sp., *Sulfolobus acidocaldarius, Clostridium histolyticum, Vibrio alginolyticus, Bacillus subtilis, Bacillus licheniformis, and Thermus thermophilus* (Mendes-Silva et al. 2020) for production of cosmeceutically active enzymes including collagenases, keratinases, superoxide dismutase, catalases, and glutathione peroxidase (Table 6.4). Advanced biotechnology processes aided with genetically engineered lactic acid bacteria are promising choices for high yield and

Table 6.3 Cosmeceutically important biosurfactants of microbial origin

Origin	Species	Bioactive compound	Cosmeceutical applications	References
Bacteria	*Pseudomonas* sp.	Viscosin, rhamnolipids	Detergent, emulsifier, and skin hydration	Chauhan et al. (2023), Moldes et al. (2020)
	Bacillus sp.	Surfactin		Ganesan and Rangarajan (2021)
Fungi	*Pseudozyma* sp., *Ustilago* sp., *Candida antarctica*	Mannosylerythritol lipid	Detergent, foaming agent, emulsifier, and skin hydration	Borah et al. (2021)
	Candida sp.	Sophorolipid		Sharma (2021)

Table 6.4 Cosmeceutically important enzymes of microbial origin

Origin	Species	Bioactive compound	Cosmeceutical applications	References
Bacteria	*Clostridium histolyticum*, *Vibrio alginolyticus*	Collagenases	Skin regeneration	Shridharani and Kennedy (2023), Vachher et al. (2021)
	Bacillus subtilis, *Bacillus licheniformis*	Keratinases	Hair removal	Rai et al. (2020), Siddiqui et al. (2021)
	Sulfolobus acidocaldarius, *Marinomonas*, *Thermus thermophilus*	Superoxide dismutase, catalases, glutathione peroxidase	Antioxidants, antiaging	Dong et al. (2023), Tan et al. (2023), Pedone et al. (2020)
Fungi	*Aspergillus*, *Penicillium*, *Rhizopus*, *Mucor*	Aspartic proteases	Treat stretch marks, scar tissues	Dhevagi et al. (2021)

stable production of these enzymes (Srivastava et al. 2020). Additionally, keratinase is also effective in enzymatic exfoliation treatments, for hair removing and hair growth delaying, and is being produced commercially by *Bacillus licheniformis, Thermoanaerobacter, Thermosipho,* and *Thermococcus.* Other gram-positive bacteria including *Lysobacter, Nesterenkonia, Kocuria,* and *Microbacterium* (Sharma and Kango 2021), and gram-negative bacteria such as *Vibrio, Xanthomonas, Stenotrophomonas, Chryseobacterium, Fervidobacterium, Thermoanaerobacter,* and *Nesterenkonia* are biofactories of keratinase enzyme (Zheng et al. 2021). Besides, fungal species belonging to *Microsporum, Epidermophyton, Trichophyton,* and *Chrysosporium* are known for their keratinophilic ability, thus helping in degrading keratin fibers (Sung et al. 2019). Together with these enzymes, particular peptides are also effective cosmeceutical ingredients, used in gels, emulsions, powders, and lotions; however, oil-soluble peptides are used in facial masks (Ahsan 2019). Synthesis of these peptides, on commercial scale, is accomplished by controlled action of proteases, majorly secreted from several *Bacillus* sp. (Cai et al. 2022; Liya et al. 2023). Also, pentapeptides and aspartic proteases from *Aspergillus*, *Penicillium*, *Rhizopus*, and *Mucor* are widely used for reducing facial wrinkles and roughness and for treating stretch marks and scar tissues (Dhevagi et al. 2021).

6.2.4 Secondary Metabolites

Secondary metabolites from microbial origin cater as important cosmeceutical ingredients due to their exclusive physiochemical properties such as antioxidant, hydrating, skin brightening, antimicrobial, collagen enhancing, exfoliating, emulsifying, antiaging, and fragrances (Pardhi et al. 2023). Different bacteria grown under

Table 6.5 Cosmeceutically important secondary metabolites of microbial origin

Origin	Species	Bioactive compound	Cosmeceutical applications	References
Bacteria	*Bacillus subtilis, Brevibacterium* sp., *Brevibacillus brevis, Microbacterium terrae*	α,β,γ-cyclodextrin	Fragrances	Pardhi et al. (2023)
	Lactobacillus	Lactic acid	Impart smoothness, exfoliator	Huang et al. (2020)
Actinobacteria	*Pseudonocardia* sp. *Actinosynnema mirum, Streptomyces avermitilis, Streptomyces lividans, Corynebacterium glutamicum, Aurantiochytrium* sp.	Mycosporine-like amino acids (MAAs)	UV and photoprotective potential protection from oxidative damage	Sánchez-Suárez et al. (2021)
Fungi	*Rhizopus* genera	Lactic acid	Skin moisture, exfoliator, biostimulatory filler	Huang et al. (2020)

controlled fermentation processes including *Bacillus subtilis, Brevibacterium* sp., *Brevibacillus brevis*, and *Microbacterium terrae* produce cyclodextrins identified for aromatic use in fragrance production. Moreover, actinobacteria, *Pseudonocardia* sp., *Actinosynnema mirum, Streptomyces avermitilis, and Streptomyces lividans*, extensively produce many mycosporine-like amino acids (MAAs) that display excellent UV and photoprotective and oxidative damage repair properties (Sánchez-Suárez et al. 2021) (Table 6.5).

6.2.5 Mycosporine-Like Amino Acids (MAA)

Among other secondary metabolites, mycosporine-like amino acids are most potent as cosmetic ingredients. Up until now, around 30 diverse MAAs have been recognized such as mycosporine-glycine, shinorine, porphyra-334, mycosporine-2-glycine, palythine, and euhalothece (Lawrence et al. 2018; Tarasuntisuk et al. 2019). MAAs are derived by specific amino acid substitutions and thus generally accumulate in the cytosol potentially combating UV-induced stress, by scattering energy as heat, without generating reactive oxidative species (ROS), thus making the organism able to survive harsh environment (Gacesa et al. 2018). The MAAs that can also be extracted from algae *Dunaliella, Arthrospira,* and *Chlorella* function as sunscreens, from *Odontella aurita* as free radical scavenging, and from *Coelastrella rubescens* have photoprotective properties. Lately, Klebsormidium is discovered to be the source of MAAs, klebsormidin A and klebsormidin B, produced under strong UV exposure, indicating them as UV sunscreens (Hartmann et al. 2020). However, very few MAAs are commercially available.

6.3 Side Effects of Synthetic Ingredients Used in Cosmetics

The addition of synthetic ingredients in cosmetics was a common practice of cosmetic companies primarily for making the product more effective keeping it stable and cost-effective. However, various formulations with synthetic pigments showed a package of side effects and were getting concerns for consumers and also the companies. For example, oxides (titanium, zinc, iron) used as pigments cause an increase in skin sensitivity, itching, redness, and eye irritation (Riley 2000; Wahab et al. 2020). Surfactants, like sodium lauryl sulfate, ammonium lauryl sulfate, olefin sulfonates, sodium stearate, and cetrimonium chloride, can cause membrane

Table 6.6 Synthetic and microbial/natural ingredient use in cosmetics

Synthetic ingredients	Microbial/natural ingredients		References
Pigments			
Titanium dioxide, iron oxides, tin oxide, and zinc oxide	Astaxanthin Phycocyanobilin Tartrazine, Zeaxanthin Quinones Beta-carotene Sporopollenin Scytonemin	Sensitive skin Redness Itching Eye irritation	Kim et al. (2017), Riley (2000), Wahab et al. (2020)
Surfactants			
Sodium lauryl sulfate Ammonium lauryl sulfate Olefin sulfonates Sodium stearate Stearic acid Cetrimonium chloride Stearalkonium chloride Sodium lauriminodipropionate disodium lauroamphodiacetate	Viscosin, rhamnolipids Surfactin Botryosphaeran Mannosylerythritol lipid Sophorolipid	Membrane damage Damage to keratinocytes Proinflammatory mediators	Hollis (1995), Velasco et al. (2022)
Exfoliators			
• Alpha-hydroxy acids • Beta-hydroxy acids • Poly-hydroxy acids	Keratinases Aspartic protease	Itching Erythema Increased skin sensitivity, epidermolysis, allergies, dermatitis, hyperpigmentation	Green et al. (2009), Heymann (1997), Wulaningsih (2023)
Moisturizers/hydrating agents			
Paraffin Squalene Dimethicone Propylene glycol Lanolin	Alginate Glucuronan Hyaluronic acid Scleroglucan Botryosphaeran	Dry, red, and cracked skin	

damage, damage to keratinocytes, hair damage, and proinflammatory effects (Velasco et al. 2022). Similarly, exfoliator and hydrating agents cause skin cracks and hyperpigmentation (Green et al. 2009; Heymann 1997; Wulaningsih 2023). With the flow of information, a common man comes to know about alternate natural ingredients that potentially can be used as natural ingredients, with fewer or no side effects associated. These alternatives provide similar beneficial, sustainable, skin-friendly, and long-lasting results with customer satisfaction (Table 6.6).

6.4 Microbe-Based Cosmetic Production and Environmental Sustainability

The production of microbe-based cosmetics involves harnessing the unique capabilities of microorganisms such as bacteria, fungi, and algae to create bioactive ingredients that are both effective and eco-friendly. This process begins with the selection of specific microbes known for their ability to produce compounds beneficial for skincare and haircare, such as hyaluronic acid, peptides, enzymes, and pigments. These microbes are cultivated under controlled conditions in bioreactors, which optimize growth factors like temperature, pH, and nutrient availability. The fermentation process, a hallmark of microbial production, allows for the large-scale synthesis of these compounds using renewable raw materials, such as agricultural waste or simple sugars. This approach minimizes reliance on nonrenewable petrochemical resources and significantly reduces the environmental footprint of the cosmetic industry. Once the desired compounds are synthesized, they are extracted and purified using sustainable techniques like membrane filtration or solvent-free processes, further contributing to their eco-friendly profile. For example, microbial fermentation can produce biodegradable surfactants, replacing traditional surfactants that often pollute waterways. Similarly, microbial pigments, derived from species like *Monascus* or *Spirulina*, are used as natural colorants in lipsticks and eye shadows, providing an alternative to synthetic dyes that are typically harmful to ecosystems. These pigments not only add aesthetic value but also offer antioxidant and antiaging properties, enhancing the functional appeal of the products.

A significant advantage of microbe-based production is its scalability and reduced resource requirements. Unlike plant-derived ingredients, which often demand extensive land, water, and chemical inputs, microbes grow rapidly in confined spaces using minimal resources. For instance, algae can be cultivated in photobioreactors with sunlight and carbon dioxide, contributing to carbon sequestration and aligning with climate action goals. Moreover, the by-products of microbial processes are often nontoxic and biodegradable, ensuring minimal waste generation and adherence to the principles of green chemistry. This eco-friendly production process directly supports the United Nations Sustainable Development Goals (SDGs), particularly those focused on responsible consumption and production, climate action, and life below water. The reduced environmental impact, combined

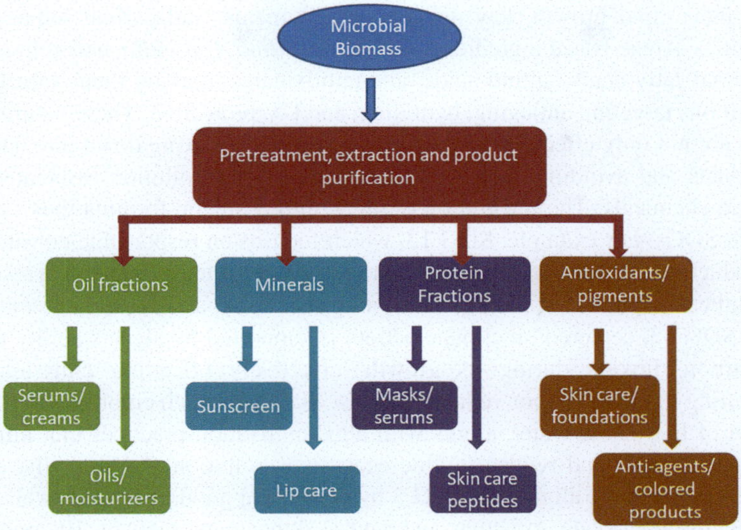

Fig. 6.3 Production of different bioactive products from microbial biomass

with the ability to create safe, effective, and innovative products, makes microbe-based cosmetics a cornerstone of the sustainable beauty industry. By embracing these processes, the cosmetic industry can significantly reduce its ecological footprint while catering to the growing demand for environmentally responsible products, paving the way for a greener future in beauty and personal care (Choi and Lee 2023; Mouro et al. 2024; Nagtode et al. 2023) (Fig. 6.3).

6.5 Microbe-Based Cosmetics: "Green Cosmetics"

Microbe-based cosmetics represent a transformative approach to sustainable beauty by integrating green chemistry principles and advancing the United Nations Sustainable Development Goals (SDGs). Utilizing microorganisms such as bacteria, fungi, and algae, these cosmetics harness bioactive compounds like enzymes, peptides, antioxidants, pigments, and polysaccharides, which are essential for skincare and haircare products. Unlike traditional cosmetic production methods that often rely on petrochemical-derived ingredients, microbe-based cosmetics adopt environmentally friendly processes such as fermentation, which use renewable resources, reduce energy consumption, and minimize hazardous by-products. Green chemistry, the foundation of these innovations, emphasizes the design of chemical products and processes that are safe, efficient, and sustainable. Microbial production exemplifies these principles by enabling the synthesis of biodegradable, renewable ingredients that lower environmental impacts and improve human health outcomes. Algae, in particular, have emerged as a cornerstone of green cosmetics due to their rapid growth, low resource requirements, and carbon sequestration capabilities. Algae-based ingredients such as *Spirulina*, *Chlorella*, and kelp are rich in essential fatty acids, amino acids, and antioxidants, making them valuable for formulations targeting antiaging, hydration, and UV protection. These natural compounds are not only effective but also biodegradable, breaking down into harmless by-products and avoiding the persistent environmental pollution associated with synthetic chemicals. The integration of algae into cosmetic formulations supports multiple SDGs. For example, SDG 12, which focuses on responsible consumption and production, is addressed by using algae to replace finite petrochemical resources and synthetic ingredients, thereby promoting renewable and sustainable manufacturing. SDG 13, centered on climate action, is supported by algae's ability to capture carbon dioxide during its growth, effectively offsetting emissions and contributing to global carbon reduction goals. SDG 14, which emphasizes the preservation of life below water, aligns with algal cultivation practices that minimize agricultural runoff and reduce marine ecosystem pollution. Additionally, algae's bioactive compounds align with SDG 3 by promoting good health and well-being, offering consumers safe, effective, and natural skincare alternatives. By embracing microbe-based cosmetics and alga-derived ingredients, the beauty industry is not only creating high-performance products but also redefining sustainability in a way that aligns with global environmental and social goals. As research and innovation in this field continue, the potential to expand the use of microbes and algae in cosmetics further strengthens the link between environmental stewardship, product efficacy, and sustainable development.

6.6 Conclusion

The rising demand for cleaner, natural, green, eco-friendly, and physiologically safe products has shifted the aim of the cosmetic industry toward better and skin-friendly product formulations. Industry efforts are focused on identifying microbially derived bioactive compounds and their production on an industrial scale using microbial diversity data, advanced biotechnology processes, and genetic enhancement. These novel biologically active molecules are rapidly replacing conventionally used synthetic ingredients such as antioxidants, antiaging, and UV-protective agents. Companies are producing these nutricosmetic products from bacterial, fungal, actinomycete, cyanobacterial, and microalgal origins. However, many of these nutricosmetic compounds may cause adverse effects that should be validated through clinical trials to study their true potential before commercialization. The shift to microbe-based cosmetics directly aligns with several United Nations Sustainable Development Goals. By focusing on sustainable production and responsible consumption, the industry is reducing dependency on synthetic and nonrenewable resources. The integration of microbial processes, such as carbon-sequestering algal cultivation, supports climate action by reducing the environmental footprint of the cosmetic industry. Furthermore, the development of biodegradable and eco-friendly cosmetics contributes to preserving life below water by minimizing marine pollution caused by conventional chemicals. As the industry continues to innovate, it strengthens its role in promoting sustainable beauty while contributing to a healthier planet.

References

Adu SA, Naughton PJ, Marchant R, Banat IM (2020) Microbial biosurfactants in cosmetic and personal skincare pharmaceutical formulations. Pharmaceutics 12(11):1099

Afroz Toma M, Rahman MH, Rahman MS, Arif M, Nazir KNH, Dufossé L (2023) Fungal pigments: carotenoids, riboflavin, and polyketides with diverse applications. J Fungi 9(4):454

Ahmadi-Ashtiani H-R, Baldisserotto A, Cesa E, Manfredini S, Sedghi Zadeh H, Ghafori Gorab M, Khanahmadi M, Zakizadeh S, Buso P, Vertuani S (2020) Microbial biosurfactants as key multifunctional ingredients for sustainable cosmetics. Cosmetics 7(2):46

Ahsan H (2019) Immunopharmacology and immunopathology of peptides and proteins in personal products. J Immunoass Immunochem 40(4):439–447

Akbari S, Abdurahman NH, Yunus RM, Fayaz F, Alara OR (2018) Biosurfactants—a new frontier for social and environmental safety: a mini review. Biotechnol Res Innov 2(1):81–90

Alipoor R, Ayan M, Hamblin MR, Ranjbar R, Rashki S (2022) Hyaluronic acid-based nanomaterials as a new approach to the treatment and prevention of bacterial infections. Front Bioeng Biotechnol 10:913912

Aman Mohammadi M, Ahangari H, Mousazadeh S, Hosseini SM, Dufossé LJB, engineering, b. (2022) Microbial pigments as an alternative to synthetic dyes and food additives: a brief review of recent studies. Bioprocess Biosyst Eng 45(1):1–12

Arshad M (2018) Cogeneration policy to contest climate changes and the future of sustainable energy. IOP Conf Ser: Earth Environ Sci 200:012050

Arshad M (2024) Sustainability of renewable energy options as compared to coal-fired power plants in Pakistan. Green Low-Carbon Econ 2(3):162–173

Arshad M, Abbas M (2018) Water sustainability issues in biofuel production. In: Perspectives on water usage for biofuels production: aquatic contamination and climate change, pp 55–76

Arshad M, Bano I, Younus M, Khan A, Rahman A (2018) Health concerns associated with biofuel production. In: Perspectives on water usage for biofuels production: aquatic contamination and climate change, pp 97–105

Balkrishna A, Agarwal V, Kumar G, Gupta AK (2018) Applications of bacterial polysaccharides with special reference to the cosmetic industry. In: Microbial bioprospecting for sustainable development, pp 189–202

Bianchet RT, Cubas ALV, Machado MM, Moecke EHS (2020) Applicability of bacterial cellulose in cosmetics—bibliometric review. Biotechnol Rep 27:e00502

Borah D, Chaubey A, Sonowal A, Gogoi B, Kumar R (2021) Microbial biosurfactants and their potential applications: an overview. In: Microbial biosurfactants: preparation, properties and applications, pp 91–116

Bukhari SNA, Roswandi NL, Waqas M, Habib H, Hussain F, Khan S, Sohail M, Ramli NA, Thu HE, Hussain Z (2018) Hyaluronic acid, a promising skin rejuvenating biomedicine: a review of recent updates and pre-clinical and clinical investigations on cosmetic and nutricosmetic effects. Int J Biol Macromol 120:1682–1695

Cai W-W, Hu X-M, Wang Y-M, Chi C-F, Wang B (2022) Bioactive peptides from skipjack tuna cardiac arterial bulbs: preparation, identification, antioxidant activity, and stability against thermal, pH, and simulated gastrointestinal digestion treatments. Mar Drugs 20(10):626

Chauhan V, Mazumdar S, Pandey A, Kanwar SS (2023) Pseudomonas lipopeptide: an excellent biomedical agent. MedComm Biomater Appl 2(1):e27

Chen W, Feng Y, Molnár I, Chen F (2019) Nature and nurture: confluence of pathway determinism with metabolic and chemical serendipity diversifies Monascus azaphilone pigments. Nat Prod Rep 36(4):561–572
Chen Z, Ni D, Zhang W, Stressler T, Mu W (2021) Lactic acid bacteria-derived α-glucans: from enzymatic synthesis to miscellaneous applications. Biotechnol Adv 47:107708
Choi KR, Lee SY (2023) Systems metabolic engineering of microorganisms for food and cosmetics production. Nat Rev Bioeng 1(11):832–857
Coker JA (2016) Extremophiles and biotechnology: current uses and prospects. F1000Research 5
da Silva RT, Bersaneti GT, Bigotto BG, Silveira VAI, Lonni AASG, Borsato D, Celligoi MAPC (2022) Development of a facial biocosmetic containing levan, almond and cinnamon oils with antioxidant and moisturizing properties. Acta Sci Biol Sci 44:e58869–e58869
Dhevagi P, Ramya A, Priyatharshini S, Geetha Thanuja K, Ambreetha S, Nivetha A (2021) Industrially important fungal enzymes: productions and applications. In: Recent trends in mycological research, vol 2: environmental and industrial perspective, pp 263–309
Dini S, Bekhit AE-DA, Roohinejad S, Vale JM, Agyei D (2024) The physicochemical and functional properties of biosurfactants: a review. Molecules 29(11):2544
Dong L, Chen Y, Gu L, Gan M, Carrier A, Oakes K, Zhang X, Dong Z (2023) Oral delivery of a highly stable superoxide dismutase as a skin aging inhibitor. Biomed Pharmacother 164:114878
Duarte AWF, de Menezes GCA, e Silva TR, Bicas JL, Oliveira VM, Rosa LH (2019) Antarctic fungi as producers of pigments. In: Fungi of Antarctica: diversity, ecology and biotechnological applications, pp 305–318
Dubessay P, Andhare P, Kavitake D, Shetty P, Ursu A, Delattre C, Pierre G, Michaud P (2021) Microbial glucuronans and succinoglycans: structures, properties, and enzymes acting about them. In: Polysaccharides of microbial origin: biomedical applications. Springer, pp 1–23
Dufossé L (2023) Natural colorants from cyanobacteria and algae. In: Handbook of natural colorants, pp 417–437
Fang N, Wang C, Liu X, Zhao X, Liu Y, Liu X, Du Y, Zhang Z, Zhang HJT (2019) De novo synthesis of astaxanthin: from organisms to genes. Trends Food Sci Technol 92:162–171
Freitas F, Alves VD, Reis MA (2015) Bacterial polysaccharides: production and applications in cosmetic industry. In: Polysaccharides. Springer, pp 2017–2043
Gacesa R, Lawrence KP, Georgakopoulos ND, Yabe K, Dunlap WC, Barlow DJ, Wells G, Young AR, Long PF (2018) The mycosporine-like amino acids porphyra-334 and shinorine are antioxidants and direct antagonists of Keap1-Nrf2 binding. Biochimie 154:35–44
Gade JV, Singh A, Jain B (2023) Current industrial-and commercial-scale applications of biomolecules. In: Handbook of biomolecules. Elsevier, pp 551–574
Ganesan NG, Rangarajan V (2021) A kinetics study on surfactin production from Bacillus subtilis MTCC 2415 for application in green cosmetics. Biocatal Agric Biotechnol 33:102001
Green BA, Ruey JY, Van Scott EJ (2009) Clinical and cosmeceutical uses of hydroxyacids. Clin Dermatol 27(5):495–501
Gupta PL, Rajput M, Oza T, Trivedi U, Sanghvi G (2019) Eminence of microbial products in cosmetic industry. Nat Prod Bioprospect 9:267–278
Halkos G, Gkampoura E-C (2021) Where do we stand on the 17 sustainable development goals? An overview on progress. Econ Anal Policy 70:94–122
Hartmann A, Glaser K, Holzinger A, Ganzera M, Karsten U (2020) Klebsormidin A and B, two new UV-sunscreen compounds in green microalgal Interfilum and Klebsormidium species (Streptophyta) from terrestrial habitats. Front Microbiol 11:499
Hashemi A, Moslemi M, Pajoum Shariati F, Delavari Amrei H (2020) Beta-carotene production within Dunaliella salina cells under salt stress condition in an indoor hybrid helical-tubular photobioreactor. Can J Chem Eng 98(1):69–74
Heymann LA (1997) The cosmetic/drug dilemma: FDA regulation of alpha-hydroxy acids. Food Drug LJ 52:357
Hollis GL (1995) Surfactants Europa: a directory of surface active agents available in Europe. Royal Society of Chemistry

Huang H-C, Lee IJ, Huang C, Chang T-M (2020) Lactic acid bacteria and lactic acid for skin health and melanogenesis inhibition. Curr Pharm Biotechnol 21(7):566–577

Ibrahim D, Nazari TF, Kassim J, Lim S-H (2014) Prodigiosin-an antibacterial red pigment produced by Serratia marcescens IBRL USM 84 associated with a marine sponge Xestospongia testudinaria. J Appl Pharm Sci 4(10):001–006

Iurciuc C, Lungu C, Martin P, Popa M (2017) Gellan. Pharmaceutical, medical and cosmetic applications. Cellulose Chem Technol 51:187–202

Kalra R, Conlan XA, Goel M (2020) Fungi as a potential source of pigments: harnessing filamentous fungi. Front Chem 8:369

Kaur N, Dey P (2022) Bacterial exopolysaccharides as emerging bioactive macromolecules: from fundamentals to applications. Res Microbiol 104024

Kaur P, Singh S, Ghoshal G, Ramamurthy PC, Parihar P, Singh J, Singh A (2022) Valorization of agri-food industry waste for the production of microbial pigments: an eco-friendly approach. In: Advances in agricultural and industrial microbiology, vol 1. Microbial diversity and application in agroindustry, pp 137–167

Keshavarz S, Azizian R, Malakootikhah J, Fathizadeh H, Hamidi M (2022) Microbial exopolysaccharides in additive manufacturing. In: Encyclopedia of green materials. Springer, pp 1–11

Kiki MJ (2023) Biopigments of microbial origin and their application in the cosmetic industry. Cosmetics 10(2):47

Kim D, Kwon S-K, Kim JF, Lee HK, Lee CH, Yim JH (2008) Red to red-the marine bacterium Hahella chejuensis and its product prodigiosin for mitigation of harmful algal blooms. J Microbiol Biotechnol 18(10):1621–1629

Kim K-B, Kim YW, Lim SK, Roh TH, Bang DY, Choi SM, Lim DS, Kim YJ, Baek S-H, Kim M-K (2017) Risk assessment of zinc oxide, a cosmetic ingredient used as a UV filter of sunscreens. J Toxic Environ Health, Part B 20(3):155–182

Kumar S, Kumar V, Nag D, Kumar V, Darnal S, Thakur V, Patial V, Singh D (2022) Microbial pigments: learning from the Himalayan perspective to industrial applications. J Ind Microbiol Biotechnol 49(5):kuac017

Kumar M, Kumar D, Garg Y, Mahmood S, Chopra S, Bhatia A (2023) Marine-derived polysaccharides and their therapeutic potential in wound healing application—a review. Int J Biol Macromol 253:127331

Lawrence K, Gacesa R, Long P, Young A (2018) Molecular photoprotection of human keratinocytes in vitro by the naturally occurring mycosporine-like amino acid palythine. Br J Dermatol 178(6):1353–1363

Lee HJ, Jeong SE, Lee S, Kim S, Han H, Jeon CO (2018) Effects of cosmetics on the skin microbiome of facial cheeks with different hydration levels. Microbiology 7(2):e00557

Li J, Guo C, Wu J (2020) Astaxanthin in liver health and disease: a potential therapeutic agent. Drug Des Devel Ther 14:2275–2285

Lin L, Xu J (2022) Production of fungal pigments: molecular processes and their applications. J Fungi 9(1):44

Liya SM, Umesh M, Nag A, Chinnathambi A, Alharbi SA, Jhanani G, Shanmugam S, Brindhadevi K (2023) Optimized production of keratinolytic proteases from Bacillus tropicus LS27 and its application as a sustainable alternative for dehairing, destaining and metal recovery. Environ Res 221:115283

Lopes FC, Tichota DM, Pereira JQ, Segalin J, de Oliveira Rios A, Brandelli A (2013) Pigment production by filamentous fungi on agro-industrial byproducts: an eco-friendly alternative. Appl Biochem Biotechnol 171:616–625

Mandal DD, Majumdar S (2023) Bacteria as biofactory of pigments: evolution beyond therapeutics and biotechnological advancements. J Biosci Bioeng

Mapari SA, Meyer AS, Thrane U, Frisvad JC (2009) Identification of potentially safe promising fungal cell factories for the production of polyketide natural food colorants using chemotaxonomic rationale. Microb Cell Factories 8:1–15

Mendes-Silva TCD, da Silva Andrade RF, Ootani MA, Mendes PVD, da Silva MRF, Souza KS, dos Santos Correia MT, da Silva MV, de Oliveira MBM (2020) Biotechnological potential of carotenoids produced by extremophilic microorganisms and application prospects for the cosmetics industry. Adv Microbiol 10(8):397–410

Moldes A, Vecino X, Rodríguez-López L, Rincón-Fontán M, Cruz J (2020) Biosurfactants: the use of biomolecules in cosmetics and detergents. In: New and future developments in microbial biotechnology and bioengineering. Elsevier, pp 163–185

Morgan S, Thomas MJ, Walstrom KM, Warrick EC, Gasper BJ (2017) Characterization of prodiginine compounds produced by a Vibrio species isolated from salt flat sediment along the Florida Gulf Coast. Fine Focus 3(1):33–51

Mouro C, Gomes AP, Gouveia IC (2024) Microbial exopolysaccharides: structure, diversity, applications, and future frontiers in sustainable functional materials. Polysaccharides 5(3):241–287

Muthusamy S, Anandharaj SJ, Kumar PS, Meganathan Y, Vo D-VN, Vaidyanathan VK, Muthusamy S (2022) Microbial pullulan for food, biomedicine, cosmetic, and water treatment: a review. Environ Chem Lett 20(5):3199–3234

Nagtode VS, Cardoza C, Yasin HKA, Mali SN, Tambe SM, Roy P, Singh K, Goel A, Amin PD, Thorat BR (2023) Green surfactants (biosurfactants): a petroleum-free substitute for sustainability—comparison, applications, market, and future prospects. ACS Omega 8(13):11674–11699

Nawaz A, Chaudhary R, Shah Z, Dufossé L, Fouillaud M, Mukhtar H, ul Haq, I. (2020) An overview on industrial and medical applications of bio-pigments synthesized by marine bacteria. Microorganisms 9(1):11

Paillié-Jiménez ME, Stincone P, Brandelli A (2020) Natural pigments of microbial origin. Front Sustain Food Syst 4:590439

Pal S, Chatterjee N, Das AK, McClements DJ, Dhar P (2023) Sophorolipids: a comprehensive review on properties and applications. Adv Colloid Interf Sci 313:102856

Pardhi DS, Rabadiya KJ, Panchal RR, Raval VH, Joshi RG, Rajput KN (2023) Cyclodextrin glucanotransferase: fundamentals and biotechnological implications. Appl Microbiol Biotechnol 107(19):5899–5907

Pedone E, Fiorentino G, Bartolucci S, Limauro D (2020) Enzymatic antioxidant signatures in hyperthermophilic archaea. Antioxidants (Basel) 9(8):703

Rai SK, Roy JK, Mukherjee AK (2020) Application of poly (vinyl alcohol)-assisted silver nanoparticles immobilized β-keratinase composite as topical antibacterial and dehairing agent. J Proteins Proteomics 11:119–134

Rather LJ, Mir SS, Ganie SA, Islam S-U, Li Q (2022) Research progress, challenges, and perspectives in microbial pigment production for industrial applications—a review. Dyes Pigments 210:110989

Riley P (2000) Colouring materials used in decorative cosmetics and colour matching. In: Poucher's perfumes, cosmetics and soaps. Springer, pp 151–165

Sajjad W, Din G, Rafiq M, Iqbal A, Khan S, Zada S, Ali B, Kang S (2020) Pigment production by cold-adapted bacteria and fungi: colorful tale of cryosphere with wide range applications. Extremophiles 24:447–473

Sánchez-Suárez J, Villamil L, Coy-Barrera E, Díaz L (2021) Cliona varians-derived actinomycetes as bioresources of photoprotection-related bioactive end-products. Mar Drugs 19(12):674

Serra M, Casas A, Toubarro D, Barros AN, Teixeira JA (2023) Microbial hyaluronic acid production: a review. Molecules 28(5):2084

Serrato RV (2024) Bacterial Alginate Biosynthesis and Metabolism. IntechOpen. https://doi.org/10.5772/intechopen.109295

Setiyono E, Adhiwibawa MAS, Indrawati R, Prihastyanti MNU, Shioi Y, Brotosudarmo THP (2020) An Indonesian marine bacterium, Pseudoalteromonas rubra, produces antimicrobial prodiginine pigments. ACS Omega 5(9):4626–4635

Sharma D (2021) Industrial applications of biosurfactants. In: Biosurfactants: greener surface active agents for sustainable future: microbial surfactants, pp 115–143

Sharma I, Kango N (2021) Production and characterization of keratinase by Ochrobactrum intermedium for feather keratin utilization. Int J Biol Macromol 166:1046–1056
Shridharani SM, Kennedy ML (2023) Real-world experience with collagenase Clostridium histolyticum-aaes for buttock and thigh cellulite: focus on administration and safety profile. Aesthet Surg J Open Forum 5:ojad014
Shuikan AM, Hozzein WN, Alzharani MM, Sandouka MN, Al Yousef SA, Alharbi SA, Damra E (2020) Enhancement and identification of microbial secondary metabolites. Extrem Microbes Metab Divers Bioprospecting Biotechnol Appl 10
Siddiqui M, Vimal A, Bhargava PJCP, Science P (2021) Biological degradation of keratin by microbial keratinase for effective waste management and potent industrial applications. Curr Protein Pept Sci 22(4):304–312
Singhvi G, Hans N, Shiva N, Dubey SK (2019) Xanthan gum in drug delivery applications. In: Natural polysaccharides in drug delivery and biomedical applications. Elsevier, pp 121–144
Solano F (2020) Photoprotection and skin pigmentation: melanin-related molecules and some other new agents obtained from natural sources. Molecules 25(7):1537
Song Y, Liu G, Li J, Huang H, Zhang X, Zhang H, Ju J (2015) Cytotoxic and antibacterial angucycline-and prodigiosin-analogues from the deep-sea derived Streptomyces sp. SCSIO 11594. Mar Drugs 13(3):1304–1316
Souto EB, Fangueiro JF, Fernandes AR, Cano A, Sanchez-Lopez E, Garcia ML, Severino P, Paganelli MO, Chaud MV, Silva AM (2022) Physicochemical and biopharmaceutical aspects influencing skin permeation and role of SLN and NLC for skin drug delivery. Heliyon 8:e08938
Srivastava B, Khatri M, Singh G, Arya SK (2020) Microbial keratinases: an overview of biochemical characterization and its eco-friendly approach for industrial applications. J Clean Prod 252:119847
Stoyneva-Gärtner M, Uzunov B, Gärtner G (2020) Enigmatic microalgae from aeroterrestrial and extreme habitats in cosmetics: the potential of the untapped natural sources. Cosmetics 7(2):27
Sung HJ, Khan MF, Kim YH (2019) Recombinant lignin peroxidase-catalyzed decolorization of melanin using in-situ generated H_2O_2 for application in whitening cosmetics. Int J Biol Macromol 136:20–26
Tan R, Lyu Y, Zeng W, Zhou J (2019) Enhancing scleroglucan production by Sclerotium rolfsii WSH-G01 through a pH-shift strategy based on kinetic analysis. Bioresour Technol 293:122098
Tan X, Guo S, Wang Y, Zhang Y, Zhang H, Lü C, Wang Q, Xu P, Ma C, Gao C (2023) Multienzyme cascade excluding costly cofactors for pyruvate production from glucose. Chem Eng J 463:142473
Tarasuntisuk S, Palaga T, Kageyama H, Waditee-Sirisattha R (2019) Mycosporine-2-glycine exerts anti-inflammatory and antioxidant effects in lipopolysaccharide (LPS)-stimulated RAW 264.7 macrophages. Arch Biochem Biophys 662:33–39
Uhliariková I, Šutovská M, Barboríková J, Molitorisová M, Kim HJ, Park YI, Matulová M, Lukavský J, Hromadková Z, Capek P (2020) Structural characteristics and biological effects of exopolysaccharide produced by cyanobacterium Nostoc sp. Int J Biol Macromol 160:364–371
Vachher M, Sen A, Kapila R, Nigam A (2021) Microbial therapeutic enzymes: a promising area of biopharmaceuticals. Curr Res Biotechnol 3:195–208
Vallejos-Almirall A, Riquelme M, Uribe E, Agurto-Muñoz C, Rosas A, Vergara C (2022) A new approach to the induction and recovery of Synechococcus leopoliensis CPD-photolyase for cosmetic applications. J Appl Phycol 34(5):2371–2382
Vega J, Bonomi-Barufi J, Gómez-Pinchetti JL, Figueroa FL (2020) Cyanobacteria and red macroalgae as potential sources of antioxidants and UV radiation-absorbing compounds for cosmeceutical applications. Mar Drugs 18(12):659
Velasco MVR, de Sá-Dias TC, Dario MF, Bedin V, Fileto MB, de Oliveira AC, de Oliveira Pinto CAS, Baby AR (2022) Impact of acid ("progressive brush") and alkaline straightening on the hair fiber: differential effects on the cuticle and cortex properties. Int J Trichology 14(6):197
Venil CK, Devi PR, Ahmad WA (2020) Agro-industrial waste as substrates for the production of bacterial pigment. pp 149–162

Verma M, Shakya S, Kumar P, Madhavi J, Murugaiyan J, Rao M (2021) Trends in packaging material for food products: historical background, current scenario, and future prospects. J Food Sci Technol 58:1–14

Wahab YA, Fatmadiana S, Naseer MN, Johan MR, Hamizi NA, Sagadevan S, Akbarzadeh O, Chowdhury ZZ, Sabapathy T, Al Douri Y (2020) Metal oxides powder technology in dielectric materials. In: Metal oxide powder technologies. Elsevier, pp 385–399

Wu J, Gu X, Yang D, Xu S, Wang S, Chen X, Wang Z (2021) Bioactive substances and potentiality of marine microalgae. Food Sci Nutr 9(9):5279–5292

Wulaningsih TI (2023) Gluconolactone in cosmetic. KESANS Int J Health Sci 2(9):730

Zampieri RM, Adessi A, Caldara F, Codato A, Furlan M, Rampazzo C, De Philippis R, La Rocca N, Dalla Valle L (2020) Anti-inflammatory activity of exopolysaccharides from Phormidium sp. ETS05, the most abundant cyanobacterium of the therapeutic euganean thermal muds, using the zebrafish model. Biomol Ther 10(4):582

Zheng M, Zheng M, Epstein S, Harnagel AP, Kim H, Lupoli TJ (2021) Chemical biology tools for modulating and visualizing Gram-negative bacterial surface polysaccharides. ACS Chem Biol 16(10):1841–1865

Zhong Q, Ma Y, Xu D, Lei P, Li S, Xu H, Qiu YJF (2023) Heterologous biosynthesis of hyaluronic acid using a new hyaluronic acid synthase derived from the probiotic Streptococcus thermophilus. Fermentation 9(6):510

Chapter 7
Cosmeceuticals from Marine Sources: Algal Products as Green Cosmetics

Iram Zovia, Annena Ijaz, Ahmad Munir, Amna Mumtaz, Muhammad Arshad, Asmara Ahmad, and Sadia Javed

Contents

I. Zovia · A. Ijaz · A. Mumtaz · A. Ahmad · S. Javed (✉)
Department of Biochemistry, Government College University Faisalabad, Faisalabad, Pakistan
e-mail: sadiajaved@gcuf.edu.pk

A. Munir
Department of Biochemistry, University of Agriculture Faisalabad, Faisalabad, Pakistan

M. Arshad
Biochemistry Section, Jhang-campus, University of Veterinary and Animal Sciences Lahore, Lahore, Pakistan

S. Javed et al. (eds.), *Sustainable Cosmeceuticals*,
https://doi.org/10.1007/978-3-031-86087-4_7

Abstract Micro- and macro-marine algae are now widely recognized as important sources of cosmeceuticals, perfectly in sync with green chemistry. These amazing organisms are adaptable to various marine habitats; they can cope with extreme and severe conditions like high-intensity UV light, high salinity, and varying temperatures. In response to such challenges, they synthesize a wide range of metabolites; bio-active peptides, vitamins, and polysaccharides that are of immense value for skincare. This shift towards marine algae is essential as more and more consumers are looking for natural cosmetics alternatives to synthetic cosmetics. Unlike harsh chemicals in traditional products, which tend to cause skin irritation and long-term health risks, algal bioactive compounds have effective products with UV protection, antioxidant properties, anti-inflammatory effects, and safety for users. Promoting these principles of green chemistry and embracing these natural compounds radically reduce the impact on the environment. Algal bioactive compounds integrated into cosmetics not only decrease reliance on synthetic ingredients but also respond to urgent ecological challenges. Algae are renewable resources that can be cultivated relatively easily with little environmental disruption, making them a great ecofriendly choice when it comes to the use of cosmetics. Recent studies have positioned them as important players in the development of innovative cosmeceuticals for common skin issues including aging, pigmentation disorders, and oxidative stress. This chapter provides a thrilling analysis of marine algae and the novel biochemistry of these microorganisms. Learning how these green wonders are going is coming up with effective skincare solutions that will revolutionize the cosmetics industry while practicing environmentally friendly approaches. This chapter aims to unveil the marvelous life of marine algae and to emphasize on their special biochemical characteristics and potentials of their valuable products in the cosmeceutical field.

Keywords Marine algae · Classification · Bioactive compounds · Applications in cosmetics

7.1 Introduction

The term "Algae" is not a taxonomic term, but rather a commonly used generic classification for all chlorophyll-containing, plant-like organisms that have photosynthesis mechanism and are not specializing land plants (De Luca et al. 2021). Algae comprise a broad range of organisms from various evolutionary groups with over 30,000 species described. These can be classified into two types including multicellular macroalgae and unicellular microalgae (microscopic algae). Though, it is crucial to remember that algae are evolutionarily diverse (Ariede et al. 2017). Phycology is the scientific study of algae in a sequential way. They have an extensive diversity of life cycles and sizes, the size ranging from microscopic *Micromonas* species to 60-m-long tremendous kelps. The marine algae have suited particularly good providers of bioactive components such as carotenoids, minerals vitamins, dietary fiber, and omega-3 fatty acids, due to prevalence in the marine ecosystem (Veluchamy and Palaniswamy 2020). The characteristics that are not observed in plants or animals have been found in them. The pigments present in them are distinct from photosynthetic pigments. Algae have diverse medicinal benefits, both internally and externally. The outstanding properties of microalgae and their capability to merely change their growth lead to the production of specific antioxidants/pigments or supplementary nutrients and secondary metabolites. Both entire algal biomass and algae extracts are included in various cosmetic formulations, primarily in skin lotions, gels body creams, soaps, and shampoos, and they demonstrate UV radiation and anti-aging properties, providing hydration and skin protection (Ariede et al. 2017). Extracts originated mostly from *Coccoid phaeodactylum tricornutum* species and filamentous microalgae, *Chlorella* species or macro-algae such as *Fucus. Porphyra, Wakame* and microalgae such as *Arthrospira platensis* and *Chlorella* are also used to extract proteins and peptides. There are several microorganisms present, including *pluvialis, Nannochloropsis gaditana, Haematococcus Isochrysis galbana, Tetraselmis chuii, Phaeodactylum tricornutum, and Dunaliella salina* (Marino et al. 2020). Many algae contain considerable quantities of high-quality oils including omega-3 and omega-6 fatty acids, which are utilized as raw material for nutritional supplements. Microalgae are the source of omega-3 fatty acids in fish. Manufacturing costs of these alternative sources may be lower than those of algae, rendering the algal product uncompetitive (Li et al. 2011). Vitamins are a fantastic example of such a product. Although microalgae have been proposed as commercial sources of vitamins such as C, E, and B12, algae-produced vitamins cannot currently compete in the market (Lordan et al. 2011).The manufacturing costs of these alternative sources may be lower than those of algae, rendering the algal product uncompetitive. Vitamins are a fantastic example of such a product. Although microalgae have been proposed as commercial sources of vitamins such as C, E, and B12, algae-produced vitamins cannot currently compete in the market (Vo and Kim 2013).

For all aquatic life they produce oxygen and supply an energy source, as well as a source of food, crude oil, and numerous industrial and pharmaceutical products for humans (Wijesinghe and Jeon 2012). Chlorophyll, astaxanthin, xanthophylls, carotene, and phycobiliproteins are pigments that are associated with photosynthetic activity, and many of them have demonstrated substantial potential as cosmetic action due to their antioxidant, anti-inflammatory, and immune-enhancing qualities (Morocho-Jácome et al. 2020).

7.2 Exploring the Versatility of Micro- and Macro-Algae

Algae, which are either can be unicellular or multicellular kind of living plants, are classified as either in macro-algae or micro-algae categories that based on their sizes and shapes. The earliest forms of life on earth are microalgae (Falkowski et al. 2004). Microalgae existed in Earth's waters when it generated around 3 billion years ago. Microalgae diversity of (eukaryotic, prokaryotic, and cyanobacteria microalgae) is vast, though it has yet to be completely investigated (Massana et al. 2006; Fehling et al. 2007). There are around 50,000 distinct categories of micro-algal classes discovered in oceans and fresh water, i.e., lakes, ponds, and rivers; only 30,000 of these species have been evaluated (Richmond 2004). Microalgae may transform solar energy by fixing CO_2 to chemical energy, with 10 times the efficiency of terrestrial plants. The recent rise in global temperature is closely related to the rise in anthropogenic carbon dioxide (CO_2) emissions during the previous century. The world's potential for utilization of CO_2 is 3.7 Gt/year, which is approximately 10% of global CO_2 emissions. According to Vooradi et al. (2018), fossil fuels are the main source of CO_2 emissions and contribute the most to this energy paradigm. Moreover, additional CO_2 emissions are released into the environment as a result of the conversion of raw materials in industries such steel, iron, cement, textile, pulp and paper, aluminum, refineries, and landfills. Approximately 75% of the total greenhouse gas emissions are from CO_2 alone. To eliminate the enormous amount of CO_2 in the atmosphere, natural processes such as forests and seas are insufficient.

According to Mata et al. (2010), green algae, especially those in the Chlorophyta class, can fix 10–50 times as much CO_2 as terrestrial plants. Algae can maintain the carbon footprint by fixing carbon in a variety of ways. Autotrophic species of marine algae fix inorganic carbon, whereas heterotrophic species can absorb organic carbon. Furthermore, certain mixotrophic algae combine the ability of autotrophy and heterotrophy to fix both organic and inorganic carbon.

Microalgae that are categorized as photo-organotrophic or photoheterotrophic have ability to absorb organic carbon by using light as an energy source. These various metabolic pathways highlight the diverse strategies algae employ for carbon fixation, making them a significant focus in carbon sequestration research. Algal biofuels have various environmental advantages over conventional biofuels, including lower GHG emissions, a smaller carbon footprint, greater energy reliability, economic viability, reduced land-use change (LUC), and the production of additional value-added goods (Anwar et al. 2020). Microalgae for commercial use are estimated to be around 5000 tons of dry matter annually (Raja et al. 2008). In the Asia-Pacific region, there are around microalgae producers are about 110 commercials with volumes ranging from 3 to 500 tons per year. Asia provides for roughly nine-tenths of algae cultivation. Among these, Taiwan, China, and India have a large number of commercially viable microalgae producers.

Haematococcus, Spirulina, Botryococcus, Nannochloris, Chlorella, Dunaliella, Porphyridium, Phaeodactylum, Tetraselmis, Chaetoceros, Isochrysis, Nitzschia, Crypthecodinium, Schizochytrium, and Skeletonema are among the commercially important microalgae (Del-Campo et al. 2007). Several types of microalgae have been found to be rich in colored high-value ingredients, which, while outstanding, have attracted little attention as possible natural forms of pigments for cosmetics.

7.3 Algal Products as Green Cosmetics

The increasing demand for organic and natural cosmetics has been fueled by growing public concern on the use of hazardous chemicals in cosmetics as well as ethical issues regarding the cosmetics industry. Natural and organic cosmetics usually refer to the "Green" attribute, which eventually designate products that specifically meet the production principles and have ecological appeal. Common claims must ensure that green cosmetic products that are developed, followed the principles of green chemistry. Despite its popularity in the cosmetics industry, the term "Green" still lacks a precise meaning. The most obvious understanding of this term is its connection with "environmentally friendly" and the use of organic/natural components (Franca and Ueno 2020).

Marine algae-derived green cosmetics are skincare and cosmetics formulated with natural ingredients and different algal extracts such as kelp, seaweed, and microalgae like chlorella and spirulina. These algae are extremely important in cosmetic formulations because they are abundant in bioactive chemicals that offer several skin benefits. Cosmetic industry uses marine algae for their moisturizing, anti-aging, and skin-nourishing properties. For example, spirulina is well-known for its high protein, vitamin, and antioxidant content, which aids in skin restoration and the fight against oxidative stress. Moreover, kelp extract soothes sensitive skin, promotes skin barrier repair, and provides hydration to irritated skin. Brown algae are the source of alginates, which are widely utilized in cosmetics because of their thickening qualities and capacity to calm sensitive skin (Pereira 2018).

Moreover, astaxanthin, a strong antioxidant present in certain algae, is well known for its anti-aging properties since it protects the skin from premature aging by scavenging free radicals (Mohiuddin 2019).

Beyond skincare, marine algae can be used in a wide range of cosmetic applications. Because of their moisturizing and strengthening qualities, they are utilized in haircare products to help dry or damaged hair. Fucoidan and phlorotannins, compounds originating from algae, have been researched for their anti-inflammatory and UV-protective properties, making them useful in anti- aging and sunscreen compositions. In addition to their uses in personal care products, these compounds also have antibacterial qualities (Dini and Laneri 2021) (Table 7.1).

Table 7.1 Algal-derived green cosmetics and their key benefits

Algae species	Bioactive extract	Product name	Company	Key benefit	Galenic form	% in Formulation	Application area	Sustainability certification	Shelf life	References
Spirulina sp.	Phycocyanin, beta-carotene, chlorophyll, SOD, xanthophyll	NutraSpir & Iraya Algae body serum	AlgaeHealth & Nykaa	Moisture the skin and hair and make it soft and look shiny, remove toxin and slow down aging of skin	Capsule, peptide (dry powder), oil	5%	Cosmetics	EcoCert	2 years	Pereira (2018)
Chlorella sp.	Chlorella growth factor, omega-3 fatty acids	Sun Chlorella Cream	Sun Chlorella USA Corporation	Anti-aging activity, UV protection, exfoliate skin and hair, moisture retention and hydration	Peptide, cream form	3%	Cosmetics	USDA Organic	1.5 years	Saini et al. (2024)
Rhodophyta	Phycoerythrin	Red Algae and Seawater Serum (Porphose)	Algae Research and supply Inc.	Moisturizing, texturing, photoprotectant, antioxidant, coloring	Powder	10%	Cosmetics	GMP Certified	3 years	Rosic (2019)
Haematococcus pluvialis	Astaxanthin	Reveal color correcting eye serum brightener	Algenist	Skin health, eye protection, smooth the eye area, treat uneven skin tone	Softgel, semiliquid	6%	Nutraceutical, Cosmetics	Non-GMO	1.5 years	Uppala (2015)
Dunaliella salina	Beta-carotene	Beta Natural Intimate Face Wash	Betadine	Antioxidant, skin health, prevent redness and irritation	Oil, gel	4%	Cosmetics	EcoCert	2 years	Uppala (2015)

Nannochloropsis gaditana	EPA omega-3	Sensitive Face Cream Sunscreen Spf 50	Apivita	Provide hydration, improve texture, and reduce irritation	Softgel	8%	Nutracosmetics	Friend of the Sea	1 year	Huang et al. (2018)
Ulva lactuca	Ulvan	Actiphyte™ Sea Lettuce GL	Lubrizol	Antimicrobial, skin hydration, anti-wrinkle, collagen synthesis and provide moisturizing effect	Gel, lotion	5%	Cosmetics	EcoCert	1.5 years	Wang et al. (2017)
Phaeodactylum tricornutum	Fucoxanthin	Sensityl	Givaudan Active Beauty	Reduce discoloration, provide smooth effect on skin, delayed the skin aging effect	Softgel	9%	Nutraceutical	Non-GMO	1 year	Guillerme et al. (2017)
Tetraselmis chuii	Caprylic/ capric triglyceride, phloretin	Algae Niacinamide Moisture Veil	Algenist	Optimizing hydration, skin protection, and preserving skin barrier	Liquid moisturizer	5%	Nutraceutical, Cosmetics	Friend of the Sea	2 years	Ahsan (2019)

7.4 Classification of Marine Algae

The classification of algae is categorized into three different groups: Chlorophyceae (Chlamydomonas, Spirogyra, and Chara)—chlorophyll a and b pigments—are found in green algae. Phaeophyceae (*Dictyota, Laminaria,* and *Sargassum)* are brown algae that predominantly occur in the sea having different kinds of pigments like carotenoids, chlorophyll A and C, and xanthophyll. Rhodophyceae (*Gracilaria, Porphyra,* and *Gelidium*)—the red pigment r-phycoerythrin—are found in red algae. The fourth form of algae is blue-green algae (Cyanobacteria), which is occasionally sometimes for seaweed. This form of algae is commonly found in household aquariums, where it quickly covers all surfaces and is known as smear algae or slime algae (Wolkers et al. 2011) (Table 7.2).

Table 7.2 Classification and pigment composition of algal species

Types	Algal species	Pigment	Metabolite	Function	References
Red algae	*Irish moss*	Phycoerythrin	Omega-3 and Omega-6 fatty acids	Pharmaceuticals food, cosmetics, and particularly as substitutes for artificial dyes Beneficial, hydrating, hair, nutrient-rich, soothing, anti-inflammatory	Arad and Yaron (1992), De Canha et al. (2020)
Red-green algae	*Phaeodactylum tricornutum*	Carotenoids	Omega-3 fatty acids	Preventing the skin from the harmful effects of ultraviolet radiation Suspending and preventing the visible signs of skin aging	Nizard et al. (2007)
Brown algae	*Fucus vesiculosus*	Fucoxanthin and polysaccharides	Fucitol, beta-carotene	Diminishes dark circles and boosts collagen production	Sun et al. (2020)
Brown algae	*Ascophyllum nodosum*	Chlorophyll-c	Fucoxanthin, alginates	Anti-aging agent, anti-wrinkle agent, soothing agent Nanotechnology, optical systems, semiconductor nanolithography, and drug delivery	Pallela et al. (2010), Pereira et al. (2020)

Table 7.2 (continued)

Types	Algal species	Pigment	Metabolite	Function	References
Green algae	*Dunaliella salina*	Chlorophyll-a, carotene chlorophyll-b, β-	Palmitic acid, linoleic acid, β-cryptoxanthin	Harnessing antioxidants for soothing and have powerful anti-inflammatory properties	Widowati et al. (2017)
Green algae	*Tetraselmis suecica*	Carotene chlorophyll-a, chlorophyll-b, β-	Palmitic acid, stearic acid and vitamin E (α-tocopherol)	Anti-oxidant, treatment of atopic dermatitis	Sansone et al. (2017)
Blue-green algae	(*Cyanobacteria*)	Chlorophyll a, xanthophyll, and phycobiliproteins		The presence of algal protein strengthens the hair and prevents breakage and broken ends	Sansone et al. (2017)

7.5 Bioactive Compounds of Marine Algae

Bioactive chemicals are substances that are biologically and physiologically active and have properties that are useful in the human body. As a result, there is a lot of potential in developing and manufacturing numerous biocompounds that could serve as functional components, such as carotenoids, fatty acids, polyphenols, phycocyanin (PC), and polyunsaturated compounds (Plaza et al. 2010) (Fig. 7.1).

Algae can be found on every continent in the world and in every biological niche accessible. As a result, they have unique features that allow them exist in the ecosystem's harsh conditions. These unique features are caused by changes in the macro- and micromolecular components of the cell, which occur in stressful situations. In addition to the macromolecules found in algae, these distinct metabolites frequently have specific capabilities and can be termed bioactive substances. Out of hundred algae species, only 25–30% have been recognized and collected. As a consequence, the pharmaceutical business has a tremendous undiscovered resource that can be utilized. A variety of therapeutically useful biochemicals are produced by microalgae that can be extracted from the plant biomass or released extracellularly into the medium (Bhagavathy et al. 2011). These microorganisms involve a wide range of bioactive substances, including proteins, polysaccharides, lipids, sterols, enzymes, vitamins, and other high-value compounds having medicinal and nutritional applications (Priyadarshani and Rath 2012). Proteins, vitamins, fatty acids, and pigments are examples of bioactive chemicals derived from microalgae that can be acquired directly from primary or secondary metabolism. These chemicals may have anti-algal, antiviral, antifungal, anti-enzymatic, or antibiotic properties (Volk 2008) (Table 7.3).

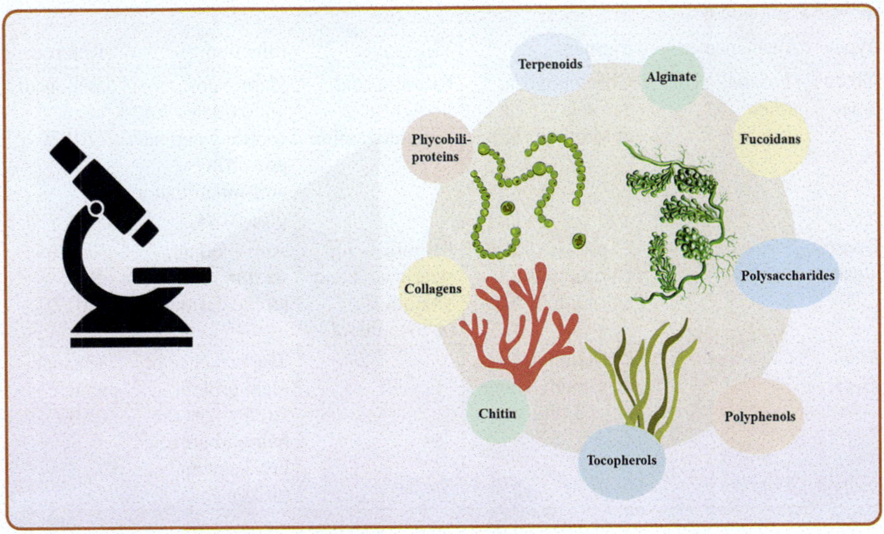

Fig. 7.1 Bioactive/natural compounds extracted from marine algae

Table 7.3 Bioactive compounds and therapeutic applications of marine algae

Bioactive compounds	Organisms	Genera	Potential applications in cosmeceuticals	References
Carotenoids	Microalgae	*Chlorella, Chlamydomonas, Dunaliella, Muriellopsis, and Haematococcus* spp.	Antioxidant stabilizers and preservative in lotion and creams for sun protection	Hallmann (2015)
Phycobiliproteins	Cynobacteria	*Rhodophyceae or Cryptophyceae*	An appealing substitute for synthetic pigments in the manufacture of colored creams and cosmetic items	Wrolstad and Culver (2012)
Collagens	Seaweeds, cynobacteria	*Ulva, Ascophyllum, Palmaria, and Porphyra*	As an anti-ageing	Sionkowska et al. (2020)
Polysaccharides	Seaweeds	*Rhodophyceae*	Used as hydrating agents in cosmetics and toiletries bath lotion and moisturizers	Usman et al. (2017)
Chitin	Microalgae	*Pithophora oedogonia*	Used in toothpastes, fragrances, and toiletries for men as well	Aranaz et al. (2018)
Fucoidans	Macro-algae, brown seaweed	*Ecklonia cava, fucus serratus*	Antiviral, anticoagulant, and antitumor properties helpful in skin protection as antioxidant anti-aging compound	Kim et al. (2018)

Table 7.3 (continued)

Bioactive compounds	Organisms	Genera	Potential applications in cosmeceuticals	References
Carrageenans	Red algae	*Agardhiella, Chondrus crispus, Eucheuma, Furcellaria, Gigartina, Hypnea, Iridaea, Sarconema, and Solieria*	Provide texture and consistency in various health and beauty products The anti-aging, antioxidant, and anticarcinogenic activities of Carrageenan are documented	Pacheco-Quito et al. (2020)
Alginate	Brown algae	*Fucus vesiculosus and Turbinaria conoides*	Outstanding biocompatibility gel-forming performance non-toxicity, biodegradability, and easy-to-process properties	Priyan Shanura Fernando et al. (2019)
Terpenoids	Microalgae	*Erythrolobus australicus*	Scavenge reactive oxygen species, free radicals that cause aging, and have anti-aging properties	Michniak-Kohn and Leonardi (2017)
Tocopherols	Marine algae	*Rhodophyceae, Phaeophyceae, Rhodophyceae, Chlorophyceae*	Tocopherols are micro-algal biotechnology products exposed to radiation; it transforms to dimers and trimers that are UV-absorbing, which act similarly to sunscreen agents	Sinbad et al. (2019)
Polyphenols	Marine macro-algae	*Fucus vesiculosus*	By single electron transfer and through hydrogen atom transfer, they act as antioxidants The phloroglucinol group, present in brown algae such as *Fucus vesiculosus*, is the most important group in the cosmetics industry	Haoujar et al. (2019)
Rutin	Blue-green algae	*Mastigocladus laminosus*	Radical scavenging activity Kill cancer cells Inhibit signaling pathway that causes cell cycle arrest	El-Baky et al. (2009)
Quercetin	Microalgae	*Microchaete tenera*	Anti-oxidative activity Anti-radical activity Arrest cell cycle by releasing p53	Singh et al. (2017)

7.6 Marine Algae as an Industry: Converting Raw Material into High-Value Products

7.6.1 Mycosporine-Like Amino Acids (MAAs)

Mycosporines are tiny compounds, i.e., water-soluble made up of an aminocyclohexenone or an aminocycloheximine ring with amino alcohol substituents or nitrogen. They are known as MAAs (mycosporine-like amino acids) when they are swapped with amino acid residues (Oren and Gunde-Cimerman 2007) (Fig. 7.2).

Many marine and freshwater species have been shown to have Mycosporine-like amino acids (MAAs) (Rosic et al. 2015). MAAs are found in a wide range of mammals, cyanobacteria, and algae. MAAs act as UV shields by absorbing and dispersing UV light as harmless heat without the risk of initiating photochemical reactions. Furthermore, MAAs have considerable antioxidant activity (Fonseca et al. 2023).

7.6.2 Polysaccharides of Marine Seaweeds: The Emerging Source of Food-, Medicine-, and Cosmeceutical-Related Products

Seaweeds are primarily composed of polysaccharides. They are composed of multiple monosaccharides linked by glycosidic linkages, with or without sulfate and other groups. They are classified as sulfated or non-sulfated. Their structural qualities provide unique functional capabilities, making them valuable additives in food processing, medicines, cosmeceuticals, and other applications (Venugopal 2019). The presence of different types of polysaccharides (matrix and storage) varies per macro-algal species. Green marine seaweeds, are high in ulvans, while and fucoidans, alginic acids (or alginates), laminarans (or laminarins), found in brown macroalgae while red seaweeds contain xylogalactans (especially in the Corallinales order), floridean starch, carrageenans, xylans, porphyran, agars and sulfated galactans (Fig. 7.3).

Recent studies have extensively characterized the important biological actions of natural polysaccharides of seaweed as well as derivatives, such as oligosaccharides

Fig. 7.2 Chemical structure of Mycosporine

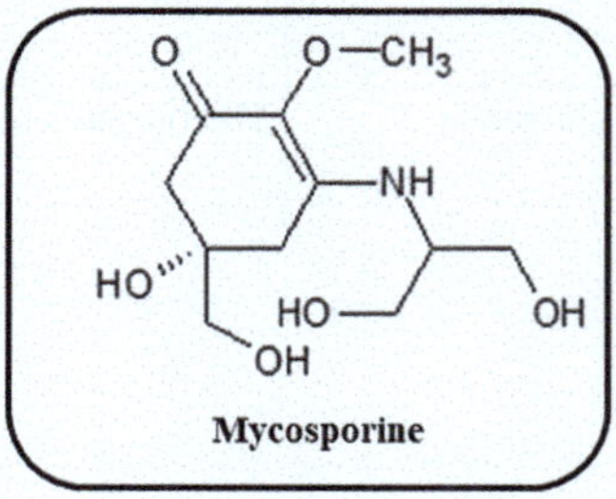

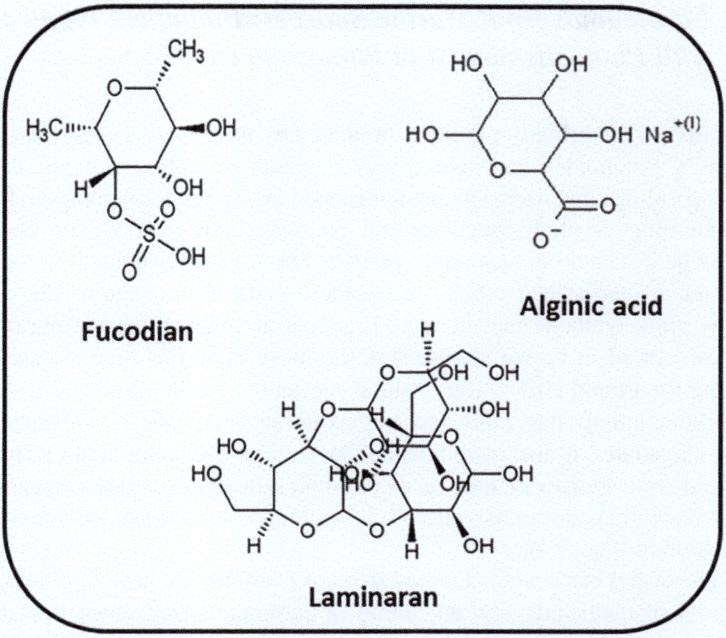

Fig. 7.3 Chemical structure of algal polysaccharides

and polysaccharides that are biochemically modified. Anti-inflammatory, anti-diabetic, anti-obesity, anti-hyperlipidemic, immunomodulatory, antioxidant, anti-cancer, antiviral, antibacterial, and gastro-protective actions have been thoroughly investigated (Hentati et al. 2020). The polysaccharides act as a potential cosmetic ingredient because of their anti-inflammatory, and antibacterial properties high water retention, as well as the fact that they are produced by a variety of marine creatures, including those from the Phaeophyta, phyla Crustacea, and Rhodophyta (Corinaldesi et al. 2017). Algae are the utmost chief marine source of polysaccharides (Pangestuti et al. 2018). Red algae-derived polysaccharides also have aesthetic advantages, such as wound healing. Polysaccharides extracted from brown seaweed have cosmetic benefits as well. Sargassum sp. isolated polysaccharides with anti-aging (elastase inhibiting ability), skin whitening (tyrosinase-inhibiting ability), antioxidant (antioxidant and scavenging abilities), water retention, and moisturizing properties and demonstrating the potential for cosmetic formulations as *Sargassum* species. Extracted polysaccharides, including fucoidan, a sulfated polysaccharide derived from brown algae are used in cosmetic products. Fucoidan is a brown algae-derived sulfated polysaccharide (Jesumani et al. 2019).

7.6.3 *Carotenoids from Marine Sources: Biological Pigments with Photochemical and Photoprotective Functions*

Carotenoids are the primary pigments produced by many plants, algae, and microorganisms. Carotenoids occur naturally as pigments with photosynthetic light-harvesting capabilities that transport accumulated solar energy into chlorophylls while also functioning as photoprotectors and photosynthetic agents. The chemicals, which can be evaluated in cyanobacteria cells but are also prominent in the organisms found in marine ecosystem, shield these cells from photo-oxidative stress caused by photosynthetic metabolism (Pagels et al. 2022). Fucoxanthin (the substantial carotenoid in brown seaweed) is the most numerous marine carotenoids, accounting for around 10% of total natural carotenoid production.

The production of astaxanthin and canthaxanthin from algae, as well as perinidin from dinoflagellates, is also incredible and increasing. Other carotenoids that originate from marine sources include fucoxanthinol, astaxanthin, peridinin, canthaxanthin and halocynthiaxanthin, as well as violaxanthin, echinenone, halocynthiaxanthin, and tunaxanthin (Fig. 7.4).

Though several carotenoids that are derived from marine have high antioxidant activity, out of them only few are found in commercialized cosmetics and sunscreens due to unsatisfactory clinical trial outcomes. Marine carotenoids' skin absorption has yet to be investigated, which is essential for demonstrating them as antioxidant ingredients in sunscreen compositions. Additionally, these carotenoids have been linked to a number of biological functions, comprising anticancer, anti-obesity, antidiabetic activity, and protective properties against inflammation and cardiovascular disease (Matsuno 2001).

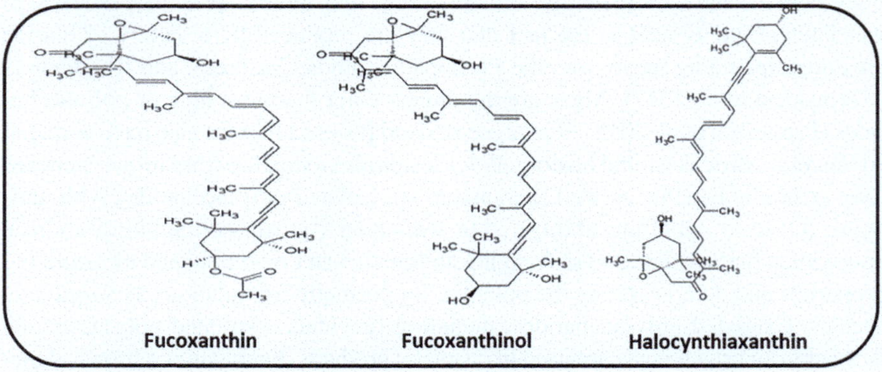

Fig. 7.4 Chemical structure of algal carotenoids

7.6.4 Polyphenolic Compounds and Their Biological Importance

Polyphenols including (flavonoids, tannins, phenolic acid, and phlorotannins) as they are secondary metabolites that have potential in a wide range of biological activities such as reproduction, photosynthesis, and cell division, as well as exhibiting both anti-melanogenesis and antioxidant properties. Polyphenolic compounds derived from marine algae such as fucofuroeckol-A, dieckol, triphlorethol-A, and phloroglucinol have photoprotective properties, particularly against UVB radiation, and hence it is sustainable to be used in sunscreens (Fig. 7.5).

Brown algae, on an average, have wider levels of phlorotannin than red and green algae (Ibañez et al. 2011; Mabeau and Fleurence 1993). These phlorotannins have numerous biological roles, comprising antioxidant, anti-proliferative, antimicrobial, anti-diabetic, anti-HIV, anti-allergic, and anti-inflammatory characteristics (Li et al. 2011). Phlorotannins, the most investigated class of polyphenolic compounds found in seaweed, in particular, exhibit a wide range of bioactivities, including antioxidant activities, UV protection, skin whitening, and anti-aging and are thus being examined for aesthetic applications (Kalasariya and Pereira 2022). They have also been shown to detoxify heavy metals and have other bioactivities that could be used in functional foods.

7.6.5 Terpenoids: Focused Natural Ingredients for Skincare and Anti-aging

Terpenoids are hydrocarbons that comprise a diverse family of natural chemicals that include both primary and secondary metabolites (Núñez-Pons et al. 2020). Terpenoids are the most common and widespread class of secondary metabolites identified in marine algae. Because of their ability to scavenge reactive oxygen species, free radicals that contribute to aging, they are considered to be anti-aging. Due

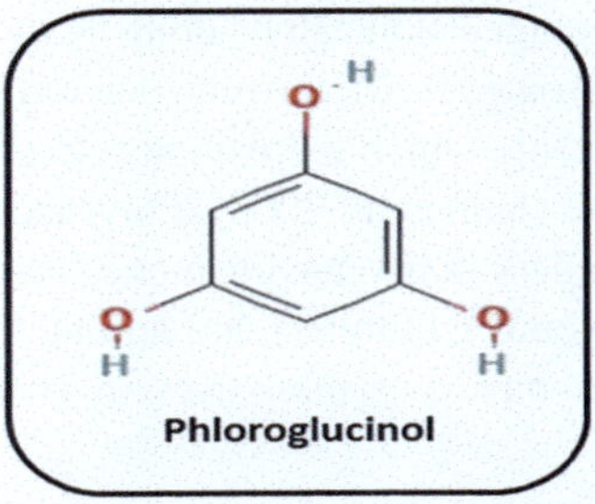

Fig. 7.5 Chemical structure of algal polyphenolic compounds (Phloroglucinol: The basic building block of Phlorotannins, which are polyphenolic components, has been found in a variety of brown algal groups and have high antioxidant activity)

to their remarkable capabilities in boosting penetration, exhibiting minimal systemic toxicity, and causing minimal discomfort, newly discovered marine terpenoids hold significant promise as a valuable resource for developing fresh antioxidant agents in cosmetic formulations. The levels of natural enzymatic and non-enzymatic antioxidants have diminished in the skin of both the elderly and those who have been exposed to solar radiation (Michniak-Kohn and Leonardi 2017).

7.7 Application of Microalgae in Derma-Cosmeceuticals: An Approach for Unlocking Therapeutic Potential

Cosmeceuticals are topical cosmetic-pharmaceutical hybrids that refer to a cosmetic product containing active ingredients that claim to offer medicinal or drug-like skin health benefits (Kligman 2000). Due to the ineptitude of artificial cosmetics, the cosmeceutical industry is progressively shifting to natural bioactive ingredients in the recent years (Smit et al. 2009). Microalgae and cyanobacteria are potentially rich sources of bioactive chemicals used in a variety of industry sectors and for human consumption in pharmaceutical, nutraceutical, and cosmetic disciplines (Tabarzad et al. 2020). These organisms have been considered therapeutic agents for numerous health conditions and other purposes in the cosmetics, food, and pharmaceutical industries due to their ability to create a variety of primary and secondary metabolites (Sathasivam et al. 2019). A number of secondary metabolites extracted from algae have been associated to skin benefits. For example, an *Arthrospira* extract can repair the indications of early skin aging, tighten the skin, and prevent striae formation; whereas a *Chlorella vulgaris* extract increases collagen production in the skin, enabling tissue regeneration and wrinkle reduction (Stolz and Obermayer 2005).

Cosmetic products that contain micro-algal bioactive compounds are in high demand particularly when combined with additional antioxidants or other bioactive chemicals, with one area of growth being treatments to protect the skin from UV damage (Pulz and Gross 2004).

Prevention of blemishers, helping seborrhea, controlling the process of inflammation and repairing the damaged skin can be done by the usage of the active ingredients obtained from microalgae. Moreover, derived active compounds from microalgae have significant role which helps in accelerating the process of healing and sustaining skin moisture (Kim et al. 2008). Polysaccharides, which are utilized as gelling agents and thickeners in diverse cosmetic formulae, as well as for moisturizing (for example, the genus Chlorella), are among the active components isolated from microalgae that have potential uses in cosmetics (Jain et al. 2005) (Fig. 7.6).

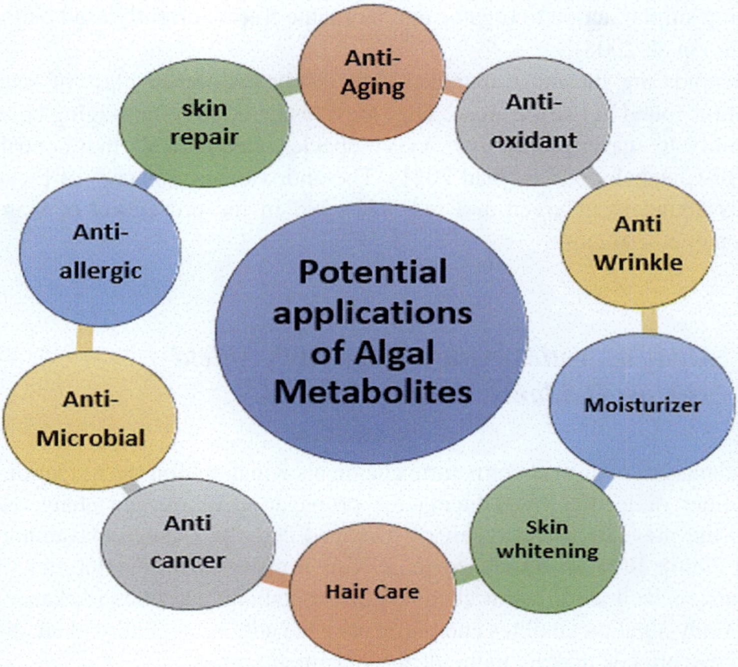

Fig. 7.6 Potential applications of algal bioactive compounds in cosmetics (Biologically active algal compounds are known to have anti-aging, anti-photo-aging, antioxidant, antimicrobial, photoprotective, skin conditioning, moisturizing, protecting, smoothing, soothing, viscosity controlling, lipolytic, and whitening properties (Agatonovic-Kustrin and Morton 2013). They can also be used in anti-cellulite and slimming treatments, as well as in the treatment of skin problems, such as tanning and pigment disorders) (Bedoux et al. 2014; Zanella and Alam 2020)

7.7.1 Algae as Nature's Anti-aging Remedy: Skin Revitalization Potential Unleashed

Skin aging is characterized by the deterioration of extracellular matrix (ECM) in both the epidermal and dermal layers, the appearance of visible indications on the skin's surface, and changes in the physical qualities of the skin. Skin aging is a complicated biological process that refers to the loss of skin elasticity, the formation of fine lines, ridges, wrinkles, and skin discoloration with increasing age (Ganceviciene et al. 2012).

The term "skin aging" is frequently associated with thinning of dermis and decreasing collagen content. The degradation of collagen caused by ultraviolet A (UVA) radiations is one of the causes of wrinkle formation (Lourenço-Lopes et al. 2020). When exposed to oxidative stress, algae can develop a number of effective defense mechanisms against free radicals and reactive oxygen species, producing compounds that can act in cosmetics against the harmful effects of UV radiations, promoting similar action to organic and inorganic filters currently used on the market (Sousa et al. 2008).

Terpenoids are the most abundant and widely distributed class of secondary metabolites found in marine algae. They are considered to be anti-aging because of their ability to scavenge reactive oxygen species, free radicals that contribute to aging (El-Chaghaby and Rashad 2021). The antioxidant properties of β-carotene, which is abundant in green and red algae, aid in the prevention of skin aging (Schagen et al. 2012).

7.7.2 Algae as Antioxidants: The Body's Shield Against Oxidants

Antioxidants are man-made or natural chemicals which inhibit the oxidation of lipids or other molecules by reducing the propagation of oxygen chain reactions through the prevention and repair of damaged cells by oxygen (Javanmardi and Kubota 2006). ROS are produced in cells as a result of ultraviolet and ionizing radiations, as well as different chemical and metabolic activities (Askarova et al. 2011). Many algal metabolites and pigments have anti-aging, antioxidant, and neuroprotective effects, making them ideal for cosmetic usage.

Numerous marine algal species have been demonstrated to have antioxidant activity, particularly *Hydroclathrus, Colpomenia, Gracilaria, Laurencia, Padina, Polysiphonia, Ahnfeltiopsis, Halymenia, and Turbinaria* (Cornish and Garbary 2010). Marine algae are rich source of metabolites, which can be used to fight against oxidative stress and delay the mechanism of skin aging. Biologically active compounds and metabolites from marine sources have gained increasing recognition these days. Mycosporine-like amino acids (MAAs), polysaccharides, sulfated polysaccharides, glucosyl glycerols, pigments, and polyphenols are among the metabolites (Berthon et al. 2017).

Porphyridium, a red microalgae species with sulfated polysaccharides, appears to be an attractive candidate to replace hyaluronic acid as a biolubricant and has antioxidant action against linoleic acid auto-oxidation (De Jesus Raposo et al. 2015). Green seaweed sulfated polysaccharides are known for their antioxidant characteristics, as they effectively resist free radicals such as superoxide, hydroxyl, and 1,1-diphenyl-2-picrylhydrazyl (DPPH) radicals (Wang et al. 2014). Meanwhile, mycosporine-like amino acids (MAAs) are naturally occurring chemical compounds found in a variety of organisms such as cyanobacteria and macro-algae. They gained a lot of attention for their functions in UV radiation shielding and scavenging reactive oxygen species (ROS), since they have been shown to attenuate both the direct and indirect negative effects of ambient UV radiation (Siezen 2011). Phenolic compound-rich algae exhibit a multitude of mechanisms, encompassing antioxidant, antiradical, anti-allergic, anti-inflammatory, and ultraviolet (UV) protective effects. Among the array of phenolic compounds derived from algae, phlorotannins stand out as particularly significant naturally occurring secondary metabolites, boasting diverse functional and bioactive properties (Heo et al. 2009). Carotenoids are often found in up to 0.2% of marine microalgae. They serve as potent antioxidants due to their capacity to quench singlet oxygen, to be oxidized and isomerized. The main sources of carotenoids are microalgae that belong to the Chlorophyceae family. They constitute important natural lipid-soluble pigments that directly protect the skin from ultraviolet light-induced photo-oxidation (Berthon et al. 2017). Trehalose is one of the best antioxidants present in numerous mushrooms such as *Auricularia auricula-judae, Lentinula edodes, Pholiota nameko,* and *Grifola frondosa*. Trehalose has a high water retention capacity as well as remarkable antioxidant properties. This makes it beneficial in cosmetics goods such as moisturizing creams.

7.7.3 The Anti-wrinkle Potential of Algae: Skin Rejuvenation as an Outcome of Natural Healing

Tanning of the skin can be caused by the continuous exposure to sunlight which enhances the amount of the melanin pigment in the skin. Ultraviolet rays from the sun play a crucial role in the production of tyrosinase, an enzyme that initiates reactions leading to the creation of melanosomes. These melanosomes eventually mature into melanin, which then transforms into keratinocytes, assisting in the skin's normal shedding process. L-tyrosine is first hydroxylated to yield 3,4-dihydroxy-L-phenylalanine and L-DOPA, which is then subjected to oxidation to produce dopaquinone. Finally, the generated dopaquinone is used to synthesize melanin. Skin pigmentation is caused by an excess of melanin, which must be regulated. Therefore, tyrosinase inhibitors are employed to facilitate the rate-limiting step in the pigmentation process (Wang et al. 2011). The use of *Fucus vesiculosus* extract aims to lessen the appearance of dark circles under the eyes by increasing the expression of heme oxygenase-1 (HO-1), a molecule that lessens heme synthesis in the skin by removing heme catabolites (Fig. 7.7).

The anti-inflammatory and antioxidant properties of this extract in topical preparations can improve the look of under-eye bags and encourage the formation of collagen, which may assist to diminish fine lines and wrinkles. Furthermore, by using cosmeceuticals products and sunscreens, it may be possible to reduce or even avoid skin aging (Sun et al. 2020). The extract from *Chlorogloeopsis* species benefits keratinous tissues by protecting them from ultraviolet radiations A and ultraviolet radiations B (UVA and UVB), which can create free radicals when exposed to UV. This defense assists in preventing photo-aging, the appearance of wrinkles, and skin sagging. Furthermore, *Nannochloropsis* algae work effectively to block the transmission of UVA and UVB light. Furthermore, the inclusion of cyanobacteria in sunscreen formulation demonstrated improved UVB-UVA absorption (290–400 nm) in comparison with a commercial formulation, as well as good absorption in the

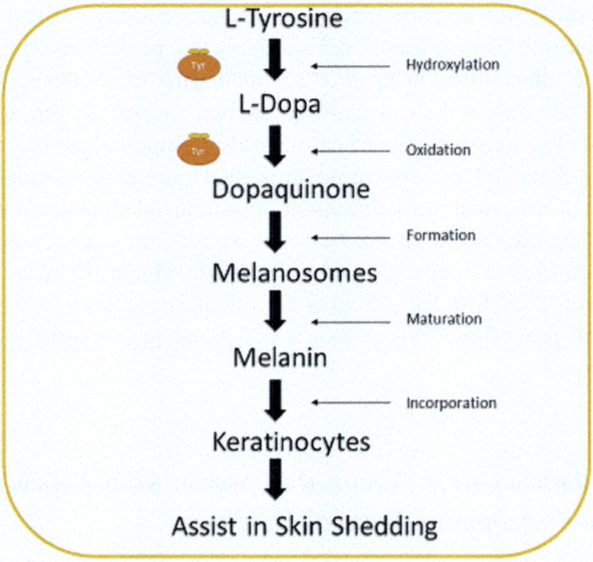

Fig. 7.7 Mechanism of melanin production

visible spectral area (400–650 nm) (Ariede et al. 2017). Various algal species produce compounds with specific chemical structures that absorb UV radiation while simultaneously inhibiting melanin synthesis (Hagino and Saito 2010).

7.7.4 Hydration from the Ocean: Algae as a Natural Moisturizer

A moisturizer is a complex mixture of different chemical substances used to improve the suppleness of the skin's epidermis. Skin that has adequate moisture is less prone to breakouts of acne and may even aid in avoiding eczema. Moisturizers are essential to maintain skin moisture, avoid dryness, minimize bruising, and minimize the appearance of wrinkles. Some ingredients such as water and specific acids such as hyaluronic acid play a major role in hydration of the skin (Bonté 2011). Cosmetic products that are enriched with algae peptides find applications in haircare and skincare, including skin lotions and moisturizers, facial creams, shampoos, rinses, hair restorers, permanent wave solutions, hair coloring agents, and body soaps or various bath products (Ariede et al. 2017). Widely recognized as the primary measure in addressing the process of skin aging, moisturization plays a crucial role in preserving the youthful aspect and flexibility of the skin, while simultaneously strengthening its defensive capabilities against external factors that pose a threat to its well-being. Hyaluronic acid, corneocytes, and regulated water transport systems all work together in the complex network of the skin to keep the body moist. Recent research revealed that aquaporins, proteins that facilitate the passage of water between cells in different epidermal layers, have been implicated in the moisturizing process. Additionally, lipid production and corneocyte shedding-related epidermal enzymes are regulated by water and pH levels. The level of skin moisture is highly related to how well it is protected from irritants and how frequently it is exposed to abrasive cleansing products (Bonté 2011).

The degradation of skin has been linked to heightened activity and enhanced expression of matrix metalloproteinases (MMPs), including hyaluronidase, collagenases, gelatinases, stromelysins, elastases, MMPs specific to membranes, and various other MMP variants (Thomas et al. 2014). These enzymes are of significant importance in the degradation of crucial structural components of the skin, including collagen and elastin, resulting in a decline in skin tautness, flexibility, and overall youthful appearance. On the other hand, there is evidence to suggest that cosmetics containing extracts derived from *Porphyra,* a kind of red algae, have notable performance in improving skin radiance as well as providing a refreshing and hydrating effect, in comparison with products without such extracts. *Porphyra* extracts possess a high concentration of bioactive substances, which contribute to the nourishment and refreshment of the skin. Consequently, these extracts hold significant value as supplementary components in skincare formulations.

Squalene, derived from *Thraustochytrium, Aurantiochytrium, and Schizochytrium*, taxa classified within the *Thraustochytriales* order, is a significant constituent within the field of cosmetic products. Squalene is widely preferred in the field of cosmetics owing to its outstanding characteristics, which include being non-toxic, non-irritating, and non-sensitizing. The substance serves a dual function, functioning both as an antistatic agent and an emollient in moisturizing creams. This results in the skin experiencing a sensation of softness, suppleness, and enough hydration (Pora et al. 2018). Exciting findings have been made by researchers concerning the potential of compounds derived from microalgae in the field of cosmetics. These compounds not only function as primary active ingredients but also serve as additives, fulfilling various roles such as stabilizers, dyes, and thickening agents (Patinier 2014). The adaptability of these organisms enables their integration into a wide range of personal care commodities, which includes facial lotions, ointments, shampoos and body cleansers. In addition to their primary functions, microalgae-based components possess further potential contributions that they can provide. The sterols found in organisms, such as plants and animals, play a crucial role in the formulation of moisturizing lotions. These sterols have the ability to effectively retain vital moisture, thereby promoting the overall well-being and appearance of the skin. Furthermore, the pigments found in microalgae, such as the keto carotenoid Astaxanthin and β-carotene, are of great importance in the skincare industry due to their remarkable antioxidant properties and potential advantages in enhancing a youthful complexion (Kaya et al. 2011).

In short, the cosmetics and skincare industries are still investigating the incredible potential of natural substances like *Porphyra* extracts, squalene, and compounds produced from microalgae. These compounds provide numerous benefits, from hydration and revitalization to antioxidant protection and general skin regeneration, in addition to helping to improve skin health. We may anticipate seeing even more cutting-edge and successful cosmetic products that harness the power of nature for healthier, more radiant skin as a result of advancement in this field.

7.7.5 Natural Radiation and Acquiring a Natural Glowing Skin: Potential of Algae in Skincare

Melanin, the pigment responsible for skin color, is essential for protecting the skin from UV light damage and halting the growth of skin cancer. However, an increased melanocyte production might result in hyperpigmentation or skin darkening. Additionally, as we age, the regulation, control, and distribution of melanocytes deteriorate, which causes dark and discolored areas to form on the skin. Tyrosinase is a pivotal enzyme involved in the process of melanin biosynthesis. The inhibition of tyrosinase is a viable approach for reducing hyperpigmentation. Compounds such as phlorotannin and 7-phloroeckol, which is derived from the marine brown seaweed Ecklonia cava, have proven the ability to effectively suppress the activity of the enzyme tyrosinase. These chemicals present a viable approach for reducing melanogenesis and serve as a potential candidate for treatments for skin lightening (Yoon et al. 2009). Prolonged exposure to sunlight induces increased melanin synthesis inside the epidermis, hence giving rise to the phenomenon of skin pigmentation known as tanning. The exposure to sunlight promotes the production of tyrosinase, an enzyme that aids biochemical events involved in the creation of melanosomes. The melanosomes undergo maturation, resulting in the production of melanin, which then undergoes differentiation into keratinocytes, so playing a role in the process of skin pigmentation. The process includes the hydroxylation of L-tyrosine, resulting in the production of 3, 4-dihydroxy-L-phenylalanine and L-DOPA. Then, oxidation occurs, leading to the generation of dopaquinone, which is then transformed into melanin. The overproduction of melanin is responsible for the coloring of the skin and requires effective regulation (Wang et al. 2011). It has been observed that natural pigments obtained from algae, namely fucoxanthin generated from brown algae such as *Laminaria japonica, Alaria, Chorda, and Macrocystis,* exhibit the ability to decrease tyrosinase activity and melanogenesis (Shimoda et al. 2010). Moreover, vitamins C and E has the ability to inhibit the skin's production of melanosomes. The protection of mammalian skin from UV radiation damage is commonly achieved by a well-established process that involves the synergistic combination of vitamins E and C. This combination operates through a mechanism that is based on nicotinamide adenine dinucleotide hydrate (NADH). *Pediastrum cruentum,* a micro-algae species recognized for its notable levels of vitamins E and C, has attracted attention as a prospective candidate in the development of cosmetic compositions aimed at melanoma prevention (Liu et al. 2015). The complex mechanisms governing the regulation and production of melanin in the skin highlight the need of implementing efficient approaches to address pigmentation concerns and protect against the detrimental impacts of ultraviolet radiation. The utilization of natural compounds derived from marine sources, in conjunction with the incorporation of vitamins C and E, presents a prospective direction for the advancement of cosmetic products that focus on enhancing skin uniformity and safeguarding against melanoma.

7.7.6 Algae: Nature-Enriched Solution for Hair Nutrition and Hair Growth

Alopecia or hair loss can be a severe dermatological and psychological issue as full head hair is an important quality of youth personality for many individual. Extracts of *Chlorococcum* species, *Thalassiosira* species, *Chaetoceros* species, *and Monodus* species have been proposed for hair loss prevention formulations because they can modulate melanogenesis in hair and human skin, improving and stimulating keratinocyte differentiation, melanocyte proliferation, and growth of human hair and hair follicles (Olsen and Iorizzo 2019). The growing role of nutrition and other natural bioactive components derived from algae are increasingly being used to treat alopecia (Almohanna et al. 2019). Oral treatment with the mixture of two algal species extracts, *L. japonica and Cistanche tubulosa,* has enhanced hair growth alopecia (Shin et al. 2015). 7-phloroeckol is a natural bioactive compound extracted from brown algae *E. cava* has good effect on hair growth (Bak et al. 2014). The biomass and bio-oil from *P. tricornutum* and C. *protothecoides* were used to introduce a new cosmetic product with long-term applications for skin and hair care (Balboa et al. 2015). Species of the microalgae Chlorella genus, which are predominantly composed of intact micro-algal cells and contain oil by dry weight, have been shown to soften and stretch both skin and hair (Ariede et al. 2017).

7.7.7 Algae as an Anticancer Agent: A New Horizon in Oncology

Cancer is the second most leading cause of death, i.e., in 2020 accounting for approximately ten million deaths (Sung et al. 2021). Oncogene mutations with predominant gain of function over genes that suppress tumors and a tumor suppressor gene dysfunction are thought to be the primary causes of cancer development (Hanahan and Weinberg 2000). Natural and synthetic medications both have advantages, but natural products are more expensive. Novel anticancer chemicals are being sought not just from microbes and terrestrial plants, but also from marine species (Dehelean et al. 2021). Carotenoids, fatty acids, alkaloids, flavonoids, phenolics, and other pharmaceutically significant bioactive substances are abundant in marine microalgae. The microalgae advantages of utilizing in drug development including their very short generation time, metabolic versatility, arable land incompetents, ability to grow in any season, and low special nutritional requirements (Martínez Andrade et al. 2018). Algal species include a variety of compounds that help to prevent cancer. Cyanobacterial species include *Microchaete, laminosus, Mastigocladus tenera, and Anabaena doliolum*. This cyanobacterial Rutin outperformed BHT and tocopherol in terms of radical scavenging activities (Singh et al. 2017). Rutin destroys cancer cells by increasing the level of ROS (reactive oxygen species) in the cells that cause oxidative stress. It has the ability to block the signaling pathways either PI3K/Akt or Ras/Raf/MAPK resulting in cell cycle arrest and, eventually, apoptosis mechanism. Apoptosis can also cause by Rutin's upregulation of Bax and downregulation of MMP-2 and Bcl-2 via activation of caspase-3. Rutin substantially reduces nephrotoxicity and doxorubicin-induced neurotoxicity in an animal model (Imani et al. 2021). Flavonol compound includes Kaempferol (3, 5, 7-trihydroxy-2-(4-hydroxyphenyl)-4H-1-benzopyran-4-one) that has been identified in the marine microalgae *Nannochloris* species (Haoujar et al. 2019). Apoptosis in ovarian cancer cells is caused by kaempferol by triggering the death receptors/FADD/Caspase-8 pathway or by increasing the apoptotic protein production, i.e., caspase-(3, -8, -9), and Bax. In G2/M phase cell cycle progression is suppressed in these cancer cells by Chk2/p21/Cdc2 and Chk2/Cdc25C/Cdc2 pathway activation (Gao et al. 2018). *Chaetomorpha* species have polyphenolic chemicals abundantly. The hydrophilic polyphenolic chemicals presence, i.e., phlorotannins, which are bipolar in nature, may act as a key antioxidant that helps out algae in oxidative stress resistance (Senthilkumar and Sudha 2012). There is a crucial link between oxidative stress and cancer prevalence. Several in vivo and in vitro investigations have demonstrated that administering exogenous antioxidants may reduce the generation of free radicals and damage to DNA and proteins, lowering the probability of developing cancer (Bonner and Arbiser 2014). Another phytochemical found in various algae species, quercetin, has high anticancer behavior. Quercetin regulates the cell cycle by increasing the levels of p53 and p21, enhancing the expression of GADD45 and Bax, while inhibiting the activity of cyclin A, cyclin B, and CDK2 inhibited (Tang et al. 2020). For nutritional supplement in medicine, Nostoc biomass frequently used due to its high fatty acid, protein, and vitamin content, has been used for medicine and as a nutritional supplement. This microalga's medical effectiveness has been illustrated by its use in the treatment of different kinds of cancer and fistulas (Temina et al. 2007).

7.7.8 Microalgae-Derived Bioactive Compounds for Anti-inflammatory Use

Inflammation is part of the immune system's natural physiological response to tissue damage that occurs as the human body attempts to deal with various pathogens, harmful chemicals, and cell injury (De Morais et al. 2015). Anti-inflammatory properties have been demonstrated by several microalgae metabolites from cyanobacteria, i.e., *Dunaliella, Chlorella, Phaeodactylum* and more (Guzman et al. 2003). Chlorophyll, carotenoids, xanthophylls, vitamins, terpenoids, saturated as well as polyunsaturated fatty acids, amino acids, alkaloids, polysaccharides, polyphenols, and acetogenins, are among the substances they produce (De Almeida et al. 2011). The abundant production of cytokines like IL (interleukin)-1beta, IL-6, and TNF (tumor necrosis factor)-alpha and PGE2 (prostaglandin E2), NO (Nitric oxide), ROS (reactive oxygen species) is all prevalent during inflammatory responses (Dray 1995). Many uses of marine algae, i.e., *Sargassum thunbergii and Sargassum fulvellum* to cure swelling, dropsy, and painful scrotum, have been documented (Kang et al. 2008). Several anti-inflammatory marine algae have been shown to modulate and target the inflammatory response and via inhibition of inflammation-mediating enzymes, i.e., inducible cyclooxygenase (COX)-2, phospholipases A2 (PLA2s) and nitric oxide synthase (iNOS), nuclear factor kB inactivation, and ROS scavenging.

7.7.8.1 Mechanisms of Anti-inflammatory Activity by Inhibition of Inflammation-Mediating Enzymes

Anti-inflammatory substance derived from algae, microalgae, and cyanobacteria works by one or more of these mechanisms. In this regard, chemokine modulation via cyclooxygenase-2 activity and expression, as well as reduced expression and activity of cell contact receptors, pro-inflammatory cytokines, and radical scavenging metabolites, is significant. These enzymes, COX-2 (cyclooxygenase) and iNOS (inducible nitric oxide synthase) for the regulation of inflammatory disorders, are thought to be important therapeutic targets. For this regard, marine algae have been identified as a possible option for modulating the generation of inflammatory mediators via the inactivation of PLA2s and iNOS (inducible nitric oxide synthase) COX-2 activity. Indeed, *Sargassum micracanthum*, the brown algae both Petalonia binghamiae and Dictyota dichotoma inhibited PGE2 and NO production in RAW 264.7 macrophages by reducing LPS-induced protein and mRNA expression of COX-2 and iNOS. Similarly, Ecklonia cava treatment declined inducible nitric oxide synthase and cyclooxygenase expression in LPS-stimulated BV2 microglia (Jung et al. 2009) (Fig. 7.8).

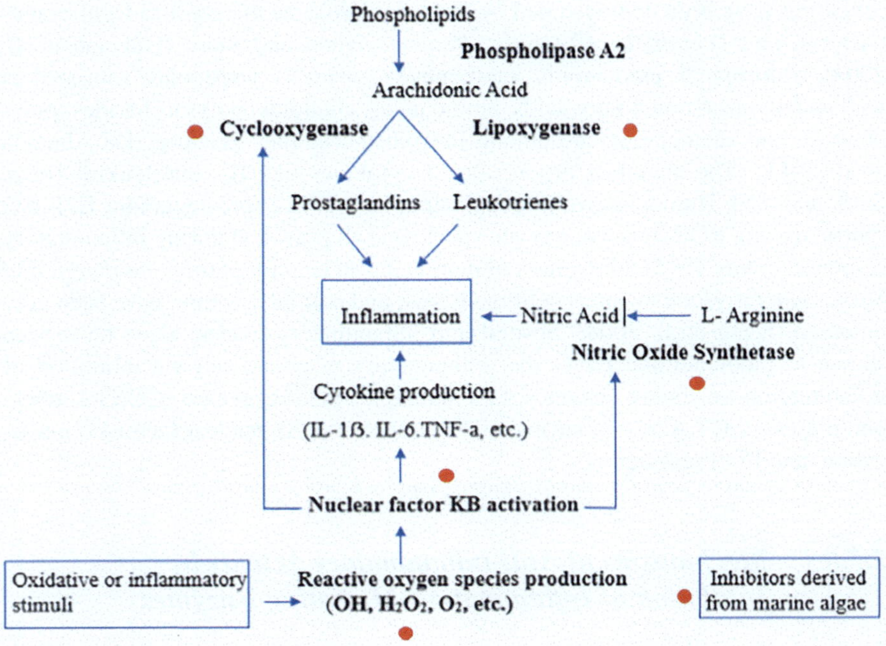

Fig. 7.8 Different metabolites from algae with anti-inflammatory activity (anti-inflammatory activity of various chemical compounds observed in microalgae and cyanobacteria has been identified)

7.7.9 Algae in Allergy Management: Potential of Nature to Combat Allergy

Marine algae have been recommended as a potential source of anti-allergic therapy. Some algae species have also shown enormous potential as cosmetics, pharmaceuticals, and medication adjuvants, with possibilities for the treatment of cough, fever, allergies, dermatitis, and other diseases (Anis et al. 2017). The largest number of anti-allergic natural compounds isolated from marine algae is polyphenols. Three chemicals isolated from Ecklonia Cava act as anti-allergic agents by reducing histamine release and disrupting with the binding of IgE and FcRI. They are also involved in cytokine production and block mast cell activation and, offering all-around anti-allergic treatment (Vo et al. 2014). Green algae have also been reported to contain anti-allergic chemicals. Compounds derived from *Eisenia arborea* have anti-allergic activity when given orally. Phlorotannins are phenolic chemicals found in brown algae such as *Eisenia* species and *Ecklonia* species that have a discrete chemical structure than terrestrial products. Phlorotannins reduce allergic inflammation, atopic dermatitis, the PCA reaction, and DMH-induced inflammation by suppressing the AA cascade, regulating the MAPK/NF-B signaling pathway, and modulating the immune system. As a result, phlorotannins could minimize allergic reactions of both acute type I and chronic type IV (Le et al. 2009) (Fig. 7.9).

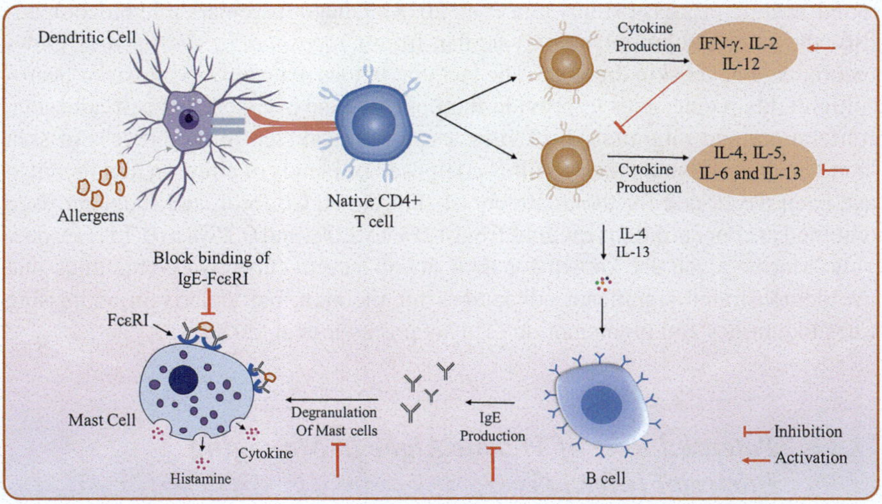

Fig. 7.9 Natural anti-allergic compounds isolated from marine algae

7.7.10 Algae: A Natural Stimulator for the Skin Regeneration

Restoration of the anatomical structure and function of the damaged skin occurs as a result of a complex series of overlapping phases known as wound repair (also known as healing) (Dhivya et al. 2015). This sophisticated biological process, which also includes tissue remodeling, collagen production, epithelialization, and other forms of cell rejuvenation, requires exact coordination (Aderibigbe and Buyana 2018). Wound healing can be divided into five overlapping phases due to its dynamic nature: hemostasis, inflammation, migration, proliferation, and maturation. Throughout history, humans have relied on medicinal plants for wound treatment due to their perceived safety, efficacy, affordability, and patient tolerance. Despite advances in pharmaceuticals, wound healing drugs remain limited in availability, affordability, and side effects. Marine algae, a traditional and functional food source, are now being explored for their potential to yield new therapeutic and cosmeceutical compounds, sparking research interest (Namvar et al. 2013).

Certain microalgae-derived substances and extracts have the potential to accelerate the renewal and healing of injured skin while also providing cosmetic advantages (Nowruzi et al. 2020). Alginate-based films can speed up the healing of wounds when mixed with other polymers because they enable the exchange of oxygen, carbon dioxide, and water vapor while protecting the wound from bacterial infections. This protective effect is especially noticeable when combined with other antibacterial agents. Sacran, a polysaccharide produced by the cyanobacterium *Aphanothece sacrum*, has been studied for its potential to treat a variety of ailments, including atopic dermatitis, skin wounds, hemorrhoids, and problems with the corneal epithelium. Additionally, the use of Sacran enhances the wound dressings Anti-inflammatory capabilities, enabling their release at an appropriate rate to assist the wound-healing process (Motoyama et al. 2018). Due to its remarkable biocompatibility, the extracellular polymer Cyanoflan from *Crocosphaera Chwakensis* shows potential as a novel skin dressing. The fact that it does not cause apoptosis or necrosis highlights its capacity to provide the optimum environment for cell adhesion, proliferation, and migration in in vitro tests, a critical step in the process of skin wound healing. Wound dressing films composed of blends of chitosan and fucoidan have been developed for the treatment of skin burns. Chitosan and fucoidan were combined at concentrations ranging from 1.0% to 2.0% and 0.25% to 0.75% respectively. Marine algae are known for their abundance of bioactive compounds that have demonstrated significant advantages for the skin, particularly in addressing issues like rashes and pigmentation (Thiyagarasaiyar et al. 2020).

7.7.11 Potential Role of Marine Algae in Managing Bacterial Infections

The cosmetic industry faces a significant challenge in maintaining microbiological purity, which can lead to infection risk and alterations in cosmetic properties. Synthetic preservatives have traditionally been used to counter this issue but may cause allergies. Consequently, there is a growing demand for natural preservatives as alternatives to synthetic ones (Senevirathne et al. 2013). Algae extracts present a viable alternative for cosmetic preservatives since they are highly effective against microbiological, environmental, and individual contaminants. Notably, phenolic substances prolong the shelf life of cosmetics by suppressing microbial growth. Phlorotannin, one of the bioactive substances obtained from algae, contributes to the antibacterial activity of the organism by binding to bacterial proteins, including enzymes and cell membranes, and inhibiting oxidative phosphorylation, which ultimately leads to cell lysis. Algal polysaccharides with glycoprotein-receptors bind to bacterial surfaces, disrupt cytoplasmic membranes, inhibit energy transfer, and ultimately lead to bacterial cell lysis (Shannon and Abu-Ghannam 2016). Bioactive substance laurinterol, which is present in *Laurencia pacifica*, shows antibacterial activities that are beneficial against infections caused on by Staphylococcus aureus, *Synechocystis species*, *Pterocladia capillacea*, *Himanthalia elongata* and *Escherichia coli* (Pimentel et al. 2017).

7.8 Algae for a Sustainable Future: Leading Biodegradation Solutions

Marine algae-derived green cosmetics mark a change in the beauty industry toward more natural and sustainable alternatives. Cosmetics makers can lessen their dependency on artificial and petroleum-based chemicals while simultaneously improving the efficacy of cosmetics by incorporating marine algae into their products. This change encourages biodiversity, strengthens the link between consumers and the environment, and promotes sustainability goals. Future of green cosmetics depends on using marine algae's untapped potential to develop novel and environmentally friendly products that meet the expanding demand in the beauty industry (Mago et al. 2023).

These algae-based components have made green cosmetics a well-liked option for customers looking for sustainable and environmentally friendly skincare products. The quick growth rates, renewability, and low freshwater and land requirements of marine algae make them sustainable resources. Their cultivation supports sustainable development along with responsible consumption by lessening the impact on the environment. Marine algae absorb CO_2 during growth, making them carbon-neutral, environmental-friendly, and safe consumable (Kadam et al. 2015). Numerous algal-based cosmetic ingredients, including pigments like astaxanthin, polysaccharides (like carrageenan or alginate), and algal oils, are naturally biodegradable. The goal of sustainability, on the other hand, is to meet present demands while maintaining ecological balance and resource efficiency for future generations. Algae are a highly sustainable raw resource due to their low energy, freshwater, and land requirements. Additionally, they absorb CO_2 during growth, making production carbon-neutral or negative. Seaweed-derived biofilms and other algae-based packaging materials improve the biodegradability of cosmetics and minimize waste in the environment. These concepts work together to promote a circular economy, minimize ecological impact, and support marine conservation through methods such as aquaculture. Algal cosmetics are a model for ecologically innovation that aligns with global sustainability goals (Arora et al. 2021).

7.8.1 Algae Innovation Advancing Sustainable Development Goals

Green cosmetics must be effective and environmentally safe at all stages of development, and environmental impact will be negligible. Ecologically safe cosmetics are therefore more suitable and align with sustainable development goals when compared with conventional cosmetics, thus supporting green strategies for their production (Hitce et al. 2018).

The 17 Sustainable Development Goals (SDGs) described by the UN are intended to promote social, environmental, and economic sustainability as well as global well-being. In particular, SDGs 3, 6, 7, 8, 9, 12, 14, and 13 address reducing the effects of waste management, water pollution, and climate change. These goals are crucial to the growth of an eco-friendly and sustainable cosmetics industry. Microalgae, bacterial cellulose, essential oils, and natural preservatives are examples of natural raw materials that can be used to reduce the risk of employing harsh chemicals while still achieving SDGs 3, 6, 12, 14, and 15. Both SDG 3 and the cosmetics industry prioritize human health and well-being, with an emphasis on personal care, body maintenance, and cleanliness (Chen et al. 2020). Hazardous ingredients that might irritate skin, trigger allergies, or cause long-term health problems, such as parabens, sulfates, and artificial perfumes, are forbidden in green cosmetics. These products promote safer lifestyles by lowering health risks for both consumers and production workers using natural and organic components. Additionally, green cosmetics generally promote ingredient sources and labeling, allowing customers to make well-informed decisions for their health. Furthermore, harmful chemicals and microplastics that can contaminate water sources and damage ecosystems are reduced by sustainable practices in green cosmetics, which contributes to public health (Cubas et al. 2022). Green cosmetics support a comprehensive approach to health, prioritize personal safety, and safeguard the environment for long-term well-being through these actions, which mainly aligns with SDG 3. Microalgae contribute to SDG 14 by maintaining biodiversity and ecological health and serving as a foundation for marine food chains. Their role in preserving aquatic environments, improving soil health, and developing farming practices emphasizes their importance in global efforts to achieve sustainability thus making an essential resource for future environmental and economic resilience. The green cosmetics industry strictly aligns with SDG 13 by significantly reducing carbon emissions, developing climate-resilient systems, and facilitating global efforts to address the climate issue. The use of biodegradable packaging and refillable containers reduces plastic waste, lowering the carbon footprint production and disposal (Sasounian et al. 2024).

7.9 Conclusion

In conclusion, cosmetics derived from marine sources, particularly the use of algae in cosmeceuticals, represent a promising frontier in the beauty and skincare industry. The attributes of algae, including their rich nutrient content, antioxidant properties, and potential for sustainable cultivation, make them valuable ingredients for innovative skincare products. As consumers increasingly seek eco-friendly and effective beauty solutions, the incorporation of marine-derived ingredients into cosmetics offers both natural and scientific advantages.

Algae-based cosmeceuticals have demonstrated remarkable potential in addressing various skin concerns, from hydration and anti-aging to protection against environmental stressors. The pigments produced by photosynthetic organisms are the primary elements of algal products. Polysaccharides, MAAs, proteins, and other algal metabolites have a wide range of activities and applications. They benefit skin health by acting as anti-aging, antioxidant, anti-inflammatory, wrinkle-reducing, and collagen-boosting agents. Furthermore, the sustainable cultivation and harvesting of algae align with growing environmental consciousness, providing an ethical choice for consumers.

7.10 Future Directions

The utilization of marine sources and the harnessing of cosmeceutical attributes of algae have opened up a world of possibilities in the cosmetics industry. The future holds tremendous potential for these eco-friendly and sustainable alternatives. Also, algae are declared natural sinks of carbon since they grab a lot of carbon from the environment during photosynthesis, hence reducing the greenhouse gases emanating from climate change. Algae-based production is sustainable and also helps reduce greenhouse gas emissions which makes it ideal for use. As consumers become more environmentally conscious, the appeal of marine-based cosmetics is set to grow, offering a guilt-free path to beauty enhancement.

Furthermore, the versatility of algae, with its abundance of beneficial compounds such as vitamins, minerals, and antioxidants, ensures its continued relevance in skincare. With ongoing research and development, we can anticipate the formulation of even more potent and specialized products that target specific skin concerns, ranging from hydration and anti-aging to protection against environmental stressors.

The potential of marine algae as a source of bioactive compounds in the cosmetics industry can be further enhanced by the use of advanced genetic engineering technologies, such as CRISPR/Cas, to improve algal strains for higher production of bioactive compounds. Polysaccharides, pigments, antioxidants, and essential fatty acids are key compounds in skincare formulations, due to their anti-aging, moisturizing, and protective properties. CRISPR/Cas technology facilitates precise modifications in the algal genome, permitting scientists to target genes responsible for the synthesis of desirable bioactive compound. For example, upregulating genes involved in the biosynthesis of carotenoids (well known for their anti-inflammatory and UV protective effects) can increase yield. Similarly, CRISPR/Cas can be used to knock out genes that produce undesirable by products and thus improve purity and efficacy of the extracted ingredients. CRISPR/Cas in addition to improving yields of known bioactive compounds can be used to introduce new traits, such as resistance to environmental stressors like salinity or temperature changes. These improvements can result in more robust algal strains that will be able to survive in myriad cultivation conditions, therefore lowering production costs and ensuring a continual supply of high-quality raw material for use in cosmetic formulations. Using CRISPR/Cas technology, the cosmetics industry can not only respond to the growing consumer demand for sustainable and effective natural ingredients, but also provide the foundation for the advancement of the development of new, innovative skincare solutions. This is in line with the industry's direction toward science-based, environmentally friendly products, and marine algae are a cornerstone of the future of cosmetic biotechnology.

Moreover, the synergy between science and nature in cosmeceuticals attributes of algae is poised to redefine beauty routines. These innovative products not only hold the promise of delivering healthier, more radiant skin but also align with the principles of sustainability and ethical consumption. As we venture into this exciting era of skincare, it is essential to maintain rigorous quality control and safety standards to ensure that consumers can fully embrace the benefits of cosmetics from marine sources and algae-based cosmeceuticals. Ultimately, these advancements are poised to transform the beauty landscape, offering consumers effective and eco-conscious choices for their skincare needs.

To ensure the sustainability of "Going Green," future cosmetics must address the following issues: the rate at which natural products are produced compared to synthetic ones; the availability of natural products due to their high cost; the shelf life of natural ingredients; the environmental impact of competing with agricultural materials; and international certification of "Green Manufactures."

References

Aderibigbe BA, Buyana B (2018) Alginate in wound dressings. Pharmaceutics 10(2):1–13

Agatonovic-Kustrin S, Morton DW (2013) Cosmeceuticals derived from bioactive substances found in marine algae. Oceanography 1(2):106

Ahsan H (2019) The significance of complex polysaccharides in personal care formulations. J Carbohydr Chem 38:213

Almohanna HM, Ahmed AA, Tsatalis JP, Tosti A (2019) The role of vitamins and minerals in hair loss: a review. Dermatol Ther 9(1):51–70

Anis M, Ahmed S, Hasan MM (2017) Algae as nutrition, medicine and cosmetic: the forgotten history, present status and future trends. World J Pharm Pharm Sci 6(6):1934–1959

Arad SM, Yaron A (1992) Natural pigments from red microalgae for use in foods and cosmetics. Trends Food Sci Technol 3:92–97

Aranaz I, Acosta N, Civera C, Elorza B, Mingo J, Castro C et al (2018) Cosmetics and cosmeceutical applications of chitin, chitosan, and their derivatives. Polymers 10(2):213

Ariede MB, Candido TM, Jacome ALM, Velasco MVR, de Carvalho JCM, Baby AR (2017) Cosmetic attributes of algae—a review. Algal Res 25:483–487

Arora K, Kumar P, Bose D, Li X, Kulshrestha S (2021) Potential applications of algae in biochemical and bioenergy sector. 3 Biotech 11(6):296

Askarova S, Yang X, Sheng W, Sun GY, Lee JM (2011) Role of Aβ-receptor for advanced glycation end products interaction in oxidative stress and cytosolic phospholipase A2 activation in astrocytes and cerebral endothelial cells. Neuroscience 199:375–385

Bak SS, Sung YK, Kim SK (2014) 7-Phloroeckol promotes hair growth on human follicles in vitro. Naunyn-Schmiedeberg Arch Pharmacol 387:789–793

Balboa EM, Conde E, Soto ML, Pérez-Armada L, Domínguez H (2015) Cosmetics from marine sources. In: Springer handbook of marine biotechnology, pp 1015–1042

Bedoux G, Hardouin K, Burlot AS, Bourgougnon N (2014) Bioactive components from seaweeds: cosmetic applications and future development. Adv Bot Res 71:345–378

Berthon JY, Nachat-Kappes R, Bey M, Cadoret JP, Renimel I, Filaire E (2017) Marine algae as attractive source to skin care. Free Radic Res 51(6):555–567

Bhagavathy S, Sumathi P, Bell IJS (2011) Green algae *Chlorococcum humicola*—a new source of bioactive compounds with antimicrobial activity. Asian Pac J Trop Biomed 1(1):S1–S7

Bonner MY, Arbiser JL (2014) The antioxidant paradox: what are antioxidants and how should they be used in a therapeutic context for cancer. Future Med Chem 6(12):1413–1422

Bonté F (2011) Skin moisturization mechanisms. Ann Pharm Fr 69(3):135–141

Chen TL, Kim H, Pan SY, Tseng PC, Lin YP, Chiang PC (2020) Implementation of green chemistry principles in circular economy system towards sustainable development goals: challenges and perspectives. Sci Total Environ 716:136998

Corinaldesi C, Barone G, Marcellini F, Dell'Anno A, Danovaro R (2017) Marine microbial-derived molecules and their potential use in cosmeceutical and cosmetic products. Mar Drugs 15(4):118

Cornish ML, Garbary DJ (2010) Antioxidants from macroalgae: potential applications in human health and nutrition. Algae 25(4):155–171

Cubas ALV, Bianchet RT, Reis IMASD, Gouveia IC (2022) Plastics and microplastic in the cosmetic industry: aggregating sustainable actions aimed at alignment and interaction with UN sustainable development goals. Polymers 14(21):4576

De Almeida CLF, Falcão HDS, Lima GRDM, Montenegro CDA, Lira NS, de Athayde-Filho PF, Batista LM (2011) Bioactivities from marine algae of the genus *Gracilaria*. Int J Mol Sci 12(7):4550–4573

De Canha MN, Steyn A, Blom van Staden A, Fibrich BD, Lambrechts IA, Denga LL, and Lall N (2020) Book review: herbal principles in cosmetics: properties and mechanisms of action. 460

De Jesus Raposo MF, De Morais AMB, De Morais RMSC (2015) Marine polysaccharides from algae with potential biomedical applications. Mar Drugs 13(5):2967–3028

De Luca M, Pappalardo I, Limongi AR, Viviano E, Radice RP, Todisco S, Vassallo A (2021) Lipids from microalgae for cosmetic applications. Cosmetics 8(2):52

de Morais MG, Vaz BDS, de Morais EG, Costa JAV (2015) Biologically active metabolites synthesized by microalgae. Biomed Res Int 2015(1):835761

Dehelean CA, Marcovici I, Soica C, Mioc M, Coricovac D, Iurciuc S, Pinzaru I (2021) Plant-derived anticancer compounds as new perspectives in drug discovery and alternative therapy. Molecules 26(4):1109

Del-Campo JA, Garcia-Gonzalez M, Guerrero MG (2007) Outdoor cultivation of microalgae for carotenoid production: current state and perspectives. Appl Microbiol Biotechnol 74:1163–1174

Dhivya S, Padma VV, Santhini E (2015) Wound dressings–a review. Biomedicine 5(4):22

Dini I, Laneri S (2021) The new challenge of green cosmetics: natural food ingredients for cosmetic formulations. Molecules 26(13):3921

Dray A (1995) Inflammatory mediators of pain. Br J Anaesth 75(2):125–131

El-Baky HHA, El Baz FK, El-Baroty GS (2009) Production of phenolic compounds from *Spirulina maxima* microalgae and its protective effects. Afr J Biotechnol 8(24):7059–7067

El-Chaghaby GA, Rashad S (2021) An overview of algae prospects in cosmeceuticals. J Egypt Women's Dermatol Soc 18(3):161

Falkowski PG, Katz ME, Knoll AH, Quigg A, Raven JA, Schofield O, Taylor FJR (2004) The evolution of modern eukaryotic phytoplankton. Science 305:354–360

Fehling J, Stoecker DK, Baldauf SL (2007) Photosynthesis and the eukaryote tree of life. In: Evolution of primary producers in the sea, pp 75–107

Fonseca S, Amaral MN, Reis CP, Custódio L (2023) Marine natural products as innovative cosmetic ingredients. Mar Drugs 21(3):170

Franca CCV, Ueno HM (2020) Green cosmetics: perspectives and challenges in the context of green chemistry. Desenvolvimento e Meio Ambiente 53:133–150

Ganceviciene R, Liakou AI, Theodoridis A, Makrantonaki E, Zouboulis CC (2012) Skin anti-aging strategies. Dermatoendocrinology 4(3):308–319

Gao Y, Yin J, Rankin GO, Chen YC (2018) Kaempferol induces G2/M cell cycle arrest via checkpoint kinase 2 and promotes apoptosis via death receptors in human ovarian carcinoma A2780/CP70 cells. Molecules 23(5):1095

Guillerme JB, Couteau C, Coiffard L (2017) Applications for marine resources in cosmetics. Cosmetics 4(3):35

Guzman S, Gato A, Lamela M, Freire-Garabal M, Calleja J (2003) Anti-inflammatory and immunomodulatory activities of polysaccharide from *Chlorella stigmatophora* and *Phaeodactylum tricornutum*. Phytother Res 17(6):665–670

Hagino H, Saito M (2010) Use of algal proteins in cosmetics. European Patent EP1433463B1. p 18

Hallmann A (2015) Algae biotechnology—green cell factories on the rise. Curr Biotechnol 4(4):389–415

Hanahan D, Weinberg RA (2000) The hallmarks of cancer. Cell 100(1):57–70
Haoujar I, Cacciola F, Abrini J, Mangraviti D, Giuffrida D, Oulad El Majdoub Y, Skali Senhaji N (2019) The contribution of carotenoids, phenolic compounds, and flavonoids to the antioxidative properties of marine microalgae isolated from Mediterranean Morocco. Molecules 24(22):4037
Hentati F, Tounsi L, Djomdi D, Pierre G, Delattre C, Ursu AV (2020) Bioactive polysaccharides from seaweeds. Molecules 25(14):3152
Heo SJ, Ko SC, Cha SH, Kang DH, Park HS, Choi YU, Jeon YJ (2009) Effect of phlorotannins isolated from *Ecklonia cava* on melanogenesis and their protective effect against photo-oxidative stress induced by UV-B radiation. Toxicol In Vitro 23(6):1123–1130
Hitce J, Xu J, Brossat M, Frantz MC, Dublanchet AC, Philippe M, Dalko-Csiba M (2018) UN sustainable development goals: how can sustainable/green chemistry contribute ? Green chemistry as a source of sustainable innovations in the cosmetic industry. Curr Opin Green Sustain Chem 13:164–169
Huang TH, Wang PW, Yang SC, Chou WL, Fang JY (2018) Cosmetic and therapeutic applications of fish oil's fatty acids on the skin. Mar Drugs 16(8):256
Ibañez E, Herrero M, Mendiola JA, Castro-Puyana M (2011) Extraction and characterization of bioactive compounds with health benefits from marine resources: macro and microalgae, cyanobacteria, and invertebrates. In: Marine bioactive compounds: sources, characterization, and applications. Springer, pp 55–98
Imani A, Maleki N, Bohlouli S, Kouhsoltani M, Sharifi S, Maleki Dizaj S (2021) Molecular mechanisms of anticancer effect of rutin. Phytother Res 35(5):2500–2513
Jain R, Raghukumar S, Tharanathan R, Bhosle NB (2005) Extracellular polysaccharide production by thraustochytrid protists. Mar Biotechnol 7:184–192
Javanmardi J, Kubota C (2006) Variation of lycopene, antioxidant activity, total soluble solids and weight loss of tomato during postharvest storage. Postharvest Biol Technol 41(2):151–155
Jesumani V, Du H, Pei P, Zheng C, Cheong KL, Huang N (2019) Unravelling properties of polysaccharides from *Sargassum* sp. as an anti-wrinkle and skin-whitening property. Int J Biol Macromol 140:216–224
Jung WK, Ahn YW, Lee SH, Choi YH, Kim SK, Yea SS, Choi IW (2009) Ecklonia cava ethanolic extracts inhibit lipopolysaccharide-induced cyclooxygenase-2 and inducible nitric oxide synthase expression in BV2 microglia via the MAP kinase and NF-κB pathways. Food Chem Toxicol 47(2):410–417
Kadam SU, Tiwari BK, O'Donnell CP (2015) Extraction, structure, and biofunctional activities of laminarin from brown algae. Int J Food Sci Technol 50(1):24–31
Kalasariya HS, Pereira L (2022) Dermo-cosmetic benefits of marine macroalgae-derived phenolic compounds. Appl Sci 12(23):11954
Kaya K, Nakazawa A, Matsuura H, Honda D, Inouye I, Watanabe MM (2011) Thraustochytrid *Aurantiochytrium* sp. 18W-13a accumulates high amounts of squalene. Biosci Biotechnol Biochem 75(11):2246–2248
Kim SK, Ravichandran YD, Khan SB, Kim YT (2008) Prospective of the cosmeceuticals derived from marine organisms. Biotechnol Bioprocess Eng 13:511–523
Kim YI, Oh WS, Song PH, Yun S, Kwon YS, Lee YJ et al (2018) Anti-photoaging effects of low molecular-weight fucoidan on ultraviolet B-irradiated mice. Mar Drugs 16(8):286
Kligman D (2000) Cosmeceuticals. Dermatol Clin 18(4):609–615
Le QT, Li Y, Qian ZJ, Kim MM, Kim SK (2009) Inhibitory effects of polyphenols isolated from marine alga *Ecklonia cava* on histamine release. Process Biochem 44(2):168–176
Li YX, Wijesekara I, Li Y, Kim SK (2011) Phlorotannins as bioactive agents from brown algae. Process Biochem 46(12):2219–2224
Liu Q, Wang Y, Cao M, Pan T, Yang Y, Mao H, Liu G (2015) Anti-allergic activity of R-phycocyanin from *Porphyra haitanensis* in antigen-sensitized mice and mast cells. Int Immunopharmacol 25(2):465–473
Lordan S, Ross RP, Stanton C (2011) Marine bioactives as functional food ingredients: potential to reduce the incidence of chronic diseases. Mar Drugs 9(6):1056–1100

Lourenço-Lopes C, Fraga-Corral M, Jimenez-Lopez C, Pereira AG, Garcia-Oliveira P, Carpena M, Simal-Gandara J (2020) Metabolites from macroalgae and its applications in the cosmetic industry: a circular economy approach. Resources 9(9):101

Mabeau S, Fleurence J (1993) Seaweed in food products: biochemical and nutritional aspects. Trends Food Sci Technol 4(4):103–107

Mago Y, Sharma Y, Thakran Y, Mishra A, Tewari S, Kataria N (2023) Next-generation organic beauty products obtained from algal secondary metabolites: a sustainable development in cosmeceutical industries. Mol Biotechnol 1–21

Marino T, Iovine A, Casella P, Martino M, Chianese S, Larocca V, Molino A (2020) From *Haematococcus pluvialis* microalgae a powerful antioxidant for cosmetic applications. Chem Eng 79:271–276

Martínez Andrade KA, Lauritano C, Romano G, Ianora A (2018) Marine microalgae with anticancer properties. Mar Drugs 16(5):165

Massana R, Terrado R, Forn I, Lovejoy C, Pedros-Alio C (2006) Distribution and abundance of uncultured heterotrophic flagellates in the world oceans. Environ Microbiol 8:1515–1522

Mata TM, Martins AA, Caetano NS (2010) Microalgae for biodiesel production and other applications: a review. Renew Sust Energ Rev 14(1):217–232

Matsuno T (2001) Aquatic animal carotenoids. Fish Sci 67(5):771–783

Michniak-Kohn B, Leonardi GR (2017) An overview about oxidation in clinical practice of skin aging. An Bras Dermatol 92:367–374

Mohiuddin AK (2019) Skin aging & modern age anti-aging strategies. Int J Clin Dermatol Res 7:209–240

Morocho-Jácome AL, Ruscinc N, Martinez RM, de Carvalho JCM, Santos de Almeida T, Rosado C, Baby AR (2020) (Bio)Technological aspects of microalgae pigments for cosmetics. Appl Microbiol Biotechnol 104:9513–9522

Motoyama K, Higashi T, Okajima MK, Kaneko T, Arima H (2018) Potential uses of sacran hydrogels as wound dressing material. J Pharm Soc Jpn 138(4):517–520

Namvar F, Mohamad R, Baharara J, Zafar-Balanejad S, Fargahi F, Rahman HS (2013) Antioxidant, antiproliferative, and antiangiogenesis effects of polyphenol-rich seaweed (*Sargassum muticum*). Biomed Res Int 2013:1–9

Nizard C, Friguet B, Moreau M, Bulteau AL, Saunois A (2007) U.S. Patent No. 7,220,417. U.S. Patent and Trademark Office, Washington, DC

Nowruzi B, Sarvari G, Blanco S (2020) The cosmetic application of cyanobacterial secondary metabolites. Algal Res 49:101959

Núñez-Pons L, Shilling A, Verde C, Baker BJ, Giordano D (2020) Marine terpenoids from polar latitudes and their potential applications in biotechnology. Mar Drugs 18(8):401

Olsen EA, Iorizzo M (2019) Hair disorders. In: Harper's textbook of pediatric dermatology, pp 2103–2138

Oren A, Gunde-Cimerman N (2007) Mycosporines and mycosporine-like amino acids: UV protectants or multipurpose secondary metabolites? FEMS Microbiol Lett 269(1):1–10

Pacheco-Quito EM, Ruiz-Caro R, Veiga MD (2020) Carrageenan: drug delivery systems and other biomedical applications. Mar Drugs 18(11):583

Pagels F, Almeida C, Vasconcelos V, Guedes AC (2022) Cosmetic potential of pigment extracts from the marine cyanobacterium *Cyanobium* sp. Mar Drugs 20(8):481

Pallela R, Na-Young Y, Kim SK (2010) Anti-photoaging and photoprotective compounds derived from marine organisms. Mar Drugs 8(4):1189–1202

Pangestuti R, Siahaan EA, Kim SK (2018) Photoprotective substances derived from marine algae. Mar Drugs 16(11):399

Pereira L (2018) Seaweeds as source of bioactive substances and skin care therapy—cosmeceuticals, algotherapy, and thalassotherapy. Cosmetics 5(4):68

Pereira L, Morrison L, Shukla PS, Critchley AT (2020) A concise review of the brown macroalga *Ascophyllum nodosum* (Linnaeus) Le Jolis. J Appl Phycol 32:3561–3584

Pimentel FB, Alves RC, Rodrigues F, Oliveira MPP (2017) Macroalgae-derived ingredients for the cosmetic industry. Cosmetics 5(1):2

Plaza M, Santoyo S, Jaime L, García-Blairsy Reina G, Herrero M, Señoráns FJ et al (2010) Screening for bioactive compounds from algae. J Pharm Biomed Anal 51(2):450–455

Pora B, Qian Y, Caulier B, Comini S, Looten P, Segueilha L (2018) U.S. Patent No. 10,087,467. U.S. Patent and Trademark Office, Washington, DC

Priyadarshani I, Rath B (2012) Commercial and industrial applications of microalgae—a review. J Algal Biomass Util 3(4):89–100

Priyan Shanura Fernando I, Kim KN, Kim D, Jeon YJ (2019) Algal polysaccharides: potential bioactive substances for cosmeceutical applications. Crit Rev Biotechnol 39(1):99–113

Pulz O, Gross W (2004) Valuable products from biotechnology of microalgae. Appl Microbiol Biotechnol 65:635–648

Raja R, Hemaiswarya S, Ashok Kumar N, Sridhar S, Rengasamy R (2008) A perspective on the biotechnological potential of microalgae. Crit Rev Microbiol 34:77–88

Richmond A (2004) Handbook of microalgal culture: biotechnology and applied phycology. Blackwell Science Ltd

Rosic NN (2019) Mycosporine-like amino acids: making the foundation for organic personalised sunscreens. Mar Drugs 17(11):638

Rosic NN, Braun C, Kvaskoff D (2015) Extraction and analysis of mycosporine-like amino acids in marine algae. In: Natural products from marine algae: methods and protocols, pp 119–129

Saini KC, Rani A, Gupta SK, Sharma P (2024) Algae a potential source in cosmetics: current status, challenges, and economic implications. Indian J Microbiol 64:1–16

Sansone C, Galasso C, Orefice I, Nuzzo G, Luongo E, Cutignano A, Ianora A (2017) The green microalga *Tetraselmis suecica* reduces oxidative stress and induces repairing mechanisms in human cells. Sci Rep 7(1):41215

Sasounian R, Martinez RM, Lopes AM, Giarolla J, Rosado C, Magalhães WV, Baby AR (2024) Innovative approaches to an eco-friendly cosmetic industry: a review of sustainable ingredients. Clean Technol 6(1):176–198

Sathasivam R, Radhakrishnan R, Hashem A, Abd Allah EF (2019) Microalgae metabolites: a rich source for food and medicine. Saudi J Biol Sci 26(4):709–722

Schagen SK, Zampeli VA, Makrantonaki E, Zouboulis CC (2012) Discovering the link between nutrition and skin aging. Dermatoendocrinol 4(3):298–307

Senevirathne WS, Kim SK, Dominguez H (2013) Functional ingredients from algae for foods and nutraceuticals. Mar Drugs 11(3):557–589

Senthilkumar P, Sudha S (2012) Antioxidant and antibacterial properties of methanolic extract of green seaweed *Chaetomorpha linum* from Gulf of Mannar: Southeast coast of India

Shannon E, Abu-Ghannam N (2016) Antibacterial derivatives of marine algae: an overview of pharmacological mechanisms and applications. Mar Drugs 14(4):81

Shimoda H, Tanaka J, Shan SJ, Maoka T (2010) Anti-pigmentary activity of fucoxanthin and its influence on skin mRNA expression of melanogenic molecules. J Pharm Pharmacol 62(9):1137–1145

Shin K, Kim TS, Kyung J, Kim D, Park D, Choi EK, Kim YB (2015) Effectiveness of the combinational treatment of *Laminaria japonica* and *Cistanche tubulosa* extracts in hair growth. Lab Anim Res 31:24–32

Siezen RJ (2011) Microbial sunscreens. Microb Biotechnol 4(1):1

Sinbad OO, Folorunsho AA, Olabisi OL, Ayoola OA, Temitope EJ (2019) Vitamins as antioxidants. J Food Sci Nutr Res 2(3):214–235

Singh DP, Prabha R, Verma S, Meena KK, Yandigeri M (2017) Antioxidant properties and polyphenolic content in terrestrial cyanobacteria. 3 Biotech 7:1–14

Sionkowska A, Adamiak K, Musiał K, Gadomska M (2020) Collagen-based materials in cosmetic applications: a review. Materials 13(19):4217

Smit N, Vicanova J, Pavel S (2009) The hunt for natural skin whitening agents. Int J Mol Sci 10(12):5326–5349

Sousa I, Gouveia L, Batista AP, Raymundo A, Bandarra NM (2008) Microalgae in novel food products. In: Food chemistry research developments, pp 75–112

Stolz P, Obermayer B (2005) Manufacturing microalgae for skin care. Cosmet Toiletries 120(3):99–106

Sun C, Xu Y, Hu N, Ma J, Sun S, Cao W, Zhao Y (2020) To evaluate the toxicity of atrazine on the freshwater microalgae Chlorella sp. using sensitive indices indicated by photosynthetic parameters. Chemosphere 244:125514

Sung H, Ferlay J, Siegel RL, Laversanne M, Soerjomataram I, Jemal A, Bray F (2021) Global cancer statistics 2020: GLOBOCAN estimates of incidence and mortality worldwide for 36 cancers in 185 countries. CA Cancer J Clin 71(3):209–249

Tabarzad M, Atabaki V, Hosseinabadi T (2020) Anti-inflammatory activity of bioactive compounds from microalgae and cyanobacteria by focusing on the mechanisms of action. Mol Biol Rep 47(8):6193–6205

Tang SM, Deng XT, Zhou J, Li QP, Ge XX, Miao L (2020) Pharmacological basis and new insights of quercetin action in respect to its anti-cancer effects. Biomed Pharmacother 121:109604

Temina M, Rezankova H, Rezanka T, Dembitsky VM (2007) Diversity of the fatty acids of the *Nostoc* species and their statistical analysis. Microbiol Res 162(4):308–321

Thiyagarasaiyar K, Goh BH, Jeon YJ, Yow YY (2020) Algae metabolites in cosmeceutical: an overview of current applications and challenges. Mar Drugs 18(6):323

Thomas NV, Manivasagan P, Kim SK (2014) Potential matrix metalloproteinase inhibitors from edible marine algae: a review. Environ Toxicol Pharmacol 37(3):1090–1100

Uppala L (2015) A review on active ingredients from marine sources used in cosmetics. SOJ Pharm Pharm Sci 2(3):1–3

Usman A, Khalid S, Usman A, Hussain Z, Wang Y (2017) Algal polysaccharides, novel applications, and outlook. In: Algae-based polymers, blends, and composites. Elsevier, pp 115–153

Veluchamy C, Palaniswamy R (2020) A review on marine algae and its applications. Asian J Pharm Clin Res 13(3):21–27

Venugopal V (2019) Sulfated and non-sulfated polysaccharides from seaweeds and their uses: an overview. ECronicon Nutr 2(2):126–141

Vo TS, Kim SK (2013) Fucoidans as a natural bioactive ingredient for functional foods. J Funct Foods 5(1):16–27

Vo TS, Ngo DH, Kang KH, Park SJ, Kim SK (2014) The role of peptides derived from *Spirulina maxima* in downregulation of FcεRI-mediated allergic responses. Mol Nutr Food Res 58(11):2226–2234

Volk RB (2008) A newly developed assay for the quantitative determination of antimicrobial (anti-cyanobacterial) activity of both hydrophilic and lipophilic test compounds without any restriction. Microbiol Res 163:161–167

Vooradi R, Bertran MO, Frauzem R, Anne SB, Gani R (2018) Sustainable chemical processing and energy-carbon dioxide management: review of challenges and opportunities. Chem Eng Res Des 131:440–464

Wang HM, Chen CY, Wen ZH (2011) Identifying melanogenesis inhibitors from *Cinnamomum subavenium* with in vitro and in vivo screening systems by targeting the human tyrosinase. Exp Dermatol 20(3):242–248

Wang L, Wang X, Wu H, Liu R (2014) Overview on biological activities and molecular characteristics of sulfated polysaccharides from marine green algae in recent years. Mar Drugs 12(9):4984–5020

Wang HMD, Li XC, Lee DJ, Chang JS (2017) Potential biomedical applications of marine algae. Bioresour Technol 244:1407–1415

Widowati I, Zainuri M, Kusumaningrum HP, Susilowati R, Hardivillier Y, Leignel V, Mouget JL (2017) Antioxidant activity of three microalgae *Dunaliella salina, Tetraselmis chuii* and *Isochrysis galbana* clone. Earth Environ Sci 55(1):012067

Wijesinghe WAJP, Jeon YJ (2012) Enzyme-assisted extraction (EAE) of bioactive components: a useful approach for recovery of industrially important metabolites from seaweeds: a review. Fitoterapia 83(1):6–12

Wolkers H, Barbosa MJ, Kleinegris DM, Bosma R, Wijffels RH, Harmsen PFH (2011) Microalgae: the green gold of the future? Large-scale sustainable cultivation of microalgae for the production of bulk commodities. Wageningen UR-Food & Biobased Research

Wrolstad RE, Culver CA (2012) Alternatives to those artificial FD&C food colorants. Annu Rev Food Sci Technol 3:59–77

Yoon NY, Eom TK, Kim MM, Kim SK (2009) Inhibitory effect of phlorotannins isolated from *Ecklonia cava* on mushroom tyrosinase activity and melanin formation in mouse B16F10 melanoma cells. J Agric Food Chem 57(10):4124–4129

Zanella L, Alam MA (2020) Extracts and bioactives from microalgae (*sensu stricto*): opportunities and challenges for a new generation of cosmetics. In: Microalgae biotechnology for food, health and high value products, pp 295–349

Chapter 8
Mushroom-Based Cosmeceuticals: Nature's Blessings to Skin Health

Asia Shaheen, Saqib Mahmood, Muhammad Zulqurnain Haider, Altaf Ahmed Simair, Naveed Sultan, Irum, Muhammad Azeem, and Muhammad Afzaal

Contents

Abstract Mushroom-derived cosmeceuticals in skincare have been gaining considerable attention due to their skin health benefits in an eco-friendly and sustainable manner. This chapter explores the other transformative ability of mushrooms in beautifying and ensuring a healthy skin. Owing to their high antioxidant content,

A. Shaheen · S. Mahmood · M. Z. Haider (✉) · N. Sultan · Irum
Department of Botany, Government College University Faisalabad, Faisalabad, Pakistan
e-mail: drmzhaider@gcuf.edu.pk

A. A. Simair
Department of Botany, Government College University, Hyderabad, Sindh, Pakistan

M. Azeem
Department of Botany, Government College University Faisalabad, Faisalabad, Pakistan

Department of Biology, College of Science, University of Bahrain, Zallaq, Bahrain

M. Afzaal
School of Science, Engineering and Environment, University of Salford, Manchester, UK

S. Javed et al. (eds.), *Sustainable Cosmeceuticals*,
https://doi.org/10.1007/978-3-031-86087-4_8

mushrooms prevent premature skin aging and the decline of collagen and elastin levels as they scavenge free radicals. Their polysaccharides also provide exceptional moisturization, which ensures a reasonable amount of water is maintained to avert dryness and dullness. For instance, shiitake mushroom extracts have been useful in lightening skin by inhibiting melanin synthesis and dealing with hyperpigmentation. While doing so, mushrooms also possess soothing and anti-inflammatory properties, making them suitable for sensitive and acne-prone skin. Protecting the skin from damage and psychological stress results in self-enhancement and stimulates collagen production within the body, which helps to smooth out fine wrinkles and rejuvenate skin. This chapter underscores the natural goodness of mushrooms and their potential to change the face of skincare by offering safe and effective means of achieving a healthy and glowing skin. Mushroom-based cosmetics align with United Nations Sustainable Development Goals (UNSDG), as these are natural, nontoxic, renewable, environmentally friendly, and sustainable sources, owing to overall good health and well-being.

Keywords Mushroom-based cosmeceuticals · Fungal beauty · Transformative skincare · Mushroom magic · Skin aging · Moisturizing · Polysaccharides

8.1 Introduction

Mushroom-based cosmetics and cosmeceutical products are getting more attention from people. In recent years, mushroom-derived fermented cosmetics and cosmeceutical products have spurred interest and excitement among people. The shift of consumers to the natural or plant-based cosmeceutical products in order to maintain healthy skin has become more noticeable due to the desire to get rid of petrochemical and synthetic cosmetic products. It has also been shown that bioactive substances contained in various mushrooms are beneficial not only in cosmetics but also contribute to the nutraceutical and pharmaceutical industries for well-being (Wu et al. 2016). Cosmeceutical can be defined as a term which is derived from the combined words of cosmetics and pharmaceutical. Also, there are many kinds of cosmetic ingredients with cosmeceutical properties, in particular drugs that could even treat skin disorders which include skin inflammation (Wu et al. 2016; Taofiq et al. 2016a, b). According to published literature on bioactive compounds obtained from mushrooms, these chemicals have gained a lot of attention in the cosmeceutical industry and are being explored for purposes other than alternative medicine and nutrition. The use of mushroom extract formulation is popularly applied in the cosmeceutical industries due to its potent effect for skin health. Used to create a variety of cosmetic products, mushroom-based formulations, including skin lightening, moisturizers, anti-aging, antioxidants, and anti-inflammatory agents are among the most common (Visvanathan et al. 2022). It would provide an interesting area of research, which is linked with establishing new species in the context of cosmetic efficacy. Sunlight lentigines, freckles, melasma, wrinkles, and dullness are among many beauty concerns that hamper selfconfidence and aesthetics in most people (Wasmeier et al. 2008; Maranduca et al. 2019).

Other factors, besides tyrosinase, that are involved in melanin synthesis include L-DOPA (L-3,4-dihydroxyphenylalanine) and DOPA quinone (L-3,4-dihydroxyphenylalanine quinone) (Kameyama et al. 1993). Liu et al. indicated some forms of matrix-degrading enzymes might be involved in skin aging too. As indicated by a proteomic study of UV-induced skin biopsies, collagen fibrils were also fragmented by upregulating gene expression of pro-inflammatory proteases (McCabe et al. 2020; Wang et al. 2019). Studies show that UV radiation increased matrix metalloprotease, which induced cleavages in ECM proteins (Quan et al. 2009). Alterations of the skin structure are attributed to aging of the skin. Skin by itself, when exposed to sunlight and other environmental stressors, produces reactive oxygen species, which are largely responsible for skin damage (Taofiq et al. 2016a, b; Sies and Jones 2020; Fussell and Kelly 2020).

Cosmetics and personal care products that are intended to be applied on the benefactor's body provide remedies in the form of enhancement and protection of one's physical characteristics. These remedies constitute a blend of both conventional and synthetic chemicals that are beneficial to the skin or other targeted area of application (Wu et al. 2016). With time, the demands of younger consumers, which referred to the use of more "green" products, have gradually altered owing to an increase in the understanding of the shortcomings of chemical products and the benefits of going natural (Kapoor et al. 2019). There was an increased interest on mushroom due to the production of cosmeceutical valuable ingredients such as proteins, lipids, several essential amino acids, glycogens, riboflavin, nicotinic acid, and other important minerals. Cultivated mushrooms have several bioactive components including polysaccharides, phenolics, terpenoids, lectins, ergosterols, and volatile organic substances in the mycelium and fruiting bodies. It is believed that these components exhibit activities including that which is anti-oxidant, anti-microbial, anti-inflammatory, anti-tyrosinase, and anti-hyaluronic acid (Meng et al. 2011; Chien et al. 2008). Biomaterials derived from mushrooms are positioned as environmentally friendly options due to their biodegradability and life cycle evaluation, providing innovative approaches to ensure a sustainable future. The ethical aspects of sustainability and environmental impact are highlighted by the use of local feedstock residues for mushroom-based biomaterials, which helps to facilitate a smooth transition to more sustainable methods across a range of businesses. The safety, sustainability, and responsible use of these novel materials for the good of society and the environment are all ethical factors in the study of mushroom biomaterials (Arshad et al. 2023). These are used in biomedicine. These materials are appropriate for drug delivery systems, tissue engineering, and medical devices because they have unique attributes like biocompatibility, biodegradability, and antibacterial activity. In contrast to traditional synthetic polymers, biomaterials derived from mushrooms are sourced responsibly and have a low environmental impact, which is in line with international sustainability objectives SDGs (Mensah et al. 2022; Srivastava et al. 2024).

8.2 Mushrooms: A Natural Skin Savior

Some of the mushrooms more commonly found in cosmeceuticals are Tremella, Grifola, Schizophyllum, Lentinula, Ganoderma, Trametes, Agaricus, Phellinus, Antrodia, Pleurotus, Cordyceps, Inonotus, Pleurotus, Coriolus, Ganoderma, Euphorbia, and Pistacia. Such diverse genera of mushrooms that produce different kinds of secondary metabolites have significant potential in the formulation of certain types of cosmetics. As mushroom extracts are increasingly applied as ingredients in skincare and hair care, there has also been a ramp-up in commercial production of mushrooms, an area worthy of mass attention because of the numerous health benefits associated with mushrooms (Mago et al. 2023). The various tissues of an organism's body consist of different cells, which proceed with biochemical reactions for the normal functioning of the multiples. On being subjected to biochemical reactions, reactive oxygen species (RNS and RSS) and free radicals are often produced. Under normal conditions, the intrinsic antioxidant pools in the body, consisting of glutathione peroxidase, catalase, and superoxide dismutase, balance the production of ROS and neutralization. But, when they are overproduced, they get depleted and cause oxidative stress upon body cells. The mushroom taken as a food component provides good quantities of bioactive compounds exhibiting antioxidant activity to help the body in meeting its antioxidant needs (Carocho and Ferreira 2013). The above-mentioned ROS-induced DNA damage, erythema, and hyperpigmentation, as well as the stimulation of fibroblast in the skin for the synthesis of matrix metalloproteinase 1 (MMP-1), which is a protease that breaks down collagen, and consequently, decreases the formation of this protein to cause photoaging skin (Masaki 2010).

Mushroom extract is an excellent source of an antioxidant and inhibitor of lipid peroxidation that is reported in the previous literature (Carocho and Ferreira 2013). The detailed investigation revealed that some species of mushrooms significantly produce diverse kinds of metabolites having significant antioxidant potential. The ethanolic extract of different edible mushrooms, including *Agaricus brasiliensis*, *A. bisporus* (Portobello), *F. velutipes*, and *A. bisporus* (Champignon), contained gallic, ferulic, gentisic, *trans*-cinnamic, catechol, and *p*-hydroxybenzoic acids, which contained high antioxidant activity. However, *p*-Hydroxybenzoic protects the skin from harmful UV radiation (Bach et al. 2019).

Yahia et al. (2017) reported that *Lactarius indigo* produced massive amounts of secondary metabolites, including myricetin, quercetin-hexoside, isoquercetin-hexoside, kaempferol, procyanidin, coumaric, *p*-coumaroylmalic, and malic acid, while *Ramaria flava* biologically produced hexosemalonate-carboxyl, catequin, which could potentially be used as powerful antioxidants in cosmeceutical products (Fig. 8.1; Table 8.1).

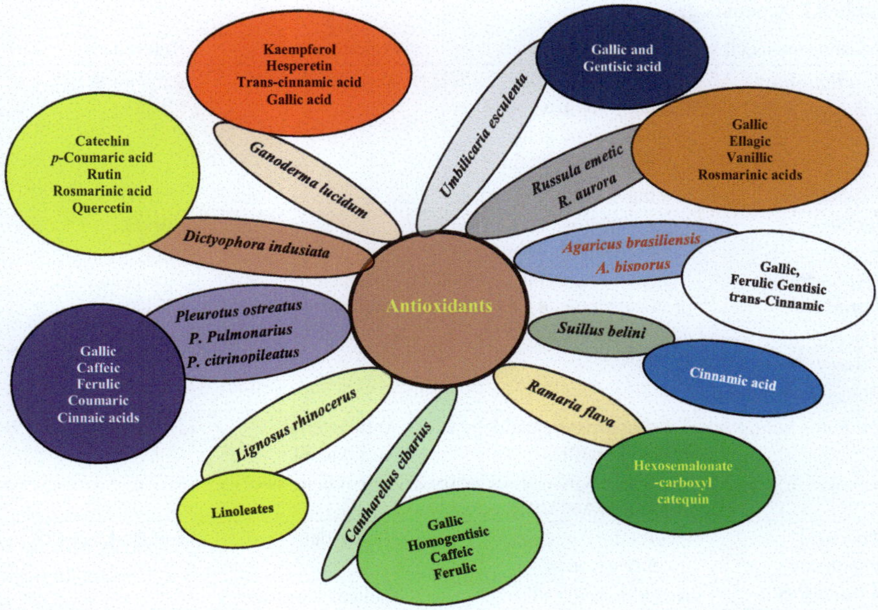

Fig. 8.1 Secondary metabolites from mushrooms used as antioxidants in cosmeceutical products

Table 8.1 List of mushroom species with potential for biosynthesis of bioactive compounds and their putative functions

Mushroom species	Bioactive compounds	Function	References
Agaricus brasiliensis	Gallic, ferulic, gentisic, *trans*-cinnamic, catecho	Antioxidant Antibiotic	Bach et al. (2019)
A. bisporus (Portobello)	*p*-Hydroxybenzoic	UV-protection	
F. velutipes *A. bisporus* (Champignon)	*p*-Coumaric	Reduced hyperpigmentation	
Auricularia polytricha	Flavonoids, phenolic acids, polysaccharides	Moisturizer	Packialakshmi et al. (2017)
A. bisporus	Polysaccharides		Nitthikan et al. (2022)
A. subrufescens	*p*-Coumaric, dihexoses	Antioxidant	Sabino-Ferrari et al. (2021)
Pleurotus ostreatus			
Agrocybe aegerita	Sinapic, chlorogenic, gallic, protocatechuic, ferulic acids	Radical scavenger	Lin et al. (2017)
Amanita crocea	Cinnamic, *p*-hydroxybenzoic, *p*-coumaric acids	Antioxidant Radical scavenger Anti-tyrosinase	Alkan et al. (2020)
H. depilatum			
N. erythropus			
Calocybe indica	Gallic, vanillin, ferulic acid, caffeic acid, and formononetin β-Carotene Linoelaic acid	Antioxidant Anti-tyrosinase Biosurfactant	Alam et al. (2019)

(continued)

Table 8.1 (continued)

Mushroom species	Bioactive compounds	Function	References
Cantharellus cibarius	Gallic, homogentisic, caffeic, ferulic, *p*-hydroxybenzoic, protocatechuic acids, pyrogallol, myricetin	Antioxidant	Palacios et al. (2011)
Ganoderma lucidum	Kaempferol, hesperetin, *trans*-cinnamic acid, gallic acid, quercetin, naringenin		Veljović et al. (2017)
Hyphodontia paradoxa *Trichaptum fuscoviolaceum* *Fomes fomentarius*	Synapic, rosmarinic, salicylic, protocatechuic, 4-OH-benzoic acid, *p*-coumaric, caffeic, syringic acids	Antibacterial Radical scavenger Antioxidant	Nowacka et al. (2015)
Inonotus obliquus	3,4-dihydroxybenzaldehyde, syringic, gallic, protocatechuic, caffeic acids, *β*-glucans, and triterpenoids	Antibacterial Antioxidant Anti-inflammatory	Hwang et al. (2019)
Laccaria amethystea *Craterellus cornucopioides* *Clitocybe maxima* *Catathelasma ventricosum* *Stropharia rugoso-annulata*	Quercetin, catechin, *p*-coumaric acid	Antioxidant	Liu et al. (2012)
	Ascorbic acid (VC)	Radical scavenger, collagen formation (anti-aging)	
	Ergosterol	Antimicrobial, antioxidant	
	Tocopherol	Moisturizer	
Lactarius indigo	Myricetin, quercetin-hexoside, isoquercetin-hexoside, kaempferol, procyanidin, coumaric, *p*-coumaroylmalic, malic, *p*-hydroxybenzoic, sinapic, *o*-chlorogenic, succinic acids	Radical scavenger Antioxidant	Yahia et al. (2017)
Lentinus edodes	*p*-Hydroxybenzoic	UV protection	Islam et al. (2016)
	Gallic	Antioxidant	
	3,4-Dihydroxybenzoic acids	Skin lightener	
Melanoleuca cognate	Benzoic, syringic, *p*-hydroxybenzoic, protocatechuic, *p*-coumaric, *trans*-cinnamic acids	UV protection	Bahadori et al. (2019)
		Antioxidant	
		Skin lightener (anti-hyperpigmentation)	
Morchella esculenta *Lepista gilva* *Leccinum scabrum* *Calvatia gigantea*	Gallic, protocatechuic, caffeic, salicylic, syringic, vanillic acids	Antioxidant	Gasecka et al. (2018)
	Ascorbic acid	Radical scavenger, collagen formation (anti-aging)	

(continued)

Table 8.1 (continued)

Mushroom species	Bioactive compounds	Function	References
Pleurotus citrinopileatus	Rutin, quercetin, gallic acid, caffeic acid	Antioxidant	Gogoi et al. (2019)
Pleurotus ostreatus	Homogentisic, *p*-coumaric, cinnamic acids		Junior et al. (2019)
Pleurotus pulmonarius	Cafeic, Gallic, Ferulic acids		Contato et al. (2020)
Ramaria flava	Hexosemalonate-carboxyl, catequin	Antioxidant	Yahia et al. (2017)
Russula aurora	Gallic, ellagic, rosmarinic, *trans*-cinnamic acids	Radical scavenger Moisturizer	Çayan et al. (2020)
Russula emetic	Protocatechuic acid, catechin, rutin, gallic, vanillic, rosmarinic acids	Antioxidant	Kaewnarin et al. (2016)
Suillus belini	*p*-Hydroxybenzoic acid	UV protection	Souilem et al. (2017)
	Protocatechuic acid	Anti-aging Anti-wrinkles	
	Cinnamic acid	Antioxidant	
Umbilicaria esculenta	Gallic acid, gentisic acid	Antioxidant	Islam et al. (2016)
	3,4-Dihydroxybenzaldehyde		
	3,4-Dihydroxybenzoic acid	Skin whitener	
Volvariella volvacea	Beta-glucan Other polysaccharide	Antioxidant Anti-tyrosinase	Sangthong et al. (2022)
Ganoderma lucidum	Triterpenoids Alkaloids Phenolic compounds Flavonoids Polysaccharides	Anti-tyrosinase Anti-melanogenesis Antioxidant Akin whitening	Ha et al. (2023)
	β-Glucans and proteins		Vaithanomsat et al. (2022)
Phallus indusiatus	Bioactive compounds	Anti-melanogenic Anti-tyrosinase Antioxidant Anti-inflammatory	Theeraraksakul et al. (2023)
Dictyophora indusiata	Catechin *p*-Coumaric acid Rutin Rosmarinic acid Quercetin Naringenin	Antioxidant Anti-inflammatory	Nazir et al. (2021)
Lignosus rhinocerus	Linoleates Ascorbic acid Dipalmitate and All-*trans*-squalene	Antioxidant Anti-inflammatory Anti-aging Anti-wrinkle	Fung et al. (2022)
Grifola fondosa	Exopolysaccharide (EPS)	Collagen formation	Kim et al. (2007)
Tricholoma matsutake	Secondary bioactive	Anti-aging, anti-collagenase, and anti-elastase	Kim et al. (2014)

8.2.1 *Mushrooms Contribution to Preventing Premature Aging*

Skin performs various functions, prevents the body from dehydration, and saves the internal organs from various pathogenic, chemical, and physical attacks. Meanwhile, it is also the largest organ and takes part in about 15% of the whole-body weight (Kendall and Nicolaou 2013). It is widely accepted that skin aging is caused by intrinsic, natural, or cellular processes that affect both the skin and other organs of the body. The basic hormonal shift associated with aging leads to collagen degradation, dryness, degeneration of elastic fiber networks, and wrinkles (Papakonstantinou et al. 2012). In addition, it has been noted that the expressions of matrix metalloproteinases, MMP-1, MMP-3, and MMP-9, increase the degradation of collagen as the cell overexpresses them (Leem 2015).

In addition, the UV radiation results in the downregulation of the production of TGF- that promotes pro-collagen production, therefore reducing the synthesis of collagen. Besides elastin and collagen, the extracellular matrix consists of hyaluronic acid, an important molecule that moisturizes the skin, protecting it from oxidative stress damage (Papakonstantinou et al. 2012). Therefore, enzyme inhibitors may be used as cosmetic ingredients that would treat the aging of the skin by restoring elasticity, increasing the contents of moisture, stimulating the synthesis of collagen, and cleansing the skin. Natural phenolic compounds have been found to scavenge ROS and so have shown interesting activity against anti-aging creams and lotions (Soto et al. 2015). Of course, various types of mushrooms can potentially yield anti-aging and anti-wrinkle compounds useful in cosmeceuticals for enhancing skin health. Ascorbic acid (Vitamin C) was incredibly produced by *Morchella esculenta*, *Lepista gilva*, *Leccinum scabrum*, and *Calvatia gigantea* (Fig. 8.2,

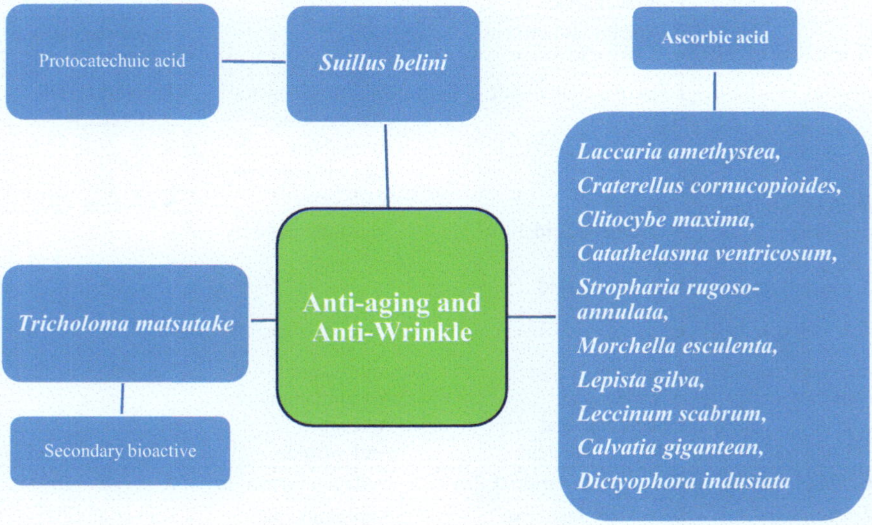

Fig. 8.2 Mushroom species reported for the synthesis of anti-aging and anti-wrinkle bioactive metabolites

Table 8.1) and other secondary metabolites (Secka et al. 2018). The combination of ascorbic acid with ferulic acid is stated to be used to work as an anti-aging product to prevent premature skin aging (Zduńska-Pęciak et al. 2022).

A detailed investigation by Souilem et al. (2017) resulted in the fact that *Suillus belini* produced various forms of secondary metabolites and phenolic compounds such as *p*-hydroxybenzoic acid, protocatechuic acid (PCA), and cinnamic acid. A further study mentioned that protocatechuic acid is significantly used in the synthesis of collagen protein and stops the skin from the aging process. PCA is not only used as an anti-aging but also highly used as an anti-wrinkle (Shin et al. 2020). Polyphenol types of extract found from *Dictyophora indusiata* were Catechin, *p*-Coumaric acid, rutin, rosmarinic acid, quercetin, naringenin, and epigallocatechin gallate. The detailed analysis of aquas of *D. indusiate* and ascorbic acid as the standard showed anti-aging results similar to each other (Nazir et al. 2021).

8.3 Polysaccharides for Moisturization

Moisturizers are the primary constituents of the SC skin barrier function, and they are also crucial in TEWL (Rawlings et al. 1994). NMFs, which are found within corneocytes and serve as very potent humectants at capturing moisture from the skin, constitute the main constituents of the SC. NMFs consist of sugars, urocanic acid, lactic acid, urea, and derivatives of amino acids like pyrrolidone carboxylic acid (PCA) (Fowler 2012). These are water-soluble chemicals produced by the degradation of filaggrin (FLG) in keratinocytes (Rawlings and Matts 2005). Water-soluble substances are generated through the breakdown of filaggrin (FLG) in keratinocytes. Aquaporin-3 (AQP3) is a protein that functions as an aquaglyceroporin channel for water or glycerol transport and has also been reported to relate to the proliferation of keratinocytes and TEWL in the epidermis (Ma et al. 2002). The edible fungus *Auricularia polytricha*, or a species of polytricha, cloud ear mushroom is widely cultivated in East Asia. The flavonoids and phenolic acids in *A. polytrichas* have been reported to possess bioactivities that include antioxidant, hypoglycemic, and anticancer properties (Packialakshmi et al. 2017). Protein, carbohydrates, and polysaccharides are also reported to be present in *A. polytrichas* (Table 8.1). Hydration of the skin and maintenance of water balance in the skin have been a long concern for the development of transparent thin films. Brown *A. bisporus* extract, however, has been proven to hold up a spectrum of biological properties, including antioxidant, anti-aging, and anti-inflammation, in addition to promising the promotion of skin barrier function and hydration (Nitthikan et al. 2022).

8.4 Shiitake Mushrooms: Skin Lightening and Hyperpigmentation Treatment

The literature review demonstrated that shiitake mushroom (*Lentinula edodes*) extract is widely used in pharmaceutical industries for manufacturing various medicines concerning anticancer, antidiabetic, and other cardiovascular diseases (Bisen et al. 2010). However, some studies reported its cosmeceutical importance. Shiitake extract possesses an immense ability to antiwrinkle, antiaging, anti-inflammatory, and antioxidant. Despite this fact, there is no real evidence in favor of skin whitening and the antipathetic role of hyperpigmentation (Seo et al. 2018; Wu et al. 2016). Several studies suggest that *L. edodes* extract may have the promise of inhibiting melanin and tyrosinase activities that directly participate in pigment formation (Hyde et al. 2010). To develop the efficacy and safety of *Lentinula*, more research needs to be conducted on edodes extract for skin whitening since the results obtained from this compound are not significantly strong and reliable. Therefore, an application of *L. edodes* extract is advised as a supplementary ingredient that adds quality and health to the skin rather than for whitening purposes alone.

Human skin color is due to the presence of dark brown pigment, which is commonly known as melanin. UV rays affect the human skin and synthesis of melanin pigment when continuously experienced the UV radiation. Overproduction of melanin pigment after exposure to ultraviolet radiation is mainly because of increased tyrosinase enzyme activity. The responsible tyrosinase enzyme catalyzes the conversion of tyrosine into dihydroxyphenylalanine (DOPA) and then DOPA oxidized into dopaquinone (Bae-Harboe et al. 2012; Meng et al. 2012).

Once it undergoes auto-oxidation, the dopaquinone changes into dopachrome, which is converted into eumelanin (brown-black pigment) in the presence of dopachrome tautomerase. Alternatively, dopaquinone can change into cysteinyl DOPA to be converted into pheomelanin in the presence of cysteine or glutathione (yellow-red pigment) (Jung et al. 2009). The main factors that cause skin hyperpigmentation include reactions to medication from drug-induced causes, autoimmune diseases, exposure to UV radiations, hormonal changes that trigger melanocyte-stimulating hormone production, genetics, hormonal therapy, and birth control pills. All these factors stimulate melanocytes to produce excess amounts of melanin, causing hyperpigmentation (Ali et al. 2015). A large number of species of mushrooms provide bioactive compounds that are antityrosinases, which reverse the darkening process of melanin in the skin (Fig. 8.2). Among the most cultivated edible mushrooms, *Pleurotus* species are of high nutritional value and health benefits, like antibacterial, antioxidant, and anti-inflammatory. The medicinal properties of these molecules have fueled optimism among scientists concerning their cosmetic applications. The extracts obtained from the fruiting body of the oyster mushroom *P. citrinopileatus* have been reported to inhibit tyrosinase activity and, consequently, suppress the synthesis of melanin (Meng et al. 2011). The remaining *Pleurotus* species, which produced high levels of tyrosinase-inhibiting metabolites, are the fungus of the basilisk, *Pleurotus ferulae* (Arshad et al. 2018; Alam et al. 2012), Ribbed Cachrys oyster mushroom *Pleurotus nebrodensis* (Inzenga) Quél. *Pleurotus ostreatus* (Alam et al. 2010; Hapsari et al. 2012) and pink Oyster Mushroom *Pleurotus salmoneostramineus* (Alam et al. 2011).

Recently, several ethanolic mycelium extracts of *Inonotus mikadoi* mushroom were studied by Park et al. (2015) for anti-tyrosinase properties in the presence of arbutin and ascorbic acid as positive controls. *Inonotus mikadoi* surprisingly reduced tyrosinase activity at 10 mg/mL. Meanwhile, the same level of concentration of *Coriolus versicolor* and *Fomitopsis* sp. also considerably reduced the tyrosinase activity. It was studied that *Inonotus obliquus* (chaga mushroom) widely used in Russia, and other European nations biologically synthesized the secondary metabolites considerably utilized as immunomodulatory and anti-cancer (Taofiq et al. 2016a, b), but ether and n-butanol extract enhance the anti-tyrosinase activity (Yan et al. 2014).

Because bioactive compounds like flavonoids, terpenoids, polysaccharides, and coumarin derivatives can control melanin synthesis by either preventing tyrosinase expression in one or more steps or by controlling melanosome distribution, they are used by the cosmetic industry (Sánchez 2017; Chang 2012). Strong anti-tyrosinase action is shown by mushrooms like *Ganoderma lucidum* (Chien et al. 2008), *Euphorbia characias* (Pintus et al. 2015), and *Pistacia atlantica* subsp. *mutica* (Arshad et al. 2022; Tayarani-Najaran et al. 2018), which are being thoroughly studied for use in skin-whitening products (Fig. 8.3, Table 8.1).

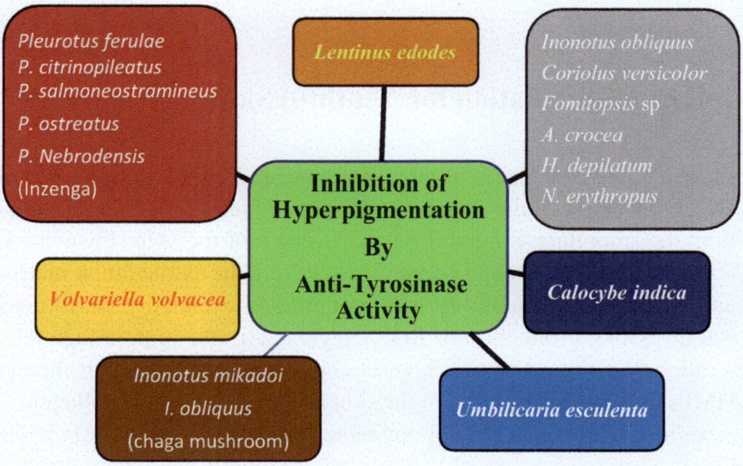

Fig. 8.3 Some macro-fungal species enable to produce anti-tyrosinase bioactive compounds

8.5 Calming and Anti-inflammatory Effects

Mushrooms and their isolated compounds, such as polysaccharides, terpenes, phenolic compounds, sterols, fatty acids, polysaccharide-protein complexes, and other bioactive metabolites, have anti-inflammatory activities as they are supposed to inhibit the production of inflammatory mediators (Taofiq et al. 2016a, b). Based on the findings of the experiment, the effect was caused by bioactive metabolites linked with mushrooms and their anti-inflammatory properties. The most investigated species included *Agaricus bisporus* Imbach, *Phellinus linteus*, *Cordyceps* species, *Antrodia camphorata*, *Pleurotus* species, and *Ganoderma lucidum*. Polysaccharides, terpenes, and phenolic derivatives obtained from the above-mentioned macrofungi were attributed to their anti-inflammatory activity (Taofiq et al. 2016a, b). As reported by Ukawa et al. (2007), the extract of *Lyophyllum decastes* inhibited the development of atopic dermatitis in mice that were topically sensitized with picryl chloride, the outcomes having been compared against controls based on severity of lesions. The extract decreased serum levels of IgE and IL-4 production. Park et al. (2015) isolated the polysaccharides of *Grifola frondosa* known as GFP. The study also elucidated the compound's potential in reducing atopic dermatitis-like skin lesions in NC/Nga mice caused by 2,4-dinitrochlorobenzene.

Due to their natural and nontoxic nature, mushroom metabolites also can be used in cosmeceutical formulations without any side effects or adverse effects (Elsayed et al. 2014). One of the anti-inflammatory compounds, which promotes the formation of healthy skin, is linoleic acid (Mago et al. 2023). *Agaricus subrufescens* has been known to produce linoleic acid (Eira et al. 2005), although other sources for this compound include *Astraeus* spp. (Kakumyan and Matsui 2009), *Cordyceps* (Yang et al. 2009), and *Pleurotus sajor-caju* (Nieto and Chegwin 2008).

8.6 Collagen Stimulation for Youthful Skin

UV irradiation causes the breakout of collagen because the most important component of the skin's extracellular matrix is collagen, which plays an important role in maintaining the smoothness, flexibility, and strength of the skin (Piwowarski et al. 2011). With age, some of its components are lost from the extracellular matrix of the skin, leading to it losing flexibility and strength and eventually causing wrinkles (Ndlovu et al. 2013). Production of ROS triggers activator protein factor 1, which activates mitogen-activated protein kinases. This leads to the uncontrolled production of MMPs that cause wrinkling of the skin and break down the collagen. For this reason, several researches are being conducted to identify natural AP-1 inhibitors that can be used in cosmeceutical preparations to inhibit the expression of MMPs (Thomas et al. 2013). Although the literature comprises various studies relating to the anti-elastase activity of plant extracts, there are very few studies about the anti-elastase and anti-collagenase activity of mushroom extracts or their metabolites (Thring et al. 2009; Moon et al. 2010; Piwowarski et al. 2011).

Kim et al. (2014) reported the mycelial extract of *Tricholoma matsutake* as possessing anti-collagenase and antielastase properties. This was done by exposing dermal fibroblast cells to *T. matsutake* extract at different concentrations from 1 to 100 g/mL for 72 h, after which the expression of type 1 collagen, MMPs, and elastase was measured. *T. matsutake* extract has anti-aging potential and inhibits the degradation of the ECM (Table 8.1).

Kim et al. (2007) researched *Grifola fondosa* mycelial extract and an isolated exopolysaccharide (EPS) that showed anti-inflammatory, anti-tyrosinase, and anti-collagenase effects. Furthermore, the extract increased collagen biosynthesis by 53%. These findings indicate that the isolated polysaccharides and the mushrooms can become useful factors in the prevention of hyperpigmentation and skin aging (Mago et al. 2023). Among these extracts of *Lignosus rhinocerus*, aquas contained linoleates, ascorbic acid dipalmitate, and all-*trans*-squalene bioactive compounds significant in improving collagen synthesis. Extract of the fungus *L. rhinocerus* may be used cosmeceutically to increase the health of the skin (Fung et al. 2022).

The human skin is made up of the epidermis, which securely linked and sustained by connective tissue to the underlying dermis. The dermal fibroblasts, through the two structural proteins collagen and elastin in the extracellular matrix (ECM), are essential for the skin's multiple defensive functions (Kim et al. 2014). Elastin, an extracellular matrix protein, makes the flexible tissues in living organisms like the aorta, lung, cartilage, elastic ligaments, and skin (Arshad 2024). A metalloproteinase enzyme, elastase, can degrade elastin although this latter protein is actually quite resistant to proteolytic degradation, the continual exposure to elastase causes damage to elastic fibers that contributes to wrinkled skin. There is accumulating evidence that the activity of elastase increases with age, thus of late, there has been a revived interest in screens of natural matrices, including plants, mushrooms, rhizomes, and marine algae, for ingredients that reduce the signs of aging on the skin and wrinkles (Onar et al. 2012). Bae et al. (2005) examined the ability of an EXP isolated from submerged mycelial cultures of *Grifola frondosa* to suppress MMP-1 expression both before and after UV A-exposure. At a concentration of 250 g/mL polysaccharides, MMP-1 mRNA levels were significantly lowered; thus, EXP may contribute to the photoaging activity of the mushroom extract (Fung et al. 2022) disclosed that bioactive compounds related to linoleates, ascorbic acid, dipalmitate, and all-*trans*-squalene present in the aqueous extract of *Lignosus rhinocerus* were significant for improving collagen synthesis. The extract of *L. rhinocerus* potentially be used in cosmeceutical products to improve skin health (Table 8.1).

8.7 Mushrooms in Skincare: A Holistic Approach

Due to their organic, healthier, and environmentally friendly characteristics, the demand for cosmetics made from these ingredients is on the rise. Cosmetics are personal care products that purify and beautify the skin (Komura et al. 2010; Wu et al. 2016). Taofiq et al.'s study (2016) indicated that this may be used as a possible treatment for cosmetological issues, including uneven tone, fine lines, and wrinkles. However, among the cosmeceutical significant contents of the extract were terpenoids present in *G. lucidum*, namely ganoderic acids C2, A, and H, and phenolics, such as protocatechuic, *p*-hydroxybenzoic, and syringic acids. It was reported that the ethanolic extracts of *Agaricus bisporus* and *Pleurotus ostreatus* are rich in essential bioactive compounds, yet they are readily degraded and oxidized. Hence, they prepared their extracts as they were microencapsulated and blended with the semi-solid base cream. Their efficacy was compared with their free-form counterparts through comparison in terms of bioactivity in vitro release, and real-time settings of up to 6 months. Formulations prepared with encapsulated forms were reported to have antimicrobial and antityrosinase activities, but not enhance the antioxidant activity of the extract to a satisfactory level with the release of the extract (Arshad et al. 2018; Taofiq et al. 2016a, b). Non-irritation on the skin of volunteers who received the placebo upon application of the moisturizing hand sanitizer developed by Lee et al. (2013) had snow mushroom polysaccharides.

8.8 Mushroom-Based Commercially Available Skincare Products

Mushroom-based skincare is an exemplary manifestation of the United Nations' Sustainable Development Goals, particularly in responsible consumption and production (Goal 12), good health and well-being (Goal 3), and life on land (Goal 15). These are excellent examples of innovative beauty products and represent true-to-life applications of ecological consciousness through sustainably cultivated and naturally derived mushrooms like reishi (*Ganoderma lucidum*), snow mushrooms (*Tremella fuciformis*), shiitake (*Lentinula edodes*), etc. These are renewable natural resources that create minimal environmental impact while fostering biodiversity and maintaining soil health. They add to bioactive compounds, which have antioxidant, anti-inflammatory properties, and can hydrate, hence creating a healthier planet and helping consumers become healthier as well through efficient alternatives to synthetic ingredients in a plant-based format to replace synthetics (Table 8.2). Such products underscore how the beauty sector can promote a greener, more sustainable future aligned with the needs of modern skincare.

8.9 Conclusion

The young generation of the entire world has developed a great interest in natural-based cosmeceutical formulations acquired from a variety of plants, algae, microbes, and macro-fungi. The mushroom-based cosmeceutical products get more interest globally as a result of the formation of different types of bioactive compounds, which form a central role in the synthesis of cosmetics products. The various species of mushrooms biochemically produce some sort of bioactive metabolites like gallic, ferulic, gentisic, *trans*-cinnamic, *p*-hydroxybenzoic, *p*-coumaric, sinapic, protocatechuic, chlorogenic, dihexoses, vanillin, β-carotene, linoelaic acid,

Table 8.2 Comprehensive list of mushroom-based skincare products aligned with sustainability and innovation

Sr #	Product name	Mushroom common name	Mushroom scientific name	Ingredients used	Maker name
1	Kora Organics Milky Mushroom Gentle Cleansing Oil	Snow mushroom	*Tremella fuciformis*	Silver ear mushroom, rose quartz	Kora Organics
2	Herbivore Pink Cloud Creamy Jelly Cleanser	Snow mushroom	*Tremella fuciformis*	Tremella mushroom, rosewater	Herbivore Botanicals
3	Youth To The People adaptogen Deep Moisture Cream	Reishi mushroom	*Ganoderma lucidum*	Reishi, ashwagandha, pro-grade peptide complex	Youth To The People
4	Tata Harper Clarifying Moisturizer	Tremella mushroom	*Tremella fuciformis*	Tremella, salicylic acid, zinc	Tata Harper
5	Murad Intensive-C Radiance Peel	Fomes mushroom	*Fomes officinalis*	Fomes extract, Vitamin C	Murad
6	Hydropeptide 5X Power Peel	Various mushroom enzymes	*Mucor rouxii* or *Mucor miehei*	Mushroom enzymes, chamomile, aloe vera	Hydropeptide
7	Drunk Elephant Protini Power Peptide Serum	Snow mushroom	*Tremella fuciformis*	Snow mushroom, peptides, lactic acid	Drunk Elephant
8	AziMD DermaBright Pads	Shiitake mushroom	*Lentinula edodes*	Kojic acid, shiitake mushroom extract	AziMD Skincare
9	Eminence Snow Mushroom Moisture Cloud Eye Cream	Snow mushroom	*Tremella fuciformis*	Tremella mushroom, botanical peptides	Eminence Organic Skin Care
10	Bliss Mighty Biome Ultra-Hydrating Moisturizer	Snow mushroom	*Tremella fuciformis*	Snow mushroom extract, probiotics	Bliss
11	Tatcha The Dewy Skin Cream	Snow mushroom	*Tremella fuciformis*	Japanese purple rice, snow mushroom extract	Tatcha
12	MicroDose Magical Moisturizer	Reishi	*Ganoderma lucidum*	Fermented reishi extract, adaptogens	FRWRD Skincare
13	Dermalogica PowerBright Dark Spot Serum	Shiitake mushroom	*Lentinula edodes*	Shiitake mushroom, niacinamide, hexylresorcinol	Dermalogica
14	Tata Harper Clarifying Moisturizer	Snow mushroom	*Tremella fuciformis*	Snow mushroom, salicylic acid	Tata Harper

(continued)

Table 8.2 (continued)

Sr #	Product name	Mushroom common name	Mushroom scientific name	Ingredients used	Maker name
15	Dr. Dennis Gross B3 Adaptive Superfoods Stress Rescue Super Serum	Maitake mushroom	*Grifola frondosa*	Maitake mushroom, niacinamide, adaptogens	Dr. Dennis Gross
16	Kypris Glow Philtre Exfoliating Mask	Snow mushroom	*Tremella fuciformis*	Snow mushroom, pomegranate enzymes	Kypris
17	Joanna Vargas Euphoria Face Mask	Snow mushroom	*Tremella fuciformis*	Snow mushroom, aloe vera, chamomile extract	Joanna Vargas
18	Volition Beauty Snow Mushroom Water Serum	Snow mushroom	*Tremella fuciformis*	Snow mushroom, watermelon extract	Volition Beauty
19	Moon Juice Beauty shroom Plumping Jelly Serum	Snow mushroom	*Tremella fuciformis*	Snow mushroom, watermelon extract, marine water	Moon Juice
20	AVYA Skincare Day Moisturizer	Snow mushroom	*Tremella fuciformis*	Snow mushroom, peptides, antioxidants	AVYA Skincare
21	Youth to the People adaptogen Deep Moisture Cream	Reishi mushroom	*Ganoderma lucidum*	Shea butter, jojoba oil, peptides	Youth to the People
22	Thayers Milky Hydrating Face Toner	Snow mushroom	*Tremella fuciformis*	Hyaluronic acid, elderflower water	Thayers
23	Kiehl's Super Multi-Corrective Cream	Chaga mushroom	*Inonotus obliquus*	Hyaluronic acid, vitamin A, nourishing agents	Kiehl's
24	Tata Harper Clarifying Moisturizer	Snow mushroom	*Tremella fuciformis*	Salicylic acid, zinc, moisturizing agents	Tata Harper
25	Beekman 1802 Mushroom Milk Eye Cream	Reishi, Lion's Mane	*Ganoderma lucidum, Hericium erinaceus*	Goat's milk, peptides, other hydrating ingredients	Beekman 1802
26	Alice Nightcap Mushroom Chocolates (Wellness)	Reishi	*Ganoderma lucidum*	Reishi, adaptogens	Alice
27	Dew Drops Mushroom Hyaluronic Acid + Vitamin C	Snow mushroom	*Tremella fuciformis*	Hyaluronic acid, vitamin C, snow mushroom	Three Ships Beauty

(continued)

Table 8.2 (continued)

Sr #	Product name	Mushroom common name	Mushroom scientific name	Ingredients used	Maker name
28	Renee Rouleau BHA Clarifying Serum	Silver ear mushroom	*Tremella fuciformis*	Beta hydroxy acid, niacinamide, silver ear mushroom	Renee Rouleau
29	Moon Juice Beauty Shroom Acid Potion	Reishi	*Ganoderma lucidum*	AHAs, BHAs, reishi extract	Moon Juice
30	Herbivore Botanicals Cloud Jelly Serum	Snow mushroom	*Tremella fuciformis*	Hyaluronic acid, strawberry extract, snow mushroom	Herbivore Botanicals
31	Dr. Andrew Weil for Origins Mega-Mushroom Skin Relief	Reishi, Chaga, Cordyceps	*Ganoderma lucidum, Inonotus obliquus, Cordyceps militaris*	Reishi extract, fermented chaga, cordyceps	Origins
32	Volition Snow Mushroom Water Serum	Snow mushroom	*Tremella fuciformis*	Tremella extract, hyaluronic acid	Volition Beauty
33	Youth to the People Adaptogen Deep Moisture Cream	Reishi, Ashwagandha	*Ganoderma lucidum*	Reishi, pentapeptide, shea butter	Youth to the People
34	Moon Juice Beauty Shroom Plumping Jelly Serum	Silver ear mushroom, Reishi	*Tremella fuciformis, Ganoderma lucidum*	Hyaluronic acid, reishi, tremella	Moon Juice
35	The Plant Base Time Stop Collagen Ampoule	Golden needle mushroom	*Flammulina velutipes*	76.52% mushroom extract, collagen peptides	The Plant Base
36	REN Ultra Comforting Rescue Mask	White mushroom	*Albatrellus ovinus*	White mushroom extract, arnica, beta-glucans	REN Clean Skincare
37	Herbivore Pink Cloud Soft Moisture Cream	Tremella	*Tremella fuciformis*	Rose hydrosol, tremella extract, vegan squalane	Herbivore Botanicals
38	Damdam Snow Mushroom Pore Cleanser	Snow mushroom	*Tremella fuciformis*	Snow mushroom, plant fibers, salt	Damdam

homogentisic, caffeic, ascorbic acid, β-carotene, and β-glucans and proteins. These bioactive metabolites show antibiotic, antioxidant, anti-inflammatory, anti-aging, anti-tyrosinase, anti-melanogenic, anti-wrinkle, and collagen formation properties. In a nutshell, mushroom-based cosmetics represent an integral combination of customary wisdom and modern science. They offer innovative solutions for contemporary beauty and wellness challenges.

These products are not only healthy for the skin via natural and bioactive ingredients but also green by the sustainability of the biodiversity of issues. Consumers going for renewable, mushroom-derived formulations are performing a direct act of reducing the environmental footprint of the beauty industry. Second, these products echo the UN SDGs through ethical practices that foster environmental stewardship and ensure healthier lives. This trend is quite growing, and it speaks toward the nature-inspired innovations that can work toward the transformational transitions bridging the gap between personal care and planetary well-being.

Further Reading

Chakraborty S, Beura M, Sharma SK, Singh A, Dahuja A, Krishnan V (2023) Lentinan, β-glucan from Shiitake (Lentinula edodes): a review on structure, conformational transition, and gastro-intestinal interaction contributing towards its antidiabetic potential. Trends Food Sci Technol 142:104224

Ciernikova S, Sevcikova A, Drgona L, Mego M (2023) Modulating the gut microbiota by probiotics, prebiotics, postbiotics, and fecal microbiota transplantation: an emerging trend in cancer patient care-reviews on cancer. Biochim Biophys Acta (BBA) 1878(6):188990

Kamagaju L, Morandini R, Bizuru E, Nyetera P, Nduwayezu JB, Stévigny C, Ghanem G, Duez P (2013) Tyrosinase modulation by five Rwandese herbal medicines traditionally used for skin treatment. J Ethnopharmacol 146(3):824–834. https://doi.org/10.1016/j.jep.2013.02.010

Lourith N, Pungprom S, Kanlayavattanakul M (2021) Formulation and efficacy evaluation of the safe and efficient moisturizing snow mushroom hand sanitizer. Abstract. J Cosmet Dermatol 20(2):554–560. https://doi.org/10.1111/jocd.13543

Montagna W (2012) The structure and function of skin. Elsevier

Mwangi RW, Macharia JM, Wagara IN, Bence RL (2022) The antioxidant potential of different edible and medicinal mushrooms. Biomed Pharmacother 147:112621. https://doi.org/10.1016/j.biopha.2022.112621

Nattaya, Lourith Sathaporn, Pungprom Mayuree, Kanlayavattanakul (2021) Formulation and efficacy evaluation of the safe and efficient moisturizing snow mushroom hand sanitizer Abstract. *Journal of Cosmetic Dermatology 20*(2) 554–560. 10.1111/jocd.v20.2 10.1111/jocd.13543

Rai M, Feitosa CM (eds) (2022) Eco-friendly biobased products used in microbial diseases. CRC Press

References

Alam N, Yoon KN, Lee KR et al (2010) Antioxidant activities and tyrosinase inhibitory effects of different extracts from Pleurotus ostreatus fruiting bodies. Mycobiology 38(4):295–301

Alam N, Yoon KN, Cha YJ et al (2011) Appraisal of the antioxidant, phenolic compounds concentration, xanthine oxidase and tyrosinase inhibitory activities of Pleurotus salmoneostramineus. Afr J Agric Res 6:1555–1563

Alam N, Yoon KN, Lee JS et al (2012) Consequence of the antioxidant activities and tyrosinase inhibitory effects of various extracts from the fruiting bodies of Pleurotus ferulae. Saudi J Biol Sci 19(1):111–118

Alam N, Sikder MM, Karim MA, Amin SMR (2019) Antioxidant and antityrosinase activities of milky white mushroom. Bangladesh J Bot 48(4):1065–1073

Ali SA, Choudhary RK, Naaz I, Ali AS (2015) Understanding the challenges of melanogenesis: key role of bioactive compounds in the treatment of hyperpigmentory disorders. J Pigment Dis 2:1–9

Alkan S, Uysal A, Kasik G, Vlaisavljevic S, Berežni S, Zengin G (2020) Chemical characterization, antioxidant, enzyme inhibition and antimutagenic properties of eight mushroom species: a comparative study. J Fungi 6(3):166

Arshad M (2024) Sustainability of renewable energy options as compared to coal-fired power plants in Pakistan. Green Low-Carbon Econ 2(3):162–173

Arshad M, Bano I, Younus M, Khan A, Rahman A (2018) Health concerns associated with biofuel production. In: Perspectives on water usage for biofuels production: aquatic contamination and climate change, pp 97–105

Arshad M, Ansari AR, Qadir R, Tahir MH, Nadeem A, Mehmood T, Alhumade H, Khan N (2022) Green electricity generation from biogas of cattle manure: an assessment of potential and feasibility in Pakistan. Front Energy Res 10:911485

Arshad M, Syed-Hassan SSA, Masood R, Ansari AR, Mumtaz A, Rahman A, Batool S, Atif FA, Muhammad A, Rehman AU (2023) Utilization of waste animal fat for sustainable biodiesel production. In: Climate changes mitigation and sustainable bioenergy harvest through animal waste: sustainable environmental implications of animal waste. Springer, pp 149–169

Bach F, Zielinski AAF, Helm CV, Maciel GM, Pedro AC, Stafussa AP et al (2019) Bio compounds of edible mushrooms: in vitro antioxidant and antimicrobial activities. LWT 107:214–220

Bae JT, Sinha J, Park JP (2005) Production of exopolysaccharide from mycelial culture of Ganoderma lucidum for application to cosmetic products. J Microbiol Biotechnol 15(2):287–293

Bae-Harboe YSC, Park HY (2012) Tyrosinase: a central regulatory protein for cutaneous pigmentation. J Invest Dermatol 132(12):2678–2680

Bahadori MB, Sarikurkcu C, Yalcin OU, Cengiz M, Gungor H (2019) Metal concentration, phenolics profiling, and antioxidant activity of two wild edible Melanoleuca mushrooms (M. cognata and M. stridula). Microchem J 150:104172

Bisen PS, Baghel RK, Sanodiya BS (2010) Lentinus edodes: a macrofungus with pharmacological activities. Curr Med Chem 17(22):2419–2430

Carocho M, Ferreira ICFR (2013) A review on antioxidants, prooxidants and related controversy: natural and synthetic compounds, screening and analysis methodologies and future perspectives. Food Chem Toxicol 51:15–25

Çayan F, Deveci E, Tel-Çayan G, Duru ME (2020) Identification and quantification of phenolic acid compounds of twenty-six mushrooms by HPLC–DAD. J Food Meas Charact 14:1690–1698

Chakraborty S, Beura M, Sharma SK, Singh A, Dahuja A, Krishnan V (2023) Lentinan, β-glucan from Shiitake (Lentinula edodes): a review on structure, conformational transition, and gastro-intestinal interaction contributing towards its antidiabetic potential. Trends Food Sci Technol 142:104224

Chang TS (2012) Natural melanogenesis inhibitors acting through the down-regulation of tyrosinase activity. Materials (Basel) 5:1661–1685

Chien CC, Tsai ML, Chen CC et al (2008) Effects on tyrosinase activity by the extracts of *Ganoderma lucidum* and related mushrooms. Mycopathologia 166:117–120

Ciernikova S, Sevcikova A, Drgona L, Mego M (2023) Modulating the gut microbiota by probiotics, prebiotics, postbiotics, and fecal microbiota transplantation: an emerging trend in cancer patient care-reviews on cancer. Biochim Biophys Acta (BBA) 1878(6):188990

Contato AG, Inácio FD, de Araújo CAV, Brugnari T, Maciel GM, Haminiuk CWI et al (2020) Comparison between the aqueous extracts of mycelium and basidioma of the edible mushroom Pleurotus pulmonarius: chemical composition and antioxidant analysis. J Food Meas Charact 14:830–837

Elsayed EA, El Enshasy H, Wadaan MA, Aziz R (2014) Mushrooms: a potential natural source of anti-inflammatory compounds for medical applications. Mediators Inflamm 2014(1):805841. https://doi.org/10.1155/2014/805841

Eira F, Da A, Kaneno R, Rodrigues Filho E, Barbisan LF, Pascholati SF, Di Piero RM, Salvadori DMF, Alves de Lima PL, Ribeiro LR (2005) Farming technology, biochemistry characterization, and protective effects of culinarymedicinal mushrooms Agaricus brasiliensis S. Wasser et al. and Lentinus edodes (Berk.) Singer: five years of research in Brazil. Int J Med Mushrooms 7(1–2):281–299

Fowler J (2012) Understanding the role of natural moisturizing factor in skin hydration. Pract Dermatol 9:36–40

Fung SY, Yap HYY, Ng ST, Tan CS (2022) In vitro inhibition of melanin formation and enhancement of collagen production by a mushroom sclerotial water extract from the tiger milk mushroom, Lignosus rhinocerus (agaricomycetes), with no skin and eye irritation. Int J Med Mushrooms 24(5):19

Fussell JC, Kelly FJ (2020) Oxidative contribution of air pollution to extrinsic skin ageing. Free Radic Biol Med 151:111–122

Gasecka M, Magdziak Z, Siwulski M, Mleczek M (2018) Profile of phenolic and organic acids, antioxidant properties and ergosterol content in cultivated and wild growing species of Agaricus. European Food Research and Technology 244(2):259–268

Gogoi P, Chutia P, Singh P, Mahanta CL (2019) Effect of optimized ultrasound-assisted aqueous and ethanolic extraction of Pleurotus citrinopileatus mushroom on total phenol, flavonoids and antioxidant properties. J Food Process Eng 42(6):e13172

Ha HT, Tran-Van H, Van Tran T, Nguyen HTN, Phan DTA (2023) Study on chemical compositions, antioxidants and intracellular anti-melanogenic activities of varieties of Ganoderma lucidum in Vietnam. Int J Food Sci Technol 58(8):4127–4135

Hapsari R, Elya B, Amin J (2012) Formulation and evaluation of antioxidant and tyrosinase inhibitory effect from gel containing the 70 % ethanolic Pleurotus ostreatus extract. Int J Med Arom Plants 2(1):135–140

Hwang AY, Yang SC, Kim J, Lim T, Cho H, Hwang KT (2019) Effects of non-traditional extraction methods on extracting bioactive compounds from chaga mushroom (Inonotus obliquus) compared with hot water extraction. LWT 110:80–84

Hyde KD, Bahkali AH, Moslem MA (2010) Fungi—an unusual source for cosmetics. Fungal Divers 43:1–9

Islam T, Yu X, Xu B (2016) Phenolic profiles, antioxidant capacities and metal chelating ability of edible mushrooms commonly consumed in China. LWT-Food Sci Technol 72:423–431

Jung E, Lee J, Huh S et al (2009) Phloridzin-induced melanogenesis is mediated by the cAMP signaling pathway. Food Chem Toxicol 47(10):2436–2440

Junior FADSC, Petrarca MH, Meinhart AD, de Jesus Filho M, Godoy HT (2019) Multivariate optimization of extraction and validation of phenolic acids in edible mushrooms by capillary electrophoresis. Food Res Int 126:108685

Kaewnarin K, Suwannarach N, Kumla J, Lumyong S (2016) Phenolic profile of various wild edible mushroom extracts from Thailand and their antioxidant properties, anti-tyrosinase and hyperglycaemic inhibitory activities. J Funct Foods 27:352–364

Kakumyan P, Matsui K (2009) Characterization of volatile compounds in Astraeus spp. Biosci Biotechnol Biochem 73(12):2742–2745. https://doi.org/10.1271/bbb.90282

Kamagaju L, Morandini R, Bizuru E, Nyetera P, Nduwayezu JB, Stévigny C, Ghanem G, Duez P (2013) Tyrosinase modulation by five Rwandese herbal medicines traditionally used for skin treatment. J Ethnopharmacol 146(3):824–834. https://doi.org/10.1016/j.jep.2013.02.010

Kameyama K, Takemura T, Hamada Y et al (1993) Pigment production in murine melanoma cells is regulated by tyrosinase, tyrosinase-related protein 1 (TRP1), DOPAchrome tautomerase (TRP2), and a melanogenic inhibitor. J Invest Dermatol 100(2):126–131

Kapoor D, Singh S, Kumar V, Romero R, Prasad R, Singh J (2019) Antioxidant enzymes regulation in plants in reference to reactive oxygen species (ROS) and reactive nitrogen species (RNS). Plant Gene 19:100182

Kendall AC, Nicolaou A (2013) Bioactive lipid mediators in skin inflammation and immunity. Prog Lipid Res 52(1):141–164

Kim JS, Kuk E, Yu KN, Kim JH, Park SJ, Lee HJ et al (2007) Antimicrobial effects of silver nanoparticles. Nanomed Nanotechnol Biol Med 3(1):95–101

Kim SY, Go KC, Song YS, Jeong YS, Kim EJ, Kim BJ (2014) Extract of the mycelium of T. matsutake inhibits elastase activity and TPA-induced MMP-1 expression in human fibroblasts. Int J Mol Med 34(6):1613–1621

Komura DL, Carbonero ER, Gracher AHP et al (2010) Structure of Agaricus spp. fucogalactans and their anti-inflammatory and antinociceptive properties. Bioresour Technol 101(15):6192–6199

Lee JS, Shin DB, Lee SM et al (2013) Melanogenesis inhibitory and antioxidant activities of Phellinus baumii methanol extract. Korean J Mycol 41(2):104–111

Leem KH (2015) Effects of Olibanum extracts on the collagenase activity and procollagen synthesis in Hs68 human fibroblasts and tyrosinase activity. Adv Sci Technol Lett 88:172–175

Lin S, Ching LT, Lam K, Cheung PC (2017) Anti-angiogenic effect of water extract from the fruiting body of Agrocybe aegerita. LWT 75:155–163

Liu YT, Sun J, Luo ZY, Rao SQ, Su YJ, Xu RR, Yang YJ (2012) Chemical composition of five wild edible mushrooms collected from Southwest China and their antihyperglycemic and antioxidant activity. Food Chem Toxicol 50(5):1238–1244

Lourith N, Pungprom S, Kanlayavattanakul M (2021) Formulation and efficacy evaluation of the safe and efficient moisturizing snow mushroom hand sanitizer. Abstract. J Cosmet Dermatol 20(2):554–560. https://doi.org/10.1111/jocd.13543

Ma T, Hara M, Sougrat R, Verbavatz JM, Verkman AS (2002) Impaired stratum corneum hydration in mice lacking epidermal water channel aquaporin-3. Journal of Biological Chemistry 277(19):17147–17153.

Mago P, Sharma R, Hafeez I et al (2023) Mushroom based cosmeceuticals: an upcoming biotechnology sector. Biosci Biotechnol Res Asia 20(2). https://doi.org/10.13005/bbra/3097

Maranduca MA, Branisteanu D, Serban DN et al (2019) Synthesis and physiological implications of melanic pigments. Oncol Lett 17(5):4183–4187

Masaki H (2010) Role of antioxidants in the skin: anti-aging effects. J Dermatol Sci 58(2):85–90

McCabe MC, Hill RC, Calderone K et al (2020) Alterations in extracellular matrix composition during aging and photoaging of the skin. Matrix Biol Plus 8:100041

Meng TX, Furuta S, Fukamizu S et al (2011) Evaluation of biological activities of extracts from the fruiting body of Pleurotus citrinopileatus for skin cosmetics. J Wood Sci 57:452–458

Meng TX, Zhang CF, Miyamoto T et al (2012) The melanin biosynthesis stimulating compounds isolated from the fruiting bodies of Pleurotus citrinopileatus. J Cosmet Dermatol Sci Appl 2(03):151

Mensah RA, Shanmugam V, Narayanan S, Renner JS, Babu K, Neisiany RE, Försth M, Sas G, Das O (2022) A review of sustainable and environment-friendly flame retardants used in plastics. Polym Test 108:107511

Moon JY, Yim EY, Lee NH, Hyun CG (2010) Screening of elastase and tyrosinase inhibitory activity from Jeju Island plants. Eurasia J Biosci 4:41–53. https://doi.org/10.5053/ejobios.2010.4.0.6

Nazir Y, Linsaenkart P, Khantham C, Chaitep T, Jantrawut P, Chittasupho C et al (2021) High efficiency in vitro wound healing of Dictyophora indusiata extracts via anti-inflammatory and collagen stimulating (MMP-2 inhibition) mechanisms. J Fungi 7(12):1100

Ndlovu G, Fouche G, Tselanyane M, Cordier W, Steenkamp V (2013) In vitro determination of the anti-aging potential of four southern African medicinal plants. BMC Complement Altern Med 13:1–7

Nieto IJ, Chegwin CA (2008) Triterpenoids and fatty acids identified in the edible mushroom Pleurotus sajor-cajú. J Chil Chem Soc 53(2). https://doi.org/10.4067/S0717-97072008000200015

Nitthikan N, Leelapornpisid P, Naksuriya O et al (2022) Potential and alternative bioactive compounds from brown Agaricus bisporus mushroom extracts for xerosis treatment. Sci Pharm 90(4):59

Nowacka N, Nowak R, Drozd M, Olech M, Los R, Malm A (2015) Antibacterial, antiradical potential and phenolic compounds of thirty-one polish mushrooms. PLoS One 10(10):e0140355

Onar HC, Yusufoglu A, Turker G, Yanardag R (2012) Elastase, tyrosinase and lipoxygenase inhibition and antioxidant activity of an aqueous extract from Epilobium angustifolium L. leaves. J Med Plants Res 6(5):716–726

Packialakshmi B, Sudha G, Charumathy M (2017) Free radical scavenging and antioxidant efficacy of ethanolic fractions from an edible medicinal mushroom—Auricularia polytricha. Int J Pharm. 7(2):108–114

Palacios I, Lozano M, Moro C, D'arrigo M, Rostagno MA, Martínez JA et al (2011) Antioxidant properties of phenolic compounds occurring in edible mushrooms. Food Chem 128(3):674–678

Papakonstantinou E, Roth M, Karakiulakis G (2012) Hyaluronic acid: a key molecule in skin aging. Dermatoendocrinol 4(3):253–258

Park HS, Hwang YH, Kim MK et al (2015) Functional polysaccharides from Grifola frondosa aqueous extract inhibit atopic dermatitis-like skin lesions in NC/Nga mice. Biosci Biotechnol Biochem 79(1):147–154

Pintus F, Spano D, Corona A, Medda R (2015) Antityrosinase activity of Euphorbia characias extracts. PeerJ 3:e1305. https://doi.org/10.7717/peerj.1305

Piwowarski JP, Kiss AK, Kozłowska-Wojciechowska M (2011) Anti-hyaluronidase and anti-elastase activity screening of tannin-rich plant materials used in traditional Polish medicine for external treatment of diseases with inflammatory background. J Ethnopharmacol 137(1):937–941

Quan T, Qin Z, Xia W, Shao Y, Voorhees JJ, Fisher GJ (2009) Matrix-degrading metalloproteinases in photoaging. J Invest Dermatol Symp Proc 14(1):20–24. https://doi.org/10.1038/jidsymp.2009.8

Rai M, Feitosa CM (eds) (2022) Eco-friendly biobased products used in microbial diseases. CRC Press

Rawlings AV, Scott IR, Harding CR, Bowser PA (1994) Stratum corneum moisturization at the molecular level. J Invest Dermatol 103(5):731–740

Rawlings AV, Matts PJ (2005) Stratum corneum moisturization at the molecular level: an update in relation to the dry skin cycle. J Invest Dermatol 124(6):1099–1110

Sabino-Ferrari AB, Azevedo de Oliveira G, Mannochio Russo H, de Carvalho Bertozo L, da Silva Bolzani V, Cunha Zied D, Farias Ximenes V, Zeraik ML (2021) Pleurotusostreatus and Agaricus subrufescens: investigation of chemical composition and antioxidant properties of these mushrooms cultivated with different handmade and commercial supplements. Int J Food Sci Technol 56(1):452–460

Sánchez C (2017) Bioactives from mushroom and their application. In: Food bioactives: Extraction and biotechnology applications, pp 23–57

Sangthong S, Pintathong P, Pongsua P et al (2022) Polysaccharides from Volvariella volvacea mushroom: extraction, biological activities and cosmetic efficacy. J Fungi 8(6):572

Seo MS, Jang YA, Lee JT (2018) The study of cosmeceutical activities from Lentinula edodes extracts and application a natural cosmetic material. J Korean Appl Sci Technol 35(4):1003–1012

Shin S, Cho SH, Park D, Jung E (2020) Anti-skin aging properties of protocatechuic acid in vitro and in vivo. J Cosmet Dermatol 19(4):977–984

Sies H, Jones DP (2020) Reactive oxygen species (ROS) as pleiotropic physiological signalling agents. Nat Rev Mol Cell Biol 21(7):363–383

Soto ML, Falqué E, Domínguez H (2015) Relevance of natural phenolics from grape and derivative products in the formulation of cosmetics. Cosmetics 2(3):259–276

Souilem F, Fernandes Â, Calhelha RC, Barreira JC, Barros L, Skhiri F et al (2017) Wild mushrooms and their mycelia as sources of bioactive compounds: antioxidant, anti-inflammatory and cytotoxic properties. Food Chem 230:40–48

Taofiq O, González-Paramás AM, Martins A, Barreiro MF, Ferreira IC (2016a) Mushrooms extracts and compounds in cosmetics, cosmeceuticals and nutricosmetics—a review. Ind Crop Prod 90:38–48

Taofiq O, Heleno SA, Calhelha RC, Alves MJ, Barros L, Barreiro MF et al (2016b) Development of mushroom-based cosmeceutical formulations with anti-inflammatory, anti-tyrosinase, antioxidant, and antibacterial properties. Molecules 21(10):1372

Tayarani-Najaran Z, Eghbali-Feriz S, Taleghani A et al (2018) Anti-melanogenesis and anti-tyrosinase properties of Pistacia atlantica mutica extracts on B16F10 murine melanoma cells. Res Pharm Sci 13:533

Theeraraksakul K, Jaengwang K, Choowongkomon K, Tabtimmai L (2023) Exploring the biological functions and antimelanogenesis of Phallus indusiatus for mushroom-based cosmetic applications. Cosmetics 10(5):121

Thomas NV, Kim SK (2013) Beneficial effects of marine algal compounds in cosmeceuticals. Mar Drugs 11(1):146–164

Thring TS, Hili P, Naughton DP (2009) Anti-collagenase, anti-elastase and anti-oxidant activities of extracts from 21 plants. BMC Complement Altern Med 9(1):1–11

Ukawa Y, Izumi Y, Ohbuchi T et al (2007) Oral administration of the extract from Hatakeshimeji (Lyophyllum decastes sing.) mushroom inhibits the development of atopic dermatitis-like skin lesions in NC/Nga mice. J Nutr Sci Vitaminol 53(3):293–296

Vaithanomsat P, Boonlum N, Chaiyana W et al (2022) Mushroom-glucan recovered from antler-type fruiting body of Ganoderma lucidum by enzymatic process and its potential biological activities for cosmeceutical applications. Polymers 14(19):4202

Veljović S, Veljović M, Nikićević N, Despotović S, Radulović S, Nikšić M, Filipović L (2017) Chemical composition, antiproliferative and antioxidant activity of differently processed Ganoderma lucidum ethanol extracts. J Food Sci Technol 54:1312–1320

Visvanathan S, Krishnamoorthy R, Sabesan GS (2022) Fungal cosmetics mushrooms in beauty care and the new age of natural cosmetics. In: Applied mycology: entrepreneurship with fungi. Springer, Cham, pp 1–37

Wang P-W, Hung Y-C, Lin T-Y, Fang J-Y, Yang P-M, Chen M-H, Pan T-L (2019) Comparison of the biological impact of UVA and UVB upon the skin with functional proteomics and immunohistochemistry. Antioxidants 8(12):569. https://doi.org/10.3390/antiox8120569

Wasmeier C, Hume AN, Bolasco G, Seabra MC (2008) Melanosomes at a glance. J Cell Sci 121(24):3995–3999. https://doi.org/10.1242/jcs.040667

Wu Y, Choi MH, Li J et al (2016) Mushroom cosmetics: the present and future. Cosmetics 3(3):22

Yahia EM, Gutiérrez-Orozco F, Moreno-Pérez MA (2017) Identification of phenolic compounds by liquid chromatography-mass spectrometry in seventeen species of wild mushrooms in Central Mexico and determination of their antioxidant activity and bioactive compounds. Food Chem 226:14–22

Yan ZF, Yang Y, Tian FH, Mao XX, Li Y, Li CT (2014) Inhibitory and acceleratory effects of inonotus obliquus on tyrosinase activity and melanin formation in B16 melanoma cells. Evid Based Complement Altern Med 2014(1). https://doi.org/10.1155/2014/259836

Yang FQ, Feng K, Zhao J, Li SP (2009) Analysis of sterols and fatty acids in natural and cultured Cordyceps by one-step derivatization followed with gas chromatography-mass spectrometry. J Pharm Biomed Anal 49(5):1172–1178

Zduńska-Pęciak K, Dębowska R, Kołodziejczak A, Rotsztejn H (2022) Ferulic acid—a novel topical agent in reducing signs of photoaging. Dermatol Ther 35(7):e15543

Chapter 9
Mushroom Fusion: Blending Science and Nature for Aesthetic Marvels

Irum, Muhammad Zulqurnain Haider, Qasim Ali, Altaf Ahmed Simair, Asia Shaheen, Naveed Sultan, and Maha Gul Zafar

Contents

Abstract Traditionally acknowledged for bioactive properties, mushrooms are now emerging as a novel focus in cosmetic industry, uncovering their potential for use in aesthetic formulations. A variety of mushrooms, as well as the nutrients they contain, have been proved to be beneficial to skin and hair. Medicinal mushrooms are becoming more popular as a possible element in cutting-edge skincare products.

Irum · M. Z. Haider (✉) · Q. Ali · A. Shaheen · N. Sultan
Department of Botany, Government College University Faisalabad, Faisalabad, Pakistan
e-mail: drmzhaider@gcuf.edu.pk

A. A. Simair
Department of Botany, Government College University, Hyderabad, Hyderabad, Sindh, Pakistan

M. G. Zafar
Biochemistry Section, Jhang-campus, University of Veterinary and Animal Sciences Lahore, Lahore, Pakistan

S. Javed et al. (eds.), *Sustainable Cosmeceuticals*,
https://doi.org/10.1007/978-3-031-86087-4_9

This chapter delves into the potential of mushroom-based aesthetic skincare, concentrating on enhancing cosmetic appearance by addressing issues like scars, skin laxity, wrinkles, moles, liver spots, excess fat, cellulite, undesired hair, skin discoloration, and spider veins. It also explores the potential of mushroom-based cosmeceuticals, and this integration aligns with support for sustainability by promoting biodiversity, enhancing sustainable agricultural practices, supporting local communities, and reducing the ecological impact of the beauty industry. Mushroom bioactive compounds such as polysaccharides, triterpenoids, polyphenols, and peptides offer specific benefits to skin health. These ingredients are pivotal in skincare formulations due to their antiaging, anti-inflammatory, and antioxidant properties. This chapter also highlights the advantages and disadvantages of combining these miraculous substances into modern cosmetic procedures, as well as paving the way for further research and improvement in this emerging biotechnology field.

Keywords Nature's pharmacy · Natural antioxidant · Anti-inflammatory · Natural antiaging · Sustainable personalized beauty · Mushroom biotechnology

9.1 Introduction

Humans have spent most of their time on Earth as hunter-gatherers, living off the natural resources of their surroundings. Since the beginning of human history, people have been eating mushrooms. Mushrooms have been eaten since 300 B.C., after Theophrastus called them "truffles." Romans considered mushrooms as "food of the Gods," while Greek soldiers used them to boost their strength before combat. It is estimated that $63 billion was spent on producing 30 million tons of mushrooms (Royse et al. 2017). The global mushroom harvest in 2013 was up dramatically from 1978, when only one million tons were harvested. Approximately $54 billion, or 54%, of all produced mushrooms are not fit for human consumption (Raut 2019).

Mushrooms have been consumed and used medicinally for centuries. Environmental stress promotes the production of secondary metabolites, which are frequently associated with medicinal properties; thus, wild mushrooms can provide a wide range of beneficial compounds (Qaderi et al. 2023). The market for functional foods and medicinal mushrooms has grown significantly over the past 10 years, and there are now a wide range and quantity of mushroom products accessible in markets all over the world (Cateni et al. 2021; Zhang et al. 2021; Singh et al. 2022; Pandey et al. 2020). In terms of their main nutritional components, mushrooms occupy a special space between meat and vegetables. For those who are concerned about their health, mushrooms are a desired alternative because they are low in calories and devoid of cholesterol. Ergosterol and important PUFAs such as linoleic, oleic, and linolenic acids are frequently found as the primary components of their lipid profile (Sande et al. 2019). Due to their comprehensive amino acid composition essential for human metabolism, mushrooms serve as a very advantageous inclusion within the progressively prevalent vegetarian and vegan dietary patterns (González et al. 2020). Mushrooms further boost their nutritional value and improve general well-being because they are an excellent source of carotenoids, lycopene, tocopherols, and vitamin C, as well as being rich in B-complex vitamins (Arshad et al. 2022; Vunduk et al. 2021).

Secondary metabolites (polyphenols) and nitrogen-containing compounds (L-ergothioneine, terpenoids, sterols, lovastatin) are some of the active constituents and metabolites found in abundance within mushrooms (Zhang et al. 2021). The bioactivity of these substances transcends the scope of age-related illness prevention and therapy, encompassing conditions such as arthritis, diabetes, dementia, cancer, atherosclerosis, vascular disorders, obesity, osteoporosis, and metabolic syndromes (Cateni et al. 2021). Mushrooms are not just good for eating; they're also ideal for the emerging cosmeceutical/mycoceutical industry.

Due to their beneficial effects on the skin and hair, they are used in a wide variety of skin and hair care products. The major constituents and important effects of mushrooms, such as moisturizing and antiaging (including whitening and antioxidant activity), will be discussed in this chapter, leading to the conclusion that mushroom extracts (and/or ingredients) are among the best options for the application of cosmetic products.

Mushrooms, in terms of morphology, have inspired numerous artists and designers. The mushroom aesthetic derives inspiration from mushrooms, fungi, toadstools, and nature, with visuals and motifs of these elements dominating. Earthy tones, such as maroon, white, yellow, and red, prevail as dominant colors, embodying the aesthetic's reverence for mushrooms and the natural world. This artistic movement has permeated social media and the Internet through Mushroomcore, Aesthetics Wiki, or Fandom, an in-depth page exploring the Mushroomcore aesthetic, covering its history, visual motifs, fashion, hobbies, and related aesthetics. Furthermore, Mushroomcore Fashion: The Mushroomcore Trend and the Pervasiveness of Fungi delves into the Mushroomcore fashion trend, focusing on its current popularity spike. Additionally, Goblincore, the fashion movement embracing "chaos, dirt, and mud," is studied, along with its relationship to Mushroomcore.

However, this chapter exclusively concentrates on the aesthetic marvel related to the effects of mushroom metabolites or extracts.

9.2 Nutritional and Medicinal Facts About Mushrooms

9.2.1 Nutritional Benefits of Mushrooms

Wild mushrooms have been used in a wide variety of ways for ages as a source of food, medicine, and cosmetics (Fig. 9.1). These days, you may get cosmetics that include mushroom extracts, whether it be in the form of a cream, lotion, ointment, serum, or facial preparation. Well-known edible mushrooms that are commercially grown on a global scale include *Agaricus bisporus*, *Lentinula edodes*, *Pleurotus*

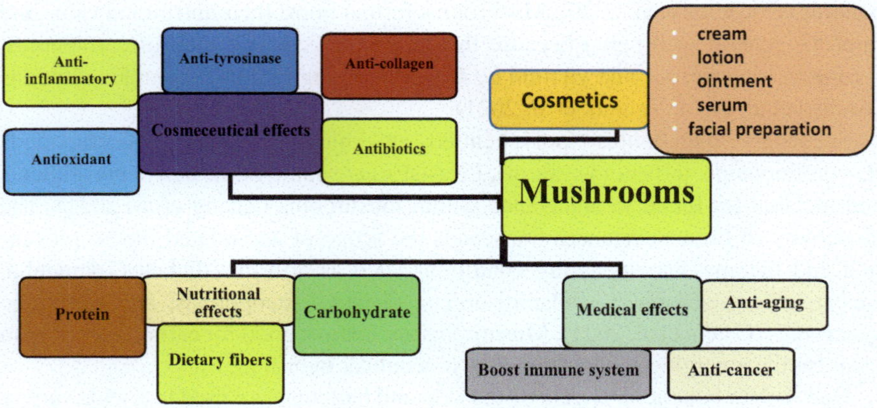

Fig. 9.1 Overview of the nutritional and medicinal impact of mushrooms

spp., and *Volvariella volvacea* due to protein, fat, carbohydrate, and dietary fiber levels. These mushrooms provide 16–31% proteins, 2–6% lipids, 51–73% carbs, and 8–12% dietary fiber, depending on their dry weight (Singh et al. 2022). Ergosterol is found in abundance in mushrooms. During the production of the fruiting body, ergosterol is converted into ergocalciferol, often known as vitamin D2 (Morales et al. 2017).

Mushrooms are great for diabetics and hypertensive individuals because of their low glycemic index, high-quality proteins and fibers, and high potassium: sodium ratio (Feeney et al. 2014). Beyond the beneficial nutritional effects, mushrooms are also gaining popularity for their capacity to alter physiological systems such as the immunological, endocrine, neurological, circulatory, and digestive systems (Plaami et al. 2001). The scientific community has concluded that mushrooms are nature's pharmacy (Mohiuddin 2021; Yadav and Negi 2021). Polyphenols, nitrogen-containing substances including ergothioneine, terpenoids, sterols, lovastatin, dietary fibers, and secondary metabolites are just some of the bioactive chemicals found in these foods. They're present in negligible numbers and have a low caloric density (Yadav and Negi 2021). Therefore, mushrooms are more than just food; they are a perfect match for an emerging cosmeceutical industry.

9.2.2 Medicinal Benefits of Mushrooms

In various East Asian countries such as China, Japan, Thailand, and Korea, it is a prevalent practice among individuals to incorporate the consumption of medicinal mushrooms into their dietary habits (Fig. 9.1). These mushrooms include *Ganoderma lucidum*, popularly known as Lingzhi or Reishi, *Agaricus blazei*, widely known as sun mushroom or himematsutake, *Lentinula edodes*, generally known as shiitake, and *Grifola frondosa*, also known as maitake or hui-shu-hua. Consuming these mushrooms has the primary purpose of boosting the immune system, inhibiting the growth of malignant cells, and assisting in post-chemotherapy recovery. The immunomodulatory characteristics of mushroom-derived dietary fibers, particularly glucans, and secondary metabolites such as polyphenols have piqued the interest of researchers in the health field (Shakoor et al. 2021). In addition to enhancing cytoprotective and antioxidant enzymes, these compounds provide protection against autoimmune disorders and reduce organ transplant rejection (Hyde et al. 2019). According to recent research, mushroom metabolites may help to attenuate coronavirus infections through suppressed viral protease activity and by preventing an inflamed cytokine storm (Rangsinth et al. 2021). Fungal immune-regulating proteins (FIPs) have been extensively studied for their prospective medical applications in the pharmaceutical sector (Liu et al. 2020). FIPs have shown antiallergic, anti-inflammatory, and anticancer bioactivity because their bioactive domains remain intact when reaching the intestines, after oral administration (Liu et al. 2020).

9.3 Aesthetic Marvels of Mushroom

The human skin regulates body temperature and water loss in addition to acting as a physical barrier against environmental hazards such as ultraviolet (UV) radiation, urban air pollution, infections, and toxins (Abdo et al. 2020). Recent years have seen explosive expansion in the cosmetic market. There has been a gradual trend toward "natural" solutions among young people, as they learn about the dangers of chemical products and the advantages of natural ones.

Cosmetology refers to the use of topical or ingestible products that renew, protect, and enhance one's appearance. Since they include a wide range of chemical components, both synthetic and natural, they have a beneficial effect on the skin or the area to which they are applied.

Cosmeceuticals are items that combine cosmetic and therapeutic properties. Topical medications like creams, lotions, and ointments with biologically active components fall under this category (Sharma 2011). In addition to improving the skin's look, vibrancy, texture, and antiaging activity, it is well known that these products contain substances that affect the skin's biological function by supplying the nutrients necessary for healthy skin. Ingredients derived from plants and microorganisms are increasingly common in skin care products. These include vitamins and minerals, as well as dairy products (Keller et al. 2014) and animal protein components (Rodrigues et al. 2015). Cosmeceuticals can be made with either naturally occurring (minerals, plant extracts, microbial products) or synthetically produced (chemically synthesized) ingredients. Mushroom-based products have become popular due to the rising demand for alternatives to synthetic chemicals. Nutritional supplements containing mushroom extracts as a functional ingredient act as a nutraceutical source that can help improve people's health in general (Wu et al. 2016).

Cosmeceutical ingredients in cosmetics have therapeutic effects against several skin diseases, including inflammation (Taofiq et al. 2016). Several cosmetic industries have utilized mushroom-based cosmetic formulations. It was asserted that mushroom extracts have cosmeceutical properties, such as moisturizing, anti-aging, antiacne, and skin-lightening effects (Visvanathan et al. 2022). *Tremella fuciformis* polysaccharide was extracted for testing, and it was found to have an excellent moisturizing effect and to suppress the development of dark spots (Yang et al. 2006). *Matsutake mycelium* extract significantly suppresses elastase activity, lowers elastin breakdown, and prevents or repairs the production of wrinkles in aging skin (Kim et al. 2014). *Grifola frondosa* ergosterols were confirmed to inhibit aging enzyme cyclooxygenase 2 (Cox-2) after being isolated from the mushroom's mycelium (Zhang et al. 2002).

A multitude of research has extensively documented the existence of a diverse array of bioactive metabolites, encompassing phenolic compounds, terpenoids, polysaccharides, lectins, steroids, glycoproteins, and lipid constituents. Numerous studies have demonstrated the significant biological functions of several mushroom extracts and their secondary metabolites. These functions encompass a range of key impacts, including antioxidant, anticancer, antibacterial, and anti-inflammatory properties. Additionally, these extracts have been found to possess immunomodulatory, antiatherogenic, and hypoglycemic roles (Taofiq et al. 2016). Very little research has been done on the effects of mushroom extracts or specific components on tyrosinase, collagenase, elastase, and hyaluronidase. The bioactivities mentioned here provide dermatologists and cosmetics-based industries a reason to be hopeful about the potential of chemicals found in mushrooms.

9.3.1 Antioxidant Activity

Oxygen is necessary for all life on Earth to survive, but it is especially important for humans. It has been related to cell death and tissue injury, which in turn has been connected to a sped-up aging process and many diseases such as diabetes, cardiovascular disease, neurodegenerative disease, and cancer (Majeed and Prakash 2021). For the body to function normally, cellular and tissue levels undergo a number of biological responses. These reactions frequently generate free radicals or species with unpaired electrons. These free radicals include reactive oxygen species (ROS), reactive nitrogen species, and reactive sulfur species. Excessive exposure to UV light causes ROS to be produced, which in turn leads to DNA damage, inflammation, hyperpigmentation which degrades collagen, and a decrease in collagen synthesis, all of which contribute to the photo-aged skin effect (Masaki 2010). The excessive generation of ROS has the potential to exhaust the endogenous antioxidant reservoir inside the body, which comprises glutathione peroxidase, catalase, and superoxide dismutase. Consequently, this phenomenon can induce oxidative stress within the cellular milieu (Carocho and Ferreira 2013). Mushrooms, as a dietary source containing a high concentration of bioactive compounds with antioxidant activity, can help the body overcome its paucity of endogenous antioxidant sources. These dietary sources of antioxidants, such as vitamin C (ascorbic acid), vitamin E (tocopherol), beta-carotene, vitamin K, flavonoids, phenolic acids, selenium, and zinc, tend to keep oxidative stress under control (Carocho and Ferreira 2013).

Extensive studies on the scavenging ability of mushrooms for ROS have uncovered the antioxidative effects of *Pleurotus citrinopileatus* and *Coriolus versicolor* (Sánchez 2017; Kozarski et al. 2015; Meng et al. 2011; Karaman et al. 2010). Mushroom extracts have been reported to have antioxidant activity in the form of radical scavengers (hydrogen peroxide), reducing power (e.g., ferric reducing antioxidant power), and lipid peroxidation inhibitors (e.g., heme degradation of peroxides, ferrous oxidation-xylenol, and thiobarbituric acid- reactive compounds) (Carocho and Ferreira 2013).

9.3.2 Anti-tyrosinase Activity

Melanin not only determines the color of skin, hair, and eyes, but it also shields the skin from harmful factors including UV light, poisonous medications, and chemicals. The increased activity of the tyrosinase enzyme is responsible for the overproduction of melanin in the epidermis when the body is overexposed to UV radiation. Tyrosinase is the enzyme that acts as the rate-limiting factor in the biosynthesis of melanin. Its primary function is to catalyze the oxidation of tyrosine, leading to the formation of dihydroxyphenylalanine (DOPA), which subsequently undergoes other enzymatic reactions to produce dopaquinone (Meng et al. 2012).

Dopaquinone is oxidized to form eumelanin (pigment with a brown-black color) in the presence of the dopachrome tautomerase. Dopaquinone can be converted to cysteinyl DOPA in the presence of cysteine or glutathione, which is a precursor of a yellow-red pigmentation, pheomelanin (Jung et al. 2009; Chang 2009).

Higher amounts of eumelanin can lead to darker skin tones, while lower levels of eumelanin are responsible for lighter skin tones. One of the numerous advantages of eumelanin is its ability to provide protection against skin cancer. The study conducted by researchers revealed a negative correlation between the presence of higher levels of eumelanin in individuals' skin and their susceptibility to developing skin cancer. By blocking the sun's heat, eumelanin aids in temperature control (Solano 2020). Hyperpigmentation issues such as melasma, post-inflammatory melanoderma, freckles, lentigo, acne scars, and age spots are caused by overabundance of melanin pigment or an abnormal distribution of melanin pigment (Chang 2012). Skin whitening products in the cosmetic sector rely heavily on the inhibition of tyrosinase activity.

As a result of the drawbacks of conventional depigmenting methods, there has been a significant increase in the study of natural products obtained from plants, mushrooms, rhizomes, and marine algae that are high in bioactive chemicals capable of treating these conditions. Many phenolic compounds have been isolated from plants with the claim that they can block melanin production at various stages of the process (Ali et al. 2015). Some examples of these chemicals are arbutin, resveratrol, ellagic acid, and gentisic acid. Mushrooms, including *Ganoderma lucidum*, *Euphorbia characias*, and *Pistacia atlantica* subsp. mutica, have exhibited robust anti-tyrosinase activity, prompting extensive research for potential applications in skin whitening products (Pintus et al. 2015; Tayarani-Najaran et al. 2018; Chien et al. 2008). *Pleurotus* species are renowned as popular mushrooms with well-established medical benefits, including anti-inflammatory qualities, antioxidants, immunomodulatory effects, antimicrobial properties, and high nutritional content. Researchers express optimism about the prospective utilization of these compounds in cosmetics, capitalizing on their therapeutic properties (Meng et al. 2011).

Park et al. (2015b) assessed the effectiveness of various ethanolic mycelium extracts in inhibiting tyrosinase. Remarkably, *Inonotus mikadoi* demonstrated high efficacy, decreasing tyrosinase activity by 46.07 ± 7.5% at a dosage of 10 mg/mL. Similarly, *Coriolus versicolor* and *Fomitopsis* sp. exhibited inhibitory effects at comparable concentrations, reducing activity by approximately 26.3% and 23.9%, respectively. The study's authors suggested that certain tyrosinase-inhibiting extracts have the potential to be developed into cosmetically beneficial functional additives. The medicinal fungus *Inonotus obliquus*, commonly referred to as Chaga mushroom, has gained popularity in Moscow and other North European countries due to its antitumor and immune-modulating capabilities (Taofiq et al. 2016).

9.3.3 Anti-inflammatory Activity

The pain, redness, and swelling that accompany an injury are all symptoms of inflammation, which is a natural physiological response. Inflammatory mediators with the potential to contribute to the pathogenesis of inflammatory disorders include tumor necrosis factor (TNF), interleukins (IL-1, IL-6, and IL-8), inducible nitric oxide synthase (iNOS), Cox-2, and intercellular adhesion molecule-1 (ICAM-1) (Taofiq et al. 2016).

Atopic dermatitis is a chronic inflammatory skin disease that manifests as intense itching, redness, and rash when provoked by a combination of physiological and environmental factors (Park et al. 2015a). According to current studies, this condition impacts 1–2% of adults and 10–20% of children. Inflammatory mediators, including TNF-α, IL-1, IL-6, and IL-8, are produced by macrophages and other cells during this process. These mediators play a crucial role in initiating inflammation (Wu et al. 2011). A study conducted by Ukawa et al. (2007) documented the inhibitory effect of *Lyophyllum decastes* (Fr.) The impact of Singer extract on the progression of atopic dermatitis is noteworthy. Moreover, because they are all-natural and have no known negative effects, these mushroom metabolites can be incorporated into cosmeceutical formulations. Linoleic acid, found in anti-inflammatory products such as Bobbi Brown's Brightening Intensive Cream and True Cosmetics' Youth Revealing Complex, promotes healthy skin growth (Taofiq et al. 2019). Linoleic acid is present in some mushrooms, including *Pleurotus sajor-caju* (Nieto and Chegwin 2008; Taofiq et al. 2019), *Astraeus* (Ferreira Da Eira et al. 2005), and *Agaricus subrufescens* (Elsayed et al. 2014).

9.3.4 Anti-collagenase and Anti-elastase Activity

The epidermis, the outermost layer of human skin, provides essential structural support. Within the skin's extracellular matrix (ECM), dermal fibroblasts play a crucial role in producing collagen and elastin, fundamental structural proteins with various protective functions (Kim et al. 2014). Elastin, an ECM protein, enhances flexibility in anatomical components like the epidermis, lungs, aorta, and cartilage. Prolonged exposure to elastases results in the degradation of elastic fibers, contributing to the loss of skin elasticity and the formation of wrinkles. Elastase (metalloproteinase enzyme) is responsible for degrading elastin. Natural derivative plants, mushrooms, rhizomes, and sea algal compounds are being used for the development of active cosmetic compounds involved in reducing skin aging and wrinkles (Onar et al. 2012). UV rays trigger collagen breakdown and speed up the aging process (Piwowarski et al. 2011). Collagen plays a crucial role in the skin's ECM, contributing to suppleness, pliability, and durability. As age progresses, levels of collagen, elastin, and hyaluronic acid in the skin's ECM decline, leading to a loss of skin elasticity and strength, leading to the appearance of wrinkles on skin (Ndlovu et al. 2013).

Kim et al. (2007) studied the antioxidant, anti-tyrosinase, and anti-collagenase activities of mycelial extract (GF-M) and an exopolysaccharide (EPS) of *Grifola frondosa*. The anti-collagenase activity was assessed by evaluating its ability to inhibit MMP-1 synthesis in UVA-irradiated human skin fibroblasts. Treatment with EPS or GF-M (100 g/mL) was found to inhibit MMP-1 expression by 20% and 40%, respectively. Additionally, EPS showed more effective inhibition of UVA-induced MMPs compared to trans-retinoic acid (tRA, a known inhibitor of UVA-induced MMPs). Moreover, the treatment can enhance collagen biosynthesis by 53%. EPS and GF-M therapy reduced approximately 25% and 17% melanogenesis in B16 murine melanoma cells, respectively. These findings suggest that mushrooms and their polysaccharides could be effective in alleviating skin aging and hyperpigmentation. Similarly, Bae et al. (2005) also identified an EPS from a water-grown *Grifola frondosa* mycelial culture. This EPS (250 g/mL) is found to involve in the suppression of UVA-induced matrix metalloproteinase 1 (MMP-1) protein production up to 61.1% in cutaneous fibroblasts. The results suggest that EPS present in mushroom extract may play an important role in controlling photo-aging effects on the skin to satisfy the human aesthetic sense.

9.3.5 Antimicrobial Activity

Human skin is housing many kinds of non-pathogenic bacteria. The distribution and density of this natural skin microflora is continuously being influenced by person's age and environmental conditions including humidity, temperature, and sebum secretions (Elsner 2006). The inflammatory skin illnesses like excessive seborrheic dermatitis, cellulitis, erysipelas, impetigo, folliculitis, furuncle, and carbuncle abscess are linked to abnormal distribution and density of skin microflora (Alsterholm et al. 2010). Mushrooms and their bioactive compounds have been studied extensively for their antimicrobial characteristics. Alves et al. (2012) found that both edible and inedible mushrooms were effective against harmful bacteria (microflora). They also analyzed the extracts of mushrooms for their bioactive compounds and antibacterial qualities. Results showed that *Lentinula edodes*, followed by species from the genera *Boletus*, *Ganoderma*, and *Lepista*, were the most interesting for their potential to fight both gram-positive and gram-negative bacteria. Mushrooms' phenolic compounds and other secondary metabolites protect them from pathogens, insects, and even UV radiation (Heleno et al. 2015). Their cell walls and membrane disruption kill invasive pathogens, as demonstrated by their antibacterial action (Ribeiro et al. 2015).

9.3.6 Hair Care Products

Hair care products aim to improve scalp health and change the hair's natural structure (Draelos 1991). Hair cosmetics include things like shampoos, conditioners, serums, sprays, waxes, gels, mousses, straighteners, bleaches, perms, and dyes (Madnani and Khan 2013). Shampoos are used to remove dirt, oil, and other debris from the hair and scalp. They come in a wide variety of formulas to address a wide range of hair and scalp conditions such as greasy hair, dandruff, and androgenic alopecia. Glucuronoxylomannan (GXM), a polysaccharide found in *tremella* mushrooms, is widely used in the cosmetic industry because of its anti-inflammatory and wound-healing characteristics (De Baets and Vandamme 2001). As a part of the haircare cosmetic composition, a patented silk film extract from *shiitake* mushrooms was included. A new product was made combining *Ganoderma lucidum* extract with three other plants and zinc extracts, with the goal of increasing hair development in humans (Meehan 2015). This product was found to be effective especially in males, since it has reduced levels of dihydrotestosterone and prostatic hyperplasia, two hormones that can inhibit hair growth.

Providing essential nutrients to hair follicles is just as important for healthy hair as it is for the rest of the body. Damage to the hair can be caused by internal factors such as a lack of essential nutrients, as well as external ones like poor diet, lack of exercise, and extended illness. Vitamin D, antioxidants, and minerals like iron, selenium, and copper can all be found in mushrooms, and these nutrients do wonders

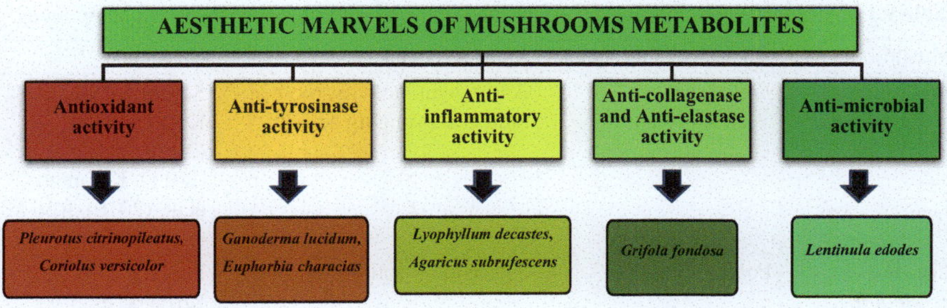

Fig. 9.2 Overview of the aesthetic marvels of the metabolite from mushrooms

for your hair by making it stronger and healthier and keeping breakage and dandruff at bay. The cosmetic industry stands to gain a lot by using mushroom extracts and/or whole mushrooms in hair care products (Arshad 2018).

The overall aesthetic properties of mushrooms are summarized with their effect and bioactive compounds obtained from various mushrooms ((Fig. 9.2); Table 9.1). It highlights the major activities of the fungal metabolite responsible for human aesthetic enhancement.

9.4 Perspective on Sustainable Development Goal (SDG)

Mushrooms, highly recognized for their therapeutic potential, offer both nutritional and medicinal benefits. Beyond their dietary importance, these play a significant role in the development of the cosmeceutical industry. These have been used in traditional skincare practices, particularly in East Asia, and their demand is rising due to the growing preference for safer, cost-effective, and toxin-free organic products. Mushroom cultivation is a sustainable agro-based microenterprise that empowers livelihoods and aligns with the framework of the Sustainable Development Goals (SDGs). As the demand for sustainability has accelerated in the past few years, many skincare brands have now incorporated medicinal mushrooms as key ingredients, ensuring ethical and sustainable source practices are followed. By incorporating mushrooms in cosmetics and skin care, both businesses and consumers will benefit. This would be multifunctional advantages of mushrooms—nourishing the skin while supporting a sustainable future (Arshad 2024; Arshad et al. 2023; Joseph et al. 2024; Mishra and Srivastava 2024; Shirur et al. 2021).

Table 9.1 Different mushrooms and their aesthetic properties with major bioactive compounds

Mushrooms	Bioactive compounds	Main activity	References
Pleurotus citrinopileatus	β-glucans	Antioxidant activity, oxygen radical scavenging activity	Sánchez (2017), Kozarski et al. (2015), Meng et al. (2011), Karaman et al. (2010), Rop et al. (2009)
Coriolus versicolor	Polysaccharides	Antioxidant activity	Sánchez (2017), Kozarski et al. (2015), Meng et al. (2011), Karaman et al. (2010)
Ganoderma lucidum	Arbutin, resveratrol, ellagic acid, and gentisic acid	Anti-tyrosinase activity	Pintus et al. (2015), Tayarani-Najaran et al. (2018), Chien et al. (2008)
Euphorbia characias	Arbutin, resveratrol, tyrosinase enzyme, ellagic acid, and gentisic acid	Anti-tyrosinase activity	Pintus et al. (2015), Tayarani-Najaran et al. (2018), Chien et al. (2008)
Pistacia atlantica subsp. mutica	Arbutin, resveratrol, ellagic acid, and gentisic acid	Anti-tyrosinase activity	Pintus et al. (2015), Tayarani-Najaran et al. (2018), Chien et al. (2008)
Pleurotus	β-glucans (pleuran) and α-glucans	Anti-tyrosinase activity, anti-inflammatory effect, antioxidants, immunomodulatory effects, and antibacterial activity	Maftoun et al. (2013), Meng et al. (2011)
Coriolus versicolor	Tyrosinase-inhibiting extracts	Anti-tyrosinase activity	Taofiq et al. (2016)
Fomitopsis sp.	Tyrosinase-inhibiting extracts	Anti-tyrosinase activity	Taofiq et al. (2016)
Inonotus obliquus	Triterpenes, polysaccharides, polyphenols, and melanin	Anti-tyrosinase activity, antitumor and immunomodulatory properties, anticancer and antitumor activities	Taofiq et al. (2016)
Lyophyllum decastes (Fr.) *Singer*	Polysaccharides	Antioxidant, anticholinesterase, antibacterial, and antidiabetic properties	Miura et al. (2002), Deveci et al. (2021)

(continued)

Table 9.1 (continued)

Mushrooms	Bioactive compounds	Main activity	References
Astraeus species	Polysaccharides, polyketides, and terpenoids	Anti-inflammatory, anticancer, and hypoglycemic efficacy	Zhang et al. (2007), De-Silva et al. (2013)
Cordyceps	Nucleosides, cyclic peptides, polyamines	Immunomodulatory, antidiabetic, anti-inflammatory, antitumor, antiproliferative	Xiao and Xiong (2013)
Grifola frondosa	Polysaccharides, proteins, peptides	Antitumor, antioxidant activity, antiviral, antibacterial effects, immunomodulation	Su et al. (2016)
Lentinula edodes	Phenolics, polysaccharides, sterols	Antioxidant, antitumor, anticancer activity	Heleno et al. (2015)
tremella	GXM	Antiallergic activities, anti-inflammatory activities	De Baets and Vandamme (2001)
Ganoderma lucidum	Triterpenic and phenolic acids, polysaccharides, terpenoids, nucleotides	Hair development in humans	Meehan (2015), Liu et al. (2020)

9.5 Conclusions and Future Perspectives

Mushrooms have gained considerable attention from the scientific and industrial communities owing to their substantial nutritional and cosmeceutical attributes. Numerous research has found that mushrooms serve as a plentiful reservoir of a diversified range of valuable natural compounds, characterized by their varied chemical composition and biological activity. Despite playing a significant role in different industries, the commercial utilization of mushrooms remains limited due to their relatively high cost and the labor-intensive processes involved. In order to enhance the accessibility of vital chemicals, it is imperative to focus on augmenting product yields and optimizing the generation of secondary metabolites in the foreseeable future. In this chapter, we mostly discussed the antioxidant, anti-inflammatory, and antibacterial benefits of mushrooms, but we also briefly discussed their anti-tyrosinase, anti-elastase, and anti-collagenase activity. Because of their high concentration of bioactive chemicals, mushrooms may 1 day be employed in cosmetics to treat a variety of skin conditions. To evaluate their possible advantages for skin health, more investigation is required to find and isolate secondary metabolites of substantial importance.

Subsequent investigations in the forthcoming years will prove advantageous in elucidating the genetic composition and enzymatic mechanisms underlying the biosynthetic pathway of different metabolites of mushrooms. Such studies will significantly contribute to the manipulation of their biosynthetic route, hence facilitating the augmentation of their production levels. The bioactive chemicals derived from mushrooms exhibit potential for future advancements in pharmaceutical research and as a dietary supplement for the prevention of human ailments. Mushrooms, which are abundant in nutrients, possess significant health benefits and have a long history of medical use in different regions worldwide. However, the product's market demand is limited due to a lack of extensive awareness and the significant costs associated with it. Furthermore, the absence of adequate quality control measures is a significant issue that raises concerns among individuals engaged in this industry and its associated disciplines. In the future, it is recommended that additional studies be performed to investigate the cosmeceutical applications of mushrooms in order to establish their effectiveness and increase public knowledge.

References

Abdo JM, Sopko NA, Milner SM (2020) The applied anatomy of human skin: a model for regeneration. Wound Med 28:100179

Ali SA, Choudhary RK, Naaz I, Ali AS (2015) Understanding the challenges of melanogenesis: key role of bioactive compounds in the treatment of hyperpigmentary disorders. J Pigment Dis 2:1–9

Alsterholm M, Karami N, Faergemann J (2010) Antimicrobial activity of topical skin pharmaceuticals? An in vitro study. Acta Derm Venereol 90(3):239–245

Alves MJ, Ferreira IC, Dias J, Teixeira V, Martins A, Pintado M (2012) A review on antimicrobial activity of mushroom (Basidiomycetes) extracts and isolated compounds. Planta Med 78:1707–1718

Arshad M (2018) Cogeneration policy to contest climate changes and the future of sustainable energy. IOP Conf Ser Earth Environ Sci 200:012050

Arshad M (2024) Sustainability of renewable energy options as compared to coal-fired power plants in Pakistan. Green Low-Carbon Econ 2(3):162–173

Arshad M, Ansari AR, Qadir R, Tahir MH, Nadeem A, Mehmood T, Alhumade H, Khan N (2022) Green electricity generation from biogas of cattle manure: an assessment of potential and feasibility in Pakistan. Front Energy Res 10:911485

Arshad M, Syed-Hassan SSA, Masood R, Ansari AR, Mumtaz A, Rahman A, Batool S, Atif FA, Muhammad A, Rehman AU (2023) Utilization of waste animal fat for sustainable biodiesel production. In: Climate changes mitigation and sustainable bioenergy harvest through animal waste: sustainable environmental implications of animal waste. Springer, Cham, pp 149–169

Bae JT, Sim GS, Lee DH, Lee BC, Pyo HB, Choe TB, Yun JW (2005) Production of exopolysaccharide from mycelial culture of Grifola frondosa and its inhibitory effect on matrix metalloproteinase-1 expression in UV-irradiated human dermal fibroblasts. FEMS Microbiol Lett 251(2):347–354

Carocho M, Ferreira IC (2013) A review on antioxidants, prooxidants and related controversy: natural and synthetic compounds, screening and analysis methodologies and future perspectives. Food Chem Toxicol 51:15–25

Cateni F, Gargano ML, Procida G, Venturella G, Cirlincione F, Ferraro V (2021) Mycochemicals in wild and cultivated mushrooms: nutrition and health. Phytochem Rev 21:339–383

Chang TS (2009) An updated review of tyrosinase inhibitors. Int J Mol Sci 10(6):2440–2475

Chang TS (2012) Natural melanogenesis inhibitors acting through the down-regulation of tyrosinase activity. Materials 5(9):1661–1685

Chien CC, Tsai ML, Chen CC, Chang SJ, Tseng CH (2008) Effects on tyrosinase activity by the extracts of Ganoderma lucidum and related mushrooms. Mycopathologia 166:117–120

De Baets S, Vandamme EJ (2001) Extracellular Tremella polysaccharides: structure, properties and applications. Biotechnol Lett 23:1361–1366

De Silva DD, Rapior S, Sudarman E, Stadler M, Xu J, Aisyah Alias S, Hyde KD (2013) Bioactive metabolites from macrofungi: ethnopharmacology, biological activities and chemistry. Fungal Divers 62:1–40

Deveci E, Çayan F, Tel-Çayan G, Duru ME (2021) Inhibitory activities of medicinal mushrooms on α-amylase and α-glucosidase-enzymes related to type 2 diabetes. S Afr J Bot 137:19–23

Draelos ZK (1991) Hair cosmetics. Dermatol Clin 9(1):19–27

Elsayed EA, El Enshasy H, Wadaan MA, Aziz R (2014) Mushrooms: a potential natural source of anti-inflammatory compounds for medical applications. Mediat Inflamm 2014:805841

Elsner P (2006) Antimicrobials and the skin physiological and pathological flora. Biofunct Textiles Skin 33:35–41

Feeney MJ, Miller AM, Roupas P (2014) Mushrooms—biologically distinct and nutritionally unique: exploring a "third food kingdom". Nutr Today 49(6):301

Ferreira Da Eira A, Kaneno R, Rodrigues Filho E, Barbisan LF, Pascholati SF, Di Piero RM, Ribeiro LR (2005) Farming technology, biochemistry characterization, and protective effects of culinary-medicinal mushrooms Agaricus brasiliensis S. Wasser et al. and Lentinus edodes (Berk.) Singer: five years of research in Brazil. Int J Med Mushrooms 7:281–299

González A, Cruz M, Losoya C, Nobre C, Loredo A, Rodríguez R, Belmares R (2020) Edible mushrooms as a novel protein source for functional foods. Food Funct 11(9):7400–7414

Heleno SA, Martins A, Queiroz MJR, Ferreira IC (2015) Bioactivity of phenolic acids: metabolites versus parent compounds: a review. Food Chem 173:501–513

Hyde KD, Xu J, Rapior S, Jeewon R, Lumyong S, Niego AGT, Stadler M (2019) The amazing potential of fungi: 50 ways we can exploit fungi industrially. Fungal Divers 97:1–136

Joseph B, Krishnakumar N, Ceasar SA, Devassy SM (2024) Role of rural women in achieving united nations sustainable development goals through mushroom agri-entrepreneurship: mushroom cultivation as a sustainable livelihood option for rural women. In: IGI Global (ed) The role of female leaders in achieving the sustainable development goals. Hershey, pp 214–231

Jung E, Lee J, Huh S, Lee J, Kim YS, Kim G, Park D (2009) Phloridzin-induced melanogenesis is mediated by the cAMP signaling pathway. Food Chem Toxicol 47(10):2436–2440

Karaman M, Jovin E, Malbaša R, Matavuly M, Popović M (2010) Medicinal and edible lignicolous fungi as natural sources of antioxidative and antibacterial agents. Phytother Res 24(10):1473–1481

Keller S, Le HY, Rödiger C, Hipler UC, Kertscher R, Malarski A, Jahreis G (2014) Supplementation of a dairy drink enriched with milk phospholipids in patients with atopic dermatitis–a double-blind, placebo-controlled, randomized, cross-over study. Clin Nutr 33(6):1010–1016

Kim SW, Hwang HJ, Lee BC, Yun JW (2007) Submerged production and characterization of Grifola frondosa polysaccharides–a new application to cosmeceuticals. Food Technol Biotechnol 45(3):295–305

Kim SY, Go KC, Song YS, Jeong YS, Kim EJ, Kim BJ (2014) Extract of the mycelium of T. matsutake inhibits elastase activity and TPA-induced MMP-1 expression in human fibroblasts. Int J Mol Med 34(6):1613–1621

Kozarski M, Klaus A, Jakovljevic D, Todorovic N, Vunduk J, Petrović P, Van Griensven L (2015) Antioxidants of edible mushrooms. Molecules 20(10):19489–19525

Liu Y, Bastiaan-Net S, Wichers HJ (2020) Current understanding of the structure and function of fungal immunomodulatory proteins. Front Nutr 7:132

Madnani N, Khan K (2013) Hair cosmetics. Indian J Dermatol Venereol Leprol 79:654

Maftoun P, Malek R, Abdel-Sadek M, Aziz R, Enshasy HE (2013) Bioprocess for semi-industrial production of immunomodulator polysaccharide Pleuran by Pleurotus ostreatus in submerged culture. J Sci Ind Res 72(11):655–662

Majeed A, Prakash L (2021) Cosmeceuticals: a revolution in the making. Sabinsa Corporation, East Windsor

Masaki H (2010) Role of antioxidants in the skin: anti-aging effects. J Dermatol Sci 58(2):85–90

Meehan K (2015) U.S. Patent No. 9,144,542. U.S. Patent and Trademark Office, Washington

Meng TX, Furuta S, Fukamizu S, Yamamoto R, Ishikawa H, Arung ET, Kondo R (2011) Evaluation of biological activities of extracts from the fruiting body of Pleurotus citrinopileatus for skin cosmetics. J Wood Sci 57:452–458

Meng TX, Zhang CF, Miyamoto T, Ishikawa H, Shimizu K, Ohga S, Kondo R (2012) The melanin biosynthesis stimulating compounds isolated from the fruiting bodies of Pleurotus citrinopileatus. J Cosmet Dermatol Sci Appl 2(3):151

Mishra M, Srivastava D (2024) Harnessing the power of medicinal mushrooms into cosmetics: a fountain of youth from nature. J Indian Bot Soc 104(4):200–209

Miura T, Kubo M, Itoh Y, Iwamoto N, Kato M, Park SR, Suzuki I (2002) Antidiabetic activity of Lyophyllum decastes in genetically type 2 diabetic mice. Biol Pharm Bull 25(9):1234–1237

Mohiuddin AK (2021) Can medicinal mushrooms fight against sars-cov-2/covid-19? J Internal Med Sci Art 2:23–24

Morales D, Gil-Ramirez A, Smiderle FR, Piris AJ, Ruiz-Rodriguez A, Soler-Rivas C (2017) Vitamin D-enriched extracts obtained from shiitake mushrooms (Lentinula edodes) by supercritical fluid extraction and UV-irradiation. Innovative Food Sci Emerg Technol 41:330–336

Nieto IJ, Chegwin A (2008) Triterpenoids and fatty acids identified in the edible mushroom Pleurotus sajor-caju. J Chil Chem Soc 53(2):1515–1517

Ndlovu G, Fouche G, Tselanyane M, Cordier W, & Steenkamp V (2013). In vitro determination of the anti-aging potential of four southern African medicinal plants. BMC complementary and alternative medicine, 13, 1–7.

Onar HC, Yusufoglu A, Turker G, Yanardag R (2012) Elastase, tyrosinase and lipoxygenase inhibition and antioxidant activity of an aqueous extract from Epilobium angustifolium L. leaves. J Med Plant Res 6(5):716–726

Pandey AT, Pandey I, Hachenberger Y, Krause BC, Haidar R, Laux P, Singh AV (2020) Emerging paradigm against global antimicrobial resistance via bioprospecting of mushroom into novel nanotherapeutics development. Trends Food Sci Technol 106:333–344

Park HS, Hwang YH, Kim MK, Hong GE, Lee HJ, Nagappan A, Kim GS (2015a) Functional polysaccharides from Grifola frondosa aqueous extract inhibit atopic dermatitis-like skin lesions in NC/Nga mice. Biosci Biotechnol Biochem 79(1):147–154

Park KM, Kwon KM, Lee SH (2015b) Evaluation of the antioxidant activities and tyrosinase inhibitory property from mycelium culture extracts. Evid Based Complement Alternat Med 2015:616298

Pintus F, Spano D, Corona A, Medda R (2015) Antityrosinase activity of Euphorbia characias extracts. PeerJ 3:e1305

Piwowarski JP, Kiss AK, & Kozłowska-Wojciechowska M (2011). Anti-hyaluronidase and anti-elastase activity screening of tannin-rich plant materials used in traditional Polish medicine for external treatment of diseases with inflammatory background. Journal of Ethnopharmacology, 137(1), 937–941.

Plaami SP, Dekker M, van Dokkum W, Ockhuizen T (2001) Functional foods: position and future perspectives (No. 2000/15). Nationale Raad voor Landbouwkundig Onderzoek, The Hague

Qaderi MM, Martel AB, Strugnell CA (2023) Environmental factors regulate plant secondary metabolites. Plants 12(3):447

Rangsinth P, Sillapachaiyaporn C, Nilkhet S, Tencomnao T, Ung AT, Chuchawankul S (2021) Mushroom-derived bioactive compounds potentially serve as the inhibitors of SARS-CoV-2 main protease: an in silico approach. J Tradit Complement Med 11(2):158–172

Raut JK (2019) Current status, challenges and prospects of mushroom industry in Nepal. Int J Agric Econ 4(4):154–160

Ribeiro AS, Estanqueiro M, Oliveira MB, Sousa Lobo JM (2015) Main benefits and applicability of plant extracts in skin care products. Cosmetics 2(2):48–65

Rodrigues F, Pimentel FB, Oliveira MBP (2015) Olive by-products: challenge application in cosmetic industry. Ind Crop Prod 70:116–124

Rop O, Mlcek J, Jurikova T (2009) Beta-glucans in higher fungi and their health effects. Nutr Rev 67(11):624–631

Royse DJ, Baars J, Tan Q (2017) Current overview of mushroom production in the world. Edible and medicinal mushrooms: technology and applications. Wiley, Hoboken, pp 5–13

Sánchez C (2017) Reactive oxygen species and antioxidant properties from mushrooms. Synth Syst Biotechnol 2(1):13–22

Sande D, de Oliveira GP, Moura MAF, de Almeida Martins B, Lima MTNS, Takahashi JA (2019) Edible mushrooms as a ubiquitous source of essential fatty acids. Food Res Int 125:108524

Shakoor H, Feehan J, Apostolopoulos V, Platat C, Al Dhaheri AS, Ali HI, Stojanovska L (2021) Immunomodulatory effects of dietary polyphenols. Nutrients 13(3):728

Sharma P (2011) Cosmeceuticals: regulatory scenario in US, Europe & India. Int J Pharm Technol 3:1512–1535

Shirur M, Barh A, Annepu SK (2021) Sustainable production of edible and medicinal mushrooms: implications on mushroom consumption. In: Climate change and resilient food systems: issues, challenges, and way forward. Springer, Cham, pp 315–346

Singh MP, Rai SN, Dubey SK, Pandey AT, Tabassum N, Chaturvedi VK, Singh NB (2022) Biomolecules of mushroom: a recipe of human wellness. Crit Rev Biotechnol 42(6):913–930

Solano F (2020) Photoprotection and skin pigmentation: melanin-related molecules and some other new agents obtained from natural sources. Molecules 25(7):1537

Su CH, Lai MN, Lin CC, Ng LT (2016) Comparative characterization of physicochemical properties and bioactivities of polysaccharides from selected medicinal mushrooms. Appl Microbiol Biotechnol 100:4385–4393

Taofiq O, Heleno SA, Calhelha RC, Alves MJ, Barros L, Barreiro MF, Ferreira IC (2016) Development of mushroom-based cosmeceutical formulations with anti-inflammatory, anti-tyrosinase, antioxidant, and antibacterial properties. Molecules 21(10):1372

Taofiq O, Rodrigues F, Barros L, Barreiro MF, Ferreira IC, Oliveira MBP (2019) Mushroom ethanolic extracts as cosmeceuticals ingredients: Safety and ex vivo skin permeation studies. Food Chem Toxicol 127:228–236

Tayarani-Najaran Z, Eghbali-Feriz S, Taleghani A, Al-Najjar H, Emami S, Rahimi H, Asili J, Hasanzadeh S (2018) Antimelanogenesis and anti-tyrosinase properties of Pistacia atlantica subsp. mutica extracts on B16F10 murine melanoma cells. Res Pharm Sci 13(6):533. https://doi.org/10.4103/1735-5362.245965

Ukawa Y, Izumi Y, Ohbuchi T, Takahashi T, Ikemizu S, Kojima Y (2007) Oral administration of the extract from Hatakeshimeji (Lyophyllum decastes sing.) mushroom inhibits the development of atopic dermatitis-like skin lesions in NC/Nga mice. J Nutr Sci Vitaminol 53(3):293–296

Visvanathan S, Krishnamoorthy R, Sabesan GS (2022) Fungal cosmetics: mushrooms in beauty care and the new age of natural cosmetics. In: Applied mycology: entrepreneurship with fungi. Springer, Cham, pp 1–37

Vunduk J, Kozarski M, Djekic I, Tomašević I, Klaus A (2021) Effect of modified atmosphere packaging on selected functional characteristics of Agaricus bisporus. Eur Food Res Technol 247:829–838

Wu G, Li L, Sung GH, Kim TW, Byeon SE, Cho JY, Park HJ (2011) Inhibition of 2, 4-dinitrofluorobenzene-induced atopic dermatitis by topical application of the butanol extract of Cordyceps bassiana in NC/Nga mice. J Ethnopharmacol 134(2):504–509

Wu Y, Choi MH, Li J, Yang H, Shin HJ (2016) Mushroom cosmetics: the present and future. Cosmetics 3(3):22

Xiao JH, Xiong Q (2013) Nucleosides, a valuable chemical marker for quality control in traditional Chinese medicine Cordyceps. Recent Pat Biotechnol 7(2):153–166

Yadav D, Negi PS (2021) Bioactive components of mushrooms: processing effects and health benefits. Food Res Int 148:110599

Yang SH, Liu HI, Tsai SJ (2006) U.S. Patent Application No. 11/095,525

Zhang Y, Mills GL, Nair MG (2002) Cyclooxygenase inhibitory and antioxidant compounds from the mycelia of the edible mushroom Grifola frondosa. J Agric Food Chem 50(26):7581–7585
Zhang M, Cui SW, Cheung PC, Wang Q (2007) Antitumor polysaccharides from mushrooms: a review on their isolation process, structural characteristics and antitumor activity. Trends Food Sci Technol 18(1):4–19
Zhang Y, Wang D, Chen Y, Liu T, Zhang S, Fan H, Li Y (2021) Healthy function and high valued utilization of edible fungi. Food Sci Human Wellness 10(4):408–420

Chapter 10
Green Biotechnological Applications in Cosmetics and Aesthetic Medicine

Muhammad Waqas Nasir, Yong Gao, Sadia Javed, and Muhammad Arshad

Contents

M. W. Nasir · Y. Gao (✉)
The First Affiliated Hospital of USTC, Division of Life Sciences and Medicine, University of Science and Technology of China, Hefei, Anhui, China
e-mail: ygao387@ustc.edu.cn

S. Javed
Department of Biochemistry, Government College University Faisalabad, Faisalabad, Pakistan

M. Arshad
Biochemistry Section, Jhang-campus, University of Veterinary and Animal Sciences Lahore, Lahore, Pakistan

S. Javed et al. (eds.), *Sustainable Cosmeceuticals*,
https://doi.org/10.1007/978-3-031-86087-4_10

Abstract Green biotechnology employs fermentation and genetics to produce active ingredients and products that can be used in cosmetics and aesthetic medicines. These products are safe and green compared to synthetic products. The recent biotechnological advances have changed the formulation, manufacturing, and effectiveness of cosmetic products and aesthetic treatments, thus improving consumer satisfaction. Biotechnology uses genetically modified organisms, cell cultures, and biomimetic methodologies to create and formulate novel cosmetics and therapies. Active ingredients like enzymes, peptides, and growth factors are helpful for skin care, hair care, and other aesthetic applications. Green biotechnology has many more applications in formulating sustainable ingredients and effective and safer high-quality ingredients for cosmetics and aesthetics. Besides this, biotechnology can also help to improve the bioavailability, stability, and delivery methods of the product, resulting in more effective and focused therapies. The use of biotechnology has been expanded to stem cell, gene, and tissue engineering therapies. Moreover, these advancements have also improved skin rejuvenation, wrinkle reduction, and scar management in a more personalized way. However, despite all these positive applications, there is still a need for rigorous testing, ingredient transparency, and regulatory compliance to ensure the safety and efficacy of the product in this rapidly developing market. This work explains the development in the cosmetic and aesthetic industry in a more detailed and comprehensive way, which can help cosmetic scientists and dermatologists to understand the more recent developments and their potential applications.

Keywords Cosmetic industry · Green technology · Biotechnology · Cosmetic ingredients · Skin care

10.1 Introduction

10.1.1 Biotechnology: Definition and Scope

Biotechnology is the use of biological systems, organisms, or processes to create goods intended to enhance human life (Bhatia and Goli 2018). According to the definition given by the Collins English Dictionary (Bhatia and Goli 2018), biotechnology is the use of living creatures, their components, or processes to create active, beneficial products and to offer services, such as waste treatment. The word biotechnology refers to a wide variety of procedures, such as using earthworms as a source of protein or genetically modifying bacteria to produce human gene products, such as growth hormones (Doran 1995).

Biotechnology can contribute to achieving sustainable development growth (SDG) rule 9, which includes innovations, construction of infrastructure, and encouraging entrepreneurship. Innovation is facilitated when industries and academic institutions work together. It also promotes sustainable growth of the economy along with productive employment. It can also create sustainable and non-toxic ingredients for various products. For example, biotechnology can create ingredients like natural ones without environmental effects.

The Golden Treasury of Science and Technology describes biotechnology as a field that relies on the regulation of biological processes for the mass production of useful chemicals. The controlled use of biological agents, such as microorganisms or cellular components, for beneficial purposes, is what biotechnology entails (Bhatia and Goli 2018). The terms "bio" and "technology" refer to the use of knowledge for practical purposes, i.e., the utilization of living things to develop or improve products (Doran 1995; Bhatia and Goli 2018). Biotechnology is also described as "the application of technology to biological processes for industrial, agricultural, and medical purposes" in The Universities Press Dictionary of Biology (Demain 2007). These definitions make it clear that biotechnology encompasses a variety of technologies that rely on knowledge obtained from recent advancements in biochemistry, cell biology, and molecular biology (Bhatia et al. 2018). These technologies are already having a significant impact on many different aspects of society, including waste management, food processing, agriculture, and medical technology (Fig. 10.1) (Zand and Narasu 2013). Biotechnology has such a wide range of uses; therefore, it is today applicable to practically all businesses. This indicates that these

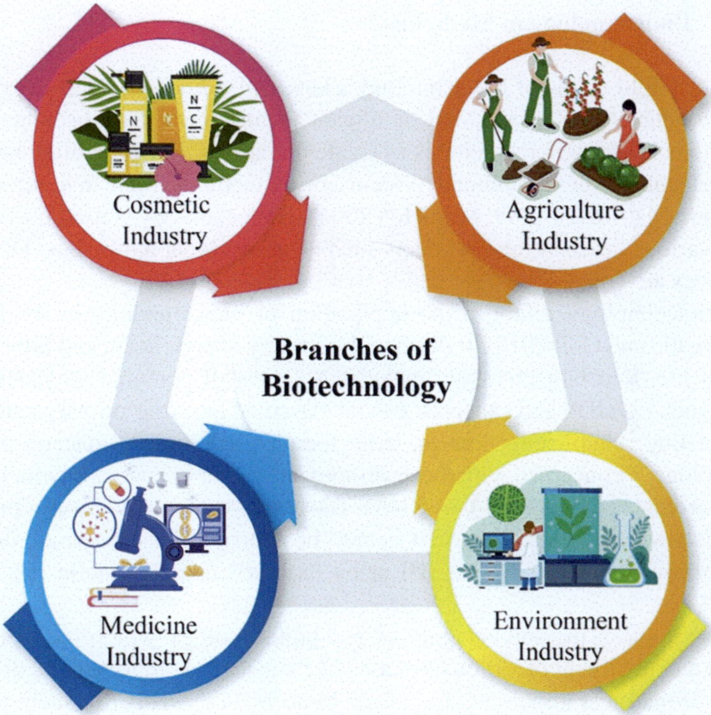

Fig. 10.1 Few branches of biotechnology

sectors can develop new or improved goods more quickly and effectively. As a result of technological advancements, pharmaceutical research will be improved, new biopharmaceuticals will be discovered and developed, certain diseases' underlying causes will be revealed, the relationship between genomic variation and drug response will be explored, and the relationship between genetics and biological function will be better understood (Zappelli et al. 2016).

Biotechnology has been widely used in the cosmetic industry in several ways. It is used to find, create, and produce ingredients for cosmetic formulations as well as to assess how these ingredients behave on the skin and how they may influence aging-related changes. Consequently, biotechnology offers a useful alternative method for creating active compounds that can delay the aging process (Zappelli et al. 2016).

Biotechnology is not a stand-alone field; it is made easier to use biological agents in technological applications by its well-known integration with allied disciplines like biochemistry, molecular biology, and microbiology. Because of this, biotechnology has emerged as a discipline with immense potential for enhancing human welfare in fields including food processing, human health, and environmental preservation. The following examples will show the major importance of this scientific area in several fields.

10.1.1.1 Biotechnology in Medicine

The medical field is one of biotechnology's key applications. Most of the research is being conducted in this area, and significant advancements have been made. The goal of biotechnology in medicine is to create therapeutic proteins, often known as biopharmaceuticals or recombinant proteins, using methods found in living systems (Galaev and Mattiasson 1999; Singleton 2004). Biotechnology is used for the diagnosis of various diseases, and products such as monoclonal antibodies, DNA, and RNA probes are created.

Red biotechnology refers to the application of biotechnology in the medical sphere (Bhatia and Goli 2018). It deals with a variety of significant and little aspects of human life. Red biotechnology includes a variety of potential medications for illnesses such as AIDS and cancer. It can be classified into four primary categories: genetic testing, pharmacogenomics, gene therapy, and biopharmaceuticals. Red biotechnology is concerned with the development of pharmaceuticals, which can be proteins or nucleic acids (DNA or RNA). Biosynthetic "human" insulin produced using rDNA technology was the first item to be approved for therapeutic use. This human insulin is often known as rHI or by its brand name Humulin (Bhatia and Goli 2018).

Gene therapy is the sub-category of red biotechnology which focuses on the detection and treatment of various diseases like cancer as well as genetic disorders. It is the second most important area of red biotechnology. This treatment includes both the modification of genes and the replacement of damaged genes. Genes are added, removed, or mutated throughout this process (Mulligan 1993; Wirth et al.

2013). Incorporating functioning genes into an unidentified genomic region to replace a damaged and defective gene is one of the most popular types of gene therapy (Clark and March 2006; Ribeil et al. 2017).

Genetic testing and pharmacogenomics are both subcategories of red biotechnology and employ individual-specific red biotechnology techniques. In pharmacogenomics, medications are created using an individual's genetic information. Numerous diseases can be diagnosed using monoclonal antibodies, and DNA and RNA probes, and pharmaceuticals like insulin and interferon can now be produced by bacteria to treat human illnesses (Phillips et al. 2004; Carr et al. 2014; Relling and Evans 2015).

10.1.1.2 Industrial Biotechnology

White biotechnology, often referred to as industrial biotechnology, is a subfield of biology that focuses on exploiting biological processes, systems, and creatures to provide long-term fixes for diverse industrial purposes (Heux et al. 2015). It entails using living beings' energy to manufacture valuable goods, chemicals, and materials in a way that is more efficient and kinder to the environment. Examples of these species' energy include bacteria and enzymes (Frazzetto 2003).

Industrial biotechnology was created to enable the large-scale microbial production of alcohol and antibiotics. Currently, genetic engineering is used to create a variety of pharmaceutical medications and substances including lactic acid, glycerin, etc. Another crucial field is protein engineering, which modifies existing proteins and enzymes to perform new functions or to do so more effectively (Drepper et al. 2006; Ulber and Sell 2007).

10.1.1.3 Environmental Biotechnology

Environmental biotechnology is a subfield of biotechnology that concentrates on employing biological systems, processes, and species to address environmental problems and provide long-term fixes. Using the ability of microbes, plants, and enzymes to remediate pollution, manage waste, and promote ecological balance, entails the use of biotechnological concepts to protect, maintain, and restore the environment (Jördening and Winter 2005; Scragg 2005).

Biotechnology is being used to address environmental issues like the depletion of natural resources for non-renewable energy, biodiversity preservation, etc. Examples of applications for bacteria include the purification of industrial effluents, the control of oil spills, the treatment of sewage, and the creation of biogas ref. For the control of insect pests and diseases, biopesticides provide an environmentally friendlier alternative to chemical pesticides (Forster and Wase 1987; Fulekar 2010; Gavrilescu 2010).

Environmental biotechnology helps to attain Sustainable Development Goals by addressing many goals. For example, SDG6 states that biotechnology aids in improving water supplies by improving sanitation. SDG 7 states that it can help by providing more clean and cost-effective energy sources. Environmental biotechnology helps to achieve SDG13, which states that biotechnology can help to fight climatic changes. SDG 11 claims that biotechnology can aid in the creation of sustainable cities. The literature claims that biotechnology can aid in achieving these goals by the replacement of chemicals and environmental pollutants with more sustainable options. It can also help in breaking down waste to produce useful materials by using less water, land, and energy (Olson et al. 2018).

10.1.1.4 Agricultural Biotechnology

Agricultural biotechnology is concerned with using scientific methods to increase agricultural production, sustainability, crops, and food quality. It involves developing crops with better characteristics, resistance to pests and diseases, and improving nutritional profiles by employing genetic engineering, molecular biology, and other biotechnological techniques. To solve issues like food security, climate change, and the sustainable use of resources, agricultural biotechnology is crucial (Altman 1997; Hoban 1998; Bennett et al. 2013).

Plant tissue culture is the most popular technique used in biotechnology. It is used to produce virus-free genetic stock and plant material, fruits, and forest trees quickly and economically. Plant tissue produces novel genetic variations through clonal variation (Phillips and Garda 2019). It is now possible to create transgenic plants with desired genes, such as herbicide resistance, disease resistance, longer shelf life, etc. Crop improvement has been speeded up with the help of molecular breeding. Molecular markers like restriction fragment length polymorphism and simple sequence repeats can be used as an instrument for the indirect selection of both qualitative and quantitative quality as well as for genotypic diversity research (Thorpe 2007; Smith 2013; Bennett et al. 2013).

Agricultural biotechnology also helps to achieve certain goals related to sustainable development. For example, SDG2 aims to end the hunger of the world. Biotechnology can help to increase the production of crops, reduce the wastage of food, increase farm yields, and improve the crops resiliency against climatic changes. SDG12 is related to responsible consumption and production. Agricultural biotechnology aids in reducing food wastage by the creation of crops that are browning and bruising resistant. It can help to increase the yield of crops along with improved quality of crops. It can also help in reducing the environmental effects produced by fertilizers.

SDG13 is related to climate action. This field of biotechnology thus helps to reduce the emissions of greenhouse gases by using genetically modified crops. It can also help in replacing conventional energy sources with cleaner fuel alternatives like biofuels. It can also help farmers to fight climatic conditions. SDG15 (life on land) can be achieved by agricultural biotechnology, which uses less farming land, thus reducing deforestation. It can help to save crops and enable precise control of pests (Tyczewska et al. 2023).

10.1.1.5 Cosmetology

By developing novel chemicals, methods, and products that improve the efficiency and safety of cosmetic formulations, biotechnology has significantly impacted the cosmetics business. Modern skin care and cosmetics have been made possible using biotechnology, or “cosmeceuticals,” in cosmetics. The study and practice of numerous cosmetic operations, treatments, and methods to improve the look and health of the skin, hair, and nails is known as cosmetology. It includes a broad range of techniques designed to enhance appearance, preserve healthy skin and hair, and increase self-confidence. The interdisciplinary area of cosmetology incorporates aspects of science, art, and practical knowledge (Millikan 2001; Bhatia and Goli 2018; Niska et al. 2018).

The ramifications of beauty for biology, society, and medicine have always captivated scientists. However, a fundamental change in the significance of scientific research as it relates to cosmetic science has been driven by an increased understanding of skin physiology. Over time, both cosmetic companies and customers have questioned the actual effectiveness of their products’ components on the skin. Before the strict rules for the examination of cosmetics, many personal care products are primarily intended to be used as decorative items with little to no actual efficacy being shown (Zappelli et al. 2016).

In many ways, biotechnology has influenced cosmetics. Biotechnology is used by the cosmetic industry to find, create, and produce the ingredients that make up cosmetic formulae as well as to assess how these ingredients operate on the skin and how they affect aging-related changes. Producers of cosmetic ingredients are subject to highly stringent manufacturing and production environment inspections, so they must assure consumer safety and repeatability of testing. Additionally, both the business and the public are increasingly aware that components made from biodegradable materials are simply healthier for the environment. Less water, soil, and/or electricity must be used, and the components must not be exposed to pesticides and other contaminants, increasing the safety of the product for the customer (Zappelli et al. 2016).

To identify ways to slow down the aging process, cosmetic corporations have spent a lot of money on molecular, genomic, and proteomic research during the past ten years. We must consider the proper integration of skin structure and functional features with the formulation’s nature, its efficacy as defined by the product’s purpose, and its safety when assessing the success of cosmetic and cosmeceutical goods (Millikan 2001; Bhatia and Goli 2018; Niska et al. 2018).

10.1.2 Sustainable Development Growth

The SDG stands for sustainable development growth goals also known as Global Goals. These goals were adopted by the UN in 2015 as a call to end poverty and hunger and protect the environment so that by 2030 all people on this planet can enjoy peace. There are 17 SDG goals. Green biotechnology can help to achieve a few goals, thus reducing carbon footprints.

SDG2 Zero Hunger: Biotechnology can help to create genetically modified crops that are not prone to pests and diseases. This leads to high production in the agricultural sector. The need to use fertilizers reduces when using biotechnological ways. This can lower carbon emissions related to agricultural methods. Loss of crops reduces due to the resistance of crops to climatic changes, which ensures food security and reduces carbon emissions (Wesseler and Zilberman 2021).

10.1.2.1 SDG 3 Good Health and Well-being

By enhancing the nutritional value of food crops, biotechnology can help to fight hunger and improve public health outcomes. Lower dependence on harmful pesticides often results in a safer food supply chain and healthier ecosystems, thus lowering carbon footprints linked to health (Wesseler and Zilberman 2021).

10.1.2.2 SDG 7 Affordable and Clean Energy

Green biotechnology converts plant resources into biofuels, which are sustainable energy sources with lower carbon emissions compared to conventional fossil fuels. The emissions of greenhouse gases can be reduced by employing biotechnological advancements to convert agricultural waste into electricity (Wesseler and Zilberman 2021).

10.1.2.3 SDG 12 Responsible Consumption and Production

The development of bioplastics and biodegradable materials can help to reduce plastic pollution and carbon footprint, which are related to their lifecycle. Green biotechnology decreases the environmental impacts by promoting recycling and sustainable resource usage (Wesseler and Zilberman 2021).

10.1.2.4 SDG 13 Climate Action

Climate change can be directly addressed by biotechnological developments that can improve plant and soil capacity to isolate carbon. Climate resilience is attributed to practices such as precision agriculture, which maximizes resource usage and reduces emissions (Wesseler and Zilberman 2021).

10.1.2.5 SDG 14 Life Below Water

By reducing the environmental effects on the aquatic environment and encouraging methods that decrease carbon emissions in food production, green biotechnology provides more promising answers for sustainable fish farming. Restoring the environment by using biotechnological methods to enhance marine contaminants can improve those ecosystems' capacity to store carbon (Wesseler and Zilberman 2021).

10.1.2.6 SDG 15 Life on Land

Genetic engineering and tissue culture are two ways that can help to restore degraded areas and save endangered or near to extinct species. Biotechnology helps in sustainable land management techniques to improve ecosystems' capacity to sequester carbon (Wesseler and Zilberman 2021).

In order to reduce carbon footprints (emissions) and achieve the SDGs, green biotechnology is important. It addresses crucial environmental issues by climate resilience, sustainable materials, and agricultural innovations. The incorporation of green biotechnology with a variety of industries not only advances short-term sustainability objectives but also builds a more robust and sustainable future for everybody. Leveraging biotechnological discoveries will be important for climatic action and sustainable development as the globe deals with urgent environmental concerns (Saeedi et al. 2013; Wesseler and Zilberman 2021).

10.1.3 History of Biotechnology in Cosmetics and Aesthetic Medicine

Cosmetics have been impacted by biotechnology in several ways. Biotechnology is used by the cosmetics industry to produce ingredients for cosmetic formulations. The cosmetic industry explores how these ingredients behave on the skin and how they may influence aging-related changes (Millikan 2001). Consequently, biotechnology offers a useful alternative method for creating active compounds that can delay the aging process (Zappelli et al. 2016).

The sole or primary purpose of cosmetic products is cleaning, perfuming, altering appearance, protecting, maintaining good condition, or correcting body flaws (Mahé 2014). Many chemical substances that have been utilized in cosmetics have the potential to have negative impacts on their consumers. Since the early 1990s, herbal cosmetics (Bijauliya et al. 2017) have gained significant importance in the cosmetic business due to these effects and the negative environmental impacts of chemical components (Sajna et al. 2015a). The use of biotechnology in aesthetic medicine and cosmetics has a long history and has seen tremendous development. The creation of novel aesthetic therapies and the composition of cosmetic items have both been transformed by advances in biotechnology (Gomes et al. 2020). The major turning points and advancements in the history of biotechnology in cosmetics and aesthetic medicine are summarized as follows:

Enzymes were first used in skin care products in the 1950s and 1960s, mostly for exfoliating and enhancing skin texture (Parisi et al. 2017). Cold creams and emulsions with natural extracts are now widely used as cosmetics (Herman et al. 2013; Harhaun et al. 2020). In the 1970s, alpha hydroxy acids (AHAs) were introduced in skin care products. They were derived from fruits and milk. AHAs were reported to have skin rejuvenating and exfoliating abilities. This was a pioneering use of substances obtained from biotechnology in cosmetic compositions (Grimes et al. 2004). In the 1980s, peptide technology was developed. Peptide-based cosmetic components were created because of an increased understanding of peptides and their function in skin health (Wintzen and Gilchrest 1996). To combat wrinkles, increase the creation of collagen, and improve skin suppleness, peptides were employed (Ferreira et al. 2020). The 1990s saw a rise in cosmetic biotechnology. The development of biotechnology made it possible to extract organic substances from plants and microorganisms for use in skin care products (Sajna et al. 2015a). Antioxidants (Kusumawati and Indrayanto 2013), plant extracts (Hoang et al. 2021), and chemicals originating from the ocean (Kijjoa and Sawangwong 2004) have become more popular in cosmetics. The chemicals originating from oceans are valuable for the advancement of biotechnology, promoting biodiversity and sustainability. By using these resources, we can develop more innovative solutions that can be beneficial to both the environment and human health. For example, phytoplankton produced by marine organisms plays an important role in carbon isolation and recycling. Similarly, extremophiles produce enzymes that can be used in various industrial processes, thus reducing the need for harsh chemicals (Arshad et al. 2014).

With stem cell extracts being employed for skin regeneration and anti-aging properties, stem cell technology came into focus in the 2000s (Prainsack et al. 2008). Hyaluronic acid was made possible by biotechnology through microbial fermentation, which helped with skin care and cosmetic procedures (de Oliveira et al. 2016). Streptococcus is the microbe that is most usually used in the industrial manufacturing of hyaluronic acid (HA) chemicals. However, this genus is known to include several human diseases; therefore, employing these bacteria to purify HA is expensive (Lu et al. 2016). To produce HA, genetically modified bacteria were investigated. *Bacillus subtilis* is one of these bacteria and is one of the most often utilized genetic engineering models, which ensures endotoxin-free goods (Chien and Lee 2007).

In the 2010s, the focus was put forward on personalization and microbiome. Personalized skin care products were made possible by genetic analysis and biotechnological techniques. The creation of items that support a healthy skin environment was made possible by our growing understanding of the skin's microbiome (Adu et al. 2020; Murphy et al. 2021). Precision and efficacy in cosmetic formulations and aesthetic procedures are being improved via biotechnology nowadays. The path of the sector is heavily influenced by ethical considerations, sustainable practices, and non-cruelty approaches. The development of biotechnology in cosmetics and aesthetic medicine may be seen as progressing from simple natural extracts to complex molecular technologies overall. The cosmetics and aesthetics sectors have benefited from biotechnological developments in terms of product efficacy and safety as well as ethical considerations and the adoption of sustainable development growth practices (Gomes et al. 2020).

10.2 Biotechnology in Cosmetic Formulations

10.2.1 Genetically Modified Organisms

In the creation of cosmetics, a wide variety of biotechnological items are utilized. The chemicals are separated into the following categories based on their molecular structure.

1. Carotenoids, terpenes, and polyphenols
2. Organic acids
3. Nitrogenous substances like amino acids
4. Vitamins and substances that resemble vitamins
5. Polysaccharides
6. Proteins and polypeptides
7. Derivatives of essential fatty acids, sterols, and lipids

10.2.1.1 Polyphenols, Terpenes, and Carotenoids

Flavonoids or polyphenols produced from plants have long been employed in cosmetic formulation. Botanical extracts high in polyphenols are frequently used in skin care products as free-radical scavengers. Examples that are most widely used include botanical extracts from green tea, pomegranate, grape seed, and coffee berry (Zillich et al. 2015). Pure phytochemicals are preferred in modern cosmetics over plant extracts since they have a greater level of action and have no negative side effects. However, in plants, these polyphenols are frequently seen to be bound to other large molecules, such as lignin (Zillich et al. 2015). One method that is frequently used to purify the phenolic chemicals from plants is microbial or enzymatic hydrolysis (Panagiotou and Olsson 2007). In addition, microbes release a variety of phenolic or terpene-like compounds with strong antioxidant properties. According to studies, topical application of cosmetic products containing microbial antioxidants is more advantageous and effective than using synthetic or plant-derived substances (Sajna et al. 2015a).

Ellagic acids are derivatives of tannin that are found in a variety of fruits, macrophytes, and fungi. In addition to being an antioxidant, studies have demonstrated that ellagic acid also has lightening properties. Some commercial products, including White Perfect Laser TM, which bills itself as a whitening lotion, include ellagic acid produced from plants (Vattem and Shetty 2005; Sajna et al. 2015a). Ellagic acid is often found in plants as ellagitannins, which need to be hydrolyzed chemically or enzymatically to get ellagic acid for usage as an active component in cosmetics. Tannase and polyphenol oxidase work together during enzymatic treatment to transform ellagitannins into ellagic acid (Hsu et al. 2019). Because of the synergistic effects of tannase and polyphenol oxidase, coculturing of *Aspergillus niger* and *Candida utilis* resulted in greater conversion of valonea tannins to ellagic acid (Shi et al. 2005; Hsu et al. 2019). With Lentinus edodes, cranberry pomace, a byproduct of the cranberry processing industry, was solid-state fermented to produce ellagic acid at a yield of 350 g/g dry weight of substrate (Vattem and Shetty 2003).

Numerous plants contain gallic acid, which is structurally 3,4,5-trihydroxy benzoic acid. Due to their ability to scavenge free radicals, gallic acid esters like propyl gallate and octyl gallate are the most often employed stabilizers or preservatives in cosmetic formulation. Utilizing fungus, often from the genera *Aspergillus*, *Penicillium*, and *Rhizopus*, bioconversion of a substrate high in tannin to gallic acid has been accomplished. It uses solid-state fermentation, which is both affordable and effective (Umakumar et al. 2012). Additionally, cosmetic products for hair straightening might include gallic acid. Gallic acid and other polyhydroxylated aromatic chemicals are used in conjunction with heat to effectively relax hair, which is a less invasive technique than chemical therapy (Sajna et al. 2015b). Gallic acid exhibits strong antibacterial action as well, making it a strong contender for anti-acne therapies (Chanwitheesuk et al. 2007).

Ferulic acid is a hydroxyl cinnamic acid used in hair products. It is a phenolic substance that is generated from plants. Most of the time, it may be found in both free form and covalently bonded to lignin in the seeds and leaves of plants. Because it quickly creates a resonance-stabilized phenoxy radical that can put an end to free-radical processes. Ferulic acid is a potent antioxidant (Graf 1992). By bio-converting lignocellulosic biomass, such as rice straw, wheat straw, and maize cob, utilizing the enzyme feruloyl esterase and the combined action of cellulolytic and xylanolytic enzymes, ferulic acid may be generated as a value-added product. Some fungi, including probiotic *Lactobacilli* and *Aspergillus* species, have the potential to be sources of feruloyl esterase (Gottschalk et al. 2010; Tomaro-Duchesneau et al. 2012). With a decrease in the synthesis of thymine dimers brought on by UV radiation, ferulic acid, vitamins C, and vitamin E have increased photo-protective effect, suggesting a possible application in sunscreen lotion composition (Lin et al. 2005).

Resveratrol has been known to have excellent antioxidant properties. Plants that are being attacked by microbes generate polyphenol known as resveratrol. When applied topically, a resveratrol-based cosmetic product demonstrated 17 times more antioxidant capacity than idebenone and protected against photoaging (Baxter 2008). Application of a retinoid and resveratrol-containing cosmetic formulation topically to the skin has been shown to increase skin radiance and reduce wrinkles. It also provided protection against photoaging. Resveratrol does have certain health benefits, but they are rather limited because of the molecule's fragility when exposed to light, air, or situations with extreme pH levels. These stressors may result in isomerization or oxidation, which lowers the compound's bioavailability and bioactivity. It is crucial to create resveratrol derivatives with improved stability, such as trans-resveratrol (Beekwilder et al. 2006; Lim et al. 2011). One of the most well-known plant secondary metabolites is the polyphenol trans-resveratrol (3,5,4′-trihydroxy-trans-stilbene). It has potent antioxidant effects and may be useful for skin anti-aging (Kiselev 2011; Pandey et al. 2015). Resveratrol is often produced by biotechnological methods, utilizing the yeasts *Saccharomyces cerevisiae* or *Pichia pastoris* (Ratz-Łyko and Arct 2019). Increased bioproduction of this substance may result from the genetic alteration of yeast and bacteria with genes encoding the trans-resveratrol production pathway's enzymes. For instance, Transderma (Sweden) developed a serum with trans-resveratrol that enhances the skin's antioxidant defenses (Donnez et al. 2009). Among antioxidants, a terpene antioxidant called astaxanthin is found in various fish, crabs, and birds. It is frequently referred to as the "king of antioxidants" since it has the highest antioxidant activity when compared to other antioxidants like lycopene, vitamin E, and vitamin A (Naguib 2000). Microbial sources of astaxanthin are from microalgae *Haematococcus pluvialis* and yeast *Xanthophyllomyces dendrorhous*. In addition to combating acne, astaxanthin and other photoconvertible carotenoids shield skin from photoaging (Olaizola 2000).

Phlorotannins are important components in cosmetic compositions. They are polyphenolic compounds derived from brown algae such as *Ecklonia cava*, *Ascophyllum nodosum*, and *Fucus vesiculosus*. Different phlorotannins generated by brown algae, including phloroglucinol, eckol, and dieckol, are not poisonous to animals and have impressive free-radical scavenging activity, making them valuable in cosmetic compositions (Pavia et al. 1999; Ahn et al. 2007; Zaragozá et al. 2008).

Many organic substances are being used in the cosmetics industry. Among them, hydroxy acids are crucial. Salicylic acid (SA) and its derivatives, hydroxy acids, and HAs may all be categorized based on their structural similarities. Glycolic acid has been produced using yeast like *Rhodotorula* sp. 3Pr-126 and *Pichia naganishii*, with yields of 35.3 and 25.1 g/L from ethylene glycol and propylene glycol, respectively (Kataoka et al. 2001). A typical preservative or antioxidant in food and cosmetic preparation is citric acid. The molasses or maize starch hydrolysate is fermented by A. niger to produce citric acid for industrial use. Propionic acid is made by fermenting sugar using Propionibacterium species. Salicylic acid, which is produced by microbes like Pseudomonas and Serratia species, makes plants more resistant to a variety of diseases (Liaud et al. 2014). Kojic acid is frequently used in food items because it stabilizes the color and flavor and inhibits enzymatic browning. There are numerous applications for Kojic acid (KA) (5-hydroxy-2-hydroxymethyl-pyran-4-one) in the food, pharmaceutical, and cosmetics industries. KA can act as a fungistatic agent by preventing the development of melanin. It can be used as a skin lightener in cosmetics (Kim et al. 2012; Saeedi et al. 2019). It is a naturally occurring pyrone produced by *Aspergillus* and *Penicillium* fungi (Kim et al. 2012). *Aspergillus oryza* creates a lot of different substances, including proteins. One of this microorganism's main secondary metabolites, KA, is one of its well-known genetic modification pathways (Chang 2009; Leyden et al. 2011;

Yamada et al. 2014). Melasma, post-inflammatory pigmentation, hyperpigmentation, and the aging of the skin are all caused by abnormal melanin synthesis (Novellino et al. 1999; Heo et al. 2010). The well-known tyrosinase enzyme inhibitor kojic acid is used to cure wrinkles, melasma, and hyperpigmentation (Niwa and Akamatsu 1991; Eisenhofer et al. 2003). KA molecules must enter the cell membrane to prevent the production of cellular tyrosinase and, subsequently, the production of melanin which is the byproduct of the tyrosinase enzyme. The administration of KA through the skin to prevent melanin formation has been researched using nano-delivery technologies. Commercially available cosmetic items including lotions, creams, and soaps already contain KA (Saeedi et al. 2019).

Numerous amino acids have tremendous promise in the pharmaceutical sector. Natural skin rejuvenators, exfoliators, and antioxidants are all functions performed by amino acids (Milbrada 1975; Mecca 1975; Shiba 1980; Sugimoto et al. 2016). The Corynebacterium species is the most well-known microbial supplier of amino acids. Using the host bacteria *Corynebacterium glutamicum* or *Escherichia coli*, recombinant DNA technology may be used to generate amino acids at large titers (Adrio and Demain 2010). Low molecular weight UV absorbing compounds like mycosporine and mycosporine-like amino acids (MAAs) are present in both prokaryotes and eukaryotes in large quantities. Because they gather a lot of MAAs when exposed to sunlight, marine algae are a source of MAA. MAA is generated from marine algae, including palythine, porphyra, and shinorine. Commercial sunblock contains *Porphyra umbilicalis*, which shields against all cellular damage brought on by UV radiation. Microalga extracts from *Aphanizomenon flos aquae* have demonstrated intriguing dermatological capabilities including wrinkle reduction, suggesting its potential for aesthetic applications (Oren and Gunde-Cimerman 2007; Scoglio et al. 2012).

Microalgae biotechnology has made significant strides; recombinant proteins and other useful substances that may be utilized in cosmetics are now produced using microalgae suspension cultures (Yan et al. 2016). Spirulina, one of these microalgae, has been utilized in cosmetics (Delsin et al. 2015). With formulations including Spirulina extract produced through biotechnological procedures, Maia Campos et al. conducted some early experiments and showed antioxidant capacity, skin compatibility, and rapid benefits on the hydration of the skin (Zappelli et al. 2016). The bioactive substance phycocyanin, which is produced by this alga, has antioxidant capabilities and reverses the effects of premature aging (Romay et al. 1998). The production of more reliable and secure formulations at a cheaper cost is made possible using Spirulina extracts in dermo-cosmetic formulations. Long-term trials, however, are required to demonstrate its anti-aging efficacy (Delsin et al. 2015). *Dunaliella salina* is another well-known microalga that is abundant in carotenoids, particularly carotene, which has anti-aging characteristics (Christaki et al. 2013). No harmful organic solvents must be utilized in the separation process of carotene for cosmetic purposes (Zappelli et al. 2016). In all living cells, there is a nitrogen molecule called betaine that has been methylated. The most effective osmotic protectant is glycine betaine. Halophiles create glycine betaine from a simple carbon source, whereas *Escherichia coli*, *Staphylococcus aureus*, *Bacillus subtilis*, *Rhizobium meliloti*, *Rhodobacter sphaeroides*, *Vibrio costicola*, and *Alcaligenes* manufacture it from choline (Kiene 1998; Nyyssölä et al. 2000). A non-ionic surfactant called betaine is used to make liquid soap and prescription shampoos (Arkin et al. 2006).

10.2.2 Cell Culture and Biomimetic Strategies

In biotechnology and scientific research, particularly in the areas of tissue engineering, regenerative medicine, and drug testing, cell culture and biomimetic techniques are crucial ideas. These methods seek to explore and create novel solutions by imitating the natural surroundings and situations of living things.

In a laboratory setting, cultivating and maintaining cells outside of their natural environment is known as cell culture. Studying cell behavior, physiology, reactions to stimuli, and interactions with other cells or substances is made possible through this process (Haycock 2011; Gordon and Amini 2021). Cell cultures are essential for several processes, such as tissue engineering (Timmins et al. 2007; Carletti et al. 2011), illness research (Druschky et al. 2001), and medication development (Hiemstra et al. 2018). Cells of many kinds, including those from humans, animals, and even plants, can be grown. Plant tissue culture is quite frequent in the pharmaceutical business, but it is not yet common in the beauty market. Single chemicals created by or isolated from microorganisms like bacteria, yeast, or microalgae are more established. Designing materials, structures, or systems that mirror the traits and operations of biological systems observed in nature is known as a "biomimetic strategy." The objective is to develop solutions that mimic the efficacy and efficiency of organic processes. Biomimetic techniques aim to create materials and constructions that closely mimic the extracellular matrix (ECM), the supporting framework around cells in tissues, in the context of tissue engineering and regenerative medicine (Fierrez-Aguilar et al. 2003a, b).

Cell culture and biomimetic techniques are both used in the cosmetics business to create new and potent products that resemble real biological interactions and processes. These methods aid in the development of cosmetics that deliver better outcomes and are more skin-friendly. These ideas are used as cosmetics in the following ways:

1. Before a cosmetic component is applied to human skin, its safety and effectiveness can be evaluated in cell cultures. In a controlled setting, this aids in finding possible allergies, irritants, or unpleasant responses (Kim et al. 2020).
2. To examine the impact of cosmetic items on human skin, researchers utilize skin models that have been culture-grown. The physiological characteristics of human skin layers may be replicated by these models, enabling precise evaluations of product penetration, irritability, and overall performance (Rattanawiwatpong et al. 2020; Gautam et al. 2021).

3. Compounds that mirror the natural components of the skin are known as biomimetic compounds, and they are frequently used in cosmetic formulations. These components can enhance hydration, barrier function, and the general look of the skin (Sutterby et al. 2020).
4. Models of hair follicles may be created using cell culture and used to test hair care products. These models aid in the research of hair strength, growth, and the results of different aesthetic procedures (Messenger 1984).
5. Cell culture models are used to determine if cosmetic goods may be poisonous or sensitive. In order to guarantee that goods are safe to use and do not hurt the skin or eyes, this is essential (Astashkina et al. 2012).
6. To ensure that active compounds are effectively absorbed into the skin for improved effects, biomimetic design techniques are used to create delivery systems that mirror the natural barrier of the skin (Sheikhpour et al. 2017).
7. Using biomimetic techniques, customized cosmetics may be made to suit a person's skin type, needs, and concerns (Sutterby et al. 2020).
8. Some cosmetic products make the claim to help with skin regeneration and restoration (Aldag et al. 2016). Utilizing substances that promote the healing processes of the skin naturally is one biomimetic strategy (Marder et al. 1993; Jayarama Reddy et al. 2013).
9. Biomimetic formulations provide solutions that improve skin health without upsetting microbial populations in an effort to respect the delicate balance of the skin's microbiome (Jayarama Reddy et al. 2013).
10. Biomimetic chemicals may be utilized to make cosmetic products feel more natural and pleasant on the skin by enhancing their texture, appearance, and sensory experience (Chen et al. 2023; Wang et al. 2023).

10.2.3 Formulation with Novel Cosmetic Ingredients

"Cosmeceuticals" are usually described as active ingredients and finished products with scientifically proven efficacy on the skin. They are now the main forces behind the cosmetic business. The cosmetics industry entered what we might refer to as "a transition phase" thanks to American dermatologist Albert Kligman, who coined the term "cosmeceutical" to describe a new type of cosmetic that combines cosmetic and pharmaceutical characteristics to satisfy consumers' demands for high efficacy (Kligman). Cosmeceuticals are skin care products that target physiological modifications to skin cells to help them seem younger and healthier rather than just being typical cosmetics. Both artificial and organic substances are utilized as active components in cosmetic formulations (Zappelli et al. 2016; Gomes et al. 2020). The following conditions must be met for a substance to be used as an active ingredient in cosmetic formulation.

1. It should be able to pass through all skin barriers and be delivered to the target location with the necessary concentration.

2. It should cause a certain mechanism of action to be triggered at the target site on human skin.
3. To demonstrate its effectiveness and safety, it must have undergone extensive clinical trials and rigorous scientific research.

Herbal cosmetics, which contain botanical extracts, have attracted consumer interest and developed into a niche in the cosmetics and personal care sector since the early 1990s because of the negative effects and environmental repercussions of chemical chemicals (Chanchal and Swarnlata 2008). A few substances with microbiological origins offer intriguing skin and hair care qualities that make them effective active ingredients in cosmetic formulations. The compounds have several functional properties that should be considered for cosmetic applications. Its biocompatibility with skin or hair is the main need. The substance shouldn't cause any unfavorable immune reactions and should be safe. A suitable downstream processing is necessary to purify the compounds with the intact structure and functional qualities since the performance of the compound is related to its structure and physiochemical properties (Pandey et al. 2015).

Reactive oxygen species (ROS), which can cause accelerated aging and skin cancer, are produced by the skin in response to exposure to UV radiation, environmental contaminants, and microbial infections (Fig. 10.2) (Takmakov et al. 2015). Antioxidants applied topically can reduce the production of ROS, which helps to delay the onset of premature aging and other symptoms of oxidative stress.

Sunscreen lotions, skin-brightening cosmetics, and anti-aging items all contain antioxidants as active components. In addition, antioxidants act as stabilizers and lengthen the shelf life of products by reducing the auto-oxidation of substances like fats and oils (Ulber and Sell 2007; Sajna et al. 2015b). The list of significant antioxidants used in cosmetics is shown in Table 10.1.

Cosmetics for skin lightening are in high demand and can be used to lighten skin tone or lessen age-related pigmentation patches. The quantity of melanin pigment in the skin affects the complexion of a person. Tyrosinase is a crucial enzyme in the melanogenesis process, which produces melanin from the amino acid tyrosine. A fairer or lighter complexion is produced by reducing melanin production, and this is seen to be visually pleasing. By preventing the tyrosinase gene from being transcribed, retinoic acid and its derivatives have an effect on the genetic level (Zolghadri et al. 2019). Arbutin and hydroquinone, two commonly used skin-lightening

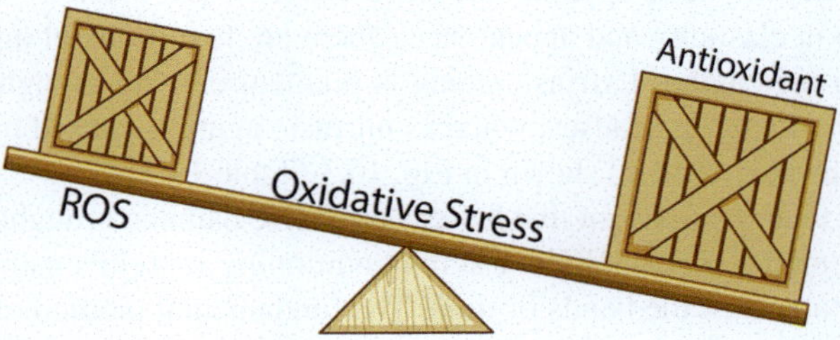

Fig. 10.2 Relationship of antioxidant and ROS

Table 10.1 Significant antioxidants are used in cosmetics

	Antioxidants
1	Vitamin A and its derivatives
2	Ascorbic acid
3	Ferulic acid
4	Phlorotannins
5	Gallic acid
6	Ellagic acid
7	Polyunsaturated fatty acids
8	Coenzyme Q
9	Astaxanthin

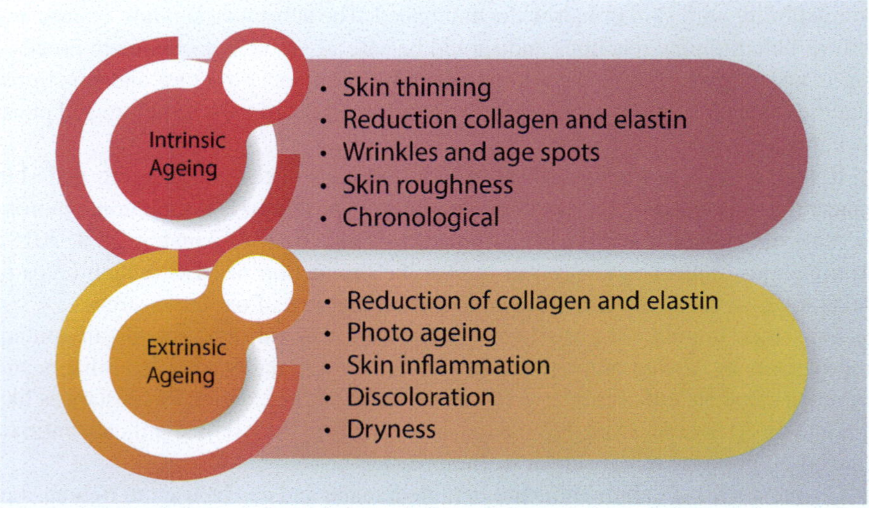

Fig. 10.3 Intrinsic and extrinsic types of aging

treatments, work by competitively blocking the tyrosinase enzyme (Huang et al. 2004). Microbial substances like kojic acid and azelaic acid have the ability to lighten skin because they inhibit the action of tyrosinase by chelating copper atoms from the enzyme's active site (Sajna et al. 2015b). Anti-aging skin care products have had the highest boom in the cosmetics sector as a result of consumers' increasing desire to keep a young appearance. Different chemical changes brought on by aging affect skin elasticity and appearance. There are two forms of aging: intrinsic aging and extrinsic aging. Extrinsic aging is premature aging brought on by solar radiation or other external causes, whereas intrinsic aging is caused by physiological aging (Sajna et al. 2015b) shown in Fig. 10.3 (Table 10.2).

In addition to an antioxidant that lessens oxidative damage brought on by aging, anti-aging cosmetics also include a skin-conditioning or rejuvenating ingredient that works to turn back the hands of time. The fundamental method of skin rejuvenation is to change collagen metabolism, which reduces the appearance of fine lines

Table 10.2 Biotechnologically derived compounds used in cosmetics and their anti-aging activity

Active ingredient	Actions
Human growth factor	Skin rejuvenation
Collagen	Antiwrinkle and skin restructuring
Ceramides	Restore lipid's structure and maintain skin barrier function
Hyaluronic acid	Skin hydrate and regenerate
Coenzyme Q10/Ubiquinol	Antioxidant function and antiwrinkle
l-ascorbic acid (vitamin C)	Antioxidant function and promote collagen synthesis
Retinol palmitate (vitamin A)	Skin rejuvenation
Pullulan	Skin tightening and skin hydration
Mycosporine-like amino acids	Photo-protection

and wrinkles (Almohideb 2021). The skin is shielded from the damaging effects of sun radiation using cosmetics with photo-protective ingredients (Baldisserotto et al. 2018). Acute sunburn causes skin damage with tanning, inflammation, erythema, and Langerhans cell damage as symptoms (Ziegler et al. 1994). Numerous microbial compounds have photo-protective properties and can lessen sunburn symptoms. In comparison to using a sun-blocking agent alone, the usage of baker's yeast extract plus sun-blocking agents such as cinnamate and benzophenone displays increased photo-protective activity (Elmets et al. 1992). Potential sources of photo-protective chemicals include marine creatures. Phlorotannins and phenol extracts from *Corallina pilulifera* block the UV-induced production of matrix metalloproteases, which are responsible for the breakdown of collagen and the development of wrinkles. The mycosporine-like amino acids (MAAs) from marine microalgae absorb UV light due to the presence of the chromophore cyclohexanone. Since MAAs can absorb UV without producing ROS, they can be employed as secure UV-blocking agents in cosmetics. Mycosporine, which inhibits UV-induced thymine dimer production and stops skin erythema, is produced by marine fungus (Pallela et al. 2010; Rastogi et al. 2010).

Hair is treated with harsh chemicals during aesthetic operations, including coloring, straightening, and hair waving, which increases the risk of damage. When you bleach or otherwise alter your hair, you cause damage to the cuticle, which gives your hair a harsh texture, less elasticity, a loss of luster, and difficulty in managing it (Bolduc and Shapiro 2001). The usage of numerous biotechnologically generated substances in modern hair care products helps to reduce hair breakage and other indicators of hair damage. Hair damage is reduced, and the loss of ceramides in damaged hair is restored when ceramide-containing hair care products are used (Morita et al. 2010).

10.3 Effective Ingredients and Enhanced Delivery

10.3.1 Peptides, Growth Factors, and Enzymes

A peptide is an amino acid chain joined together by peptide bonds. Peptides fall somewhere between usual small molecules and bigger proteins in terms of complexity (Rose et al. 1985). Naturally occurring peptides have a number of biological roles, such as signaling (Antonova et al. 2012) and regulatory molecules (Alix and Blanc-Potard 2009), including defense (Marmiroli and Maestri 2014), immunity (Zasloff 2007), growth (Xiao et al. 2015), and homeostasis (Muniz et al. 2012).

Recently, the dermatology and cosmeceutical industry has developed stable and synthetic peptides that have a role in matrix synthesis, pigmentation, and inflammation (Zhang and Falla 2009). Oligopeptides have just lately been studied for their potential function in influencing biological activity within the skin. Etanercept and cyclosporine are the two most often utilized peptides in dermatology (Reddy et al. 2012; Bruno et al. 2013). The usage of peptides has recently been expanded to include cosmeceuticals, a term coined by Dr. Kligman to describe substances that fall in between pure cosmetics and pharmaceuticals (Kligman 2000). Cosmeceutical products provide a biological activity in support of cosmetic claims (Zhang and Falla 2009).

Growth factors are proteins that influence cellular proliferation, differentiation, and growth under specific circumstances (Crommelin et al. 2013). To treat the issues associated with aging skin, a number of growth factors and cytokines have been utilized alone or in combination. After more than 2 months, research employing these growth factors in combination significantly reduced wrinkle scores compared to placebo. Due to their crucial function in skin care, human growth factors are regarded as remarkable molecules in the cosmetics business. The potential role that these molecules may play in the restoration of skin structures has sparked intense research in the involvement of growth factors in wound healing and regeneration (Husein el Hadmed and Castillo 2016).

Human epidermal growth factor (hEGF) has been discovered to be useful in treating wrinkles, age spots, and freckles as well as accelerating the healing process (Walsh 2013; Eissazadeh et al. 2017). Numerous hosted systems, such as *E. coli* and *S. cerevisiae*, have produced hEGF (Ptitsyn and Al'tman 1999). As a result of the hEGF cytoplasm's propensity to form inclusion bodies, which can be quickly broken down by proteases when *E. coli* is utilized as the host, the yield obtained in this case is insufficient for industrial needs. Due to the additional production procedures needed to release HEGF from inclusion bodies, the overall production cost ultimately rises. The amount of HEGF produced by prokaryotic systems is also less than that of eukaryotic systems. Therefore, the growth factor can be produced in vast quantities using eukaryotic systems, such as P. pastoris (Eissazadeh et al. 2017). Heterologous produced EGF in *E. coli* is offered for sale by Skin Actives (USA) as a skin conditioner (Pandey et al. 2015).

Different kinds of enzymes are employed in the cosmetics industry to create formulations that speed up biochemical skin processes and prevent skin aging (Pandey et al. 2015). Enzymes are found in living things and catalyze several biochemical processes that are essential to life (Eissazadeh et al. 2017). These enzymes are also in charge of defending the skin from environmental threats like UV radiation and free radicals. Enzyme use in cosmetics offers a certain biochemical pathway that is more advantageous and improves the functionality of the skin (Sim et al. 2003). Superoxide dismutase (SOD) is one of these enzymes; through its mode of action, it guards against damage from free radicals and other toxic contaminants (Pandey et al. 2015). Reactive oxygen species (ROS) and reactive nitrogen species (RNS) are formed by UV radiation, other radiation, and normal cellular metabolism. SOD enzymes regulate these species' levels, limiting their potential toxicity and regulating cellular processes that are controlled by their signaling functions (Wang et al. 2018). SOD is regarded as an anti-aging enzyme since it aids in the removal of ROS in humans, which has been proven to contribute to skin degeneration due to ROS produced in metabolic pathways. Additionally, SOD enhances skin suppleness, preserves the essential keratin structure, and gives the skin a silky sensation (Younus 2018). *S. cerevisiae* can be genetically modified to produce SOD (Levin 2005).

10.3.2 Hair Care, Skin Care, and Other Aesthetic Applications

Biotechnology has significantly improved many industries, including the maintenance of hair. The development of products that support hair growth and restoration has been made possible through biotechnology (Tsigos et al. 2000; Tinoco et al. 2022). Growth factors, stem cells, and gene therapies have all been investigated by researchers as potential means of enhancing hair growth (Ohyama and Vogel 2003; Park et al. 2010; Won et al. 2010). These developments may eventually result in more efficacious therapies for ailments like male and female pattern baldness. Biotechnology has aided in the development of sophisticated anti-hair loss remedies. For instance, some biotechnological strategies involve locating and focusing on certain chemicals or processes that cause hair loss. In comparison to conventional treatments, this might lead to therapies that are more focused and efficient (Castro et al. 2023). Genetic testing and analysis methods that can shed light on a person's genetic propensity for hair-related issues have been developed because of biotechnology. This data can be utilized to customize hair care products and treatments, providing more specialized answers to address issues (Cheng and Bayliss 2008; Westgate et al. 2013).

Biotechnology has made it possible to extract, synthesize, and modify bioactive substances from natural sources. These components, including proteins, peptides, and chemicals with plant origins, can be added to hair care products to increase their potency and nourish the hair and scalp (Park et al. 2010). The use of nanotechnology in hair care can improve the distribution of active chemicals to the hair and scalp. Nanotechnology is a subset of biotechnology. Ingredients can be encapsulated and shielded by nanoparticles, enabling their effective absorption and focused activity (Rosen et al. 2015; Santos et al. 2022). Biotechnology has the potential to alter the texture of hair. The invention of products that change hair texture, making it more manageable, straight, or curly based on personal preferences, may result from research into the molecular structure of hair fibers (Kessler-Becker 2016). Nanotechnology helps to achieve the SDG goals by addressing many issues. It can improve healthcare by enabling focused and targeted drug delivery and early detection of diseases. It can improve the quality and number of crops by controlling pests. It can improve the purification methods of water (Arshad et al. 2023).

This has sparked the creation of novel products and treatments that make use of biological processes and components to enhance skin function, look, and health (Iobst et al. 2016). Using bioactive substances, peptides, enzymes, and other biological components, biotechnology in skin care aims to provide effective and focused treatments for a range of skin issues (Stolz and Obermayer 2005).

Biotechnology makes it possible to extract and create bioactive substances from natural sources. These ingredients, which include vitamins, antioxidants, and botanical extracts, are added to skin care products to offer advantages including improved skin texture, hydration, and defense against environmental stresses (Inoue 2016). Active compounds that are good for the skin are produced through microbial fermentation methods. For instance, fermented probiotics and prebiotics can regulate the skin's microbiota and enhance general skin health (Stolz and Obermayer 2005; Iobst et al. 2016). The sustainable sources of prebiotics are onions, leeks, garlic, bananas, and whole grains that are rich in inulin and fructooligosaccharides, which are prebiotic fibers. Foods like kefir, kimchi, and sauerkraut are rich and sustainable sources of probiotics. The major advantages of fermentation-based skin care manufacturing are shown in Fig. 10.4.

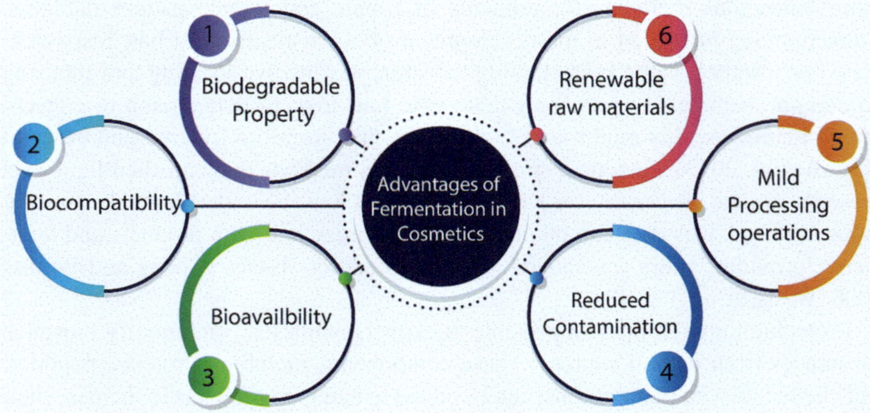

Fig. 10.4 Benefits of fermentation in cosmetic industries

In order to create particular bioactive substances, plant cells are grown in a controlled environment using plant cell culture procedures. This makes it possible to sustainably extract uncommon or difficult-to-find plant-based compounds that have anti-aging, calming, or brightening properties (Rawlings 2016). In order to encourage exfoliation, boost product penetration, and improve skin texture, skin care products often contain enzymes obtained from microorganisms. In comparison to physical exfoliants, enzymes are kinder and can target skin issues. Biotechnology makes it possible to produce biomimetic substances that resemble the organic components of healthy skin. These components can promote general skin health, retain moisture, and aid in restoring the function of the skin barrier (Inoue 2016). Products that enhance the skin's natural DNA repair mechanisms are made using biotechnology, helping to fight the effects of aging brought on by environmental damage. Certain substances can promote collagen formation and cell regeneration, resulting in skin that is smoother and firmer (Lad 2016). By modifying materials at the nanoscale, nanotechnology can improve their characteristics. Nanotechnology can enhance component delivery in skin care products, allowing for greater skin penetration and improved outcomes (Lad 2016). The creation of customized skin care products that cater to a person's unique skin demands has been made possible through biotechnology. The development of treatments that address issues including aging, sensitivity, and pigmentation is assisted by DNA analysis and other evaluations (Rawlings 2016).

While remarkable improvements in skin care have been made thanks to biotechnology, it's crucial to remember that not all items with the designation "biotech" work equally well. It is advised to seek trusted businesses, speak with skin care experts, and consider unique skin sensitivities and preferences while thinking about biotech-based skin care products. The ideal biotech-based skin care regimen for your unique needs can be determined by consulting a dermatologist or skin care expert (Stolz and Obermayer 2005).

In the discipline of aesthetics, which focuses on enhancing and improving people's appearances through different non-surgical and minimally invasive methods, biotechnology has found several applications. Here are some ways that biotechnology is used in cosmetic procedures:

10.3.2.1 Dermal Fillers

A type of cosmetic procedure called dermal fillers is used to improve facial features, add volume, and lessen the visibility of wrinkles and creases. They are frequently used to reduce the appearance of aging or to produce more harmonious facial features. Dermal fillers are injectable materials that increase volume, lift drooping regions, and smooth out wrinkles and creases beneath the skin (Funt and Pavicic 2013). Hyaluronic acid or other biocompatible materials are used in advanced dermal fillers developed by biotechnology to restore volume and smooth wrinkles. These fillers are made to deliver results that seem natural with the least amount of discomfort and downtime (Tezel and Fredrickson 2008; Ballin et al. 2015).

Dermal fillers can be divided into a variety of categories, but for the purpose of discussing dermal filler complications, it may be most useful to divide them into biodegradable and nonbiodegradable fillers, as well as particulate and nonparticulate fillers. Collagen and hyaluronic acid (HA) fillers are examples of moderate-duration biodegradable fillers. Because these fillers are swiftly reabsorbed by the body, they provide rather transient aesthetic benefits. In both Europe and the USA (Zielke et al. 2008), HA derivatives are the most often used biodegradable fillers. The effects of these biodegradable fillers last for 6–18 months, depending on the source, degree of cross-linking, concentration, and particle size of each substance. A foreign body reaction caused by the nonbiodegradable fillers encourages the fibroblastic deposition of collagen surrounding the nonabsorbable microspheres (Hilinski and Cohen 2009).

10.3.2.2 Botulinum Toxin (Botox) Treatments

Botulinum toxin, also known as botox, is a neurotoxic protein used for a variety of therapeutic and aesthetic objectives. It is made by the bacteria Clostridium botulinum and utilized in very tiny, carefully regulated quantities for medical and aesthetic procedures. Although they can have medicinal use, botox injections are most used for aesthetic purposes (Naumann et al. 2006).

Botox injections, which are generated from the bacterium Clostridium botulinum, are frequently used to temporarily relax facial muscles and minimize the appearance of fine lines and wrinkles. Botulinum toxin can now be produced in pure, risk-free versions for aesthetic purposes thanks to biotechnology (Jankovic and Schwartz 1991; Binder et al. 2000).

10.3.2.3 Platelet-Rich Plasma (PRP) Therapy

A medical procedure called platelet-rich plasma (PRP) therapy uses a concentrated type of the patient's blood to encourage tissue regeneration and repair. Platelets, growth factors, and other advantageous components are extracted from blood to create PRP, which is then administered intravenously to specific body parts. PRP therapy uses a patient's blood to separate platelets and growth factors that are then injected into the skin to promote collagen synthesis and enhance skin tone and texture. These platelets can be effectively isolated and concentrated using biotechnology techniques (Middleton et al. 2012; Le et al. 2018; Emer 2019). PRP treatment is becoming more and more used in several medical specialties, including orthopedics, dermatology, and cosmetic surgery. PRP treatment is used to treat osteoarthritis, tendinitis, and ligament sprains among other musculoskeletal conditions. To encourage healing and minimize inflammation, it is frequently injected directly into the afflicted region (Dhillon et al. 2012). PRP is used in dermatology to rejuvenate the skin and treat problems including fine lines, wrinkles, and scars. It can be injected or administered topically to promote collagen formation and enhance skin texture (Le et al. 2018). PRP treatment is utilized to stimulate hair follicles and encourage hair growth. PRP can be used with other procedures like micro-needling or dermal fillers to boost outcomes and increase skin quality (Cervelli et al. 2014; Emer 2019). Although PRP therapy is usually regarded as safe since it uses the patient's blood, outcomes might vary depending on the patient, the ailment being treated, and the healthcare provider's level of skill.

10.3.2.4 Stem Cell Therapies

Stem cells are used in stem cell therapy, a medical strategy to cure or prevent illnesses and injuries. Embryos, fetal tissue, and adult tissues can all serve as a source of stem cells. To rejuvenate the skin and encourage the generation of new, healthy cells, stem cell-based therapies harness the regenerative characteristics of stem cells. To cultivate and use stem cells for cosmetic purposes, biotechnology is crucial (Trehan et al. 2017; Aggarwal et al. 2020; Pérez-Rivero and López-Gómez 2023). In several investigations, stem cells are found to have improved dermal thickness and density of collagen fibers in elderly mice (Kim et al. 2011) as well as to reduce wrinkles (Kim et al. 2009).

10.3.2.5 Cutting-Edge Laser and Light-Based Technologies

Medical dermatology and aesthetics have undergone a revolution thanks to cutting-edge laser and light-based technology. These technologies provide cutting-edge solutions for a variety of skin issues, from medical operations to aesthetic procedures. Biotechnology helps to create cutting-edge laser and light-based technologies for skin resurfacing, hair removal, and the treatment of pigmentation problems. These techniques precisely target skin structures, improving cosmetic results. Biotechnology is utilized in tissue engineering to create tissues or constructions for cosmetic surgeries, such as artificial skin substitutes for grafting or aids in wound healing (Fang et al. 2017; CAN YOU 2021).

10.3.2.6 Cryolipolysis

The non-invasive fat reduction method known as cryolipolysis (CoolSculpting) uses controlled cooling to target and destroy fat cells. Cryolipolysis, sometimes known as "fat freezing," is a non-surgical cosmetic technique that involves subjecting stubborn localized fat to controlled chilling (Jalian and Avram 2013). The idea behind cryolipolysis is that fat cells are more susceptible to cold temperatures than the tissues around them. By carefully controlling freezing, fat cells may be targeted and destroyed without causing injury to surrounding tissues or the skin. The development of cooling technologies and the refinement of treatment regimens both include biotechnology (Avram and Harry 2009; Derrick et al. 2015).

10.3.2.7 Non-surgical Skin Tightening Treatments

Skin tightening treatments include those that use ultrasonic or radiofrequency energy to tighten the skin. To give the skin a firmer and more youthful appearance, these technologies are intended to increase collagen synthesis. These procedures tighten the skin using a variety of technologies that encourage the creation of collagen and tissue remodeling. People searching for a minimally invasive alternative to surgical procedures like facelifts frequently seek them out. Biotechnology is employed to produce these devices (Malone et al. 2021).

10.3.2.8 Thread Lifts

To lift and tighten sagging tissue, biocompatible threads are inserted under the skin. A thread lift, also known as a thread facelift or a non-surgical facelift, is a cosmetic procedure designed to lift and tighten the skin of the face and neck using specialized threads (Savoia et al. 2014). This procedure is considered a minimally invasive alternative to a traditional surgical facelift. Thread lifts are often chosen by individuals who want to achieve some degree of facial rejuvenation without undergoing surgery. Biotechnology aids in the creation of threads that encourage tissue integration and provide results that look natural (Tong and Rieder 2019).

10.3.2.9 Gene Therapy and Growth Factors

In the world of medical and scientific research, growth factors and gene therapy are two separate strategies that both attempt to address various health issues and promote tissue repair or regeneration. New biotechnological methods involve the introduction of genes through gene therapy that support tissue regeneration and young looks. Additionally, biotechnologically derived growth factors can improve skin renewal and wound healing (Crommelin et al. 2013; Kokare 2013). Gene therapy includes adding or changing genes to cells to treat or improve certain features. Gene therapy is still a new and unproven field in the field of cosmetics. Targeting genes associated with skin health and aging is the notion behind gene therapy in cosmetics, which aims to improve certain traits (Macer et al. 1995; Delhove et al. 2020). Growth factors are naturally occurring proteins that are essential for tissue regeneration, cell differentiation, and growth. Growth factors are utilized in cosmetics to make products that promise to stimulate skin renewal and anti-aging benefits. These goods frequently have growth elements that come from organic materials or are biotechnologically created. They are said to increase the formation of collagen, enhance skin texture, and lessen the visibility of fine lines and wrinkles. Growth factors can be added to micro-needling treatments, serums, and creams (Aldag et al. 2016).

10.3.2.10 Microbiome Restoration

The act of purposefully repairing and rebalancing the microbial communities that live in different regions of the human body, notably the gut, is known as microbiome restoration. The collection of bacteria, viruses, fungi, and other microbes that reside in and on our bodies and are vital to our health and well-being is known as the human microbiome. Due to its effects on metabolism, immunity, digestion, and even possible effects on mental health, the gut microbiome has drawn a lot of research (Dixit et al. 2021). The equilibrium of the gut microbiome can be upset by elements including nutrition, lifestyle, drugs, and environmental exposures, which may result in health problems. The microbiome of the skin is essential for preserving skin health and beauty. In order to enhance aesthetics, biotechnology research is examining how to balance and restore the microbiome of the skin. The use of nanotechnology in aesthetics involves creating nanoparticles for targeted drug delivery, boosting topical therapy effectiveness, and enhancing aesthetic results (Daliri et al. 2018).

10.3.3 Enhanced Bioavailability, Stability, and Delivery Methods

Enhancing the bioavailability, stability, and distribution of active substances is crucial in the cosmetics sector to guarantee that products deliver the desired advantages. The following are some popular techniques and tools for achieving these objectives:

10.3.3.1 Embedding and Microsite- and Nanotechnology

Incorporating microscopic and nanoscopic structures, devices, or particles into materials or systems to accomplish certain capabilities or advantages is known as embedding micro and nanotechnology in a variety of sectors, including cosmetics and medicine. Encapsulation is the process of encasing active chemicals inside tiny carriers, including liposomes, nanoparticles, or microcapsules (Arshad 2024). These carriers increase the stability of the substances and shield them from deterioration. As a result of improved substance penetration into the skin and hair because of nanotechnology, bioavailability can be increased. Working with materials and structures at exceedingly small sizes, frequently on the order of micrometers (millionths of a meter) or nanometers (billionths of a meter), is the focus of micro- and nanotechnology (Lv et al. 2005; Shoukat and Khan 2021).

10.3.3.2 Lipid Nanoparticles and Liposomes

In the realm of biotechnology, lipid nanoparticles and liposomes are cutting-edge delivery technologies, notably in medicinal, cosmetic, and other applications. These lipid (fat)-based structures have the capacity to encapsulate and distribute different compounds, including medications, active chemicals, and genetic material, to certain sites (Müller et al. 2002). Lipid bilayers form the spherical vesicles known as liposomes, which can contain both hydrophilic and hydrophobic substances. They boost penetration, increase component stability, and enable controlled release. Within their watery core or lipid bilayers, they can enclose molecules that are both hydrophobic and hydrophilic. Drugs that are hydrophobic can be made more bioavailable by being dissolved in liposomes. Liposomes can control the release of medications, resulting in therapeutic benefits that last longer. Liposomal administration can reduce medication exposure to healthy tissues, which lessens negative effects (Ahmad 2021; Fan et al. 2021).

10.3.3.3 Solid Lipid Nanoparticles (SLNs) and Nanostructured Lipid Carriers (NLCs)

These lipid-based nanoparticles provide both hydrophilic and lipophilic components greater stability, controlled release, and increased bioavailability. Solid lipid nanoparticles, or SLNs, provide better drug loading capacity, controlled release, and improved stability. SLNs are used in pharmaceuticals, cosmetics, and nutraceutical industries. A more flexible lipid matrix is produced by NLCs, a variant of SLNs that employ a combination of solid and liquid lipids. Higher drug loading, better drug release patterns, and increased stability are all made possible by this structure (Müller et al. 2002).

10.3.3.4 Nanoemulsions Vs. Microemulsions

These are colloidal systems made up of microscopic oil and water droplets that have surfactants stabilizing them. They make substances more soluble and aid in their more efficient delivery. Nanoemulsions are thermodynamically prone to separation and are unstable. However, because of the diminished gravitational forces acting on the tiny droplets, their small droplet size might offer increased stability (Boonme et al. 2009). Another benefit is the small droplet's large surface area, which enables efficient delivery of the active to the skin (Sharma and Sarangdevot 2012). To enhance medication solubility, bioavailability, and targeted distribution, nanoemulsions are employed in drug delivery. They are also used in skin care products to improve sensory qualities and the penetration of active components. Microemulsions may sustain stability for long periods without phase separation because they are thermodynamically stable. Microemulsion can increase medication solubility and controlled release (Chanchal and Swarnlata 2008).

10.3.3.5 Delivery Systems Based on Polymers

The progressive release of active substances from matrices made of polymers ensures prolonged effects and improves stability. Polymer-based delivery methods are extremely important in several sectors, including the pharmaceutical, cosmetic, and agricultural ones. These systems encapsulate and distribute active components, medications, nutrients, and other substances to sites using polymers as carriers. Polymers have benefits including focused distribution, controlled release, and preservation of delicate substances (Van Gheluwe et al. 2021). Biocompatible or biodegradable polymers are used to create polymer nanoparticles, which are nanoscale transporters. In order to use hydrophobic or hydrophilic substances for gene therapy, medication delivery, or aesthetic purposes. Nanoparticles made of chitosan and poly(lactic-co-glycolic acid) are two examples (Arshad and Abbas 2018). Amphiphilic block copolymers self-assemble into polymeric micelles in aqueous liquids. Drugs that are hydrophobic can be encapsulated, which increases their solubility and bioavailability. Micelles are utilized in cosmetics and medicine delivery (Costa and Santos 2017).

10.3.3.6 Hydrogels

Hydrogels, also known as hydrophilic gels, are typical pharmaceutical delivery systems used to provide medications to the skin. Gels are semisolid formulations that are clear and translucent. They have a high solvent-to-gelling agent ratio. Due to their biocompatibility, excellent spreading and adhesion to the application site, and lengthy residence duration on the site of action, hydrogels have been employed for a very long time in the sectors of dermatology and cosmetology (Van Gheluwe et al. 2021). A polymeric channel system makes up the core of a hydrogel, which is created by physically or chemically cross-linking homopolymers or copolymers (Sood et al. 2016), which causes swelling when exposed to an aqueous environment as shown in Fig. 10.5.

10.3.3.7 Techniques for pH Adjustment and Formulation

Pharmaceutical, cosmetic, and food sectors, among others, all depend on pH correction and formulation. Optimizing product stability, effectiveness, and safety requires proper pH regulation. The stability and solubility of components can be affected by

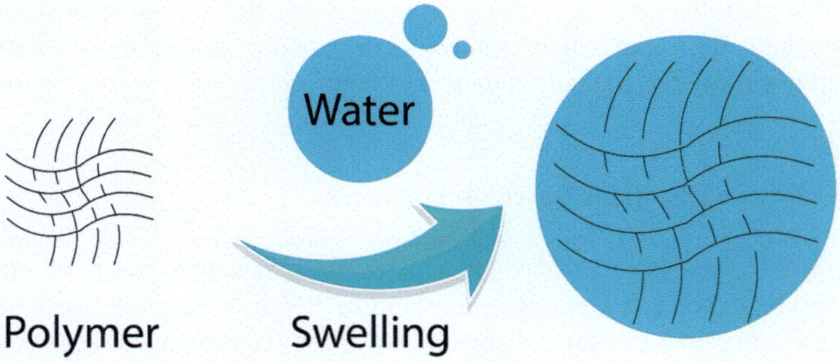

Fig. 10.5 Hydrophilic gel formation by swelling polymers in water

changing the pH of formulations. The stability and dispersion of ingredients can be improved by formulation processes including microemulsion and complicated coacervation (Vemula et al. 2010; Mobarak et al. 2014). Some of the techniques that are used for pH adjustment are buffer solutions, the addition of acid bases, buffering salts, and chelating agents. Solutions known as buffers prevent pH shifts caused by the addition of acids or bases. They are made up of either a weak base and its corresponding conjugated acid or vice versa. For compositions to stay within the required pH range, buffers are crucial (Gokarn et al. 2008). To change the pH, one can also add bases (such as NaOH or KOH) or acids (such as citric acid or lactic acid) directly. Excessive doses must be avoided since they may impair the skin's tolerance and the stability of the substance. To keep the pH steady, buffering salts are added to formulations. Citrate salts and phosphate salts are two examples of substances that can regulate and stabilize pH. The activity of metal ions that could catalyze processes resulting in pH shifts or product degradation can be controlled with the use of chelating agents (Khan and Singh 2016).

10.3.3.8 Penetration Facilitators

Penetration facilitators are chemicals that aid in the absorption and penetration of active compounds through the skin, mucous membranes, and other biological barriers. They are sometimes referred to as penetration enhancers or permeation enhancers. These enablers can improve the delivery of medications, cosmetic components, and other bioactive substances to the targets they are intended for. To improve molecular permeability, penetration facilitators alter the skin's or membranes' barrier characteristics. These are substances that make it easier for active chemicals to cross the skin barrier. They can be added to formulations to increase bioavailability (Katz et al. 2015; Arct et al. 2016). As technology develops, these strategies and tactics are always improving. Combining these methods allows cosmetic scientists and researchers to create products that efficiently distribute active chemicals, maintain stability over time, and provide consumers with better advantages.

10.4 Aesthetic Medicine and Biotechnology

10.4.1 Tissue Engineering

Biotechnology and cosmetic medicine both benefit greatly from the use of tissue engineering. It provides ground-breaking methods for increasing physical attractiveness, correcting aging-related changes, and raising aesthetic standards generally. Here are some applications of tissue engineering in biotechnology and aesthetic medicine:

10.4.1.1 Dermal Fillers and Soft Tissue Augmentation

Dermal fillers are cosmetic substances that are injected into the skin to improve facial features, increase volume, and smooth wrinkles. They are frequently utilized for soft tissue augmentation to produce a more rested look, address face asymmetry, and restore youthful features (Thoma et al. 2009). In aesthetic medicine, dermal fillers are a common non-surgical approach for treating aging symptoms and improving face looks. Injectable dermal fillers made of biocompatible materials or biomaterials coupled with the patient's cells are produced using tissue engineering procedures (Funt and Pavicic 2013).

10.4.1.2 Skin Rejuvenation

The term "skin rejuvenation" describes a range of methods and treatments used to enhance the look, feel, and general health of the skin. The skin changes as we age, showing indications of aging such as loss of elasticity, fine lines, wrinkles, uneven pigmentation, and more (Aldag et al. 2016). Treatments for skin rejuvenation concentrate on these issues to provide a more young and revitalized appearance. Tissue engineering plays a role in the creation of cutting-edge skin rejuvenation procedures (Gomes et al. 2020). By extracting and processing the patient's own cells for re-injection, procedures including platelet-rich plasma (PRP) treatment and autologous fat transfer stimulate collagen formation and improve skin texture (Emer 2019).

10.4.1.3 Hair Restoration

The term "hair restoration" refers to a variety of methods and practices used to cure hair loss and encourage hair growth. Alopecia, the medical term for hair loss, can be brought on by genetics, hormonal changes, illnesses, and aging. Treatments for hair restoration attempt to promote natural hair growth and enhance the look of balding or thinning regions. Tissue engineering has a part to play in hair restoration procedures. To restore hair follicles and encourage organic hair growth in those who are experiencing hair loss, this procedure uses stem cells, growth factors, and biomaterials (Bolduc and Shapiro 2001; Rosen et al. 2015; Sajna et al. 2015a; Kessler-Becker 2016).

10.4.1.4 Lip Augmentation

A cosmetic treatment called lip augmentation tries to improve the lips' size, shape, and fullness. People who want larger lips or who want to enhance the symmetry and dimensions of their lips frequently choose this option. Various surgical and non-surgical methods, including lip augmentation, are available. Tissue engineering is used in aesthetic medicine to provide long-lasting, natural-looking lip augmentation procedures. The lips get volume and contour via injectable lip fillers composed of biocompatible materials (Byrne and Hilger 2004; Sarnoff et al. 2008). Techniques for lip augmentation are dermal fillers, lip implants, hydroxyapatite fillers, and botox.

Dermal fillers are one of the most popular and non-invasive procedures for lip enhancement. Fillers made of hyaluronic acid (HA) are frequently utilized for this. The body naturally contains HA, which draws in and holds onto moisture to give lips their plumpness and hydration. To give the lips more volume and improve their look, filler is injected into them. Results are usually transient and might persist for a few months until the filler progressively degrades (Monheit and Coleman 2006). To improve the size and contour of the lips, artificial materials, such as silicone or GORE-TEX, are surgically inserted into the lips. The only drawback to this approach is that it necessitates surgery (Segall and Ellis 2007).

In comparison to certain other fillers, those that include hydroxyapatite particles can produce effects that stay longer. They increase collagen synthesis, which prolongs the effects of the augmentation (Eversole et al. 2013). The muscles surrounding the lips can be relaxed with injectables such as neuromodulators (e.g., botox), which results in a smoother look and subtle augmentation (Votto et al. 2021).

10.4.1.5 Reducing Wrinkles

In the world of skin care and aesthetics, minimizing wrinkles is a frequent objective. In addition to being impacted by variables including sun exposure, genetics, lifestyle, and skin care practices, wrinkles are a normal part of aging. While wrinkles cannot be fully removed, several techniques and treatments can help lessen their visibility and encourage smoother, more youthful-looking skin. Tissue engineering also helps to create cutting-edge anti-aging remedies that target wrinkles and fine lines. The use of biomaterials or biologically active substances is a technique for promoting collagen formation and enhancing skin suppleness (Imokawa 2008; Proksch et al. 2014).

10.4.1.6 Customized Implants

In the realm of cosmetic surgery, customized implants are also utilized to accentuate and enhance physical traits. These implants are made to fit the patient's preferred anatomical features and look. While cosmetic operations are optional and generally intended to improve appearance, personalized implants may be able to help patients achieve outcomes that are both natural-looking and unique to them. Tissue engineering techniques can be used to manufacture customized implants for body and face reshaping. These implants can be made to meet a person's specific anatomy and aesthetic objectives (Leordean et al. 2015; Bai et al. 2019).

10.4.1.7 Personalized Cosmetics

Cosmetics that are made specifically for each user's needs, tastes, and traits are known as personalized cosmetics. These products develop formulas that target requirements and provide the best outcomes by considering variables including skin type, skin issues, color preferences, and lifestyle. Technology advancements, data analysis, and the rising desire for individualized beauty products have led to the development of personalized cosmetics (Wang et al. 2004). Tissue engineering and biotechnology enable the creation of customized cosmetic products that are tailored to a person's unique skin type, needs, and concerns.

10.4.1.8 Scar Revision

Medical and cosmetic surgery called scar revision aims to minimize the visible effects of scars left behind by accidents, surgeries, burns, or other conditions. Scar revision treatments attempt to reduce their visibility and improve the overall aesthetics of the skin. Scars can vary in size, shape, color, and texture. Revision of scars can be done in several ways, depending on the scar's nature and features. Techniques for tissue engineering are employed to provide therapies that reduce the visibility of scars. Engineered tissues support the formation of healthy tissue over damaged regions and aid in the healing process (Garg et al. 2014). The use of tissue engineering in aesthetic medicine presents ethical issues and necessitates adherence to regulatory standards to guarantee patient safety and the moral utilization of human cells and resources. Tissue engineering, biotechnology, and cosmetic medicine can be combined to provide less intrusive, more natural-looking therapies that improve beauty while putting the needs of patients first. Tissue engineering is probably going to play an increasingly bigger part as technology develops in altering the field of biotechnology and cosmetic medicine (Alster and Zaulyanov-Scanlon 2007; Thomas and Somenek 2012).

10.4.2 Stem Cell Therapies

Due to its potential to repair tissues, expedite healing, and improve several aspects of beauty and health, stem cell treatments have drawn considerable attention in both the biotechnology and aesthetics areas. Because they may develop into numerous cell types, stem cells, which are undifferentiated cells, are useful for tissue regeneration and repair (Trounson et al. 2011).

Skin texture improvement, wrinkle reduction, and collagen stimulation are all achieved by the use of stem cells in face rejuvenation procedures (Chen et al. 2012; Jeong et al. 2015; Semsarzadeh and Khetarpal 2022). Therapies based on stem cells can improve the skin's general appearance and give it a more youthful appearance. By encouraging the formation of hair follicles, stem cells are essential for hair repair. In patients with hair loss, stem cell therapy can awaken dormant hair follicles and promote the creation of new, natural hair (Egger et al. 2020). Stem cells are being researched for possible uses in breast augmentation operations. To improve breast volume and form, adipose-derived stem cells can be employed. Stem cell treatments are being studied for scar reduction by encouraging the regeneration of healthy tissue and reducing the visibility of scars arising from injuries or surgical operations. Stem cell-based therapies also encourage the development of new skin cells and support tissue healing that can help regenerate aging or damaged skin (Branski et al. 2009).

The process of tissue engineering, in which stem cells are seeded onto scaffolds to produce functioning tissues for transplantation or study, is essential. Applications for these created tissues include drug testing and regenerative medicine (Hovatta et al. 2010; Hafizi et al. 2012). Stem cells are being investigated as a possible source for developing substitute organs in the laboratory for organ transplantation. This could assist in addressing the lack of organ donors for transplants (Li et al. 2021). To investigate the processes of illnesses and evaluate prospective therapeutic approaches, disease-specific cell lines made from stem cells are utilized. Stem cells are also being used to screen and test novel medications for effectiveness and safety during drug development. Neurodegenerative illnesses, cardiovascular ailments, and autoimmune conditions are among the diseases for which stem cell treatments are being researched.

Stem cells can be utilized to build patient-specific models for drug testing and therapy development that are customized to a person's genetic makeup and illness profile (Kaji and Leiden 2001; Trounson and McDonald 2015). It is crucial to remember that while stem cell treatments show great potential, their clinical use requires careful ethical deliberation, regulatory approval, and extensive scientific study to guarantee their success. Furthermore, the field of stem cell research and therapy is continually developing, and new developments keep influencing how stem cells are used in biotechnology and aesthetics.

10.4.3 Gene Therapies

Gene treatments provide novel ways to address genetic diseases, improve physical beauty, and increase general health, and they have the potential to revolutionize both biotechnology and aesthetics. To treat or prevent illnesses, fix genetic mutations, and affect cellular activities, gene therapy involves changing the genetic information within cells (Chulpanova et al. 2022).

Collagen and elastin are two proteins crucial for young skin. They can be stimulated via gene treatments. Gene treatments can enhance skin texture, elasticity, and general attractiveness by introducing genes that promote the production of these proteins. Gene treatments focus on hair follicles to promote hair growth and stop hair loss. These treatments try to restore hair density and thickness by providing genes that enhance hair follicle activity and inhibit elements causing hair loss. By adding genes that control the synthesis of melanin, gene treatments may provide alternatives to improve or alter skin pigmentation. This could make it possible to modify skin tone under control. Gene therapies are being investigated as potential treatments for genetic illnesses brought on by gene mutations. These treatments try to address the disease's underlying cause by introducing functioning copies of the damaged gene (Hengge and Bardenheuer 2004; March et al. 2018). By delivering genes that cause apoptosis (cell death) or inhibit tumor development, gene treatments can specifically target cancer cells. The specificity and efficacy of cancer treatments might be improved by these approaches. Gene treatments can also improve the immune system's capacity to recognize and destroy infections or cancer cells. Immune system modifications, such as CAR-T cell therapy, enable more effective targeting and cancer cell eradication (Amer 2014; Das et al. 2015). For uncommon illnesses brought on by genetic mutations, gene treatments are now being researched. These treatments try to treat symptoms while restoring normal cellular function. Gene therapies can be used to create cells to make certain antibodies, perhaps providing cures for autoimmune illnesses or infectious diseases. The use of gene therapies presents ethical issues, including worries about unexpected consequences and access to these medicines (Zhang et al. 2007).

Regulatory organizations are crucial in guaranteeing the security and efficacy of gene treatments. Applications of these novel strategies in biotechnology, health, and aesthetics are being shaped by continuing developments in gene therapy research.

10.4.4 Effective and Personalized Aesthetic Treatments

Effective and individualized cosmetic treatments are made to accommodate unique preferences, issues, and traits while providing the best outcomes. These procedures make use of medical expertise, technological developments, and patient-centered strategies to improve appearance and increase self-confidence. The following describes how efficient and unique cosmetic treatments are accomplished:

1. Aesthetic professionals do in-depth consultations with patients to learn about their objectives, medical history, skin type, and particular problems. The treatment plan will be adapted to the patient's particular needs thanks to this personalized approach (Vitagliano et al. 2023).
2. Modern methods and technology are used in aesthetic procedures to improve results while minimizing pain and recovery time. Examples include minimally invasive procedures, ultrasound-based therapies, and laser therapies (Fabi 2015).
3. Practitioners create personalized treatment programs based on the consultation that target certain issues like wrinkles, volume loss, pigmentation, or skin texture.
4. Synergistic outcomes can be achieved by combining different therapies. For instance, a holistic approach to face rejuvenation can be achieved by combining injectables with laser treatment (Mirza and Khatri 2017).
5. A customized application of dermal fillers and neuromodulators (like botox) can be used to improve facial characteristics while retaining a natural look (Binder et al. 2000; Funt and Pavicic 2013).
6. Specific skin issues, such as acne scars, pigmentation, or fine wrinkles, can be addressed with customized skin treatments including chemical peels, microdermabrasion, and laser resurfacing (Fabi et al. 2019).
7. In order to develop a strategy that produces results that seem natural, personalized hair restoration therapies take into account variables including hair type, hair loss pattern, and individual objectives (Verret 2009; Jördening and Winter 2005).
8. The ability to adapt therapies to a person's genetic profile and predispositions is made possible by advances in biotechnology. Practitioners may now employ genetic and biomarker analyses (Ulber and Sell 2007).
9. The patient's skin type, issues, and objectives may all be taken into consideration while creating a customized skin care program. These goods can improve the outcomes of cosmetic procedures (Graham and Logan 2004).
10. For controlling expectations and obtaining gratifying results, it is essential to inform patients about the therapies, anticipated results, and aftercare (Graham and Logan 2004).
11. Maintenance is frequently needed after successful treatments to maintain outcomes. To guarantee continued benefits, practitioners develop individualized plans for follow-up sessions (Bai et al. 2019).
12. Aesthetic professionals give ethical treatment choices a priority and take patients' psychological and emotional well-being into account (Newburger and Caplan 2006).
13. A customized strategy takes into account the psychological needs of the patient, ensuring that they have healthy self-perception and reasonable aspirations (Atiyeh et al. 2020).
14. Aesthetic professionals collaborate closely with patients to create the optimal treatment plan by going through possibilities, listening to worries, and discussing solutions (Atiyeh et al. 2020).

Combining medical skill, technological advancements, and patient preferences to produce results that are pleasing and natural-looking is the secret to efficient and individualized aesthetic treatments. The specific traits and objectives of each person are taken into account to create a customized strategy that improves their physical attractiveness and general well-being.

10.5 Future Considerations and Challenges

10.5.1 Biotechnology and Ethical Concerns

Like other biotechnology-related industries, cosmetic biotechnology (cosmeceuticals) involves several ethical questions. The intersection of biotechnological developments with aesthetic surgeries and goods makes ethical questions especially pertinent. The following are some moral issues with using biotechnology in cosmetology (Atiyeh et al. 2020):

1. When novel biotechnology components or processes are introduced without being thoroughly tested for safety and potential negative impacts on human health, ethical questions are raised.
2. To assess the safety and efficacy of some biotechnological compounds, animal testing may be necessary. The handling of animals and the possibility of needless suffering raise ethical questions (Gross and Tolba 2015).
3. Informed consent raises ethical concerns since patients receiving cosmetic operations cannot completely comprehend the complicated scientific nature of biotechnology therapies (Hovatta et al. 2010).
4. Consumers may experience disappointment or discontent if they have unreasonable expectations about the degree of improvement that may be made as a result of the use of biotechnology in cosmetology (Hovatta et al. 2010).
5. There may be gaps in our understanding of the long-term implications of some biotechnological treatments, raising ethical questions concerning the introduction of new goods or practices.
6. When biotechnological cosmetic procedures are only available to those who can afford them, there are ethical issues that might exacerbate socioeconomic disparities and aesthetic standards (Newburger and Caplan 2006).
7. Biotechnological developments may promote false body images and enforce unreachable beauty standards, which can have a severe impact on mental health and self-esteem.
8. Advertising that makes false or excessive promises about the advantages of biotechnology cosmetic goods might prey on consumer weaknesses and create moral questions regarding advertising truth (Newburger and Caplan 2006).
9. Genetic testing for customized cosmetic items might lead to privacy issues and ethical dilemmas about the security of genetic data.

10. Cosmetics containing biotechnological elements may have long-term environmental effects, such as bioaccumulation or ecological disturbances, posing ethical questions regarding sustainability (Atiyeh et al. 2020).
11. In order to avoid promoting unattainable beauty standards, biotechnological cosmetic procedures must encourage psychological well-being and body acceptance (Saniotis 2007).

Collaboration between industry stakeholders, regulatory agencies, practitioners, ethicists, and consumers is necessary to address these ethical issues in cosmetology. To make sure that biotechnological breakthroughs in cosmetology prioritize safety, consumer well-being, and ethical standards, transparent labeling, clear communication, rigorous testing, and adherence to ethical principles are crucial.

10.5.2 Ensuring Rigorous Testing and Ingredient Transparency

To prioritize customer safety, foster confidence, and uphold ethical standards, thorough testing and ingredient openness are required in the cosmetology industry. Here's how to accomplish these objectives:

1. Encourage independent study and clinical studies to assess the security and effectiveness of novel cosmetic components and treatments (Saraf and Kaur 2010).
2. To guarantee objectivity and accuracy, seek third-party validation and verification of testing results.
3. Conduct long-term research to determine the possible impacts of cosmetic treatments and products over a lengthy period of time (Juhász and Marmur 2014).
4. Provide thorough and unambiguous labeling that lists all components, their concentrations, and any possible allergies.
5. Don't use "proprietary blends" that make it difficult to identify the precise ingredients and their proportions.
6. Provide easy access to ingredient information on the business website so that customers may look up ingredients before making a purchase (Pauwels and Rogiers 2007).
7. Make sure that you adhere to the rules and regulations established by the appropriate regulatory agencies, including the FDA (U.S. Food and Drug Administration) or the EU Cosmetics Regulation (Wallack 2019).
8. Before putting cosmetics on the market, thoroughly evaluate the safety of the components and products.
9. Look for certification from unbiased groups that focus on conducting cosmetic safety tests, including the Environmental Working Group (EWG) or the Cosmetic Ingredient Review (CIR) (Elder 1984; Houlihan 2008).

10. Encourage peer evaluation and publishing of research findings on substances and methods used in cosmetics in respected scientific publications.
11. Give customers thorough product information, including any possible advantages and hazards, to help them make wise decisions.
12. Create a readily accessible ingredient dictionary outlining each item's function and any potential side effects.
13. Establish internal channels so that staff members or people may voice their concerns about risky or unethical behavior.
14. Ensure that those who come forward with information about wrongdoing are shielded from reprisals and that a full investigation is conducted (Hagel III and Rayport 1997).
15. Collaboration with peers in the industry to exchange best practices for in-depth testing and ingredient transparency.
16. Launch public awareness efforts to inform the public about the value of open ingredient disclosure and thorough testing.
17. Create channels for customers to submit feedback and report negative responses to aid in the identification of possible safety concerns (Takao and Sando 2010).

The cosmetics business may take proactive measures to guarantee customer safety, encourage transparency, and preserve ethical standards by putting these concepts into practice. Such actions not only assist customers but also help cosmetic firms and the sector maintain long-term success and reputation.

10.5.3 Regulatory Safety and Compliance

For cosmetic treatments and products to be secure, efficient, and up to standards, regulatory safety and compliance are essential. International regulatory organizations establish standards and criteria to protect consumer health and limit harm. Here is how the area of cosmetology manages regulatory safety and compliance (Millikan 2001; Gagliardi and Dorato 2007; Schneider 2009; Wallack 2019):

1. Under the Federal Food, Drug, and Cosmetic Act, the FDA (U.S. Food and Drug Administration) is responsible for overseeing cosmetics in the country. It regulates product claims, ingredient safety, and labeling (Barnes 2013).
2. In the European Union, the Cosmetics Regulation specifies criteria for cosmetic product safety, ingredient disclosure, and labeling (Henkler et al. 2012).
3. The Cosmetic Regulations of Health Canada in Canada specify the standards for product testing, ingredient labeling, and safety (Gagliardi and Dorato 2007).
4. Different nations each have their own regulatory bodies in charge of regulating the compliance, efficacy, and safety of cosmetics.
5. Cosmetic items must go through safety testing to guarantee that, when used as intended or under circumstances that are reasonable to expect, no dangers to consumers exist.

6. The safety, possible allergies, and compatibility of cosmetic ingredients are all assessed.
7. Lists of compounds that are banned or forbidden from usage in cosmetic goods are kept by regulatory organizations owing to safety concerns.
8. Cosmetic items must identify all their components in descending order of concentration so that consumers are aware of what's within.
9. Claims regarding a product's advantages that are untrue, deceptive, or overstated are not permitted on labels or in promotional materials.
10. Manufacturers and distributors must include their contact information on labels so that customers may get in touch with them with questions or complaints.
11. To guarantee consistent product quality, safety, and hygiene throughout manufacturing, cosmetic producers must adhere to GMP regulations.
12. Regulatory agencies urge users of cosmetic items to report negative reactions or side effects.
13. Cosmetic makers are frequently forced to notify regulatory authorities of unfavorable incidents.
14. Before releasing brand-new cosmetic items into the market, certain regulatory organizations require producers to inform them.
15. When a cosmetic product is discovered to be dangerous, regulatory organizations have processes in place for starting product recalls.
16. To make sure that production facilities are adhering to safety and quality requirements, regulatory agencies may check them.
17. To promote international trade and provide uniform safety standards, efforts are being undertaken to harmonize cosmetic rules (Millikan 2001; Gagliardi and Dorato 2007; Schneider 2009; Wallack 2019).

It is essential for preserving customer confidence and the general integrity of the cosmetics sector that safety criteria are complied with. The highest safety and compliance standards should be followed by cosmetology practitioners, distributors, and manufacturers. These parties should keep up with regulatory requirements.

10.6 Conclusion

Biotechnology is a field that depends on the regulation of processes to produce useful chemicals. Biotechnology is also defined as the application of technology for industrial, agricultural, cosmetics, and medicinal purposes. In the cosmetic industry, biotechnology has been used for many purposes. It has been used to produce ingredients for the formulations of cosmetics. It also studies the effects of ingredients on skin and some other age-related factors. Biotechnology has many branches including its use in the agricultural, medicinal, environmental, and cosmetic industries. The history of biotechnology explains major turning points in cosmetics and aesthetic medicine. According to this, enzymes were first introduced as skin care products in the 1950s and 1960s. Alpha hydroxy acids and peptide-based products were introduced in the 1970s and 1980s, respectively. In the 1990s, organic extracts gained importance in skin care products. Stem cell extracts for anti-aging products came in the 2000s. Moreover, in the 2010s, personalized cosmetics gained much importance in the cosmetic industry.

The use of biotechnology in cosmetic formulations explains the modified organisms, cell culture biomimetic strategies, and novel cosmetic ingredient formulations. Where cell culture is setting and maintaining cells outside their natural environment to study the reaction and interactions to external stimuli. While biomimetic strategy is the designing of materials that mirror the properties of natural organisms. Cell culture and biomimetic techniques are both used in the cosmetics industry to create and design new and useful products that correlate with real biological interactions and processes. Novel cosmetics ingredient formulations include arbutin and hydroquinone. Important acids may include kojic acid and azelaic acid. Green biotechnology also plays a crucial role in the advancement of Sustainable Development Goals (SDGs) by providing new solutions to different environmental challenges. It enhances the yields of crops using genetically modified organisms which are more resistant to pests and different diseases to increase food production and decrease poverty.

Moreover, the use of effective ingredients like peptides, growth factors, and enzymes was explained in detail. Hair care, skin care, and other aesthetic applications were elaborated. Enhanced bioavailability, stability, and delivery methods were also explained. Peptides have their role in immunity, growth, pigmentation, and inflammation. Growth factors are important for the proliferation of cells and growth under specific conditions. In hair care, biotechnology has supported the growth and restoration of hair, reducing baldness. Growth factors, stem cells, and gene therapies have been studied to enhance hair growth. Biotechnology has also aided in the development of sophisticated anti-hair loss remedies. Skin care has benefited greatly from biotechnology's breakthroughs. Different techniques including dermal fillers, botox, PRP therapy, stem cell therapy, cutting-edge laser, cryolipolysis, thread lifts, tightening treatments, and microbiome restoration have been explained in detail for skin care. For stability, bioavailability, and delivery methods, embedding, micro, and nanotechnologies have been discussed in detail. The role of liposomes, solid lipid nanoparticles (SLNs), nanoemulsions, microemulsions, polymer-based delivery, and hydrogels for delivery is also explained in detail.

The fourth section deals with aesthetic medicine and biotechnology where detailed analysis of tissue engineering, stem cell, and gene therapies have been reported. This section also explains the effective and personalized treatments that can be made to accommodate unique traits and preferences with the best outcomes.

The fifth section deals with future challenges and considerations, including ethical concerns, testing and ingredient transparency, regulatory safety, and compliance. Safety criteria should be complied with to preserve customer confidence and integrity. The highest safety and compliance standards should be followed by cosmetology practitioners, distributors, and manufacturers. These parties should keep up with regulatory requirements.

References

Adrio J-L, Demain AL (2010) Recombinant organisms for production of industrial products. Bioeng Bugs 1:116–131

Adu SA, Naughton PJ, Marchant R, Banat IM (2020) Microbial biosurfactants in cosmetic and personal skincare pharmaceutical formulations. Pharmaceutics 12

Aggarwal S, Sardana C, Ozturk M, Sarwat M (2020) Plant stem cells and their applications: special emphasis on their marketed products.: 3 Biotech 10:291

Ahmad J (2021) Lipid nanoparticles based cosmetics with potential application in alleviating skin disorders. Cosmetics 8:84

Ahn G-N, Kim K-N, Cha S-H et al (2007) Antioxidant activities of phlorotannins purified from Ecklonia cava on free radical scavenging using ESR and H2O2-mediated DNA damage. Euro Food Res Technol 226:71–79

Aldag C, Nogueira Teixeira D, Leventhal PS (2016) Skin rejuvenation using cosmetic products containing growth factors, cytokines, and matrikines: a review of the literature. Clin Cosmet Investig Dermatol 9:411–419

Alix E, Blanc-Potard A-B (2009) Hydrophobic peptides: novel regulators within bacterial membrane. Mol Microbiol 72:5–11

Almohideb M (2021) Management patterns for skin aging among Saudi dermatologists: a questionnaire-based study. J Family Med Prim Care 10:4525

Alster T, Zaulyanov-Scanlon L (2007) Laser scar revision: a review. Dermatologic Surg 33:131–140

Altman A (1997) Agricultural biotechnology. CRC Press

Amer MH (2014) Gene therapy for cancer: present status and future perspective. Mol Cell Ther 2:1–19

Antonova Y, Arik AJ, Moore W et al (2012) In: Gilbert LIBT-IE (ed) 2—Insulin-like peptides: structure, signaling, and function. Academic Press, San Diego, pp 63–92

Arct J, Oborska A, Mojski M, Chudzicki M (2016) Proper selection of cosmetic ingredients as a key factor in skin penetration of flavonoids–in vivo studies. Pol J Cosmetol 19:59–63

Arkin M, Zeevi A, Avram N, Uzan R, Shilo-Volin H, Hollander E, Buriakovsky O (2006) Pharmaceutical and cosmeceutical wash-off mousse shampoo compositions, processes for preparing the same and uses thereof. U.S. Patent Application 11/194,582

Arshad M (2024) Sustainability of renewable energy options as compared to coal-fired power plants in Pakistan. Green and Low-Carbon Economy 2(3):162–173

Arshad M, Abbas M (2018) Water sustainability issues in biofuel production. Perspectives on Water Usage for Biofuels Production: Aquatic Contamination and Climate Change, 55–76

Arshad M, Adil M, Sikandar A, Hussain T (2014) Exploitation of meat industry by-products for biodiesel production: Pakistan's perspective. Pak J Life Soc Sci 12:120–125

Arshad M, Syed-Hassan SSA, Masood R, Ansari AR, Mumtaz A, Rahman A, Batool S, Atif FA, Muhammad A, Rehman AU (2023) Utilization of waste animal fat for sustainable biodiesel production. In: Climate changes mitigation and sustainable bioenergy harvest through animal waste: sustainable environmental implications of animal waste. Springer, pp 149–169

Astashkina A, Mann B, Grainger DW (2012) A critical evaluation of in vitro cell culture models for high-throughput drug screening and toxicity. Pharmacol Ther 134:82–106

Atiyeh BS, Rubeiz MT, Hayek SN (2020) Aesthetic/cosmetic surgery and ethical challenges. Aesthetic Plast Surg 44:1364–1374

Avram MM, Harry RS (2009) Cryolipolysis™ for subcutaneous fat layer reduction. Laser Surg Med 41:703–708

Bai L, Gong C, Chen X et al (2019) Additive manufacturing of customized metallic orthopedic implants: materials, structures, and surface modifications. Metals 9:1004

Baldisserotto A, Buso P, Radice M et al (2018) Moringa oleifera leaf extracts as multifunctional ingredients for "natural and organic" sunscreens and Photoprotective preparations. Molecules 23:12

Ballin AC, Brandt FS, Cazzaniga A (2015) Dermal fillers: an update. Am J Clin Dermatol 16:271–283
Barnes S (2013) The new China Food and Drug Administration: key changes in Chinese food, drugs, medical devices and cosmetics regulation. FDLI Update, p 40
Baxter RA (2008) Anti-aging properties of resveratrol: review and report of a potent new antioxidant skin care formulation. J Cosmet Dermatol 7:2–7
Beekwilder J, Wolswinkel R, Jonker H et al (2006) Production of resveratrol in recombinant microorganisms. Appl Environ Microbiol 72:5670–5672
Bennett AB, Chi-Ham C, Barrows G et al (2013) Agricultural biotechnology: economics, environment, ethics, and the future. Annu Rev Environ Resour 38:249–279
Bhatia S, Goli D (2018) History, scope and development of biotechnology. Intro Pharma Biotechnol 1:1–61
Bhatia S, Naved T, Sardana S (2018) Introduction to pharmaceutical biotechnology, vol 2. IOP Publishing
Bijauliya RK, Alok S, Kumar M et al (2017) A comprehensive review on herbal cosmetics. Int J Pharma Sci Res 8:4930–4949
Binder WJ, Brin MF, Blitzer A et al (2000) Botulinum toxin type a (BOTOX) for treatment of migraine headaches: an open-label study. Otolaryngol Head Neck Surg 123:669–676
Bolduc C, Shapiro J (2001) Hair care products: waving, straightening, conditioning, and coloring. Clin Dermatol 19:431–436
Boonme P, Junyaprasert VB, Suksawad N, Songkro S (2009) Microemulsions and nanoemulsions: novel vehicles for whitening cosmeceuticals. J Biomed Nanotechnol 5:373–383
Branski LK, Gauglitz GG, Herndon DN, Jeschke MG (2009) A review of gene and stem cell therapy in cutaneous wound healing. Burns 35:171–180
Bruno BJ, Miller GD, Lim CS (2013) Basics and recent advances in peptide and protein drug delivery. Ther Dev 4:1443–1467
Byrne PJ, Hilger PA (2004) Lip augmentation. Facial Plast Surg 20:31–38
CAN YOU (2021) 'Magic wands' for Laser Tech. Quantum
Carletti E, Motta A, Migliaresi C (2011) In: Haycock JW (ed) Scaffolds for tissue engineering and 3D cell culture BT—3D cell culture: methods and protocols. Humana Press, Totowa, NJ, pp 17–39
Carr DF, Alfirevic A, Pirmohamed M (2014) Pharmacogenomics: current state-of-the-art. Genes 5:430–443
Castro AR, Portinha C, Logarinho E (2023) The booming business of hair loss. Trends Biotechnol 41:731
Cervelli V, Garcovich S, Bielli A et al (2014) The effect of autologous activated platelet rich plasma (AA-PRP) injection on pattern hair loss: clinical and histomorphometric evaluation. Biomed Res Int 20:14
Chanchal D, Swarnlata S (2008) Novel approaches in herbal cosmetics. J Cosmet Dermatol 7:89–95
Chang T-S (2009) An updated review of tyrosinase inhibitors. Int J Mol Sci 10:2440–2475
Chanwitheesuk A, Teerawutgulrag A, Kilburn JD, Rakariyatham N (2007) Antimicrobial gallic acid from Caesalpinia mimosoides Lamk. Food Chem 100:1044–1048
Chen L, Xiao Z, Meng Y et al (2012) The enhancement of cancer stem cell properties of MCF-7 cells in 3D collagen scaffolds for modeling of cancer and anti-cancer drugs. Biomaterials 33:1437–1444
Chen J, Fan Y, Dong G et al (2023) Designing biomimetic scaffolds for skin tissue engineering. Biomater Sci 11:3051–3076. https://doi.org/10.1039/D3BM00046J
Cheng AS, Bayliss SJ (2008) The genetics of hair shaft disorders. J Am Acad Dermatol 59:1–22
Chien L-J, Lee C-K (2007) Enhanced hyaluronic acid production in Bacillussubtilis by coexpressing bacterial hemoglobin. Biotechnol Prog 23:1017–1022
Christaki E, Bonos E, Giannenas I, Florou-Paneri P (2013) Functional properties of carotenoids originating from algae. J Sci Food Agr 93:5–11

Chulpanova DS, Shaimardanova AA, Ponomarev AS et al (2022) Current strategies for the gene therapy of autosomal recessive congenital ichthyosis and other types of inherited ichthyosis. Int J Mol Sci 23:2506

Clark JR, March JB (2006) Bacteriophages and biotechnology: vaccines, gene therapy and antibacterials. Trends Biotechnol 24:212–218

Costa R, Santos L (2017) Delivery systems for cosmetics—from manufacturing to the skin of natural antioxidants. Powder Technol 322:402–416

Crommelin DJA, Sindelar RD, Meibohm B (2013) Pharmaceutical biotechnology: fundamentals and applications. Springer Science & Business Media

Daliri EB-M, Tango CN, Lee BH, Oh D-H (2018) Human microbiome restoration and safety. Int J Med Microbiol 308:487–497

Das SK, Menezes ME, Bhatia S et al (2015) Gene therapies for cancer: strategies, challenges and successes. J Cell Physiol 230:259–271

de Oliveira JD, Carvalho LS, Gomes AMV et al (2016) Genetic basis for hyper production of hyaluronic acid in natural and engineered microorganisms. Microb Cell Factories 15:119

Delhove J, Osenk I, Prichard I, Donnelley M (2020) Public acceptability of gene therapy and gene editing for human use: a systematic review. Human Gene Ther 31:20–46

Delsin SD, Mercurio DG, Fossa MM, Maia Campos P (2015) Clinical efficacy of dermocosmetic formulations containing spirulina extract on young and mature skin: effects on the skin hydrolipidic barrier and structural properties. Clin Pharmacol Biopharm 4:2

Demain AL (2007) Reviews: the business of biotechnology. Indust Biotechnol 3:269–283

Derrick CD, Shridharani SM, Broyles JM (2015) The safety and efficacy of cryolipolysis: a systematic review of available literature. Aesthet Surg J 35:830–836

Dhillon RS, Schwarz EM, Maloney MD (2012) Platelet-rich plasma therapy—future or trend? Arthritis Res Ther 14:219

Dixit K, Chaudhari D, Dhotre D et al (2021) Restoration of dysbiotic human gut microbiome for homeostasis. Life Sci 278:119622

Donnez D, Jeandet P, Clément C, Courot E (2009) Bioproduction of resveratrol and stilbene derivatives by plant cells and microorganisms. Trends Biotechnol 27:706–713

Doran PM (1995) Bioprocess engineering principles. Elsevier

Drepper T, Eggert T, Hummel W et al (2006) Novel biocatalysts for white biotechnology. Biotechnol Healthc 1:777–786

Druschky A, Herkert M, Radespiel-Tröger M et al (2001) Critical illness polyneuropathy: clinical findings and cell culture assay of neurotoxicity assessed by a prospective study. Intensive Care Med 27:686–693

Egger A, Tomic-Canic M, Tosti A (2020) Advances in stem cell-based therapy for hair loss. Cell R4—repair, replacement, regeneration, & reprogramming 8:12

Eisenhofer G, Tian H, Holmes C et al (2003) Tyrosinase: a developmentally specific major determinant of peripheral dopamine. FASEB J 17:1248–1255

Eissazadeh S, Moeini H, Dezfouli MG et al (2017) Production of recombinant human epidermal growth factor in Pichia pastoris. Braz J Microbiol 48:286–293

Elder RL (1984) The cosmetic ingredient review—a safety evaluation program. J Am Acad Dermatol 11:1168–1174

Elmets CA, Vargas A, Oresajo C (1992) Photoprotective effects of sunscreens in cosmetics on sunburn and Langerhans cell photodamage. Photodermatol Photoimmunol Photomed 9:113–120

Emer J (2019) Platelet-rich plasma (PRP): current applications in dermatology. Skin Therapy Lett 24:1–6

Eversole R, Tran K, Hansen D, Campbell J (2013) Lip augmentation dermal filler reactions, histopathologic features. Head Neck Pathol 7:241–249

Fabi SG (2015) Noninvasive skin tightening: focus on new ultrasound techniques. Clin Cosmet Investig Dermatol 8:47–52

Fabi SG, Joseph J, Sevi J et al (2019) Optimizing patient outcomes by customizing treatment with microfocused ultrasound with visualization: gold standard consensus guidelines from an expert panel. J Drugs Dermatol 18:426–432
Fan Y, Marioli M, Zhang K (2021) Analytical characterization of liposomes and other lipid nanoparticles for drug delivery. J Pharm Biomed Anal 192:113642
Ferreira MS, Magalhães MC, Sousa-Lobo JM, Almeida IF (2020) Trending anti-aging peptides. Cosmetics 7:91
Fang Z, Cai H, Chen G, Qu R (2017) Light-Based Science. Nature Photonics 11:681
Fierrez-Aguilar J, Ortega-Garcia J, Garcia-Romero D, Gonzalez-Rodriguez J (2003a) A comparative evaluation of fusion strategies for multimodal biometric verification BT - audio- and video-based biometric person authentication. In: Kittler J, Nixon MS (eds) . Springer, Berlin Heidelberg, Berlin, Heidelberg, pp 830–837
Fierrez-Aguilar J, Ortega-Garcia J, Gonzalez-Rodriguez J (2003b) Fusion strategies in multimodal biometric verification. In: 2003 international conference on multimedia and expo. ICME '03. Proceedings (cat. No.03TH8698). P III–5
Forster CF, Wase DAJ (1987) Environmental biotechnology. p. 453pp
Frazzetto G (2003) White biotechnology: the application of biotechnology to industrial production holds many promises for sustainable development, but many products still have to pass the test of economic viability. EMBO Rep 4:835–837
Fulekar MH (2010) Environmental biotechnology. CRC Press
Funt D, Pavicic T (2013) Dermal fillers in aesthetics: an overview of adverse events and treatment approaches. Clin Cosmet Investig Dermatol 6:295–316
Gagliardi L, Dorato S (2007) General concepts and cosmetic legislation. Analysis of cosmetic products; Amparo, S, Alberto, C, Eds 3–28
Galaev IY, Mattiasson B (1999) 'Smart'polymers and what they could do in biotechnology and medicine. Trends Biotechnol 17:335–340
Garg S, Dahiya N, Gupta S (2014) Surgical scar revision: an overview. J Cutan Aesthet Surg 7:3
Gautam R, Yang S, Maharjan A et al (2021) Prediction of skin sensitization potential of silver and zinc oxide nanoparticles through the human cell line activation test. Front Toxicol 3:649666
Gavrilescu M (2010) Environmental biotechnology: achievements, opportunities and challenges. Dyn Biochem Process Biotechnol Mol Biol 4:1–36
Gokarn YR, Kras E, Nodgaard C et al (2008) Self-buffering antibody formulations. J Pharm Sci 97:3051–3066
Gomes C, Silva AC, Marques AC et al (2020) Biotechnology applied to cosmetics and aesthetic medicines. Cosmetics 7
Gordon J, Amini S (2021) General overview of neuronal cell culture BT—neuronal cell culture: methods and protocols. In: Amini S, White MK (eds) . Springer US, New York, NY, pp 1–8
Gottschalk LMF, Oliveira RA, da Silva Bon EP (2010) Cellulases, xylanases, β-glucosidase and ferulic acid esterase produced by Trichoderma and aspergillus act synergistically in the hydrolysis of sugarcane bagasse. Biochem Eng J 51:72–78
Graf E (1992) Antioxidant potential of ferulic acid. Free Radic Biol Med 13:435–448
Graham K, Logan J (2004) Using the Ottawa model of research use to implement a skin care program. J Clin Nurs 19:18–26
Grimes PE, Green BA, Wildnauer RH, Edison BL (2004) The use of polyhydroxy acids (PHAs) in photoaged skin. Cutis 73:3–13
Gross D, Tolba RH (2015) Ethics in animal-based research. Eur Surg Res 55:43–57
Hafizi M, Bakhshandeh B, Soleimani M, Atashi A (2012) Exploring the enkephalinergic differentiation potential in adult stem cells for cell therapy and drug screening implications. In Vitro Cell Dev Biol Animal 48:562–569
Hagel J III, Rayport JF (1997) The coming battle for customer information. McKinsey Q 64
Harhaun R, Kunik O, Saribekova D, Lazzara G (2020) Biologically active properties of plant extracts in cosmetic emulsions. Microchem J 154:104543

Haycock JW (2011) 3D cell culture: a review of current approaches and techniques BT—3D cell culture: methods and protocols. In: Haycock JW (ed) . Humana Press, Totowa, NJ, pp 1–15
Hengge UR, Bardenheuer W (2004) Gene therapy and the skin. Am J Med Genet C Semin Med Genet. Wiley Online Library:93–100
Henkler F, Tralau T, Tentschert J et al (2012) Risk assessment of nanomaterials in cosmetics: a European union perspective. Arch Toxicol 86:1641–1646
Heo S-J, Ko S-C, Kang S-M et al (2010) Inhibitory effect of diphlorethohydroxycarmalol on melanogenesis and its protective effect against UV-B radiation-induced cell damage. Food Chem Toxicol 48:1355–1361
Herman A, Herman AP, Domagalska BW, Młynarczyk A (2013) Essential oils and herbal extracts as antimicrobial agents in cosmetic emulsion. Indian J Microbiol 53:232–237
Heux S, Meynial-Salles I, O'Donohue MJ, Dumon C (2015) White biotechnology: state of the art strategies for the development of biocatalysts for biorefining. Biotechnol Adv 33:1653–1670
Hiemstra PS, Grootaers G, van der Does AM et al (2018) Human lung epithelial cell cultures for analysis of inhaled toxicants: lessons learned and future directions. Toxicol In Vitro 47:137–146
Hilinski JM, Cohen SR (2009) Soft tissue augmentation with ArteFill. Facial Plast Surg 25:114–119
Hoang HT, Moon J-Y, Lee Y-C (2021) Natural Antioxidants from Plant Extracts in Skincare Cosmetics: Recent Applications, Challenges and Perspectives. Cosmetics. 8(4), p.106
Hoban TJ (1998) Trends in consumer attitudes about agricultural biotechnology. AgBioForum 1.1:3–7
Houlihan J (2008) Comments for public meeting on 'International Cooperation on Cosmetics Regulations (ICCR) Preparations.' Grupo de Trabajo Ambiental
Hovatta O, Stojkovic M, Nogueira M, Varela-Nieto I (2010) European scientific, ethical, and legal issues on human stem cell research and regenerative medicine. Stem Cells 28:1005–1007
Hsu C-C, Chao Y-Y, Wang S-W, Chen Y-L (2019) Polyethylenimine-capped silver nanoclusters as fluorescent sensors for the rapid detection of ellagic acid in cosmetics. Talanta 204:484–490
Huang S-C, Lin C-C, Huang M-C, Wen K-C (2004) Simultaneous determination of magnesium ascorbyl phosphate, ascorbyl glucoside, kojic acid, arbutin and hydroquinone in skin whitening cosmetics by HPLC. J Food Drug Anal 12:1
Husein el Hadmed H, Castillo RF (2016) Cosmeceuticals: peptides, proteins, and growth factors. J Cosmet Dermatol 15:514–519
Imokawa G (2008) Recent advances in characterizing biological mechanisms underlying UV-induced wrinkles: a pivotal role of fibrobrast-derived elastase. Arch Dermatol Res 300:7–20
Inoue S (2016) Biotechnology in skin care (II): Moisturization. In: Biotechnology in personal care. CRC Press, pp 157–186
Iobst S, Santhanam U, Weinkauf RL (2016) Biotechnology in skin care (I): overview. Biotechnol Pers Care:141–156
Jalian H, Avram MM (2013) Cryolipolysis: a historical perspective and current clinical practice. Semin Cutan Med Surg 32(1):31–34
Jankovic J, Schwartz K (1991) Botulinum toxin treatment of tremors. Neurology 41:1185 LP – 1185
Jayarama Reddy V, Radhakrishnan S, Ravichandran R et al (2013) Nanofibrous structured biomimetic strategies for skin tissue regeneration. Wound Repair Regen 21:1–16
Jeong JH, Fan Y, You GY et al (2015) Improvement of photoaged skin wrinkles with cultured human fibroblasts and adipose-derived stem cells: a comparative study. Plast Reconstr Surg 68:372–381
Jördening H-J, Winter J (2005) Bacterial metabolism in wastewater treatment systems. Environmental biotechnology: concepts and applications.
Juhász MLW, Marmur ES (2014) A review of selected chemical additives in cosmetic products. Dermatol Ther 27:317–322
Kaji EH, Leiden JM (2001) Gene and stem cell therapies. Jama 285:545–550
Kataoka M, Sasaki M, Hidalgo A-RGD et al (2001) Glycolic acid production using ethylene glycol-oxidizing microorganisms. Biosci Biotechnol Biochem 65:2265–2270

Katz LM, Dewan K, Bronaugh RL (2015) Nanotechnology in cosmetics. Food Chem Toxicol 85:127–137
Kessler-Becker D (2016) Biotechnology in hair care (I): overview. Biotechnol Pers Care:271–290
Khan AD, Singh L (2016) Various techniques of bioavailability enhancement: a review. J Drug Deliv Therap 6:34–41
Kiene RP (1998) Uptake of choline and its conversion to glycine betaine by bacteria in estuarine waters. Appl Environ Microbiol 64:1045–1051
Kijjoa A, Sawangwong P (2004) Drugs and cosmetics from the sea. Mar Drugs 2:73–82
Kim W-S, Park B-S, Park S-H et al (2009) Antiwrinkle effect of adipose-derived stem cell: activation of dermal fibroblast by secretory factors. J Dermatol Sci 53:96–102
Kim J-H, Jung M, Kim H-S et al (2011) Adipose-derived stem cells as a new therapeutic modality for ageing skin. Exp Dermatol 20:383–387
Kim JH, Chang P-K, Chan KL et al (2012) Enhancement of commercial antifungal agents by Kojic acid. Int J Mol Sci 13:13867–13880
Kim HJ, Jung MS, Hur YK, Jung AH (2020) A study on clinical effectiveness of cosmetics containing human stem cell conditioned media. Biomed Dermatol 4:9
Kiselev KV (2011) Perspectives for production and application of resveratrol. Appl Microbiol Biotechnol 90:417–425
Kligman AM (2000) Cosmetics: a dermatologist looks to the future: promises and problems. Dermatol Clin 18:699–709
Kokare C (2013) Pharmaceutical biotechnology fundamentals and application. Nirali Prakashan
Kusumawati I, Indrayanto G (2013) Chapter 15—natural antioxidants in cosmetics. In: Atta-ur-Rahman BT-S in NPC (ed). Elsevier, pp 485–505
Lad R (2016) Biotechnology in skin care (IV): skin pigmentation, lightening, and darkening. In: Biotechnology in personal care. CRC Press, pp 225–250
Le ADK, Enweze L, DeBaun MR, Dragoo JL (2018) Current clinical recommendations for use of platelet-rich plasma. Curr Rev Musculoskelet Med 11:624–634
Leordean D, Dudescu C, Marcu T et al (2015) Customized implants with specific properties, made by selective laser melting. Rapid Prototyp J 21:98–104
Levin ED (2005) Extracellular superoxide dismutase (EC-SOD) quenches free radicals and attenuates age-related cognitive decline: opportunities for novel drug development in aging. Curr Alzheimer Res 2:191–196
Leyden JJ, Shergill B, Micali G et al (2011) Natural options for the management of hyperpigmentation. J Eur Acad Dermatol Venereol 25:1140–1145
Li J, Thomson AW, Rogers NM (2021) Myeloid and mesenchymal stem cell therapies for solid organ transplant tolerance. Transplantation 105:e303–e321
Liaud N, Giniés C, Navarro D et al (2014) Exploring fungal biodiversity: organic acid production by 66 strains of filamentous fungi. Fung Biol Biotechnol 1:1
Lim CG, Fowler ZL, Hueller T et al (2011) High-yield resveratrol production in engineered Escherichia coli. Appl Environ Microbiol 77:3451–3460
Lin F-H, Lin J-Y, Gupta RD et al (2005) Ferulic acid stabilizes a solution of vitamins C and E and doubles its photoprotection of skin. J Invest Dermatol 125:826–832
Lu JF, Zhu Y, Sun HL et al (2016) Highly efficient production of hyaluronic acid by streptococcus zooepidemicus R42 derived from heterologous expression of bacterial haemoglobin and mutant selection. Lett Appl Microbiol 62:316–322
Lv Y-G, Deng Z-S, Liu J (2005) 3-D numerical study on the induced heating effects of embedded micro/nanoparticles on human body subject to external medical electromagnetic field. IEEE Trans Nanobioscience 4:284–294
Macer DRJ, Akiyama S, Alora AT et al (1995) International perceptions and approval of gene therapy. Human Gene Ther 6:791–803
Mahé A (2014) The practice of skin-bleaching for a cosmetic purpose in immigrant communities. J Travel Med 21:282–287

Malone CH, Walters N, Stroh R, Munavalli G (2021) New technologies in skin tightening. Curr Otorhinolaryngol Rep 9:422–435

March OP, Reichelt J, Koller U (2018) Gene editing for skin diseases: designer nucleases as tools for gene therapy of skin fragility disorders. Exp Physiol 103:449–455

Marder SR, Gorman CB, Tiemann BG, Cheng LT (1993) Stronger acceptors can diminish nonlinear optical response in simple donor-acceptor polyenes. J Am Chem Soc 115:3006–3007

Marmiroli N, Maestri E (2014) Plant peptides in defense and signaling. Peptides 56:30–44

Mecca SB (1975) Allantoin glycine complex U.S. Patent 3,927,021

Messenger AG (1984) The culture of dermal papilla cells from human hair follicles. Br J Dermatol 110:685–689

Middleton KK, Barro V, Muller B et al (2012) Evaluation of the effects of platelet-rich plasma (PRP) therapy involved in the healing of sports-related soft tissue injuries. Iowa Orthop J 32:150

Milbrada EJ (1975) Dyeing human hair with oxidation dyes and arginine or a protamine protein. U.S. Patent 3,861,868

Millikan LE (2001) Cosmetology, cosmetics, cosmeceuticals: definitions and regulations. Clin Dermatol 19:371–374

Mirza FN, Khatri KA (2017) The use of lasers in the treatment of skin cancer: a review. J Cosmet Laser Ther 19:451–458

Mobarak DH, Salah S, Elkheshen SA (2014) Formulation of ciprofloxacin hydrochloride loaded biodegradable nanoparticles: optimization of technique and process variables. Pharm Dev Technol 19:891–900

Monheit GD, Coleman KM (2006) Hyaluronic acid fillers. Dermatol Ther 19:141–150

Morita T, Kitagawa M, Yamamoto S et al (2010) Glycolipid biosurfactants, mannosylerythritol lipids, repair the damaged hair. J Oleo Sci 59:267–272

Müller RH, Radtke M, Wissing SA (2002) Solid lipid nanoparticles (SLN) and nanostructured lipid carriers (NLC) in cosmetic and dermatological preparations. Adv Drug Deliv Rev 54:S131–S155

Mulligan RC (1993) The basic science of gene therapy. Science 260:926–932

Muniz L, Knosp C, Yeretssian G (2012) Intestinal antimicrobial peptides during homeostasis, infection, and disease. Front Immunol 3:17

Murphy B, Hoptroff M, Arnold D et al (2021) In-vivo impact of common cosmetic preservative systems in full formulation on the skin microbiome. PLoS One 16:e0254172

Naguib YMA (2000) Antioxidant activities of Astaxanthin and related carotenoids. J Agric Food Chem 48:1150–1154

Naumann M, Albanese A, Heinen F et al (2006) Safety and efficacy of botulinum toxin type a following long-term use. Eur J Neurol 13:35–40

Newburger AE, Caplan AL (2006) Taking ethics seriously in cosmetic dermatology. Arch Dermatol 142:1641–1642

Niska K, Zielinska E, Radomski MW, Inkielewicz-Stepniak I (2018) Metal nanoparticles in dermatology and cosmetology: interactions with human skin cells. Chem Biol Interact 295:38–51

Niwa Y, Akamatsu H (1991) Kojic acid scavenges free radicals while potentiating leukocyte functions including free radical generation. Inflammation 15:303–315

Novellino L, Napolitano A, Prota G (1999) 5,6-Dihydroxyindoles in the Fenton reaction: a model study of the role of melanin precursors in oxidative stress and Hyperpigmentary processes. Chem Res Toxicol 12:985–992

Nyyssölä A, Kerovuo J, Kaukinen P et al (2000) Extreme halophiles synthesize betaine from glycine by methylation. J Biol Chem 275:22196–22201

Ohyama M, Vogel JC (2003) Gene delivery to the hair follicle. J Investig Dermatol Symp Proc. Elsevier 8:204–206

Olaizola M (2000) Commercial production of astaxanthin from Haematococcus pluvialis using 25,000-liter outdoor photobioreactors. J Appl Phycol 12:499–506

Olson KR, Gao Y, Arif F et al (2018) Metabolism of hydrogen sulfide (H2S) and production of reactive sulfur species (RSS) by superoxide dismutase. Redox Biol 15:74–85

Oren A, Gunde-Cimerman N (2007) Mycosporines and mycosporine-like amino acids: UV protectants or multipurpose secondary metabolites? FEMS Microbiol Lett 269:1–10
Pallela R, Na-Young Y, Kim S-K (2010) Anti-photoaging and Photoprotective compounds derived from marine organisms. Mar Drugs 8:1189–1202
Panagiotou G, Olsson L (2007) Effect of compounds released during pretreatment of wheat straw on microbial growth and enzymatic hydrolysis rates. Biotechnol Bioeng 96:250–258
Pandey A, Höfer R, Taherzadeh M et al (2015) Industrial biorefineries and white biotechnology. Elsevier
Parisi N, Matts PJ, Lever R et al (2017) Hexamidine salts–applications in skin health and personal care products. Int J Cosmet Sci 39:361–365
Park B-S, Kim W-S, Choi J-S et al (2010) Hair growth stimulated by conditioned medium of adipose-derived stem cells is enhanced by hypoxia: evidence of increased growth factor secretion. Biomed Res J 31:27–34
Pauwels M, Rogiers V (2007) Database search for safety information on cosmetic ingredients. Regul Toxicol Pharmacol 49:208–216
Pavia H, Toth G, Åberg P (1999) Trade-offs between phlorotannin production and annual growth in natural populations of the brown seaweed Ascophyllum nodosum. J Ecology 87:761–771
Pérez-Rivero C, López-Gómez JP (2023) Unlocking the potential of fermentation in cosmetics: a review. Fermentation 9:463
Phillips GC, Garda M (2019) Plant tissue culture media and practices: an overview. In Vitro Cell Dev Biol Plant 55:242–257
Phillips KA, Veenstra DL, Ramsey SD et al (2004) Genetic testing and pharmacogenomics: issues for determining the impact to healthcare delivery and costs. Am J Manag Care 10:425–432
Prainsack B, Geesink I, Franklin S (2008) Stem cell technologies 1998–2008: controversies and silences. Sci Cult 17:351–362
Proksch E, Schunck M, Zague V et al (2014) Oral intake of specific bioactive collagen peptides reduces skin wrinkles and increases dermal matrix synthesis. Skin Pharmacol Physiol 27:113–119
Ptitsyn LR, Al'tman IB (1999) Extracellular production of recombinant human epidermal growth factor (hEGF) in Escherichia coli cells. Bioorg Khim 25:923–929
Rastogi RP, Richa, Sinha RP et al (2010) Photoprotective compounds from marine organisms. J Ind Microbiol Biotechnol 37:537–558
Rattanawiwatpong P, Wanitphakdeedecha R, Bumrungpert A, Maiprasert M (2020) Anti-aging and brightening effects of a topical treatment containing vitamin C, vitamin E, and raspberry leaf cell culture extract: a split-face, randomized controlled trial. J Cosmet Dermatol 19:671–676
Ratz-Łyko A, Arct J (2019) Resveratrol as an active ingredient for cosmetic and dermatological applications: a review. J Cosmet Laser Ther 21:84–90
Rawlings AV (2016) Biotechnology in skin care (III): skin aging. In: Biotechnology in personal care. CRC Press, pp 187–224
Reddy B, Jow T, Hantash BM (2012) Bioactive oligopeptides in dermatology: part I. Exp Dermatol 21:563–568
Relling MV, Evans WE (2015) Pharmacogenomics in the clinic. Nature 526:343–350
Ribeil J-A, Hacein-Bey-Abina S, Payen E et al (2017) Gene therapy in a patient with sickle cell disease. N Engl J Med 376:848–855
Romay C, Armesto J, Remirez D et al (1998) Antioxidant and anti-inflammatory properties of C-phycocyanin from blue-green algae. Inflamm Res 47:36–41
Rose GD, Glerasch LM, Smith JA (1985) In: Anfinsen CB, Edsall JT, Richards FMBT-A in PC (eds) Turns in peptides and proteins. Academic Press, pp 1–109
Rosen J, Landriscina A, Friedman AJ (2015) Nanotechnology-based cosmetics for hair care. Cosmetics 2:211–224
Saeedi M, Anafcheh M, Ghafouri R, Hadipour NL (2013) A computational investigation of the electronic properties of octahedral Al n N n and Al n P n cages (n= 12, 16, 28, 36, and 48). Struct Chem 24:681–689

Saeedi M, Eslamifar M, Khezri K (2019) Kojic acid applications in cosmetic and pharmaceutical preparations. Biomed Pharmacother 110:582–593

Sajna KV, Gottumukkala LD, Sukumaran RK, Pandey A (2015a) In: Pandey A, Höfer R, Taherzadeh M et al (eds) Chapter 18—White biotechnology in cosmetics. Elsevier, Amsterdam, pp 607–652

Sajna KV, Gottumukkala LD, Sukumaran RK, Pandey A (2015b) White biotechnology in cosmetics. In: Industrial biorefineries & white biotechnology. Elsevier, pp. 607–652

Saniotis A (2007) Changing ethics in medical practice: a Thai perspective. Indian J Med Ethics 4:24–25

Santos JS, Barradas TN, Tavares GD (2022) Advances in nanotechnology-based hair care products applied to hair shaft and hair scalp disorders. Int J Cosmet Sci 44:320–332

Saraf S, Kaur CD (2010) Phytoconstituents as photoprotective novel cosmetic formulations. Phcog Rev 4:1

Sarnoff DS, Saini R, Gotkin RH (2008) Comparison of filling agents for lip augmentation. Aesthet Surg J 28:556–563

Savoia A, Accardo C, Vannini F et al (2014) Outcomes in thread lift for facial rejuvenation: a study performed with happy lift™ revitalizing. Dermatol Ther 4:103–114

Schneider SB (2009) Global cosmetic regulations? A long way to go…. In: Global regulatory issues for the cosmetics industry. Elsevier, pp 189–233

Scoglio S, Canestrari F, Benedetti S, Zolla L (2012) Extracts of Aphanizomenon Flos Aquae and nutritional, cosmetic and pharmaceutical compositions containing the same. U.S. Patent 8,337,858

Scragg AH (2005) Environmental biotechnology. OXFORD University Press, New York

Segall L, Ellis DAF (2007) Therapeutic options for lip augmentation. FPCS Clin North Am 15:485–490

Semsarzadeh N, Khetarpal S (2022) Rise of stem cell therapies in aesthetics. Clin Dermatol 40:49–56

Sharma S, Sarangdevot K (2012) Nanoemulsions for cosmetics. Int J Adv Res Pharm Bio Sci 1:408–416

Sheikhpour M, Barani L, Kasaeian A (2017) Biomimetics in drug delivery systems: a critical review. J Control Release 253:97–109

Shi B, He Q, Yao K et al (2005) Production of ellagic acid from degradation of valonea tannins by aspergillus Niger and Candida utilis. J Chem Technol Biotechnol Int Res Process Environ Clean Technol 80:1154–1159

Shiba K (1980) Composition for permanent waving. U.S. Patent 4,218,435

Shoukat R, Khan MI (2021) Carbon nanotubes: a review on properties, synthesis methods and applications in micro and nanotechnology. Microsyst Technol 27:1–10

Sim YC, Nam Y, Shin Y-H et al (2003) Proteolytic enzyme conjugated to SC-glucan as an enzymatic transdermal drug penetration enhancer. Die Pharmazie-An Int J Pharm Sci 58:252–256

Singleton P (2004) Bacteria in biology, biotechnology and medicine. John Wiley & Sons

Smith RH (2013) Plant tissue culture: techniques and experiments. academic press

Sood N, Bhardwaj A, Mehta S, Mehta A (2016) Stimuli-responsive hydrogels in drug delivery and tissue engineering. Drug Deliv 23:748–770

Stolz P, Obermayer B (2005) Manufacturing microalgae for skin care. Cosmetics Toiletries 120:99–106

Sugimoto T, Haraya N, Saito M (2016) Moisturizer and cosmetic agent containing same. U.S. Patent 9,456,971

Sutterby E, Thurgood P, Baratchi S et al (2020) Microfluidic skin-on-a-Chip models: toward biomimetic artificial skin. Small 16:2002515

Takao Y, Sando KM (2010) Products and evaluation device of cosmetics for UV protection-AIST commercialization based on regional collaboration that combines the current strategic logic, and an intermediary's experience and trial-and-error approach. Synthesiology. English Edition 3:140–150

Takmakov P, Ruda K, Scott Phillips K et al (2015) Rapid evaluation of the durability of cortical neural implants using accelerated aging with reactive oxygen species. J Neural Eng 12:26003
Tezel A, Fredrickson GH (2008) The science of hyaluronic acid dermal fillers. Cosmet Laser Ther 10:35–42
Thoma DS, Benić GI, Zwahlen M et al (2009) A systematic review assessing soft tissue augmentation techniques. Clin Oral Implants Res 20:146–165
Thomas JR, Somenek M (2012) Scar revision review. Arch Facial Plast Surg 14:162–174
Thorpe TA (2007) History of plant tissue culture. Mol Biotechnol 37:169–180
Timmins NE, Scherberich A, Früh J-A et al (2007) Three-dimensional cell culture and tissue engineering in a T-CUP (tissue culture under perfusion). J Tissue Eng 13:2021–2028
Tinoco A, Martins M, Cavaco-Paulo A, Ribeiro A (2022) Biotechnology of functional proteins and peptides for hair cosmetic formulations. Trends Biotechnol 40:591–605
Tomaro-Duchesneau C, Saha S, Malhotra M et al (2012) Lactobacillus fermentum NCIMB 5221 has a greater ferulic acid production compared to other ferulic acid esterase producing Lactobacilli. Int J Probiot Prebiot 7:23–32
Tong LX, Rieder EA (2019) Thread-lifts: a double-edged suture? A comprehensive review of the literature. Dermatologic Surg 45:931–940
Trehan S, Michniak-Kohn B, Beri K (2017) Plant stem cells in cosmetics: current trends and future directions. Future Sci OA 3:FSO226
Trounson A, McDonald C (2015) Stem cell therapies in clinical trials: progress and challenges. Cell Stem Cell 17:11–22
Trounson A, Thakar RG, Lomax G, Gibbons D (2011) Clinical trials for stem cell therapies. BMC Med 9:1–7
Tsigos I, Martinou A, Kafetzopoulos D, Bouriotis V (2000) Chitin deacetylases: new, versatile tools in biotechnology. Trends Biotechnol 18:305–312
Tyczewska A, Twardowski T, Woźniak-Gientka E (2023) Agricultural biotechnology for sustainable food security. Trends Biotechnol 41:331–341
Ulber R, Sell D (2007) White biotechnology. Springer
Umakumar G, Naiman AM, Samuel John K, Hajera T (2012) Bio-conversion studies on gallic acid production from Chebulic myrobalan and Emblic myrobalan by aspergillus Niger mtcc 281 and Rhizopus oryzae mtcc 1987. Int J Pharma Biosci 3:146–155
Van Gheluwe L, Chourpa I, Gaigne C, Munnier E (2021) Polymer-based smart drug delivery systems for skin application and demonstration of stimuli-responsiveness. Polymers 13
Vattem DA, Shetty K (2003) Ellagic acid production and phenolic antioxidant activity in cranberry pomace (Vaccinium macrocarpon) mediated by Lentinus edodes using a solid-state system. Process Biochem 39:367–379
Vattem DA, Shetty K (2005) Biological functionality of Ellagic acid: a review. J Food Biochem 29:234–266
Vemula VR, Lagishetty V, Lingala S (2010) Solubility enhancement techniques. Int J Pharm Sci Rev Res 5:41–51
Verret DJ (2009) Patient guide to hair loss & hair restoration. DJ Verret
Vitagliano T, Garieri P, Lascala L et al (2023) Preparing patients for cosmetic surgery and aesthetic procedures: ensuring an optimal nutritional status for successful results. Nutrients 15:352
Votto SS, Read-Fuller A, Reddy L (2021) Lip augmentation. Oral Maxillofac Surg Clin North Am 33:185–195
Wallack G (2019) Rethinking FDA'S regulation of cosmetics. Harv J Legis 56:311
Walsh G (2013) Pharmaceutical biotechnology: concepts and applications
Wang Y-F, Chuang Y-L, Hsu M-H, Keh H-C (2004) A personalized recommender system for the cosmetic business. Expert Syst Appl 26:427–434
Wang Y, Branicky R, Noë A, Hekimi S (2018) Superoxide dismutases: dual roles in controlling ROS damage and regulating ROS signaling. J Cell Biol 217:1915–1928
Wang J, Liu X, Li R, Fan Y (2023) Biomimetic strategies and technologies for artificial tactile sensory systems. Trends Biotechnol 41:7

Wesseler JHH, Zilberman D (2021) Biotechnology, bioeconomy, and sustainable life on land. In: Transitioning to sustainable life on land. MDPI, pp. 415–435
Westgate GE, Botchkareva NV, Tobin DJ (2013) The biology of hair diversity. Int J Cosmet Sci 35:329–336
Wintzen M, Gilchrest BA (1996) Proopiomelanocortin, its derived peptides, and the skin. J Invest Dermatol 106:3–10
Wirth T, Parker N, Ylä-Herttuala S (2013) History of gene therapy. Gene 525:162–169
Won CH, Yoo HG, Kwon OS et al (2010) Hair growth promoting effects of adipose tissue-derived stem cells. J Dermatol Sci 57:134–137
Xiao H, Shao F, Wu M et al (2015) The application of antimicrobial peptides as growth and health promoters for swine. J Anim Sci Biotechnol 6:19
Yamada R, Yoshie T, Wakai S et al (2014) Aspergillus oryzae-based cell factory for direct kojic acid production from cellulose. Microb Cell Factories 13:71
Yan N, Fan C, Chen Y, Hu Z (2016) The potential for microalgae as bioreactors to produce pharmaceuticals. Int J Mol Sci 17:6
Younus H (2018) Therapeutic potentials of superoxide dismutase. Int J Health Sci 12:88
Zand M, Narasu M (2013) A review article biotechnology applications in medicine. Intl Res J Appl Basic Sci 4:2557–2563
Zappelli C, Barbulova A, Apone F, Colucci G (2016) Effective active ingredients obtained through biotechnology. Cosmetics 3:39
Zaragozá MC, López D, Sáiz PM et al (2008) Toxicity and antioxidant activity in vitro and in vivo of two Fucus vesiculosus extracts. J Agric Food Chem 56:7773–7780
Zasloff M (2007) Antimicrobial peptides, innate immunity, and the normally sterile urinary tract. J Am Soc Nephrol 18:7
Zhang L, Falla TJ (2009) Cosmeceuticals and peptides. Clin Dermatol 27:485–494
Zhang Q, Chen G, Liu X, Qian Q (2007) Monoclonal antibodies as therapeutic agents in oncology and antibody gene therapy. Cell Res 17:89–99
Ziegler A, Jonason AS, Leffellt DJ et al (1994) Sunburn and p53 in the onset of skin cancer. Nature 372:773–776
Zielke H, Wölber L, Wiest L, Rzany B (2008) Risk profiles of different injectable fillers: results from the injectable filler safety study (IFS study). Dermatologic Surg 34:326–335
Zillich OV, Schweiggert-Weisz U, Eisner P, Kerscher M (2015) Polyphenols as active ingredients for cosmetic products. Int J Cosmet Sci 37:455–464
Zolghadri S, Bahrami A, Hassan Khan MT et al (2019) A comprehensive review on tyrosinase inhibitors. J Enzyme Inhib Med Chem 34:279–309

Chapter 11
Hair Care Science and the Role of Cosmeceuticals

Shazia Abrar, Naveed Anjum, Sarosh Iqbal, Fareeha Arooj, and Muhammad Arshad

Contents

Abstract Hair care and styling cosmetics are active in the hair care industry. Different categories of hair care cosmetics include hair cleaning and grooming or styling cosmetics. A third category of hair care cosmetics includes products that preserve hair integrity and help treat scalp and hair problems such as dandruff and dermatitis. The active ingredients present in the cosmeceuticals have multifunctional effects on hair, such as cleansing conditioning, coloring, preserving, etc. The main purpose of cosmeceuticals is to improve hair health, growth, and dermatological problems such as alopecia and preserve the integrity of the hair. Different types of cosmeceuticals containing multifunctional therapeutic agents from plant extracts

S. Abrar (✉) · N. Anjum · S. Iqbal · F. Arooj
Department of Applied Chemistry, Government College University Faisalabad, Faisalabad, Pakistan
e-mail: shaziaabrar@gcuf.edu.pk

M. Arshad
Biochemistry Section, Jhang-campus, University of Veterinary and Animal Sciences Lahore, Lahore, Pakistan

S. Javed et al. (eds.), *Sustainable Cosmeceuticals*,
https://doi.org/10.1007/978-3-031-86087-4_11

have been formulated. These types of cosmeceuticals represent sustainable, non-toxic, eco-friendly hair care products that meet the standards of sustainable development goal of good health and well-being along with ensuring safety to the environment by releasing biodegradable chemicals. In this chapter, we will discuss the chemistry and structure of hair, types of hair care cosmetics, sustainable hair cosmeceuticals, and their applications. Different types of hair cosmeceuticals such as shampoos, conditioners, natural hair dyes, and hair growth-promoting formulations will be discussed.

Keywords Hair care science · Eco-cosmeceuticals for hairs · Herbal extracts · Sustainability in hair care · Non-toxic hair cosmetics

11.1 Introduction

Cosmetics with biologically active chemicals claiming therapeutic or drug-like effects are known as cosmeceuticals (Arshad 2024; Dias 2015). Cosmeceuticals are applied to treat various skin disorders and enhance hair's appearance and nourishment. Just like skin grooming cosmetics, hair cosmeceuticals include active ingredients that affect the biological, structural, and mechanical properties of hair. The purpose of hair cosmeceuticals is to enhance hair strength and attractiveness by providing necessary nutrients (Trüeb 2001).

Humans can change their appearance by altering their hair length, color, and style to look how they want. The first step in hair grooming is to cleanse the hair using cleansing agents, i.e., shampooing. Traditionally shampoos are used for cleaning hair and scalp, and the advanced formulas are tailored to provide additional health benefits along with cleansing such as treatment of dandruff, hair strengthening, and oily hair therapy (Rathod et al. 2020). Cosmetics for hair work on the portion of the hair shaft that extends past the scalp's surface. Various hair and scalp treatments are available, and most of them are applied as coatings of cosmeceuticals. The most commonly used hair care cosmeceutical products are shampoos, conditioners, hair strengthening, and dyes that are applied to improve hair appearance. This chapter will detail the structure and chemical composition of hair, types of hair care products, and sustainable hair care cosmeceuticals.

11.2 Hair Structure and Chemical Composition

It is important to understand the structure of hair before elaborating on the function and composition of hair care products. Hair is a polymeric biomaterial, generated inside the hair follicles, consisting of keratin protein, lipids, water, and pigments (Fig. 11.1). Keratin forms the major portion of the hair, accounting for 60–95% of the hair's total weight. Keratin is an insoluble, resistant protein, composed of

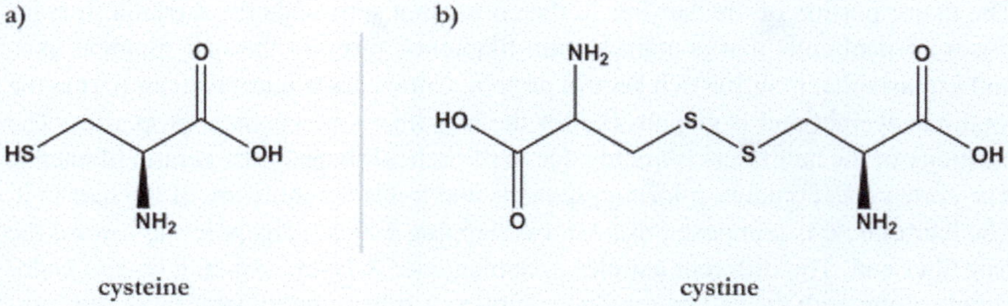

Fig. 11.1 Structure of cysteine and cystine

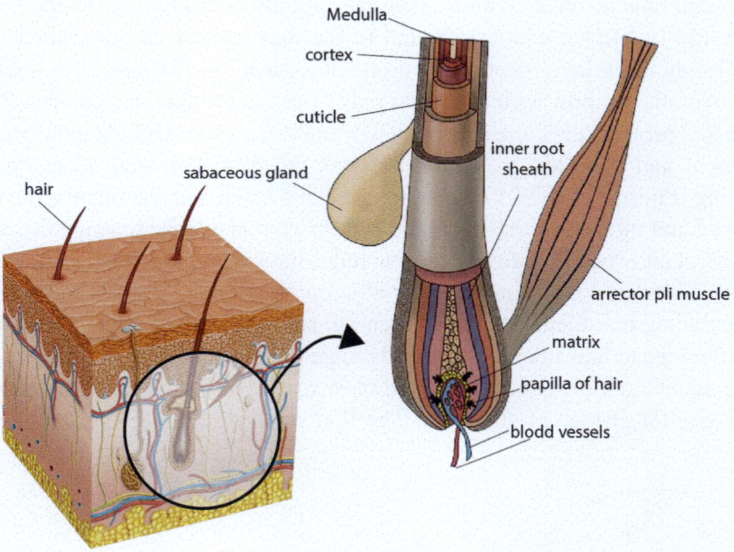

Fig. 11.2 Composition of hair fiber

peptide chains of tyrosine, cysteine, and glycine amino acids (Robbins and Robbins 2012). The sulfur-rich amino acid cysteine is responsible for establishing the strong disulfide linkages between keratin fibers, forming the cysteine units. Other than disulfide linkages, other bonds such as peptide bonds, hydrogen bonding, and van der Waals forces are also present in the keratin fibers. The resulting structure of keratin protein gives hair, its shapes, structure, and mechanical properties (Fraser et al. 1988).

The hair shaft is composed of three concentric layers, the inner cortex, the medulla, and the outer cuticle layers, respectively. A representation of the hair structure is given in Fig. 11.2.

The central part of the hair fiber is the medulla, which is composed of the nucleus of the central cells and their air spaces. This part is more apparent in thick hairs and contributes a little to the total hair mass and its mechanical properties (Jones 2001).

The major portion of the hair lies in the cortex that surrounds the medulla. It comprises cortical cells that contain keratin filaments, forming the hair filament axis, and a matrix that is sulfur-rich keratin protein. Cortex contributes a major part to the total hair weight and is responsible for the hair fiber's mechanical properties. The elasticity of the hair fibers is mainly due to the helical shape of the keratin filaments. The cortex also contains coloring pigments and holds the moisture of the hair in it. The hair cuticle is composed of 6–10 overlapping cuticle cells pointing toward the hair fiber end. The different cuticles are epicuticle, A-layer, exo-cuticle, and endocuticle along with the cell membrane complex. A schematic illustration of the complex structure of the cuticle cell is given in Fig. 11.3.

The epicuticle is covered by a lipid bilayer giving a hydrophobic character to the cuticle membrane also called an F layer. It is mainly composed of 18-methyl eicosanoid acid (18-MEA), which is bound to the hair keratin via thioester linkages. This hydrophobic F layer provides a protective barrier to the hair, thus decreasing the electrostatic friction within hair strands. The cuticle has an important barrier function and protects the cortex giving hair a smooth texture. Cuticle damage caused by physical and chemical treatment exposes the hair fiber and accelerates hair weathering (Phillips et al. 2017). After continuous harsh hair treatments, the F layer is removed and ultimately entire cuticle is destroyed and hair breakage takes place. The physical and chemical destruction includes harsh chemical hair treatments, bad hair care habits, and exposure to UV radiation (Lee et al. 2014). Chemical treatments including hair bleaching, straightening, perming, and relaxing can cause significant damage to hair fibers. Physical damage to hair results from the use of harsh cleaning agents, poor hair combing, and blow drying at high temperatures (Arshad and Abbas 2018; Bloch et al. 2019). The first stage of damage is the loss of hair

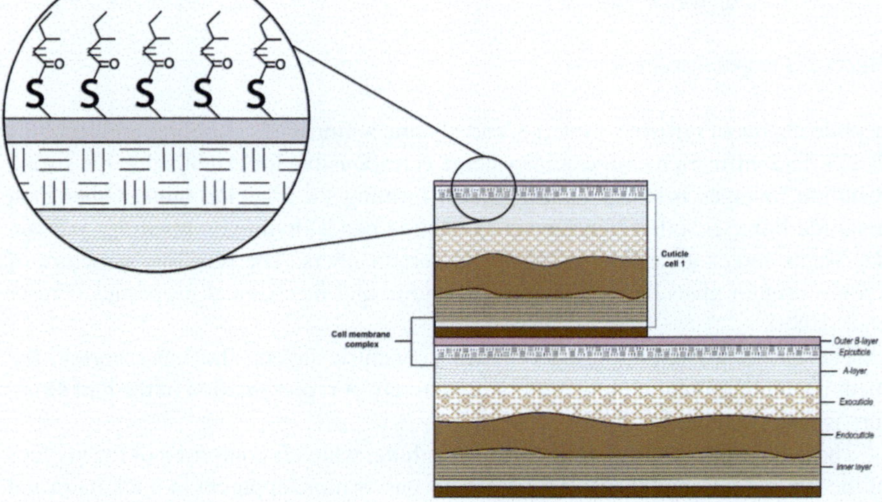

Fig. 11.3 Structure of hair cuticle

gloss and smoothness and then hair loses its elasticity and strength. After complete damage, the hair gets broken. Therefore, older hair is more prone to ultimate damage than newly formed hairs. Near the scalp, hair has a completely covered surface with overlapping scales of cuticle cells while the hair part toward the tip is more damaged. The degree of hair damage depends on personal hair care habits and the degree of exposure to chemical treatments.

11.2.1 Chemical Interactions Within the Hair Cortex

The chemical interactions in the hair cortex responsible for its mechanical strength include a range of intermolecular forces such as hydrogen bonding, hydrophobic interactions, ionic interaction, and disulfide bonds (Feughelman 2018). These chemical interactions are due to reactive groups in the keratin protein structure. The weakest interaction is hydrogen bonding most frequent in the hair and easily broken due to water. Ionic bonds are present between acidic and basic groups in the structure and are relatively stable in aqueous environments. Disulfide linkages are present between two cysteine amino acids bridging the keratin chains and giving strength to hair fiber. These bonds are the targets to achieve permanent hair modifications, causing the stretching of alpha keratin fiber when broken. These varieties of interaction in the polymeric chains provide a high resistance to keratin fibers.

11.3 Role of Cosmeceuticals in Hair Care

Nowadays, delineation of cosmetics from pharmaceuticals has become more complex through the development of cosmetics with physiologically active ingredients, i.e., cosmeceuticals. By far the most frequently applied form of cosmetic hair treatment is shampooing (Thompson et al. 2023; Trüeb 2001). Shampoos are primarily used for cleansing hair and some specialized shampoos are applied for treating scalp problems such as dandruff, oily hair, androgenic alopecia, and other scalp surface problems (Schwartz et al. 2013). These shampoos contain one or more specific active ingredients that have clinical effectiveness. These types of hair treatments should not be harmful to scalp and hair. Hair cosmeceuticals include a range of products such as conditioning agents, hair growth stimulants, hair nourishment oils, serums, etc. (Madnani and Khan 2013). Hair conditioning agents are formulated to bring softness and shine to hair and reduce static charge on the hairs that minimize the friction between hair fibers. The most commonly used conditioning agents are quaternized cationic compounds, cationic polymer, silicons, hydrolyzed proteins, and long chain fatty acids (Fernández-Peña and Guzmán 2020). Some special ingredients, such as anti-dandruff agents and antifungal compounds, can also be added. Conditioning agents have short-term contact with hair and hair growth stimulants cannot have a good effect on hair growth. A cosmetic agent such as 2, diamino-pyrimidine-3-oxide is used as a topical stimulant for hair growth. It acts via the prevention of inflammation and perifollicular fibrosis. Recently, minoxidil 5% has been approved in the United States to promote scalp hair growth in men, promoting a new treatment option for androgenetic alopecia (Gorecki 1988).

11.4 Classification of Hair Cosmeceuticals

Hair cosmetics are classified according to their active ingredients, mechanism of action, and type of applications. Following are various classes of hair cosmetics that may also act as cosmeceutical-containing active agents.

11.4.1 Shampoo

Shampoos are cleansing formulations for hair to remove oil, dirt, skin particles, dandruff, and environmental pollutants. All the shampoos are formulated as a solution of detergents and other soluble additives for hair cleansing, conditioning, lubrication, and treating specific scalp conditions. The purpose of shampoos is to cleanse the scalp, remove dirt and oils from the scalp and hair, and make them look beautiful, fragranced, and easy to comb. A wide range of the synthetic, medicated, and herbal shampoos are available; however, the popularity and demand of herbal shampoo are increasing because of the belief that these formulas are sustainable and have natural origin, safe, and non-toxic (Al Badi and Khan 2014). This meets the sustainable development goal of good health and wellbeing. The shampoo formula comprises 10–15% surfactant and 70–80% of water and rest is other additives and active agents. The high water content makes them easy to apply and increases their spreadability. Different types of shampoos are formulated according to the hair type and nature, and extent of damage. Baby shampoos are formulated to give non-irritating, mild cleansing with the use of amphoteric surfactants (O'Brien et al. 1994). Dry shampoos contain powdered ingredients such as starch, silica, and talc that physically absorb the sebum from hairs. Contact allergy due to shampoos is rare, if it occurs it's mainly due to the presence of preservatives, and fragrances (Yim et al. 2014). Shampoos are specially formulated for allergic scalp and don't contain allergic substances such as fragrances. The primary surfactants used in the non-herbal shampoo are anionic such as lauryl and laureth sulfates. Laureth sulfates are not hard for hair and cleanse hair better than lauryl sulfates. For cosmetic purposes, several surfactants are combined in a given formulation. Additional ingredients are added to improve conditioning effects, easy delivery of active agents, and improvement of product aesthetic properties. The purpose of adding synthetic surfactants is to obtain uniform foaming and cleansing; however, their regular use can cause hair and scalp dryness, and hair loss and occasionally result in irritation to the eyes and skin. During the last few decades, the use of natural ingredients in cosmetics has increased dramatically. Herbal shampoos are safe alternatives to synthetic ones that offer multiple health benefits to consumers along their with sustainable and non-toxic nature (Patidar 2018). Therefore, the addition of herbal extracts and oils in the hair cleansing products is according to the sustainable development goal number 3, which deals with good health and wellbeing. Many medicinal plants and herbs are available that have been reported for their hair-beneficial effects. The demand for herbal cosmetics is on the rise due to the belief that herbal products are safe, sustainable, and eco-friendly with minimum side effects and multiple beneficial effects. The crude extracts, powdered forms, and purified active agents of medicinal plants have been used for hair care formulations (Kolekar et al. 2021). A range of active phytochemicals of plants such as vitamins, phyto-hormones, polyphenols, enzymes, fruit acids, plant proteins, glycosides, and essential oils have been considered a multifunctional active constituent of cosmetic formulations (Bakr et al. 2019). The outer covering of *Spindus mukorossi* also called soap nut or reetha, the dried pods of Sheekakai (*Acacia concinna*), and fruits of Amla (*Phyllanthus emblica*) have been used traditionally in the sub-continents for washing the hair and scalp (Rashid et al. 2020; Wei et al. 2021). Several herbal shampoos are available in the market that contain essential oils and herbal extracts. A list of the plants and herbs that can be used in hair cleansing is given in Table 11.1. These extracts produce rich lather and foam when shaken with water due to the high concentration of saponin. They are also rich sources of phytochemicals and vitamins that have multifunctional effects such as anti-dandruff properties (Sharma et al. 2003). They also promote hair growth along with strengthening the hair (Patel et al. 2015). Saponins are natural surfactants that help remove the excess sebum and dirt from hair without causing any adverse effects. They also show strong antimicrobial properties and important ingredients in cosmetics. The use of herbal cosmetics is associated with health and beauty. The herbal shampoo has been developed using saponins of *Acanthophyllum squarrosum* (Aghel et al. 2007). In this formulation, the synthetic component like alkanolamide was replaced with a natural foam stabilizing agent. The formulation was further studied for its physical and rheological properties along with stability. In another formulation, polyherbal antidandruff shampoo was developed in which extract of *Rosmarinus officinalis*, *Nigella sative*, *Ficus bengalensis*, *Vetiverla zizanioides*, and citrus lemon along with oil of *Melaleuca leucadendron* was used (Guzmán and Lucia 2021). The formulation was effective due to synergistic effects like anti-fungal, anti-inflammatory, and immunostimulatory actions of the ingredients. The formulation has been studied for its anti-dandruff activity (Sharma et al. 2013). The extracts of *Acacia conica*, *Lawsonia inermis*, *Sapindus laurifolia*, and *Hibiscus rosa sinesis* were added in the formulation to enhance the effectiveness of shampoo as an antidandruff cleanser (Halith et al. 2009). A shampoo cream was developed with extract (Shivanand) and was found effective when compared with standard ketoconazole-containing shampoo. Another hair and scalp issue is the head lice parasites, which is a major public health problem in remote areas (Mohamed et al. 2022). The anti-lice agents are pesticides that can be harmful to the skin and hair. Most commonly used anti-lice products contain pyrethrins, permethrin, malathion, DDT, and lindane, which are harmful for patients with allergic reactions,

Table 11.1 Plants for hair care cosmetics: a brief introduction to active agents and their function

Plant scientific name	General name	Reported functions/uses	Active compounds	References
Berberis vulgaris	Barberry	Used in as hair growth stimulant	Berberine (alkaloid), isoquinoline alkaloids	Semalty et al. (2011)
Trigonella foenum-graecum	Fenugreek	Used as a hair cleansing and conditioning agent	Saponins (including diosgenin), alkaloids (trigonelline), flavonoids	Noudeh et al. (2011)
Eucalyptus sp.	Eucalyptus	Used as an antidandruff agent	Eucalyptol (1,8-cineole), flavonoids, tannins	Selvakumar (2012)
Larrea divaricata	Creosote bush	Used as an antidandruff agent in hair tonics	Lignans (larreatin, larreic acid), flavonoids	Moreno et al. (2020)
Lawsonia inermis	Henna	Conditioning, coloring agents in hair coloring cosmetics	Lawsone (2-hydroxy-1,4-naphthoquinone)	Semwal et al. (2014)
Pilocarpus jaborandi	Jaborandi	Used in shampoos and tonics as a hair growth-stimulating agent	Pilocarpine (alkaloid), tannins, flavonoids	Patel et al. (2015)
Citrus limon	Lemon	Used in shampoo as a cleanser, antidandruff, and conditioning agent	Citral, limonene, flavonoids (hesperidin, eriocitrin), ascorbic acid	Segaran et al. (2020)
Urtica dioica	Nettles	Used in shampoos as growth stimulates	Silica, iron, chlorophyll, flavonoids (quercetin), formic acid	Sahraie-Rad et al. (2015)
Rosmarinus officinalis	Rosemary	Condition agent	Rosmarinic acid, carnosic acid, camphor, 1,8-cineole, ursolic acid	González-Minero et al. (2020)
Santalum album	Sandalwood	As an antidandruff and antimicrobial agent	Sandalwood oil (Santalol), coumarins, xanthones	Kumar et al. (2019)
Quillaja saponaria	Soapbark	Used in shampoo, for itchy scalp or dandruff	Saponins (Quillaja saponin), flavonoids	Barve et al. (2016)
Saponaria officinalis	Soapwort	As an antidandruff agent and cleansing agent	Saponins (Soapwortin), flavonoids	Oleszek and Hamed (2010)

(continued)

Table 11.1 (continued)

Plant scientific name	General name	Reported functions/uses	Active compounds	References
Artemisia abrotanum	Southernwood	Acts as a hair growth stimulant	Thujone, flavonoids, essential oils (terpenoids)	Oleszek and Hamed (2010)
Melaleuca alternifolia	Tea tree	Its oil is effectively used in antidandruff shampoos	Terpinen-4-ol, gamma-terpinene, 1,8-cineole	Kinjuit and Surugau (2019)
Juglans nigra	Walnut	Decoction of leaves is added to shampoo, used as black hair dye	Juglone, tannins, polyphenols	Shao (2022)
Hamamelis virginiana	Witch hazel	Used in shampoos as an anti-dandruff agent	Tannins (Hamamelitannin), flavonoids, essential oils	Trüeb (2014)
Achillea millefolium	Yarrow	Used in shampoos and hair preparations, stimulates hair growth	Achilleine, chamazulene, flavonoids (luteolin, apigenin), tannins	Becker et al. (2016)
Phaseolus vulgaris	Bean flower	Effectively used in shampoos	Proteins (including enzymes), saponins, polysaccharides	Oakenfull (1981)
Nigella sativa	Black cumin	Used in shampoos for cleansing	Thymoquinone, nigellidine, saponins	Sudhir et al. (2016)
Vetiveria zizanoides	Ushir	Used in shampoos as cooling and antibacterial agent	Vetiverol, alkaloids, flavonoids	Rajput and Kumar (2020)
Sapindus saponaria	Soap nut	Used in shampoos as a cleanser and hair growth regulator	Saponins (sapindoside), triterpenoids	Cavaletti et al. (2023)
Amaranthus spinosus	Velvet flower	Used as a hair conditioner	Saponins, flavonoids, tannins	Sattar et al. (2024)
Schleichera oleosa	Kusum tree	Used in shampoos as a hair growth promoter	Saponins, triterpenoids, fatty acids (oleic acid, linoleic acid)	Bhatia et al. (2013)
Eclipta alba	Bhringaraja	Used for hair loss treatment	Wedelolactone, ecliptins, flavonoids	Roy et al. (2008)
Withania somnifera	Ashwagandha	Acts as an antidandruff agent	Withanolides, alkaloids (somniferine), saponins, fatty acids	Powar and Nitave (2020)
Eugenia jambos	Rose apple	Stimulates the scalp and regeneration of damaged hair.	Tannins, flavonoids, eugenol	Ranghoo-Sanmukhiya et al. (2019)

(continued)

Table 11.1 (continued)

Plant scientific name	General name	Reported functions/uses	Active compounds	References
Melaleuca leucadendron	Cajeput	Used in shampoos as an antidandruff agent.	1,8-Cineole, terpinen-4-ol, alpha-pinene	Barbosa et al. (2013)
Azadirachta indica	Neem	Used in shampoos as an antimicrobial agent	Azadirachtin, nimbidin, triterpenoids (nimbin, nimbolide)	Heukelbach et al. (2006)
Ocimum sanctum	Holy basil	Used in shampoo as an antimicrobial agent	Ursolic acid, eugenol, Triterpenoids (oleanolic acid)	Halith et al. (2009)
Cinnamomum camphora	Camphor	Used in shampoos as an antidandruff agent	Camphor, cinnamaldehyde, terpenoids	Park et al. (2019)
Cedrus atlantica	Cedarwood	Used in antidandruff shampoos and to prevent hair loss	Cedrol, triterpenoids, lignans	Abelan (2022)
Juniperus communis	Juniper berry	Cleanser and anti-infective	Alpha-pinene, beta-pinene, limonene	Uniyal et al. (2013)
Mentha piperita	Peppermint oil	Used in shampoos for its cooling properties	Menthol, menthone, flavonoids	Wulandari et al. (2022)
Thymus vulgaris	Thyme	Used in shampoos as an antimicrobial agent	Thymol, carvacrol, flavonoids	Paloukopoulou et al. (2022)
Phyllanthus emblica	Amla	Promotes hair growth, prevents premature graying, and controls dandruff	Phenolic compounds	Rahman et al. (2024)

asthma, skin diseases, and nervous system damage (Jolley et al. 1991). The most effective, healthy, and sustainable alternative is herbal anti-lice shampoos developed with ingredients that act against lice. These herbal ingredients include Paw Paw, tea tree oil, seed extract of *Azadirachta indica,* and thymol extracts (Arora et al. 2011). Neem extract shampoo was found effective more than permethrin-based shampoo (Leite and Maia Campos 2018).

11.4.2 Hair Conditioner

Chemical and environmental exposure to hair may cause tangling, dryness, and breakage. This effect is associated with weakening linkages between the 18-methyl eicosanoic acid (18-MEA) and cuticle cell membrane, making hair more hydrophilic and prone to damage. The damaged cuticle exposes the cortex and may cause hair fracture and breakage. However, the damaged cuticle can be partially repaired using hair conditioning treatments that enable the covering and filling of damaged cuticle parts, reduce friction between fibers, and maintain hair moisture level (da Gama et al. 2018; Fernandes et al. 2023). The most commonly used approaches for hair conditioning formulation involve the interaction of mixtures of macromolecular complexes such as synthetic or natural polymers, proteins, and polyelectrolytes with the surface of hair fiber. Different polycations and silicons are also used in the conditioning formulations. Hair condition formulations also increase hair softness, shine, and manageability and reduce frizz. The main purpose of using conditioner is to make combing easy and less damaging to hair and scalp by decreasing the friction between hair strands. This is achieved by neutralizing hair's negative charge with a positive charge and lubricating the hair shaft by reducing its hydrophilic character (Arshad 2024; Fernández-Peña and Guzmán 2020). Cationic polymeric materials are soluble in water at low concentrations but the silicon-based components are incompatible with the formulation. The use of conditioner is very much the same as the use of oils for reducing hair friction. Coconut oil is considered the best conditioning oil as its molecules penetrate the hair shaft and lessen friction along with minimum protein loss (Jaafar et al. 2019). Different types of conditioning formulations are based on oils, viscous liquids, creams, lotions, sprays, and gels. A new direction toward the formulation of eco-friendly and sustainable formulation is the development of plant-based conditioning agents (Isnard et al. 2019). The most common of these are the extracts of soy, rice, aloe vera, jojoba, and quinoa (Centini et al. 2023; Choi et al. 2014). The application of natural ingredients in hair conditioners meets the sustainable development goal of sustainable consumption and production, in addition to maintaining good health and wellbeing. Plant-based extracts are also rich sources of plant proteins that can penetrate the hair shaft, making it strong and enabling moisture retention and less susceptible to damage (Camargo et al. 2022). The most sustainable, bioavailable, and eco-friendly source of hair conditioning treatment is the use of vegetable oils. The use of mineral oil, sunflower, and coconut oil helps in preventing hair damage and breakage and thus reduces protein loss (Rele and Mohile 2003). Coconut oil is a rich source of fatty acids such as lauric acid that can penetrate the hair cuticle and cortex. Several other vegetable oils such as argan oil, mustard oil, and olive oils have been studied for their beneficial effect on hair and scalp (Loch et al. 2022). The organic oils are rich source of polyphenols and antioxidants. These have therapeutic properties such as anit-microbial, anti-inflammatory, and hair growth-stimulating effects.

11.4.3 Hair Color Products

In modern society, people want to change their hair color and style according to the fashion and thanks to a range of color shades, available in the market. Hair dyes are also used to restore the original hair pigmentation and cover the grey hairs due to aging effects. Hair dyes are classified as temporary, semi-permanent, and permanent. The difference between these types is their extent of penetration into the hair fiber mediated by alkaline pH (Da França et al. 2015). Temporary hair dyeing is obtained by-products in the form of lotions, creams, or shampoos and the color stays on the hair for a short period of time. Temporary hair dyes have high molecular weight due to which they deposit on the cuticle instead of penetrating the hair fiber and are washed away and removed easily (Rust and Schlatter 2022). Permanent hair colors allow brightening the darkening of the hair with little resistance to external factors for more than 4 weeks. Permanent hair colors offer a wide range of color shades. Their work is based on the penetration of color molecules into the hair shaft and then molecules are oxidized and attached to the cuticle and cortex. The modern trend of consumer desire for eco-friendly natural colors is healthier and sustainable. As compared to oxidative hair dyes, plant-based natural colorants are considered more compatible and less toxic to consumers and the environment (Cui et al. 2023). A range of coloring plants such as henna leaves, Chinese gallnut, walnut, turmeric, heartwood, and gromwell roots have been used for their hair color properties since ancient times (Cui et al. 2022). Besides the hair dyeing properties, natural colorants are therapeutically active and exhibit multifunctional effects such as ant-microbial, ant-oxidant, and anti-inflammatory effects (Cui et al. 2023; Li et al. 2022a, b). Although hair dyes offer several health benefits, their applications are restricted by various disadvantages such as time-consuming extraction procedures, poor color stability, reproducibility, and color fastness, when compared to synthetic hair dyes. In a research study, the dyeing efficiency and ultraviolet fastness of the natural dyes were assessed for various dye-yielding plants(Padfield and Landi 1966). Henna as a major hair coloring plant has been investigated for its dyeing stability (Cox 1938). The waste product from walnuts contains juglone, which is a coloring agent with good light fastness, and has been studied for its coloring properties (Beiki et al. 2018). In the Chinese cosmetic industry, bluish-black floor shade of hair dye obtained from Chinese gallnuts contains active colorant gallotannis when mordanted with ferrous salts (Arshad et al. 2018; Tang et al. 2020). Similarly, Brazilian wood is also important for dyeing properties and has been studied for its dyeing behavior (Indrawati et al. 2017). Plant-based colorants can be applied as hair dye either directly or in combination with mordant. Direct natural hair colorants have narrow hair color shades and poor fatness to light and sun. When applied with mordant, these colorants showed improved color fastness (Cui et al. 2022). The metallic mordants are used such as metallic salts, i.e., ferrous sulfate, ferrous gluconate, alum and copper sulfate (Cui et al. 2022). The choice of mordant affects the yield and color fastness of dyed hairs.

11.4.4 Hydrolyzed Proteins

One of the most efficient hair-strengthening treatments is the application of cosmetics containing hydrolyzed protein either from plants or animal sources (Sinkiewicz et al. 2018; Weathersby and McMichael 2013). The hydrolyzed amino acids are positively charged and have low molecular weight in the range of 1 kDa. These amino acids can penetrate into the hair shaft, neutralize the positive charge of the hair fiber and restore the natural moisture level of the hair. The deposition of the amino acids is achieved via heat, which leads to more stable and permanent deposition in the cuticle cell membrane.

11.4.5 Hair Growth Promoters

Natural extracts have been used since ancient times to regulate hair growth and also preserve hair integrity and structure by providing basic nutrients required for hair growth (Patel et al. 2015). Medicinal plants have been studied for their applications in the hair care industry. Various herbal extracts and oils have been investigated for their ability to promote hair growth in hair-falling conditions such as alopecia. These active agents work under different mechanisms such as growth factors- cytokines, 5α-reductase inhibitory effect, apoptosis, sonic hedgehog signaling, and cell cycle progression (Shapiro and Price 1998). Saw palmetto extracts belonging to a class of palm tree berries have saw-shaped leaves. The extracts of saw palmettos and beta-sitosterol have been used to enhance hair growth in androgenetic patients (Evron et al. 2020). This is due to its effect as a non-selective and competitive inhibitor of type 1 and II 5-α-reductase. This property is used in treating prostatic conditions, i.e., benign prostatic hyperplasia. At the level of hair follicle, saw palmetto affects the synthesis of DHT and also inhibits the update and binding ability of DHT to androgenetic receptors (Sudeep et al. 2023). Extracts of *Ginkgo biloba* have been used traditionally for stimulating follicular proliferation and inhibiting apoptosis. The extract is a mixture of glycosides such as kaempferol, terpenes, and quercetin (Arshad et al. 2014; Li et al. 2022a, b). Aloe vera extract has been used as a traditional Asian therapy for its therapeutic properties such as antimicrobial, antioxidant, and anti-inflammatory (Indriaty et al. 2018). Aloe gel is a rich source of nutrients such as aloin, alkylchromone, amino acids, minerals, vitamins, mucopolysaccharides, and hydroxyquinine glycoside. It is therefore shown to induce angiogenesis and promote hair growth by delivering active nutrients and oxygen for the functioning of hair follicles (Tyagi et al. 2023). *Vitis vinifera* extract commonly known as grapevine is incorporated in cosmeceuticals due to its good antioxidant capacity (Srisuk et al. 2023). The extract contains multifunctional therapeutic compounds such as flavonoids, fruit acids, stilbenes, essential fatty acids, tocopherols, and phenylacrylic acid. Its hair growth promoting effect is due to the presence of proanthocyanidins in its seed extract, which stimulate hair follicle cells by proliferation. The effect was studied in vivo as proanthocyanidin-induced change in the hair cycle was observed (Takahashi et al. 2001). One of the fastest-growing applications of herbal extract is the use of green tea extracts in the cosmeceutical applications. Green tea is derived from *camellia sinesis* leaves and contains polyphenolic compounds that modulate biochemical pathways for cell proliferation (Koch et al. 2019). The most dominant polyphenol in the tea extract is *Epigallocatechin gallate*

(EGCG), which is a potent antioxidant. it has been shown to inhibit the II 5-α-reductase (Arshad et al. 2022; Esfandiari and Kelly 2005). Some unsaturated fatty acids such as γ-linolenic acid inhibit 5-α-reductase activity. Other fatty acids such as arachidonic acid, linoleic acid, and α-linolenic acid also have inhibitory action for 5-α-reductase. The fatty acid extract of *Lygodii spora* as Japanese fern is known for its anti-5-α-reductase activity (Krishnamoorthi and Kumaran 2024).

11.5 Conclusion

Hair cosmeceuticals have been developed for the improvement in the biological, chemical, and mechanical properties of hair fibers. These hair care products are applied in the form of coatings creams, solutions, serums, etc. A range of hair care products containing cosmeceutical agents are hair cleansing products such as shampoos, conditioners, strengthening and growth stimulants, and nourishment oils. Natural extracts from plants have been used for the formulation of multifunctional hair coloring products. A range of coloring plants such as henna leaves, Chinese gallnut, walnut, turmeric, heartwood, and gromwell roots have been used for their hair color properties since ancient times. The demand for herbal cosmetics is on the rise due to the belief that herbal products are safe, sustainable, and eco-friendly with minimum side effects and multiple beneficial effects, according to the sustainable development goal of good health and wellbeing. The outer covering of *Spindus mukorossi* also called soap nut or reetha, the dried pods of Sheekakai (*Acacia concinna*), and fruits of Amla (*Phyllanthus emblica*) have been used traditionally in the sub-continents for washing hair and scalp. One of the most efficient hair-strengthening treatments is the application of cosmetics containing hydrolyzed protein from plants or animal sources. Herbal extracts and oils have been used as hair conditioning formulations. Vegetable oils and extracts have been used as therapeutic agents for preventing hair damage and controlling protein loss.

References

Abelan US (2022) Potential use of essential oils in cosmetic and dermatological hair products: a review. J Cosmet Dermatol 21(4):1407–1418

Aghel N, Moghimipour E, Raeis DN (2007) Formulation of a herbal shampoo using total saponins of Acanthophyllum squarrosum. Iran J Pharm Res 6(3):167–172

Al Badi K, Khan SA (2014) Formulation, evaluation and comparison of the herbal shampoo with the commercial shampoos. Beni-Suef Univ J Basic Appl Sci 3(4):301–305

Arora P, Nanda A, Karan M (2011) Shampoos based on synthetic ingredients vis-a-vis shampoos based on herbal ingredients: a review. Int J Pharm Sci Rev Res 7(1):41–46

Arshad M (2024) Sustainability of renewable energy options as compared to coal-fired power plants in Pakistan. Green Low-Carbon Econ 2(3):162–173

Arshad M, Abbas M (2018) Water sustainability issues in biofuel production. In: Perspectives on water usage for biofuels production: aquatic contamination and climate change. Springer, Cham, pp 55–76

Arshad M, Adil M, Sikandar A, Hussain T (2014) Exploitation of meat industry by-products for biodiesel production: Pakistan's perspective. Pak J Life Soc Sci 12:120–125

Arshad M, Bano I, Younus M, Khan A, Rahman A (2018) Health concerns associated with biofuel production. In: Perspectives on water usage for biofuels production: aquatic contamination and climate change. Springer, Cham, pp 97–105

Arshad M, Ansari AR, Qadir R, Tahir MH, Nadeem A, Mehmood T, Alhumade H, Khan N (2022) Green electricity generation from biogas of cattle manure: an assessment of potential and feasibility in Pakistan. Front Energy Res 10:911485

Bakr RO, Amer RI, Fayed MA, Ragab TI (2019) A completely polyherbal conditioning and antioxidant shampoo: a phytochemical study and pharmaceutical evaluation. J Pharm Bioallied Sci 11(2):105–115

Barbosa LCA, Silva CJ, Teixeira RR, Meira RMSA, Pinheiro AL (2013) Chemistry and biological activities of essential oils from Melaleuca L. species. Agric Conspec Sci 78(1):11–23

Barve K, Dighe A, Barve K, Dighe A (2016) Hair shampoo. In: The chemistry and applications of sustainable natural hair products. Springer, Cham, pp 25–35

Becker LC, Bergfeld WF, Belsito DV, Hill RA, Klaassen CD, Liebler DC, Marks JG Jr, Shank RC, Slaga TJ, Snyder PW (2016) Safety assessment of Achillea millefolium as used in cosmetics. Int J Toxicol 35(3):5S–15S

Beiki T, Najafpour GD, Hosseini M (2018) Evaluation of antimicrobial and dyeing properties of walnut (Juglans regia L.) green husk extract for cosmetics. Color Technol 134(1):71–81

Bhatia H, Kaur J, Nandi S, Gurnani V, Chowdhury A, Reddy PH, Vashishtha A, Rathi B (2013) A review on Schleichera oleosa: pharmacological and environmental aspects. J Pharm Res 6(1):224–229

Bloch L, Goshiyama AM, Dario MF, Escudeiro C, Sarruf FD, Velasco MVR, Valente NYS (2019) Chemical and physical treatments damage Caucasian and Afro-ethnic hair fibre: analytical and image assays. J Eur Acad Dermatol Venereol 33(11):2158–2167

Camargo FB, Minami MM, Rossan MR, Magalhaes WV, Porto Ferreira VT, Maia Campos PMBG (2022) Prevention of chemically induced hair damage by means of treatment based on proteins and polysaccharides. J Cosmet Dermatol 21(2):827–835

Cavaletti J, Prando W, Ribeiro E, Valladão D, d. S. (2023) Development and stability of intimate soap formulations using Sapindus saponaria L. extract as a natural surfactant. Braz J Biol 83:e276940

Centini M, Signori G, Francescon F, Tsuno F, Oguro T, Anselmi C (2023) Rice derivatives in hair protecting. Cosmetics 10(6):163

Choi J-S, Jeon M-H, Moon W-S, Moon J-N, Cheon EJ, Kim J-W, Jung SK, Ji Y-H, Son SW, Kim M-R (2014) In vivo hair growth-promoting effect of rice bran extract prepared by supercritical carbon dioxide fluid. Biol Pharm Bull 37(1):44–53

Cox H (1938) Hair dyes. I. The chemistry and analysis of henna. Analyst 63(747):397–404

Cui H, Xie W, Hua Z, Cao L, Xiong Z, Tang Y, Yuan Z (2022) Recent advancements in natural plant colorants used for hair dye applications: a review. Molecules 27(22):8062

Cui H, Cai R, Hua Z, Tang Y (2023) Plant colorants for natural hair coloration: dyeing optimization and photostability assessment. Sustain Chem Pharm 36:101285

Da França SA, Dario MF, Esteves VB, Baby AR, Velasco MVR (2015) Types of hair dye and their mechanisms of action. Cosmetics 2(2):110–126

da Gama RM, França-Stefoni SA, Sá-Dias TC, Bedin V, Baby AR, Velasco MVR (2018) Protective effect of conditioner agents on hair treated with oxidative hair dye. J Cosmet Dermatol 17(6):1090–1095

Dias MFRG (2015) Hair cosmetics: an overview. Int J Trichology 7(1):2–15

Esfandiari A, Kelly AP (2005) The effects of tea polyphenolic compounds on hair loss among rodents. J Natl Med Assoc 97(8):1165

Evron E, Juhasz M, Babadjouni A, Mesinkovska NA (2020) Natural hair supplement: friend or foe? Saw palmetto, a systematic review in alopecia. Skin Appendage Disord 6(6):329–337
Fernandes C, Medronho B, Alves L, Rasteiro MG (2023) On hair care physicochemistry: from structure and degradation to novel biobased conditioning agents. Polymers 15(3):608
Fernández-Peña L, Guzmán E (2020) Physicochemical aspects of the performance of hair-conditioning formulations. Cosmetics 7(2):26
Feughelman M (2018) Morphology and properties of hair. In: Hair and hair care. Routledge, London, pp 1–12
Fraser R, MacRae T, Sparrow L, Parry D (1988) Disulphide bonding in α-keratin. Int J Biol Macromol 10(2):106–112
González-Minero FJ, Bravo-Díaz L, Ayala-Gómez A (2020) Rosmarinus officinalis L.(Rosemary): an ancient plant with uses in personal healthcare and cosmetics. Cosmetics 7(4):77
Gorecki DK (1988) Minoxidil. In: Analytical profiles of drug substances, vol 17. Elsevier, Amsterdam, pp 185–219
Guzmán E, Lucia A (2021) Essential oils and their individual components in cosmetic products. Cosmetics 8(4):114
Halith SM, Abirami A, Jayaprakash S, Karthikeyini C, Pillai KK, Firthouse P (2009) Effect of Ocimum sanctum and Azadiracta indica on the formulation of antidandruff herbal shampoo powder. Pharm Lett 1(2):68–76
Heukelbach J, Oliveira FA, Speare R (2006) A new shampoo based on neem (Azadirachta indica) is highly effective against head lice in vitro. Parasitol Res 99:353–356
Indrawati T, Syahrin A, Irpan I (2017) Preparation of demipermanent and semipermanent hair dyes gels from ethanol extract of Caesalpinia sappan L. using carbomer as gelling agent. AIP Conf Proc 1862:030080
Indriaty S, Indrawati T, Taurhesia S (2018) Formulation and test activities of hair tonic with a combination of the aqueous extracts of aloe vera (Aloe vera L.) and licorice (Glycyrrhiza glabra L.). Pharmaciana 8(1):33–42
Isnard MD, Costa GMDA, Campos PMBGM (2019) Development of hair care formulations based on natural ingredients. Int J Phytocosmet Nat Ingredients 6(1):9
Jaafar SM, Khalid RM, Othaman R (2019) Coconut oil based microemulsion formulations for hair care product application. Sains Malaysiana 48(3):599–605
Jolley J, Kennedy J, Miller A (1991) A comparison of two insecticidal shampoos in the treatment of head louse infection. J R Soc Health 111(3):90–91
Jones LN (2001) Hair structure anatomy and comparative anatomy. Clin Dermatol 19(2):95–103
Kinjuit H, Surugau N (2019) Formulation and evaluation of hair shampoo containing tea tree (Melaleuca alternifolia) oil and virgin coconut (Cocos nucifera) oil. J Phys Conf Ser 1358:012022
Koch W, Zagórska J, Marzec Z, Kukula-Koch W (2019) Applications of tea (Camellia sinensis) and its active constituents in cosmetics. Molecules 24(23):4277
Kolekar YS, Tamboli FA, More HN, Mulani SA, Mali NP (2021) Medicinal plants used in cosmetics for skin and hair care. Int J Pharm Chem Anal 8(2):36–40
Krishnamoorthi R, Kumaran A (2024) Future aspects of plant derived bioactive metabolites as therapeutics to combat benign prostatic hyperplasia. J Ethnopharmacol 330:118207
Kumar GR, Chandrashekar B, Rao MS, Ravindra M, Chandrashekar K, Soundararajan V (2019) Pharmaceutical importance, physico-chemical analysis and utilisation of Indian sandalwood (Santalum album Linn.) seed oil. J Pharmacogn Phytochem 8(1):2587–2592
Lee Y, Kim YD, Pi L, Lee SY, Hong H, Lee WS (2014) Comparison of hair shaft damage after chemical treatment in A sian, W hite E uropean, and A frican hair. Int J Dermatol 53(9):1103–1110
Leite MGA, Maia Campos PM (2018) Photoprotective effects of a multifunctional hair care formulation containing botanical extracts, vitamins, and UV filters. Photochem Photobiol 94(5):1010–1016

Li Y, Sheng Y, Liu J, Xu G, Yu W, Cui Q, Lu X, Du P, An L (2022a) Hair-growth promoting effect and anti-inflammatory mechanism of Ginkgo biloba polysaccharides. Carbohydr Polym 278:118811
Li N, Wang Q, Zhou J, Li S, Liu J, Chen H (2022b) Insight into the progress on natural dyes: sources, structural features, health effects, challenges, and potential. Molecules 27(10):3291
Loch C, Souza PDC, Frigieri I, Isaac V, Chiavacci LA, Oshiro JA, Chiari-Andréo BG (2022) Development and characterization of highly structured rinse-off conditioners containing vegetable oils. J Dispers Sci Technol 43(9):1283–1290
Madnani N, Khan K (2013) Hair cosmetics. Indian J Dermatol Venereol Leprol 79:654
Mohamed JMM, Elhassan GO, Khan J, Nachiya R, Kayarohanam S, Janakiraman A (2022) Studies on poly herbal powder shampoo for the treatment of pediculosis capitis and pityriasis capitis infestations. Int J Appl Pharm 14:127–134
Moreno MA, Zampini IC, Isla MI (2020) Antifungal, anti-inflammatory and antioxidant activity of bi-herbal mixtures with medicinal plants from Argentinean highlands. J Ethnopharmacol 253:112642
Noudeh GD, Sharififar F, Khazaeli P, Mohajeri E, Jahanbakhsh J (2011) Formulation of herbal conditioner shampoo by using extract of fenugreek seeds and evaluation of its physicochemical parameters. Afr J Pharm Pharmacol 5(22):2420–2427
O'Brien K, Basketter D, Jones P, Dixit M (1994) An in vitro study of the eye irritation potential of new shampoo formulations. Toxicol In Vitro 8(2):257–261
Oakenfull D (1981) Saponins in food—a review. Food Chem 7(1):19–40
Oleszek W, Hamed A (2010) Saponin-based surfactants. In: Surfactants from renewable resources. Wiley, Hoboken, pp 239–249
Padfield T, Landi S (1966) The light-fastness of the natural dyes. Stud Conserv 11(4):181–196
Paloukopoulou C, Govari S, Soulioti A, Stefanis I, Angeli A, Matheeussen A, Capasso C, Cos P, Supuran CT, Karioti A (2022) Phenols from Origanum dictamnus L. and Thymus vulgaris L. and their activity against Malassezia globosa carbonic anhydrase. Nat Prod Res 36(6):1558–1564
Park BY, Ryu MJ, Park BY, Ryu MJ (2019) Assessment on the biological activities of mint, cinnamon and mixed extract and stability of the shampoo. Asian J Beauty Cosmetol 17(2):247–256
Patel S, Sharma V, Chauhan S, Thakur M, Dixit VK (2015) Hair growth: focus on herbal therapeutic agent. Curr Drug Discov Technol 12(1):21–42
Patidar K (2018) Polyherbal anti-dandruff shampoo: basic concept, benefits, and challenges. Asian J Pharm 12(3):S849–S858
Phillips TG, Slomiany WP, Allison R (2017) Hair loss: common causes and treatment. Am Fam Physician 96(6):371–378
Powar MAD, Nitave SA (2020) A review: polyherbal antidandruff hair oil. World J Pharm Res 10:440–457
Rahman SA, Ramli SI, Rahman SA (2024) Formulation and evaluation of halal hair growth promoting shampoo containing Centella asiatica and Phyllanthus emblica. Curr Trends Biotechnol Pharm 18(3):125–135
Rajput M, Kumar N (2020) Medicinal plants: a potential source of novel bioactive compounds showing antimicrobial efficacy against pathogens infecting hair and scalp. Gene Rep 21:100879
Ranghoo-Sanmukhiya VM, Chellan Y, Govinden-Soulange J, Lambrechts IA, Stapelberg J, Crampton B, Lall N (2019) Biochemical and phylogenetic analysis of Eugenia and Syzygium species from Mauritius. J Appl Res Med Aromat Plants 12:21–29
Rashid K, Raj V, Kumar P, Nishad K (2020) Hair care promising herbs: a review. Pharm Res 10:677–688
Rathod S, Mali S, Shinde N, Aloorkar N (2020) Cosmeceuticals and beauty care products: current trends with future prospects. Res J Top Cosmet Sci 11(1):45–51
Rele AS, Mohile RB (2003) Effect of mineral oil, sunflower oil, and coconut oil on prevention of hair damage. J Cosmet Sci 54(2):175–192
Robbins CR, Robbins CR (2012) Chemical composition of different hair types. In: Chemical and physical behavior of human hair. Springer, Cham, pp 105–176

Roy R, Thakur M, Dixit V (2008) Hair growth promoting activity of Eclipta alba in male albino rats. Arch Dermatol Res 300(7):357–364

Rust RC, Schlatter H (2022) Hair dyes. In: Cosmetic dermatology: products and procedures. Wiley, Hoboken, pp 309–319

Sahraie-Rad M, Izadyari A, Rakizadeh S, Sharifi-Rad J (2015) Preparation of strong antidandruff shampoo using medicinal plant extracts: a clinical trial and chronic dandruff treatment. Jundishapur J Nat Pharm Prod 10(4):e21517

Sattar M, Saeed F, Afzaal M, Rasheed A, Asif A, Sharif S, Hussain M, Asad Ur Rehman H, Raza MA, Munir H (2024) An overview of the nutritional and therapeutic properties of amaranth. Int J Food Prop 27(1):263–272

Schwartz J, Bacon R, Shah R, Mizoguchi H, Tosti A (2013) Therapeutic efficacy of anti-dandruff shampoos: a randomized clinical trial comparing products based on potentiated zinc pyrithione and zinc pyrithione/climbazole. Int J Cosmet Sci 35(4):381–387

Segaran RA, Ng CH, Al-Dhalli S (2020) Antifungal activity of successive extracts of Citrus limon peel for an antidandruff shampoo formulation. Int J Med Toxicol Legal Med 23(1):59–68

Selvakumar P (2012) Studies on the antidandruff activity of the essential oil of Coleus amboinicus and Eucalyptus globulus. Asian Pac J Trop Dis 2:S715–S719

Semalty M, Semalty A, Joshi GP, Rawat MSM (2011) Hair growth and rejuvenation: an overview. J Dermatol Treat 22(3):123–132

Semwal RB, Semwal DK, Combrinck S, Cartwright-Jones C, Viljoen A (2014) Lawsonia inermis L.(henna): ethnobotanical, phytochemical and pharmacological aspects. J Ethnopharmacol 155(1):80–103

Shao M (2022) Juglans: Tannin Ink and magic formula for hair dyeing. Chin Med Cult 5(1):72–76

Shapiro J, Price VH (1998) Hair regrowth: therapeutic agents. Dermatol Clin 16(2):341–356

Sharma L, Agarwal G, Kumar A (2003) Medicinal plants for skin and hair care. Indian J Tradit Knowl 2:62–68

Sharma S, Upadhyay U, Upadhyay SU, Patel T, Trivedi P (2013) Herbal armamentarium for the culprit dandruff. Int J Phytopharm Res 4(1):23–28

Sinkiewicz I, Staroszczyk H, Śliwińska A (2018) Solubilization of keratins and functional properties of their isolates and hydrolysates. J Food Biochem 42(2):e12494

Srisuk P, Khunkitti W, Khawee C, Yarovaya L (2023) Grape skin and seed extracts as a potential natural solution for hair loss: a bioactivity evaluation. Songklanakarin J Sci Technol 45(5):578–584

Sudeep HV, Rashmi S, Jestin TV, Richards A, Gouthamchandra K, Shyamprasad K (2023) Oral and topical administration of a standardized saw palmetto oil reduces hair fall and improves the hair growth in androgenetic alopecia subjects–a 16-week randomized, placebo-controlled study. Clin Cosmet Investig Dermatol 16:3251–3266

Sudhir S, Deshmukh V, Verma H (2016) Nigella sativa seed, a novel beauty care ingredient: a review. Int J Pharm Sci Res 7(8):3185

Takahashi T, Kamimura A, Yokoo Y, Honda S, Watanabe Y (2001) The first clinical trial of topical application of procyanidin B-2 to investigate its potential as a hair growing agent. Phytother Res 15(4):331–336

Tang Y, Yang S, He W, Liu L, Zhang Z (2020) Stabilization of Chinese gallnut (Galla Chinensis) tannins by spray-drying microencapsulation for natural hair coloring. Fibers Polymers 21:1283–1292

Thompson CJ, Ainger N, Starck P, Mykhaylyk OO, Ryan AJ (2023) Shampoo science: a review of the physiochemical processes behind the function of a shampoo. Macromol Chem Phys 224(3):2200420

Trüeb R (2001) The value of hair cosmetics and pharmaceuticals. Dermatology 202(4):275–282

Trüeb RM (2014) North American virginian witch hazel (Hamamelis virginiana): based scalp care and protection for sensitive scalp, red scalp, and scalp burn-out. Int J Trichology 6(3):100–103

Tyagi N, Tyagi A, Rastogi R, Singh B, Nagarajan K (2023) Unlocking radiance: the dynamic duo of flaxseeds and aloe vera in hair mask for bioactive brilliance. Proteins 22:29–22

Uniyal V, Saxena S, Bhatt R (2013) Screening of some essential oils against Trichosporon species. J Environ Biol 34(1):17

Weathersby C, McMichael A (2013) Brazilian keratin hair treatment: a review. J Cosmet Dermatol 12(2):144–148

Wei M-P, Qiu J-D, Li L, Xie Y-F, Yu H, Guo Y-H, Yao W-R (2021) Saponin fraction from Sapindus mukorossi Gaertn as a novel cosmetic additive: extraction, biological evaluation, analysis of anti-acne mechanism and toxicity prediction. J Ethnopharmacol 268:113552

Wulandari D, Sopyan I, Ginaris RP, Fathurrohim MF, Maya I (2022) Potential of essential oil as anti-dandruff in scalp treatment preparations. FITOFARMAKA J Ilmiah Farmasi 12(2):156–168

Yim E, Baquerizo Nole KL, Tosti A (2014) Contact dermatitis caused by preservatives. Dermatitis 25(5):215–231

Chapter 12
Analytical Methods in Cosmetics

Muhammad Azeem, Saqib Mahmood, Asia Shaheen, Aisha Tariq, Iahtasham Khan, and Muhammad Arshad

Contents

M. Azeem
Department of Biology, College of Science, University of Bahrain, Zallaq, Bahrain

S. Mahmood (✉) · A. Shaheen
Department of Botany, Government College University Faisalabad, Faisalabad, Pakistan
e-mail: saqibmahmood@gcuf.edu.pk

A. Tariq
Department of Nutritional Sciences, Government College University, Faisalabad, Pakistan

I. Khan
Epidermology and Public Health Section, Jhang-Campus, University of Veterinary and Animal Sciences, Lahore, Pakistan

M. Arshad
Biochemistry Section, Jhang-Campus, University of Veterinary and Animal Sciences Lahore, Lahore, Pakistan

S. Javed et al. (eds.), *Sustainable Cosmeceuticals*,
https://doi.org/10.1007/978-3-031-86087-4_12

Abstract The cosmeceutical industry relies on a diverse range of advanced analytical methods to ensure the quality, safety, and efficacy of its products. This chapter explores key techniques, including chromatographic methods such as high-performance liquid chromatography (HPLC), gas chromatography (GC), and thin-layer chromatography (TLC), which are essential for separating, identifying, and quantifying bioactive compounds and contaminants in cosmetic formulations. Spectroscopic methods, including UV-visible (UV-Vis) spectroscopy, Fourier transform infrared (FTIR) spectroscopy, and nuclear magnetic resonance (NMR) spectroscopy, offer critical insights into structural characterization and ingredient interactions. Microscopic and imaging techniques, such as scanning electron microscopy (SEM), transmission electron microscopy (TEM), and confocal microscopy, provide detailed visualization of particle morphology and nanoscale interactions within formulations. Thermal analysis methods, such as thermogravimetric analysis (TGA) and differential scanning calorimetry (DSC), evaluate ingredient stability and product shelf life, while microbiological techniques like enzyme-linked immunosorbent assay (ELISA) and polymerase chain reaction (PCR) ensure microbial safety. These analytical methods address the unique challenges of natural cosmeceuticals, including ingredient variability, degradation susceptibility, and compliance with global regulatory frameworks.

In addition to these applications, the discussed techniques align with several sustainable development goals (SDGs). By ensuring the safety and efficacy of products, they contribute to **SDG 3 (good health and well-being)**, safeguarding consumer health. Innovations in analytical approaches drive progress in cosmeceutical formulation and production, supporting **SDG 9 (industry, innovation, and infrastructure)**. Green analytical methods, which reduce hazardous waste and conserve resources, promote **SDG 12 (responsible consumption and production)**, while the sustainable sourcing and analysis of natural ingredients aid in conserving biodiversity, aligning with **SDG 15 (life on land)**. Emerging trends, such as green analytical chemistry, high-throughput screening, and nanotechnology, are further shaping the future of cosmeceutical analysis, enhancing efficiency, precision, and sustainability. This chapter provides a comprehensive overview of these methods, highlighting their application in natural cosmeceutical products and their critical role in advancing a safer, greener, and more sustainable cosmeceutical industry.

Keywords Techniques · Cosmetics · Chromatography · Spectrometry · Microscopy · Sustainability · Green chemistry

12.1 Introduction

The cosmetics industry extensively relies on analytical techniques to ensure the quality, safety, and efficacy of its products. These analytical methods are crucial in meeting customer expectations, complying with stringent regulatory requirements, and validating the purity and performance of ingredients (Bandaranayake 2006; El-Otmani et al. 2024). As consumer interest shifts toward natural cosmetics and cosmeceuticals—products that blend therapeutic and cosmetic benefits—the demand for advanced, precise, and innovative analytical techniques has grown exponentially. This chapter aims to explore the fundamental concepts, instruments, and methodologies central to cosmetic analysis, with a particular focus on their application in the evaluation of natural cosmeceuticals.

Analytical techniques in cosmetic science serve two primary objectives: regulatory compliance and quality assurance. Regulatory compliance involves ensuring adherence to safety, efficacy, and labeling standards set by global frameworks such as the European Union Cosmetics Regulation (EC 1223/2009) and US FDA Guidelines (European Commission 2009; U.S. Food and Drug Administration 2024; Ferreira and Lores 2022). Quality assurance encompasses evaluating raw materials, formulations, and finished products to meet defined benchmarks for stability, performance, and safety. Together, these efforts are foundational in the development and market success of safe and effective cosmetic products (Zakaria 2013).

The growing complexity of natural cosmeceutical formulations, which often include plant-based extracts, essential oils, and bioactive compounds, has driven the adoption of sophisticated analytical tools. Chromatographic techniques, such as high-performance liquid chromatography (HPLC), gas chromatography (GC), and thin-layer chromatography (TLC), are crucial for separating, quantifying, and identifying components. Spectroscopic methods like UV-visible (UV-Vis), Fourier transform infrared (FTIR), and nuclear magnetic resonance (NMR) spectroscopy provide valuable insights into the structural and compositional aspects of formulations. Microscopic and imaging techniques, including scanning electron microscopy (SEM), transmission electron microscopy (TEM), and confocal microscopy, offer detailed visualization of particle morphology and nanoscale interactions.

This chapter also explores advanced techniques, including thermal analysis methods such as thermogravimetric analysis (TGA) and differential scanning calorimetry (DSC), which are used to evaluate product stability. Additionally, microbiological techniques like enzyme-linked immunosorbent assay (ELISA) and polymerase chain reaction (PCR) ensure the microbial safety of products. Emerging trends, such as green analytical chemistry, nanotechnology, and real-time analysis tools, are transforming the field by promoting sustainability, efficiency, and innovation.

Efforts to harmonize regulatory frameworks across different regions have further underscored the importance of analytical methods. Initiatives by international organizations, such as the International Cooperation on Cosmetics Regulation (ICCR), the Organization for Economic Cooperation and Development (OECD), and the International Organization for Standardization (ISO), aim to align safety and efficacy standards, reduce trade barriers, and encourage innovation (International Cooperation on Cosmetics Regulation, 2022; Organisation for Economic Co-Operation and Development 2022; International Organization for Standardization, 2022). These collaborative endeavors highlight the need for robust analytical methods to navigate regulatory complexities while fostering market growth.

The emphasis on sustainable analytical practices aligns closely with SDG 12 (responsible consumption and production) and SDG 9 (industry, innovation, and infrastructure) (United Nations 2015). Green analytical chemistry, in particular, supports SDG 12 by minimizing the environmental impact of cosmetic production through eco-friendly solvents, reduced waste generation, and energy-efficient processes (United Nations 2015). Similarly, the integration of advanced technologies and collaborative regulatory efforts fosters innovation and infrastructure development, contributing to SDG 9 (United Nations 2015). By adopting sustainable practices and supporting global regulatory harmonization, the cosmetics industry can ensure that its products are safe, effective, and environmentally responsible, addressing consumer demand while contributing to global sustainability goals.

By exploring the key analytical techniques and their applications, this chapter provides a comprehensive overview of how these methods contribute to the advancement of natural cosmeceuticals. It bridges the gap between traditional approaches and modern innovations, emphasizing their role in ensuring product safety, authenticity, and efficacy while meeting the growing consumer demand for sustainable and high-quality cosmetics (Zakaria 2013).

12.2 Prerequisites for Cosmetics Analysis

To ensure accurate and reliable results, certain essential conditions must be met before applying analytical techniques. These include a thorough understanding of the scientific principles underlying analytical methods, knowledge of equipment requirements, familiarity with regulatory frameworks, and a comprehensive grasp of cosmetic formulations.

12.3 Understanding Cosmetic Formulations

Cosmetic products are complex mixtures of components, including active agents, excipients, preservatives, and fragrances. Natural cosmeceutical formulations may also contain bioactive ingredients such as vitamins, minerals, essential oils, and plant extracts. Analysts must understand the roles and interactions of these components to accurately select the appropriate techniques for separation and identification.

A solid understanding of formulation chemistry is essential, focusing on key elements such as:

Chemical Composition: The polarity, reactivity, and molecular structure of the components.

Functional Properties: The intended effects of the product, such as anti-aging benefits, UV protection, or moisturizing properties.

Interactions: Possible reactions between natural and synthetic substances that could affect product stability or efficacy (Sharma et al. 2023).

12.4 Regulatory Frameworks

Global and regional regulatory standards provide a framework for analytical testing in cosmetics. These regulations define permissible limits for pollutants, allergens, and active ingredients while prioritizing consumer safety. Analysts must be familiar with key regulatory guidelines, including:

European Union Cosmetics Regulation (EC 1223/2009): Establishes safety requirements for cosmetic products sold within the EU market (Rujido-Santos et al. 2022; Pistollato et al. 2021; Mishra and Kurmi 2023; Union 2009; Ferreira and Lores 2022).

US FDA Guidelines: Encompasses good manufacturing practices (GMP), safety requirements, and labeling standards for cosmetic products (U.S. Food and Drug Administration 2024; Ferreira and Lores 2022).

ISO Standards for Cosmetics: Provides internationally recognized standards for analytical testing and quality control, ensuring consistency and reliability in cosmetic testing (Mishra and Kurmi 2023; International Organization for Standardization 2022; Lores et al. 2016).

Additional analysis may be required for natural cosmeceuticals to validate claims such as "organic" or "natural." Compliance with certifications like **ECOCERT** or **COSMOS** often necessitates comprehensive analytical evaluations. These assessments ensure that product ingredients meet the stringent criteria for natural and organic certification, thereby enhancing consumer trust and supporting market claims (Mishra and Kurmi 2023).

12.5 Instrumentation and Laboratory Setup

Well-equipped laboratories with appropriate instruments are essential for accurate cosmetic analysis. All equipment must be calibrated and validated according to standard operating procedures (SOPs), a responsibility that lies with the analysts. The following tools are commonly used:

Chromatographic Systems: Thin-layer chromatography (TLC), gas chromatography (GC), and high-performance liquid chromatography (HPLC) are employed for the identification and separation of components.

Spectroscopic Instruments: Nuclear magnetic resonance (NMR), infrared (IR), and ultraviolet-visible (UV-Vis) spectroscopy are used for quantitative analysis and structural examination.

Microscopy Techniques: Methods such as confocal microscopy and scanning electron microscopy (SEM) are used to analyze particle size, shape, and surface morphology.

To ensure reproducible results, it is critical to maintain a controlled laboratory environment with effective measures for managing temperature, humidity, and contamination (Rico et al. 2023a, b; Lores et al. 2016).

12.6 Sampling and Sample Preparation

Appropriate sample preparation and representative sampling are essential prerequisites for reliable analysis. Variability in raw materials, such as plant extracts, can present challenges for the analysis of natural cosmeceuticals. To address this, the following key steps are involved:

Sampling Methods: Ensuring uniformity and preventing contamination during sample collection.

Extraction Techniques: Separating active compounds using methods such as solvent extraction, distillation, or supercritical fluid extraction (Lores et al. 2016).

Stabilization: Preventing the degradation of sensitive substances during analysis to maintain the integrity of the sample.

These steps are critical for obtaining accurate, reproducible, and meaningful analytical results.

12.7 Understanding of Ingredients

Analyzing natural cosmeceuticals requires a solid understanding of botanical and natural product chemistry. A key aspect of this process is phytochemistry, which involves identifying and characterizing the chemical composition of plant-based substances such as alkaloids, flavonoids, terpenoids, and polyphenols.

Key considerations in the analysis of natural cosmeceuticals include:

Adulteration Detection: Identifying synthetic contaminants or unauthorized additives present in natural products.

Batch Variability: Addressing differences in the quality of natural ingredients caused by factors such as cultivation practices, geographical location, and seasonal changes (Rico et al. 2023a, b; Aziz et al. 2022).

These factors are critical for ensuring the consistency, authenticity, and safety of natural cosmeceuticals.

12.8 Data Analysis and Reporting

To derive meaningful insights from analytical outcomes, analysts must possess strong skills in data interpretation. This process involves several key components:

Quantitative Analysis: Measuring the concentration of pollutants or active substances in a sample.

Qualitative Analysis: Determining the presence or absence of specific substances within a formulation.

Statistical Evaluation: Ensuring the accuracy and precision of results through methods such as regression analysis and standard deviation.

Effective communication of findings is essential for supporting regulatory compliance and guiding decisions in product development. Clear, accurate reporting ensures that stakeholders can rely on the data for quality control and safety assessments.

12.9 Appropriate Selection of Analytical Techniques for Natural Cosmeceuticals

The growing demand for safe, sustainable, and natural products has significantly driven the growth of the cosmetics and cosmeceuticals sector. These products often incorporate bioactive substances, essential oils, and plant-based extracts. Selecting appropriate analytical methods is critical to ensuring the effectiveness, safety, and quality of natural cosmeceuticals while adhering to environmentally conscious practices (Prado and Rostagno 2022).

Incorporating the principles of green chemistry is essential in this context. Analytical techniques that utilize eco-friendly solvents, minimize waste, and optimize energy efficiency are increasingly favored (Anastas & Warner 2000). For example, supercritical fluid extraction and microwave-assisted extraction are preferred for isolating bioactive compounds due to their reduced environmental impact and superior efficiency (Aili et al. 2024). Additionally, methods that prioritize biodegradability and non-toxic reagents contribute to sustainable practices, ensuring that the production and testing processes align with environmental and consumer safety standards.

The choice of analytical techniques directly influences product characterization, stability assessment, and regulatory compliance. Techniques such as high-performance liquid chromatography (HPLC) and gas chromatography (GC) provide precise separation and quantification of complex natural compounds while maintaining accuracy in identifying biodegradable and non-toxic formulations (Kiani et al. 2022). Spectroscopic methods like Fourier transform infrared (FTIR) spectroscopy and nuclear magnetic resonance (NMR) spectroscopy offer detailed insights into the structural composition of ingredients, aiding in the design of safe and sustainable cosmeceuticals (Liu et al. 2023).

Moreover, adopting non-invasive and real-time analytical tools, such as near-infrared (NIR) spectroscopy and Raman spectroscopy, further supports sustainability. These methods allow on-site quality assessment and reduce the need for extensive sample preparation, thereby lowering chemical usage and environmental impact (Evangeline et al. 2024).

The integration of green chemistry principles with advanced analytical techniques ensures that natural cosmeceuticals meet stringent safety and quality requirements while minimizing harm to the environment. By emphasizing sustainability, biodegradability, and non-toxicity, the cosmetics industry can address consumer demand for ethical and eco-friendly products, contributing to a healthier planet and promoting responsible production practices (United Nations 2015).

12.10 Importance of Analytical Techniques in Natural Cosmeceuticals

Analytical procedures are essential to ensure that cosmeceutical products meet quality standards and perform as intended (Prado and Rostagno 2022). Unlike synthetic cosmetics, natural cosmeceuticals often have complex formulations with multiple bioactive components. Their analysis requires techniques capable of detecting, measuring, and characterizing these compounds amidst a variety of natural substances (Sathyaseelan et al. 2024).

Regulatory agencies demand reliable evidence of a product's safety and efficacy, which necessitates the use of accurate and validated methods. Analytical techniques also help businesses maintain batch-to-batch consistency, a critical factor in building consumer trust (Yingling & Onel 2013; Mukherjee et al. 2015) (Fig. 12.1).

12.11 Common Analytical Techniques in Cosmeceutical Analysis

The key analytical methods in cosmeceutical analysis include chromatography for the separation and identification of components, spectroscopy for structural analysis, microscopy for visualizing particle size and morphology, mass spectrometry for detecting trace contaminants, and microbial analysis for ensuring product safety. Each technique plays a vital role in product characterization, stability assessment, and regulatory compliance, thereby supporting the development of high-quality natural cosmeceuticals. These methods are also closely linked to the sustainable development goals (SDGs), including SDG 3 (good health and well-being), SDG 9 (industry, innovation, and infrastructure), SDG 12 (responsible consumption and production), and SDG 15 (life on land), as they promote safety, innovation, sustainability, and biodiversity conservation. The details are as follows.

12.12 Chromatographic Techniques

Chromatography is a key component of analytical chemistry in cosmeceuticals, offering precise techniques for separating and identifying complex mixtures. It plays a vital role in assessing raw materials, monitoring manufacturing processes,

Fig. 12.1 Factors of analytical technique selection in natural cosmeceutical research

and ensuring that final products meet consumer and regulatory requirements. The primary chromatographic methods used in cosmeceuticals are high-performance liquid chromatography (HPLC), gas chromatography (GC), and thin-layer chromatography (TLC), each serving distinct analytical needs. These methods support the separation, identification, and quantification of active ingredients, pollutants, and preservatives, ensuring the quality, performance, and safety of natural cosmeceuticals (Rico et al. 2023a, b; Mildau 2018; Mohamed 2015; Lores et al. 2016). Furthermore, these analytical tools contribute to **SDG 12 (responsible consumption and production)** by promoting sustainable practices in the production and quality assurance of natural products.

12.13 High-Performance Liquid Chromatography (HPLC) in Cosmetics

HPLC is a versatile and widely used method for cosmeceutical analysis. It operates on the principle of partitioning components between a mobile phase (a liquid solvent) and a stationary phase (often silica-based) (Baranowska & Wojciechowska 2013). This technique is particularly effective for analyzing heat-sensitive substances, such as peptides, vitamins, and botanical extracts (Ngoc et al. 2023).

HPLC is frequently used to quantify active components in plant extracts, including alkaloids, polyphenols, and flavonoids. For instance, it is commonly employed to measure ascorbic acid (Vitamin C) in skin serums. Since environmental factors like light, heat, and oxygen can degrade cosmetic formulations, HPLC plays a vital role in stability testing. It ensures that active ingredients maintain their efficacy throughout the product's shelf life. HPLC is also crucial for detecting contaminants such as pesticide residues in plant-sourced raw materials, ensuring the safety and purity of the final product. Additionally, it supports formulation optimization by characterizing the interactions between active components and excipients, thereby guaranteeing consistent product performance (Mohamed 2015; Maia et al. 2007).

HPLC is also widely used in sunscreen analysis, where it is employed to assess the stability and effectiveness of UV-filtering agents in cosmetic formulations. This ensures the efficacy of sun-protection products, supporting regulatory compliance and consumer safety. Such innovations directly align with **SDG 3 (good health and well-being)** by ensuring safe and effective formulations that protect against harmful UV radiation.

12.14 Gas Chromatography (GC) in Cosmetics

Gas chromatography (GC) is a powerful technique used to separate and analyze volatile compounds. In this method, samples are vaporized and passed through a capillary column, where they are separated based on their boiling points and interactions with the stationary phase. GC is widely used for analyzing essential oils, fragrance components, and other volatile substances in cosmeceuticals.

The combination of GC with mass spectrometry (GC-MS) significantly enhances its ability to detect and identify unknown compounds in complex mixtures. This makes it indispensable for analyzing fragrance ingredients in creams and perfumes, where the identification and quantification of aroma compounds are critical. GC is also crucial for ensuring the consistency and regulatory compliance of essential oils, which are widely used in natural cosmetics for their therapeutic properties.

Additionally, GC plays a vital role in detecting volatile contaminants, such as residual solvents from extraction processes, which may compromise product safety. GC-MS is often used to compare the chemical profiles of natural oils and plant extracts against reference standards, helping to verify the origin, purity, and authenticity of raw materials (Sharmeen et al. 2021; Chaintreau 2007). These efforts contribute to **SDG 9 (industry, innovation, and infrastructure)** by promoting innovation in analytical techniques to ensure high standards in cosmeceutical production.

12.15 Thin-Layer Chromatography (TLC) in Cosmetics

Thin-layer chromatography (TLC) is a simple, cost-effective method used primarily for the preliminary screening of bioactive compounds in cosmeceuticals. In this method, a sample is applied to a stationary phase, which consists of a thin layer of silica or alumina coated on a plate made of glass or plastic. The sample components are separated as they migrate with the mobile phase, allowing for visual identification.

TLC is frequently used for the rapid screening of phytochemicals, such as alkaloids, flavonoids, and terpenoids. Although it is less sensitive than HPLC or GC, it remains a valuable tool for routine quality control, especially in the early stages of analysis. TLC is instrumental in ensuring the uniformity of raw materials by comparing the chromatographic profiles of different batches. This method is also essential for detecting synthetic adulterants in natural cosmeceuticals, as it can distinguish components based on their polarity.

Furthermore, when combined with bioautography, TLC becomes a powerful tool for evaluating the antimicrobial properties of cosmetic compounds. This approach is particularly useful for assessing the bioactivity of plant extracts used in natural formulations (Kowalska and Sajewicz 2022). This capability supports **SDG 12 (responsible consumption and production)** by encouraging the use of natural, sustainable bioactive ingredients in cosmetic formulations.

12.16 Integration of Chromatographic Techniques

The integration of chromatographic techniques enhances analytical efficiency and accuracy in modern cosmeceutical analysis. For instance, TLC can be used as a preliminary screening method to identify potential bioactive compounds, while HPLC or GC can provide more detailed quantification and validation. This integrated approach ensures a comprehensive examination of raw materials, intermediate formulations, and final products. By focusing on plant-based bioactive compounds, these methods contribute to SDG 15 (Life on Land) by supporting the sustainable utilization of biodiversity resources. Ensuring the traceability and authenticity of natural ingredients helps promote conservation and sustainable sourcing of raw materials, reducing the overharvesting of endangered plant species and promoting sustainable agricultural practices.

By facilitating the detection of prohibited substances, supporting accurate labeling of active ingredients, and verifying product claims, chromatographic methods play a vital role in regulatory compliance. Regulatory bodies, such as the FDA and the European Commission, rely on these methods to approve and oversee cosmeceutical products (European Commission 2009; U.S. Food and Drug Administration 2024; Ferreira and Lores 2022). By ensuring compliance with international standards and promoting sustainable practices, these methods contribute to achieving SDG 12 (responsible consumption and production), SDG 9 (industry, innovation, and infrastructure), and SDG 15, advancing responsible production, fostering innovation, and supporting environmental conservation.

12.17 Advancements in Chromatographic Techniques

The future of chromatographic analysis in cosmeceuticals is being shaped by advancements such as ultra-performance liquid chromatography (UPLC), multidimensional chromatography, and hyphenated methods like GC-MS and LC-MS. These technologies offer faster, more accurate, and environmentally friendly approaches to analyzing natural cosmeceuticals.

Ultra-Performance Liquid Chromatography (UPLC): This technique enhances the speed and sensitivity of HPLC, allowing for faster separation of bioactive components.

Multidimensional Chromatography: This method combines two or more chromatography systems (like HPLC-HPLC or GC-GC) to improve the separation of complex mixtures.

Hyphenated Techniques: Techniques like GC-MS and LC-MS provide enhanced analytical capabilities, allowing for the detection and identification of trace contaminants in complex cosmeceutical matrices.

These developments provide natural cosmeceutical analysis techniques that are faster, more precise, and eco-friendly. As consumer demand for sustainable and natural cosmetics continues to grow, chromatographic techniques will play an increasingly vital role in ensuring product safety, quality, and transparency. The continued development of these techniques will drive innovation, enhance regulatory compliance, and build consumer trust in natural cosmeceuticals.

12.18 Mass Spectrometry (MS) in Cosmeceuticals

Mass spectrometry (MS) is a highly effective technique for identifying and quantifying chemical components in cosmeceuticals. When combined with chromatographic methods such as gas chromatography-mass spectrometry (GC-MS) and liquid chromatography-mass spectrometry (LC-MS), it enables precise analysis of complex mixtures. This combination allows for the separation, detection, and identification of multiple compounds within a single analysis.

MS is particularly useful for the detection of trace contaminants, such as pesticide residues and heavy metals, which may be present in raw materials or final products. Its high sensitivity and accuracy make it an indispensable tool for ensuring product safety, maintaining quality control, and meeting regulatory requirements (Rico et al. 2023a, b). Furthermore, by safeguarding consumer health and ensuring that cosmeceuticals are free from harmful substances, MS supports **SDG 3 (good health and well-being)**. Additionally, its role in quality control and minimizing environmental risks associated with contaminants aligns with **SDG 12 (responsible consumption and production)**. By enabling the sustainable and safe use of natural resources, it also contributes to **SDG 15 (life on land)**.

12.19 Spectroscopic Techniques in Cosmeceuticals

Spectroscopy provides non-destructive methods for the qualitative and quantitative analysis of cosmeceuticals. It enables the identification of functional groups, structural characterization, and determination of ingredient concentrations. The most widely used spectroscopic techniques in cosmeceuticals include UV-visible (UV-Vis) spectroscopy, Fourier transform infrared (FTIR) spectroscopy, and nuclear magnetic resonance (NMR) spectroscopy. Each of these techniques plays a vital role in quality control, stability testing, and regulatory compliance.

12.20 UV-Visible (UV-Vis) Spectroscopy in Cosmeceuticals

UV-Vis spectroscopy is one of the most commonly used techniques for the quantitative analysis of pigments, dyes, and sunscreens in cosmeceutical products. It measures light absorption in the ultraviolet (190–400 nm) and visible (400–800 nm) regions of the electromagnetic spectrum. This technique works by detecting the absorption of specific wavelengths of light by molecules, corresponding to electronic transitions within the molecules.

UV-Vis spectroscopy is widely used to quantify the concentration of active ingredients, such as synthetic colorants and natural pigments derived from plants (e.g., carotenoids, flavonoids, and anthocyanins). These pigments contribute to the visual appeal of cosmetics and are associated with antioxidant properties. UV-Vis spectroscopy plays a crucial role in ensuring regulatory compliance and consistent product quality.

This method is especially valuable in assessing the effectiveness of sunscreen formulations, which rely on ingredients like zinc oxide, titanium dioxide, and organic filters (e.g., oxybenzone and avobenzone) to protect the skin from UV radiation. UV-Vis spectroscopy measures the absorption of UV light to determine the sun-protection factor (SPF) of sunscreens, helping formulators maintain regulatory compliance and optimize product performance (Ambruso et al. 2023; Guerra et al. 2018; Butnariu 2016).

Additionally, UV-Vis spectroscopy is useful in tracking the stability of cosmeceutical products. By analyzing changes in the absorbance profile of a product over time, researchers can identify degradation or chemical interactions, ensuring the stability of active ingredients throughout the product's shelf life.

12.21 Fourier Transform Infrared (FTIR) Spectroscopy in Cosmeceuticals

FTIR spectroscopy is an essential tool for identifying chemical bonds and functional groups in cosmeceuticals. This technique works by measuring the absorption of infrared light, which causes molecular vibrations unique to specific functional groups. As a result, FTIR generates a "chemical fingerprint" of the sample, making it useful for characterizing raw materials, finished products, and potential contaminants.

FTIR spectroscopy is widely used in the analysis of oils, waxes, and essential oils frequently found in natural cosmeceuticals. By comparing the FTIR spectra of different samples, manufacturers can verify the authenticity of raw materials and detect possible adulteration. This is crucial for natural products, where synthetic adulterants may be present.

Another important application of FTIR in cosmeceuticals is the analysis of emulsions, which are the basis for many creams and lotions. FTIR can track changes in the interactions between water, oil, and surfactants, providing valuable insights into the stability of emulsions. This is particularly important for natural formulations that use plant-based emulsifiers, as the quality of these emulsifiers can vary significantly.

FTIR also facilitates the analysis of packaging materials to ensure compatibility with cosmetic formulations. It can detect the leaching of packaging materials into the product, reducing the risk of contamination. The development of attenuated total reflectance (ATR) technology has further improved FTIR's sensitivity and resolution, enabling more precise analysis of raw materials, active ingredients, and packaging materials (Ribechini et al. 2011; Pasieczna-Patkowska and Olejnik 2013; Angrish et al. 2020).

12.22 Nuclear Magnetic Resonance (NMR) Spectroscopy in Cosmeceuticals

Nuclear magnetic resonance (NMR) spectroscopy is an advanced analytical method used to investigate the molecular structure and dynamics of chemical compounds. NMR relies on the magnetic properties of atomic nuclei, particularly hydrogen (^{1}H) and carbon (^{13}C), to generate detailed information on the structure of organic molecules.

In the cosmeceutical industry, NMR spectroscopy is widely used to characterize bioactive compounds in plant extracts, such as phenolics, terpenoids, and alkaloids. These bioactives contribute to the anti-aging, anti-inflammatory, and skin-brightening properties of cosmeceuticals. NMR helps identify and quantify these compounds, ensuring uniformity and effectiveness in product formulations. By supporting the development of safe and effective products that promote skin health, NMR aligns with **SDG 3 (good health and well-being)**.

NMR is also essential for analyzing lipids, which are common ingredients in natural cosmetics. Vegetable oils, butters, and waxes are frequently used as moisturizers and emollients. NMR spectroscopy can analyze the degree of unsaturation and oxidative stability of these lipids, ensuring they meet the required standards for cosmetic formulations. This is particularly crucial for cold-pressed oils, which are prized in natural cosmeceuticals but are more susceptible to degradation. By safeguarding the quality of plant-based ingredients, NMR contributes to **SDG 12 (responsible consumption and production)** and **SDG 15 (life on land)**, promoting sustainable sourcing and use of biodiversity resources.

Another key application of NMR is the analysis of complex formulations containing bioactives, surfactants, and preservatives. NMR can detect interactions among these components, helping formulators identify potential issues that may affect stability or product performance. Moreover, NMR provides unmatched precision in detecting impurities and confirming the authenticity of natural components. For instance, NMR can be used to detect synthetic additives or identify the adulteration of essential oils with cheaper alternatives. These capabilities are vital for maintaining consumer trust and meeting regulatory requirements, further supporting **SDG 9 (industry, innovation, and infrastructure)** by fostering technological advancements in the cosmeceutical industry.

Advanced NMR techniques, such as high-resolution magic angle spinning (HRMAS) and two-dimensional (2D) NMR, allow for the analysis of solid samples and entire formulations (Ribechini et al. 2011). These innovations not only enhance product quality but also ensure sustainable and ethical practices throughout the product life cycle, aligning with global efforts to achieve the SDGs.

12.23 Integrating Spectroscopic Methods in Cosmeceutical Development

The development of natural cosmeceuticals presents unique challenges due to the variability of plant-based raw materials, sensitivity to environmental factors, and the absence of synthetic stabilizers commonly used in conventional cosmetics. Spectroscopic techniques, such as UV-visible (UV-Vis) spectroscopy, Fourier transform infrared (FTIR) spectroscopy, and nuclear magnetic resonance (NMR) spectroscopy, are indispensable tools for addressing these complexities. These methods provide critical insights into the structural and functional properties of cosmetic formulations, ensuring their quality, stability, and efficacy (Evangeline et al. 2024).

For example, **UV-Vis spectroscopy** is widely applied in the quantification of active ingredients such as flavonoids, anthocyanins, and carotenoids, which are commonly used as natural pigments or antioxidants in cosmetic formulations. It also plays a crucial role in assessing the efficacy of sunscreen agents like titanium dioxide and zinc oxide by evaluating their ultraviolet absorption profiles (Rico et al. 2023a, b). On the other hand, **FTIR spectroscopy** enables the identification of functional groups in raw materials, such as ester bonds in plant oils or hydroxyl groups in humectants, making it essential for verifying the purity and authenticity of natural ingredients (Al-Harrasi et al. 2022). Additionally, FTIR is highly effective in monitoring the stability of emulsions and detecting degradation products in cosmetic formulations (Rico et al. 2023a, b).

NMR spectroscopy further complements these techniques by providing detailed structural information on complex bioactive compounds, such as terpenoids, alkaloids, and phenolic acids, which are often used for their anti-aging, anti-inflammatory, and skin-brightening properties (Rico et al. 2023a, b). For instance, NMR can be employed to analyze the degree of unsaturation in natural oils or detect the adulteration of essential oils with synthetic substitutes. When combined with chemometric analysis, NMR aids in interpreting large datasets, enabling the prediction of bioactive efficacy and interactions within multi-component formulations (Al-Harrasi et al. 2022).

Spectroscopic techniques are not only instrumental in quality control and formulation optimization but also play a pivotal role in regulatory compliance. Regulatory agencies require robust evidence of product safety and efficacy, especially for natural cosmeceuticals marketed with claims of being organic or free from harmful chemicals (European Commission 2009). For instance, compliance with frameworks such as the European Union Cosmetics Regulation (EC 1223/2009) often necessitates the use of validated spectroscopic methods to detect contaminants, quantify active ingredients, and ensure labeling accuracy (Rico et al. 2023a, b).

By providing precise and reliable data, these techniques enable manufacturers to verify the authenticity and quality of their products, ensuring they meet both consumer expectations and regulatory standards. Spectroscopic methods thus not only enhance the scientific rigor of cosmeceutical development but also foster consumer trust by supporting claims of naturality, sustainability, and efficacy.

12.24 Advancements in Spectroscopic Techniques

While UV-Vis, FTIR, and NMR spectroscopy have revolutionized cosmeceutical analysis, future developments are expected to expand their potential. Emerging technologies, such as portable NMR devices, miniaturized spectrometers, and machine learning algorithms, will increase the accessibility and efficiency of these methods. Real time on-site analysis of raw materials and final products will reduce reliance on centralized laboratories, accelerating the formulation process.

However, there are also challenges associated with the use of spectroscopic techniques in cosmeceuticals. Due to the complexity of natural formulations, data interpretation can be difficult, and small-scale firms may not be able to afford the sophisticated equipment required, such as NMR spectrometers. Collaboration among manufacturers, regulatory agencies, and researchers is essential to overcome these obstacles. Standardizing analytical procedures, offering affordable equipment, and providing training in spectroscopic techniques are key to supporting the broader adoption of these methods.

Spectroscopic techniques, such as UV-Vis, FTIR, and NMR, are essential for the analysis of natural cosmeceuticals. These methods enable the characterization of bioactive components, optimize formulations, and ensure the safety and quality of products. As the demand for sustainable and natural cosmetics continues to rise, spectroscopic techniques will play an increasingly critical role in the development of innovative and consumer-friendly cosmeceuticals.

12.25 Microscopic and Imaging Techniques in Cosmeceuticals

The rapid growth of the natural cosmeceutical industry has driven the need for advanced analytical techniques to ensure product effectiveness, safety, and quality. Microscopic and imaging techniques are essential for tracking product stability, understanding the physicochemical properties of raw materials, and ensuring compliance with regulatory standards. These methods offer detailed visual and structural analysis of cosmeceutical formulations, allowing researchers to examine surface morphology, particle size, and the distribution of active ingredients. Key imaging and microscopic techniques used in cosmeceuticals include scanning electron microscopy (SEM), transmission electron microscopy (TEM), and confocal microscopy, each serving a specific analytical purpose.

12.26 Scanning Electron Microscopy (SEM) in Natural Cosmeceuticals

Scanning electron microscopy (SEM) is a widely used imaging method for examining surface morphology and nanoscale structures. By directing a focused electron beam onto the surface of a sample, SEM produces secondary and backscattered electrons, which are captured to create high-resolution images. This technique provides detailed information on the texture, composition, and structure of samples.

In cosmeceuticals, SEM is commonly used to visualize the surface morphology of plant-based extracts, microcapsules, and nanoparticles used in formulations. For instance, SEM allows for the examination of encapsulated active ingredients, such as vitamins or essential oils, to ensure proper encapsulation and structural integrity. It is also employed to analyze the dispersion of pigments in natural colorants, ensuring uniform distribution.

Another significant application of SEM is in the analysis of the effects of cosmeceutical products on the skin. SEM images of treated skin samples can reveal changes in surface moisture, pore size, or texture, providing insight into the product's effectiveness. This is particularly relevant for products targeting skin hydration, anti-aging, or texture enhancement. Moreover, SEM is a valuable tool for quality control, as it can detect impurities, contamination, or biofilms that may compromise product safety and stability (Belostozky et al. 2019; Rico et al. 2023a, b).

12.27 Transmission Electron Microscopy (TEM) for Nanostructural Analysis

Transmission electron microscopy (TEM) is a powerful imaging technique that provides atomic-level insights into the internal structure of materials. Unlike SEM, which focuses on surface imaging, TEM allows for the visualization of internal structures by passing electrons through an ultra-thin sample.

TEM is widely used in cosmeceutical research to characterize nanoparticles, liposomes, and nanofibers. These nanostructures are often used as delivery systems to enhance the penetration and bioavailability of active ingredients. TEM enables the visualization of particle size, shape, and homogeneity, which are crucial for ensuring the stability and effectiveness of these delivery systems.

Another application of TEM is in understanding how natural bioactive compounds interact with skin cells. For instance, it can visualize the uptake of nanoparticles by skin cells, providing insight into how bioactive ingredients penetrate the skin. This capability is particularly valuable for products with cellular-level claims, such as anti-aging or regenerative treatments. Additionally, TEM plays a vital role in characterizing liposomal delivery systems, which are used to encapsulate and protect active ingredients (Ribechini et al. 2011; Shanbhag et al. 2019).

12.28 Confocal Microscopy for 3D Imaging

Confocal microscopy is a non-invasive imaging technique that provides high resolution, three-dimensional images of biological and non-biological materials. By using a pinhole to eliminate out-of-focus light and illuminating samples with lasers, confocal microscopy generates clear and detailed images.

In cosmeceutical analysis, confocal microscopy is widely used to assess the penetration of active ingredients into the skin. By labeling active components with fluorescent markers, researchers can track the distribution and depth of penetration within skin layers. This is particularly relevant for anti-aging creams, moisturizers, and serums designed to reach the dermal or subdermal layers.

Confocal microscopy also plays a crucial role in visualizing the interaction of cosmeceutical ingredients with skin cells. For instance, it can provide evidence of how moisturizers are deposited on the skin surface or how emollients form protective coatings. Additionally, it is employed to assess the stability and uniformity of formulations, such as emulsions, by analyzing the homogeneity of active ingredients within the formulation. This method is widely used to ensure the consistent distribution of encapsulated bioactive compounds, thereby supporting the production of high-quality cosmeceutical products (Krombholz and Lunter 2020; Förster et al. 2011).

12.29 Comparative Applications and Advantages of Microscopic Techniques

Although SEM, TEM, and confocal microscopy operate on different principles, they often work in tandem to provide a comprehensive analysis of cosmeceuticals. SEM is best suited for surface morphology studies, while TEM offers unmatched resolution for internal structure analysis at the atomic level. Confocal microscopy, on the other hand, provides non-invasive, three-dimensional imaging, making it ideal for studying biological interactions and penetration of active ingredients into the skin.

Each method has distinct advantages:

SEM is frequently used for routine analysis because of its simplicity and adaptability.

TEM provides atomic-level insights into nanoparticles, offering unmatched resolution for nanostructural analysis.

Confocal Microscopy is particularly valuable for studying biological interactions, as it allows the visualization of dynamic processes in real time.

The selection of the appropriate technique depends on the specific analytical objective. For instance, confocal microscopy is preferred for penetration studies, while TEM may be required for structural analysis when characterizing a novel nanoparticle-based delivery system. Integrating these techniques enables a more comprehensive understanding of cosmeceutical formulations, enhancing both development and quality assurance.

12.30 Advancing Quality and Innovation of Microscopy in Natural Cosmeceuticals

Microscopic and imaging techniques are at the forefront of cosmeceutical research, playing a pivotal role in enhancing the quality, safety, and efficacy of products. These methods enable researchers to validate delivery systems, optimize formulations, and explore innovative materials. As the demand for natural cosmeceuticals increases, the ability to scientifically validate claims related to sustainability, efficacy, and product safety becomes more critical.

Confocal microscopy, TEM, and SEM are not only analytical tools but also drivers of creativity and innovation. They allow manufacturers to visualize product interactions, such as ingredient penetration into the skin, or track the structural integrity of encapsulated bioactives. The insights provided by these imaging techniques support the development of high-quality cosmeceutical products that meet the needs of well-informed consumers.

By employing these techniques, researchers and manufacturers can substantiate marketing claims, such as "deep penetration" or "nanotechnology-enhanced absorption." This scientific support boosts consumer confidence and strengthens regulatory compliance. As technological advancements continue, improvements such as high-resolution imaging, faster processing, and the miniaturization of microscopy equipment are likely to expand the accessibility of these techniques, even for smaller-scale manufacturers.

Microscopic and imaging techniques, such as **SEM**, **TEM**, and **confocal microscopy**, play an essential role in the analysis and development of cosmeceuticals. These methods allow for the visualization of surface morphology, internal nanostructures, and three-dimensional interactions between active ingredients and skin cells. By integrating these techniques, researchers can comprehensively analyze cosmeceutical formulations, ensuring product safety, efficacy, and quality. SEM is widely used for surface analysis, TEM provides insight into internal nanostructures, and confocal microscopy tracks the penetration and distribution of active ingredients within the skin.

The role of microscopic techniques extends beyond analysis, as they also drive innovation in delivery systems and formulation design. With the growing demand for sustainable and natural cosmeceuticals, these techniques are vital for validating product claims and ensuring regulatory compliance. As new imaging technologies emerge, the capacity for real-time analysis, higher resolution, and increased accessibility will support further advancements in the field of natural cosmeceuticals.

12.31 Thermal Analysis in Cosmeceuticals

Thermal analysis techniques are essential for evaluating the stability and compatibility of components in cosmeceuticals. The most commonly used methods are **thermogravimetric analysis (TGA)** and **differential scanning calorimetry (DSC)**, which provide valuable insights into the thermal behavior of raw materials, intermediates, and final products.

Thermogravimetric analysis (TGA) measures changes in the mass of a sample as it is heated, providing data on thermal degradation, moisture content, and volatile components. This technique is essential for assessing the stability of natural ingredients and identifying decomposition points that may affect the shelf life of cosmeceutical products.

Differential scanning calorimetry (DSC) determines the heat flow associated with thermal transitions, such as melting, crystallization, and glass transition temperatures. It is used to evaluate the compatibility of excipients with active ingredients and identify changes in crystalline structure. This is particularly useful for creams, gels, and emulsions, where the stability of emulsifiers, surfactants, and bioactives can affect product consistency and performance (Bezerra et al. 2017).

Both TGA and DSC are indispensable tools for ensuring the thermal stability of cosmeceuticals, supporting product development, storage condition optimization, and shelf-life prediction.

12.32 Examination of Microbes in Cosmeceuticals

Ensuring microbiological safety is crucial for cosmeceutical products, as microbial contamination can compromise product quality, stability, and consumer safety. To prevent contamination, manufacturers use a range of microbiological examination techniques, including **enzyme-linked immunosorbent assays (ELISA)**, **polymerase chain reaction (PCR)**, and **plate count methods**.

ELISA is a rapid and specific method for detecting microbial antigens in cosmetic products. It is widely used to identify specific bacteria, fungi, and yeast in raw materials and finished products.

Polymerase Chain Reaction (PCR) is a highly sensitive technique that amplifies specific DNA sequences to detect microbial contamination at extremely low levels. PCR is essential for the rapid identification of pathogens in probiotic formulations, creams, and lotions.

Plate Count Methods are conventional techniques that involve culturing microorganisms on growth media to quantify microbial contamination. This method is still widely used for routine quality control due to its simplicity and cost-effectiveness.

Additionally, **next-generation sequencing (NGS)** is an emerging technique for analyzing the microbiome of probiotic-based cosmeceuticals. NGS provides detailed insights into the microbial community within products, ensuring product consistency, efficacy, and compliance with regulatory requirements (Tuzimski and Petruczynik 2020; Rohman et al. 2020; Drouet et al. 2018).

By employing these microbial testing methods, manufacturers can ensure that cosmeceutical products meet stringent microbial limits, providing consumers with safe, effective, and contamination-free products.

12.33 Emerging Trends in Analytical Techniques for Cosmeceuticals

As the cosmeceutical industry evolves, so do the analytical techniques used to ensure product quality, efficacy, and regulatory compliance. Emerging trends such as **green analytical chemistry**, **high-throughput screening**, **omics technologies**, and **real-time analysis** are transforming the way cosmeceuticals are tested and analyzed. These innovative approaches aim to increase efficiency, reduce environmental impact, and support personalized cosmetic solutions.

1. **Green Analytical Chemistry**: Green analytical chemistry emphasizes sustainability by reducing the use of hazardous chemicals, minimizing waste, and promoting energy-efficient instrumentation. The cosmetics industry is increasingly adopting these practices to align with global sustainability goals. Analytical techniques that use smaller sample sizes, safer solvents, and energy-efficient equipment are now preferred. For example, miniaturized chromatographic systems and solvent-free extraction methods have become more common (Mohamed 2015). By adopting green analytical methods, cosmeceutical manufacturers can reduce their environmental impact and support eco-friendly product claims.
2. **High-Throughput Screening**: High-throughput screening (HTS) techniques enable the rapid analysis of multiple samples, significantly increasing efficiency in quality control and research. Automation systems, robotic chromatography, and microplate-based assays are now being used to handle large sample volumes simultaneously. HTS methods reduce analysis time and cost, accelerating the formulation development process. This is especially useful in stability studies and the rapid evaluation of raw material variability (Georgousaki et al. 2015). The ability to screen large datasets in a short time frame enhances decision-making and speeds up product development.
3. **Omics Technologies**: The incorporation of **proteomics**, **metabolomics**, and **genomics** into cosmeceutical analysis is opening new possibilities for personalized cosmetic products. **Proteomics** studies proteins to understand the impact of bioactive ingredients on skin structure and function, while **metabolomics** analyzes small molecules to monitor product stability and identify potential degradation products. **Genomics** provides insights into how cosmetic ingredients

affect gene expression, enabling the development of personalized skincare products based on individual genetic profiles (Gaur et al. 2023). Omics technologies help manufacturers design customized cosmetics, such as products tailored to a specific skin type or genetic profile, enhancing consumer satisfaction.

4. **Real-Time Analysis**: Real-time analytical techniques enable on-site quality evaluation and process monitoring during manufacturing. Portable devices such as **near-infrared (NIR) spectroscopy** and **Raman spectroscopy** are increasingly being used to perform real time, non-destructive analysis of raw materials and finished products. These methods help manufacturers maintain batch-to-batch consistency and reduce production downtime (Blanco et al. 2007). On-site analysis allows manufacturers to address quality issues in real time, reducing the risk of product recalls and ensuring faster production cycles.

12.34 Advancing Quality and Innovation in Thermal, Microbial, and Emerging Approaches in Natural Cosmeceuticals

The use of cutting-edge analytical techniques is essential for ensuring the safety, efficacy, and regulatory compliance of natural cosmeceuticals. **Thermal analysis**, **microbial testing**, and **emerging trends** such as green analytical chemistry, high-throughput screening, and omics technologies are transforming the industry. While traditional techniques such as spectroscopy and chromatography remain essential, innovative technologies are reshaping the way manufacturers approach cosmeceutical development.

These emerging techniques offer numerous benefits:

Sustainability: Green analytical methods reduce the environmental impact of cosmeceutical production.

Efficiency: High-throughput screening and real-time analysis reduce the time and cost of testing.

Customization: Omics technologies support the development of personalized skincare solutions.

Safety: Microbial testing ensures that products meet stringent microbiological safety requirements.

By leveraging these analytical innovations, cosmeceutical manufacturers can enhance product quality, ensure regulatory compliance, and meet consumer demand for sustainable, effective, and personalized products.

Thermal analysis, microbial examination, and emerging analytical techniques are revolutionizing the field of cosmeceuticals. **TGA** and **DSC** play vital roles in ensuring the stability and compatibility of components, while **microbial testing** using ELISA, PCR, and NGS guarantees the microbiological safety of products. Emerging trends, such as **green analytical chemistry**, **high-throughput**

screening, **omics technologies**, and **real-time analysis**, are transforming how cosmeceuticals are developed, produced, and tested.

These advancements not only improve analytical efficiency but also promote sustainability and personalization. By adopting modern techniques, manufacturers can enhance product quality, optimize production, and meet the growing demand for eco-friendly and personalized cosmetics. As regulatory scrutiny increases and consumer expectations evolve, these advanced analytical methods will remain crucial to maintaining trust, ensuring safety, and driving innovation in the cosmeceutical industry (Table 12.1).

12.35 Quantitative Bioassays and In Vitro Testing

12.35.1 Immunoassays

Immunoassays can be performed in two primary configurations:

1. Non-competitive (sandwich) configuration: This method is ideal for detecting macromolecules such as pathogens, bacteria, and allergenic proteins.
2. Competitive configuration: This approach is typically used for identifying small molecules, such as antibiotics, pesticides, and mycotoxins.

12.35.2 Detection Principles

Immunoassays rely on various detection principles, particularly optical and electrochemical methods, depending on the labels used in the assay. Common labels for probe assembly through conjugation include enzymes, colloidal gold, fluorescent nanoparticles, and antibodies/antigens. The intensity of the signal generated by these labels corresponds to the quantity of target molecules present. This interaction between the probes and the target molecules in the sample provides quantitative results for the analysis (Fig. 12.2).

12.35.3 Enzyme-Linked Immunosorbent Assays (ELISA)

Methods utilizing immunoassays for general applications in clinical-diagnostic analyses are categorized under ELISA (enzyme-linked immunosorbent assay). In these methods, enzymes such as alkaline phosphatase (ALP), horseradish peroxidase (HRP), or glucose oxidase (GOD) are commonly used as tags. These enzymes catalyze reactions with specific substrates, generating a detectable signal.

Table 12.1 Advantages and limitations of analytical techniques in cosmeceutical analysis

Technique	Advantages	Limitations	References
Chromatography			
HPLC	High sensitivity and specificity for identifying and measuring active substances like vitamins, antioxidants, and preservatives Suitable for thermally labile, non-volatile chemicals Suitable for a variety of samples, including hydrophilic and hydrophobic ones	Requires sample and solvent preparation, which might take a lot of time Equipment is costly and needs operators with specialized knowledge Derivatization may be necessary for some samples in order to increase detectability	Rico et al. (2023a, b), Jones (2024), Dong (2013), Tripathi (2023)
Gas chromatography (GC)	High resolution for volatile substances like perfumes and essential oils Mass spectrometry (GC-MS) compatibility for detecting unknown compounds Quick examination of volatile substances	Incompatible with thermally unstable or non-volatile substances For some analytes, significant sample preparation and derivatization are necessary Limited use in aqueous matrices that don't require extraction	Rico et al. (2023a, b), Tranchida et al. (2017), Zeng (2022), Sharmeen et al. (2021), Chaintreau (2007)
Thin-layer chromatography (TLC)	A quick, easy, and affordable screening tool Perfect for qualitative examination of components such as small molecule actives and dyes Minimal equipment is needed	Reduced sensitivity and resolution in contrast to GC or HPLC Limited ability to quantify	Radoičić et al. (2024), Kowalska and Sajewicz (2022)
Spectroscopic techniques			
Ultraviolet-visible (UV-vis) spectroscopy	Helpful for quantitative evaluation of sunscreens, and pigments Quick and simple to use, requiring little sample preparation Economical and non-destructive	*Only substances having chromophores (groups that absorb light) are allowed* *Inadequate selectivity when examining complex compositions* *Absorption spectra that overlap and cause interference*	Rico et al. (2023a, b), Ambruso et al. (2023), Rajasekar et al. (2024), Erbas and Soylak (2022), Li et al. (2019)

(continued)

Table 12.1 (continued)

Technique	Advantages	Limitations	References
Infrared (IR) spectroscopy	Good at spotting functional groupings in both raw materials and final goods Non-destructive and little sample preparation is needed Helpful in assessing the substances' purity	*Components in complicated combinations are challenging to measure* *Low sensitivity in contrast to other techniques* *Requires spectral interpretation expertise*	Agatonovic-Kustrin et al. (2020), Bonnier et al. (2020), Sharma et al. (2021), Bruce et al. (2024)
Nuclear magnetic resonance (NMR) spectroscopy	Offers comprehensive structural details regarding organic substances Sample preparation is needed	High maintenance and instrumentation costs It shows a greater likelihood of overlapping peaks and low resolution It takes experience to interpret spectra	Ribechini et al. (2011), Emwas et al. (2019), Bifulco et al. (2007), Monakhova and Diehl (2016)
Microscopic and imaging techniques			
Scanning electron microscopy (SEM)	High-resolution imaging of formulations' particle sizes and surface morphology Ideal for researching powders, emulsions, and coatings	For non-conductive samples, conductive materials must be applied to the sample Time-consuming and costly	Klang et al. (2013), Lorenz et al. (2010), Belostozky et al. (2019), Goldstein et al. (2014), Sun et al. (2022)
Transmission electron microscopy (TEM)	Offers extremely high-resolution pictures for examining objects at the nanoscale, such as liposomes and nanoparticles Perfect for emulsion and nano-emulsion structural analysis	Requires extremely thin samples, which might be difficult to produce Operation and interpretation are expensive and demand a high level of knowledge	Sun et al. (2022), Hiba (2023), Wang et al. (2023), Peddie et al. (2022), Rizvi et al. (2021)
Confocal microscopy	Makes it possible to visualize cosmetic formulas and how they interact with the layers of the skin in three dimensions Non-invasive and appropriate for research on in vitro models or living tissues Can be used in conjunction with fluorescence to monitor the distribution of active ingredients	Reduced resolution in contrast to electron microscopy Expensive maintenance and equipment	Ribechini et al. (2011), Förster et al. (2011), Jonkman et al. (2020), Kubota et al. (2021), Roy et al. (2013)

(continued)

Fig. 12.2 Comprehensive safety testing parameters for cosmetic products (Zorić et al. 2022)

Depending on the type of enzymatic substrate used, different detection principles can be employed, including colorimetric, chemiluminescent, or electrochemiluminescent detection. The CL-ELISA method, in particular, enables specific, rapid, semi-quantitative, and quantitative detection of Chloramphenicol (CAP) in cosmetics. This approach contributes to quality control and regulatory compliance by ensuring the accurate monitoring of CAP in cosmetic formulations (Fig. 12.3).

12.36 Titration Methods

Titration is one of the quantitative chemical analytical techniques used in determining the concentration of a solute in a solution. Applying titration can be made in the analysis of natural extracts or formulations to establish the potency of active ingredients using bioassays. Researchers can determine the amount of an active component through a reaction endpoint by adding titrant stepwise to a sample. This, therefore, allows accurate quantification of the active component, essential for product efficacy and safety.

Titration is an important analytical technique in the cosmetics industry to determine the concentration of different components in cosmetic formulations. It can be

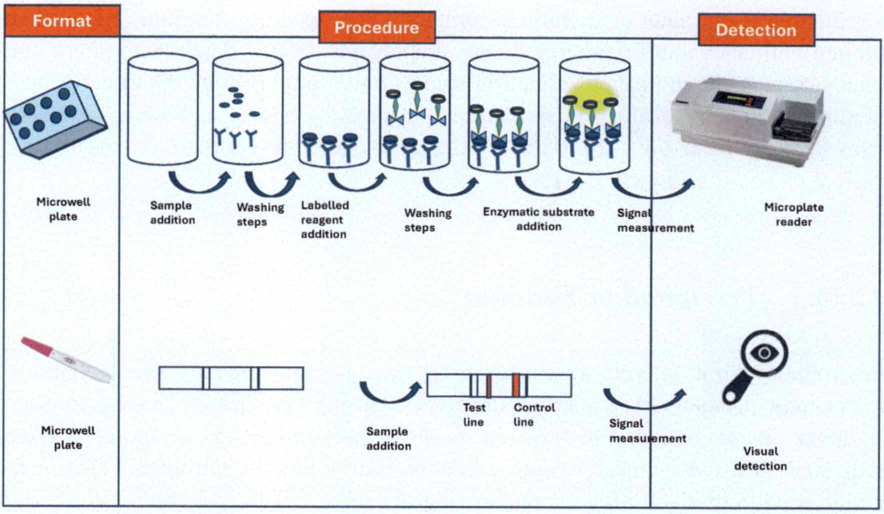

Fig. 12.3 Scheme of the major differences among the traditional immunoassay format approaches based on bench-top instrumentation used in an equipped laboratory setting (top line) and a rapid and portable platform exploiting the Lateral flow immunoassay (LFIA) principle for on-site applications (bottom line) (Zangheri et al. 2021)

particularly useful in the determination of active ingredients, including UV filters, preservatives, and other chemical compounds. Some of the most commonly used titration methods relevant to cosmetics analysis are given below: (Rico et al. 2023a, b).

12.36.1 Acid-Base Titration

Acid-base titration is a fundamental method used to measure the concentration of acidic or basic components in cosmetic formulations, such as creams, lotions, and shampoos. This method involves reacting an acid with a base (or vice versa) to identify the equivalence point, typically using pH indicators like phenolphthalein or bromothymol blue. Acid-base titrations are crucial for determining product pH, ensuring safety, and skin compatibility.

12.36.2 Redox Titration

Redox titration is a commonly used method for determining metal oxides in cosmetics, such as titanium dioxide (TiO_2), which is widely incorporated into various sunscreen products. In this method, titanium (IV) is reduced to titanium (III) using a reducing agent, such as metallic aluminum. The resulting titanium (III) is then titrated with a standard oxidizing agent, such as iron (III) sulfate, using potassium thiocyanate as the indicator. Validation studies have demonstrated the effectiveness of this method by evaluating numerous sunscreen products, achieving recovery rates ranging from 96% to 105% relative to inductively coupled plasma atomic emission spectroscopy (ICP-AES).

12.36.3 Precipitation Titration

Precipitation titration is commonly applied to quantify specific ions, such as halides, in cosmetic products like cleansers or serums. During this titration, a reagent, such as silver nitrate, is added to form an insoluble precipitate. The endpoint is often indicated by a color change using indicators like potassium chromate. This technique is particularly useful for monitoring the ionic balance in formulations and ensuring compliance with quality standards.

12.36.4 Complexometric Titration

Complexometric titration is an advanced method used to measure metal ions, such as calcium and magnesium, present in cosmetics like mineral powders or skincare products. A chelating agent, such as Ethylenediamine tetraacetic acid (EDTA), is used as the titrant, forming stable complexes with metal ions. Indicators like Eriochrome Black T are employed to detect the endpoint. This method helps in monitoring trace metal concentrations, which is essential for product stability and safety.

12.36.5 Conductometric Titration

Conductometric titration is a specialized technique that measures the ionic properties of cosmetic solutions, such as surfactant-containing formulations. By tracking changes in electrical conductivity during the reaction, this method eliminates the need for visual indicators. It is especially advantageous for complex mixtures where traditional titration methods may not be effective, ensuring precise control of ionic concentrations.

12.37 Cell-Based Assays for Efficacy and Safety

The safety assessment of cosmetic ingredients is an evolving field, focusing on the development of alternative approaches that comply with regulatory requirements while avoiding animal testing. Advances in vitro, in chemico, and in silico methodologies have enabled significant progress in assessing ingredient safety without the use of animals. This shift has been largely driven by the European Union's Cosmetics Regulation (EU No. 1223/2009), which bans the marketing of cosmetic ingredients and finished products tested on animals.

However, challenges remain in assay development. Factors such as the color and turbidity of test substances can interfere with signal measurements, including absorbance and fluorescence. Additionally, certain compounds exhibit properties like metal ion chelation and redox cycling, which can complicate assay interpretations. Non-specific interactions, such as protein aggregation, may also result in false positives or misinterpretation as mechanism-specific inhibition. Therefore, the inclusion of appropriate assay controls is critical to ensure the accuracy and reliability of results.

The following sections highlight cell-based assays used to evaluate cosmetic ingredients for specific biological activities:

- Anti-acne activity
- Hair growth-promoting activity
- Anti-aging and skin-rejuvenating activity
- Anti-psoriatic activity
- Wound-healing activity
- Anti-skin-hyperpigmentation activity

With the growing consumer preference for natural and sustainable products, cosmetics derived from herbal ingredients have gained significant attention. These natural components are evaluated for their potential biological activities, contributing to the development of effective and safe cosmetic formulations. Each section provides examples of herbal ingredients with demonstrated efficacy in specific applications, showcasing their role in enhancing cosmetic product performance (Singh et al. 2022) (Fig. 12.4).

12.38 Green Analytical Techniques

Adopting green analytical techniques in the cosmetics industry represents a significant shift toward sustainable practices. By implementing eco-friendly methods for analyzing natural ingredients, companies can enhance product safety while reducing their environmental footprint. As research in this field progresses, these techniques are expected to play a pivotal role in ensuring that cosmetic products meet both efficacy and sustainability standards.

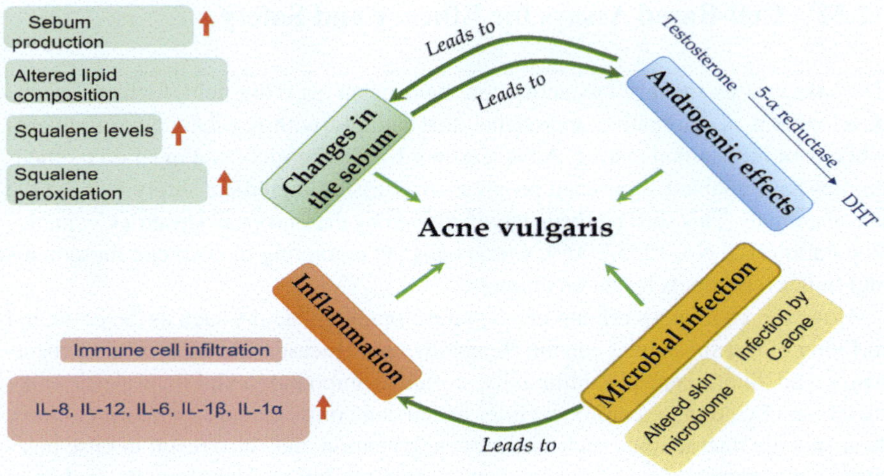

Fig. 12.4 Key components of the pathophysiology of acne vulgaris. Changes in the microbiome and infection by *C. acne* lead to the initiation of acne vulgaris. Although inflammation is beneficial for fighting against the infection, it can lead to aggravation of acne. In addition to this, androgenic factors can influence sebum quality and quantity and dihydrotestosterone (DHT). Changes in the sebum quality and quantity can create a conducive atmosphere for microbes during acne. Cell-based model systems to evaluate anti-acne ingredients take these components of pathology into consideration. The "up arrow" marked in red color indicates an increase in expression (Singh et al. 2022)

The cosmetics industry is increasingly embracing green analytical techniques to ensure the safety and effectiveness of natural ingredients while minimizing environmental impact. These eco-friendly methods prioritize reducing the use of hazardous solvents, lowering energy consumption, and minimizing waste generation during the analysis of cosmetic products. Below are some of the key green analytical techniques and their applications in cosmetic analysis:

12.39 Principles of Green Analytical Chemistry (GAC)

Green analytical chemistry (GAC) emphasizes sustainability and environmental responsibility in analytical processes. Its core principles include:

Minimization of Hazardous Chemicals: Replacing traditional organic solvents with safer, eco-friendly alternatives.

Reduction of Waste: Adopting techniques that produce minimal waste and encourage recycling.

Energy Efficiency: Utilizing methods that operate at ambient temperatures and require less energy.

Improved Safety: Ensuring that analytical processes are safe for both human health and the environment (Fig. 12.5).

12.40 Eco-Friendly Methods for Natural Ingredients

The cosmetics industry is transitioning to greener analytical techniques for natural ingredients due to consumer demand for sustainable products and regulatory pressures. These methods aim to reduce the environmental impact while enhancing the safety and efficacy of cosmetic formulations. Below are key eco-friendly methods and their applications in cosmetics analysis.

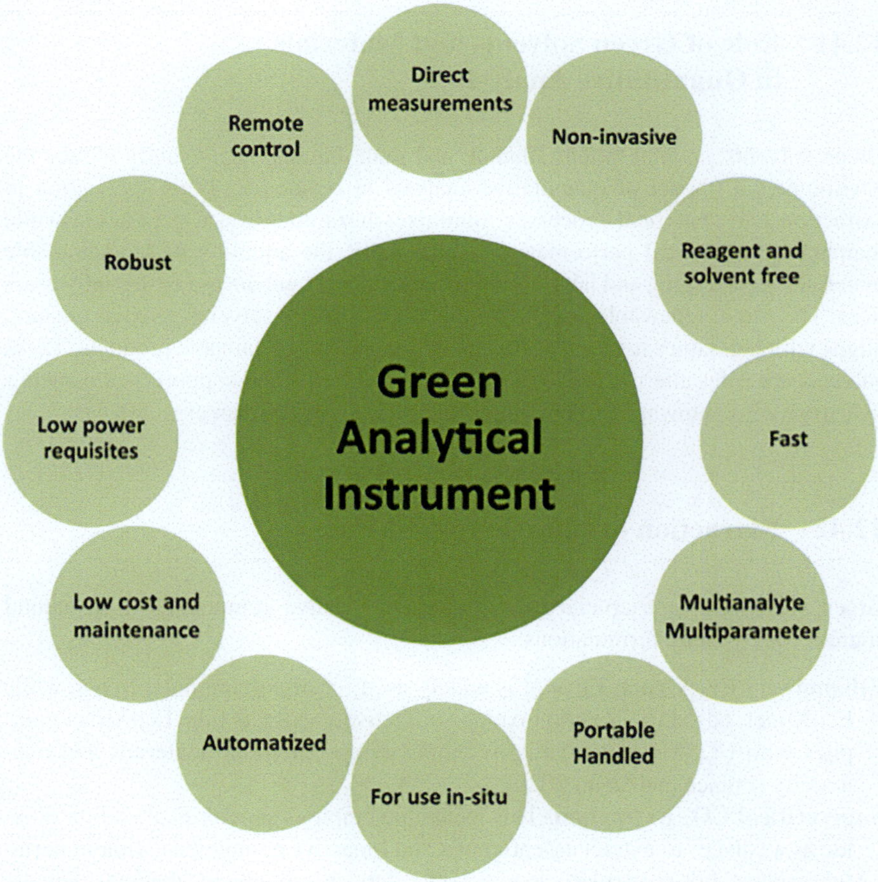

Fig. 12.5 Key features of a green analytical instrument (Armenta et al. 2023)

The need for sustainable practices has driven the adoption of green analytical methods in the analysis of natural ingredients used in cosmetology. Techniques such as supercritical fluid extraction (SFE) and microwave-assisted extraction minimize the use of solvents and energy while maximizing the yield and purity of bioactive compounds. These methods not only adhere to the principles of green chemistry but also improve the sustainability profile of cosmetic products (Barba-Ostria et al. 2022).

The integration of eco-friendly analytical techniques represents a significant advancement toward sustainability in the cosmetics industry. By embracing these methods, manufacturers can meet consumer expectations for safe and effective products while aligning with global sustainability goals. As research in this area evolves, these techniques are expected to play a vital role in shaping the future of cosmetic formulations (Siahaan et al. 2022).

12.41 Role of Green Solvents and Materials in Quantitative Analysis

Green solvents, such as water, ethanol, and ionic liquids, significantly reduce the environmental impact of quantitative analysis in cosmetics. Their application in extraction and analytical processes minimizes hazardous waste generation while maintaining analytical performance. Additionally, the adoption of biodegradable materials in packaging and laboratory processes further enhances eco-friendly practices. The use of renewable and environmentally friendly solvents, such as ethanol, aligns with regulatory norms and consumer demand for sustainable products. These solvents are safer alternatives to petrochemical solvents and support the cosmetics industry's efforts toward creating greener formulations (Usman et al. 2023).

12.42 Extraction of Natural Ingredients

Green solvents play a crucial role in extracting active compounds from natural sources for cosmetic formulations:

Ethanol and EcoXtract: Ethanol is widely used for dissolving polar lipids, while EcoXtract, based on 2-methyloxolane, effectively extracts oils. EcoXtract complies with COSMOS standards by emphasizing renewable materials and non-toxicity (Franca and Ueno 2020).

Supercritical CO_2 Extraction: This technique employs supercritical carbon dioxide as a solvent to extract essential oils and bioactive compounds without harmful residues. It is particularly beneficial for delicate compounds that may degrade under heat or traditional solvents (Usman et al. 2023).

12.43 Non-Invasive Analytical Technologies

The cosmetics industry increasingly employs non-invasive analytical technologies to ensure the safety and efficacy of formulations without harming the skin or altering product integrity. These methods enable real-time monitoring of cosmetic ingredients and their interactions in biological systems.

Technologies such as Raman spectroscopy and near-infrared spectroscopy are widely used to assess skin properties and monitor product interactions without invasive sampling. These techniques prioritize consumer safety while adhering to ethical standards in cosmetic testing, allowing manufacturers to innovate while meeting regulatory requirements (Duffy et al. 2017).

12.44 Regulatory Requirements and Compliance

12.44.1 Analytical Methods for Regulatory Approval

Regulatory agencies require robust analytical methods to ensure the safety and efficacy of cosmetic products before market approval. Standardized procedures, such as those from the International Organization for Standardization (ISO), help manufacturers develop reliable testing methods to meet regulatory expectations. Following these standards is crucial for obtaining certifications and building consumer trust.

12.44.2 Overview of Regulatory Framework

The **European Union** (EU), under **Regulation** (**EC**) **No. 1223/2009**, has established comprehensive requirements for cosmetic products, including safety assessments and good manufacturing practices (GMP). This regulation mandates rigorous testing of ingredients and finished products to ensure safety before market entry. By enforcing strict safety standards, the regulation aligns with **SDG 3** (**good health and well-being**) by protecting consumers from harmful substances and ensuring safe cosmetic products. Additionally, GMP requirements support **SDG 12** (**responsible consumption and production**) by promoting efficient resource utilization and minimizing waste during production. More details can be found at https://single-market-economy.ec.europa.eu/sectors/cosmetics/scientific-and-technical-assessment_en.

In the **United States** (US), the **Modernization of Cosmetics Regulation Act** (**MoCRA**), enforced by the FDA, emphasizes ingredient transparency and safety assessments. This regulatory framework ensures that cosmetics meet rigorous safety and transparency standards, aligning with **SDG 9** (**industry, innovation, and**

infrastructure) by fostering innovation in sustainable cosmetic formulations. Furthermore, safety assessments and transparent labeling practices contribute to **SDG 12**, ensuring responsible production and consumption. For more information, visit https://www.fda.gov/cosmetics.

China's **Cosmetic Supervision and Administration Regulations (CSAR)** require ingredient registration in the **Inventory of Existing Cosmetic Ingredients in China (IECIC)**, with strict safety evaluations for new ingredients to ensure consumer protection. These regulations support **SDG 3**, safeguarding consumer well-being by preventing harmful substances in products, and **SDG 15 (life on land)** by promoting sustainable use of natural resources and biodiversity conservation. Details are available at https://www.nmpa.gov.cn/.

In **Japan**, cosmetics are regulated by the **Ministry of Health, Labour, and Welfare (MHLW)**, which classifies products into general cosmetics and quasi-drugs, each with specific regulatory requirements. Rigorous testing ensures product safety and aligns with **SDG 3** by prioritizing consumer health and well-being. By focusing on quality control and sustainable production practices, the framework also supports **SDG 12**. Additional information can be accessed at https://www.mhlw.go.jp/english/.

This global regulatory approach highlights the commitment to aligning cosmetic safety and efficacy with sustainability goals, ensuring consumer trust and environmental protection.

12.45 Market Surveillance and Compliance Testing

In-Market Control: National authorities inspect and sample cosmetic products to ensure compliance with safety standards (Ferreira et al. 2022).

Official Methods of Analysis: The European Network of Official Cosmetics Control Laboratories (OCCLs) develops validated methods to detect prohibited or restricted substances (https://www.edqm.eu/en/occls-analytical-methods-guidance-documents-and-studies).

12.46 Conclusion

The field of cosmeceutical analysis is undergoing a transformative evolution driven by advancements in analytical methodologies, green chemistry principles, and innovative technologies. The integration of quantitative bioassays, such as ELISA, alongside cell-based assays, has enhanced the industry's ability to evaluate the safety, efficacy, and biological activities of natural ingredients. These techniques not only ensure high standards of product performance but also align with global sustainability goals by emphasizing eco-friendly and resource-efficient processes.

As natural and complex formulations become increasingly prevalent, the demand for advanced analytical methods continues to grow. The adoption of green analytical techniques, such as supercritical fluid extraction and non-invasive technologies, has demonstrated a significant reduction in environmental impact while maintaining the quality and safety of cosmetic products. These efforts underscore the industry's commitment to **Sustainable Development Goals (SDGs)**, particularly in areas such as responsible consumption and production (SDG 12) and the sustainable use of natural resources (SDG 15).

Emerging trends, including nanotechnology, high-throughput screening, and automation, are paving the way for a new era in cosmeceutical analysis. These innovations promise increased precision, efficiency, and scalability in the development and evaluation of cosmetic products. Additionally, regulatory frameworks worldwide are adapting to support these advancements, ensuring consumer safety and fostering sustainable practices in the industry.

Looking forward, the cosmeceutical sector must continue to embrace multidisciplinary approaches, fostering collaboration between researchers, manufacturers, and regulatory bodies. By prioritizing sustainability, technological innovation, and rigorous safety assessments, the industry is well-positioned to deliver solutions that are not only beneficial to consumers but also contribute positively to the environment and society as a whole. This dynamic progression marks a promising future for cosmeceutical analysis, ensuring that it remains both scientifically robust and ethically responsible.

References

Agatonovic-Kustrin S, Ristivojevic P, Gegechkori V, Litvinova TM, Morton WD (2020) Essential oil quality and purity evaluation via ft-ir spectroscopy and pattern recognition techniques. Appl Sci 10(20):7294

Aili Q, Cui D, Li Y, Zhige W, Yongping W, Minfen Y et al (2024) Composing functional food from agro-forest wastes: selectively extracting bioactive compounds using supercritical fluid extraction. Food Chem 139848

Al-Harrasi A, Bhatia S, Behl T, Aldawsari MF, Kaushik D, Chigurupati S (2022) Recent advances in natural products science. CRC Press

Ambruso K, Park HY, Riley KR (2023) Analysis of the active ingredients in sunscreen: a multi-week experiment for the analytical chemistry laboratory. J Chem Educ 100(12):4845–4852

Anastas PT, & Warner JC (2000). Green chemistry: theory and practice. Oxford University Press.

Angrish A, Kumar R, Chauhan R, Sharma V (2020) On the IR spectroscopy and chemometric based rapid and non-destructive method for the investigation of sunscreen stains: application in forensic science. Spectrochim Acta A Mol Biomol Spectrosc 242:118708

Armenta S, Esteve-Turrillas FA, Garrigues S, de la Guardia M (2023) Green analytical chemistry: concepts, evolution, and recent developments. Green Approaches for Chemical Analysis, 1–37

Aziz AA, Nordin FNM, Zakaria Z, Abu Bakar NK (2022) A systematic literature review on the current detection tools for authentication analysis of cosmetic ingredients. J Cosmet Dermatol 21(1):71–84

Bandaranayake WM (2006) Quality control, screening, toxicity, and regulation of herbal drugs. Modern phytomedicine: turning medicinal plants into drugs, 25–57

Baranowska I, Wojciechowska I (2013) The determination of preservatives in cosmetics and environmental waters by HPLC. Pol J Environ Stud 22:6

Barba-Ostria C, Carrera-Pacheco SE, Gonzalez-Pastor R, Heredia-Moya J, Mayorga-Ramos A, Rodríguez-Pólit C et al (2022) Evaluation of biological activity of natural compounds: current trends and methods. Molecules 27(14):4490

Belostozky A, Bretler S, Kolitz-Domb M, Grinberg I, Margel S (2019) Solidification of oil liquids by encapsulation within porous hollow silica microspheres of narrow size distribution for pharmaceutical and cosmetic applications. Mater Sci Eng C 97:760–767

Bezerra GSN, Pereira MAV, Ostrosky EA, Barbosa EG, de Moura MDFV, Ferrari M et al (2017) Compatibility study between ferulic acid and excipients used in cosmetic formulations by TG/DTG, DSC and FTIR. J Therm Anal Calorim 127:1683–1691

Bifulco G, Dambruoso P, Gomez-Paloma L, Riccio R (2007) Determination of relative configuration in organic compounds by NMR spectroscopy and computational methods. Chem Rev 107(9):3744–3779

Blanco M, Alcalá M, Planells J, Mulero R (2007) Quality control of cosmetic mixtures by NIR spectroscopy. Anal Bioanal Chem 389:1577–1583

Bonnier F, Miloudi L, Henry S, Bertrand D, Tauber C, Perse X et al (2020) Quantification of low-content encapsulated active cosmetic ingredients in complex semi-solid formulations by means of attenuated total reflectance-infrared spectroscopy. Anal Bioanal Chem 412:159–169

Bruce KA, Arnold DC, Sauzier G, Lewis SW (2024) Considerations and future perspectives for the vibrational spectroscopic analysis of forensic cosmetic evidence. Wiley Interdiscip Rev Forensic Sci 6(6):e1533

Butnariu M (2016) Methods of analysis (extraction, separation, identification and quantification) of carotenoids from natural products. J Ecosyst Ecograp 6(2):1–19

Chaintreau A (2007) 6.2. Analytical methods to determine potentially allergenic fragrance-related substances in cosmetics. Anal Cosmet Prod 106:257

International Cooperation on Cosmetics Regulation (2022) About ICCR. https://www.iccr-cosmetics.org/about-us. Accessed 1 April 2022

Dong, M. W. (2013). The essence of modern HPLC: Advantages, limitations, fundamentals, and opportunities. LCGC Asia Pacific, 16(3), 24–29.

Drouet S, Garros L, Hano C, Tungmunnithum D, Renouard S, Hagège D et al (2018) A critical view of different botanical, molecular, and chemical techniques used in authentication of plant materials for cosmetic applications. Cosmetics 5(2):30

Duffy E, De Guzman K, Wallace R, Murphy R, Morrin A (2017) Non-invasive assessment of skin barrier properties: investigating emerging tools for in vitro and in vivo applications. Cosmetics 4(4):44

El-Otmani N, Zeouk I, Hammani O, Zahidi A (2024) Analysis and quality control of bio-actives and herbal cosmetics: the case of traditional cooperatives from Fes-Meknes region. Trop J Nat Prod Res 8:5

Emwas AH, Roy R, McKay RT, Tenori L, Saccenti E, Gowda GN et al (2019) NMR spectroscopy for metabolomics research. Meta 9(7):123

Erbas Z, Soylak M (2022) Determination of rhodamine B by UV–vis spectrophotometry in cosmetics after microextraction by using heat-induced homogeneous liquid–liquid extraction method. J Iran Chem Soc 19(9):3935–3942

European Commission (2009) Regulation (EC) No 1223/2009 of the European Parliament and of the Council on cosmetic products. Retrieved from https://eur-lex.europa.eu

Evangeline WP, Saranya E, Rajalakshmi E, Murugan M, Mahalakshmi S, Ramya V et al (2024) Advancements of Raman spectroscopy in cosmetics and dermatology. S Afr J Bot 167:122–129

Ferreira S, Lores M (2022) Global regulatory frameworks in cosmeceutical analysis: ensuring safety, efficacy, and consumer protection. Int J Cosmet Sci 44(1):10–20

Ferreira M, Matos A, Couras A, Marto J, Ribeiro H (2022) Overview of cosmetic regulatory frameworks around the world. Cosmetics 9(4):72

Förster M, Bolzinger MA, Ach D, Montagnac G, Briançon S (2011) Ingredients tracking of cosmetic formulations in the skin: a confocal Raman microscopy investigation. Pharm Res 28:858–872

Franca CCV, Ueno HM (2020) Green cosmetics: perspectives and challenges in the context of green chemistry. Desenvolvimento e Meio Ambiente 53:133–150

Gaur P, Rana P, Pathania D, Shukla S, Suvirya S, Shukla SV, Chopra D (2023) Role of omics approach in the toxicity/safety study of cosmetics. In: Skin 3-D models and cosmetics toxicity. Springer Nature Singapore, Singapore, pp 211–225

Georgousaki K, DePedro N, Chinchilla A, Aliagiannis N, Vicente F, de Castro P et al (2015) High throughput screening of microbial biodiversity for the discovery of novel cosmeceutical agents. Panta Med 10

Goldstein A, Soroka Y, Frušić-Zlotkin M, Popov I, Kohen R (2014) High resolution SEM imaging of gold nanoparticles in cells and tissues. J Microsc 256(3):237–247

Guerra E, Llompart M, Garcia-Jares C (2018) Analysis of dyes in cosmetics: challenges and recent developments. Cosmetics 5(3):47

Hiba P (2023) Exploring exoplanets at the nanoscale: a novel approach through transmission electron microscopy. Acceleron Aerospace J 1(3):54–57

International Organization for Standardization (2022) Technical Committees, ISO/TC 217, Cosmetics. https://www.iso.org/committee/54974.html. Accessed 22 June 2022

Jones, W. (2024). GCMS vs HPLC. Celebration of Scholarship, Shawnee State University. https://digitalcommons.shawnee.edu/cos/2024/monday/3

Jonkman J, Brown CM, Wright GD, Anderson KI, North AJ (2020) Tutorial: guidance for quantitative confocal microscopy. Nat Protoc 15(5):1585–1611

Kiani H, Aznar R, Poojary MM, Tiwari BK, Halim R (2022) Chromatographic techniques to separate and identify bioactive compounds in microalgae. Front Energy Res 10:904014

Klang V, Valenta C, Matsko NB (2013) Electron microscopy of pharmaceutical systems. Micron 44:45–74

Kowalska T, Sajewicz M (2022) Thin-layer chromatography (TLC) in the screening of botanicals–its versatile potential and selected applications. Molecules 27(19):6607

Krombholz R, Lunter D (2020) A new method for in-situ skin penetration analysis by confocal Raman microscopy. Molecules 25(18):4222

Kubota R, Tanaka W, Hamachi I (2021) Microscopic imaging techniques for molecular assemblies: electron, atomic force, and confocal microscopies. Chem Rev 121(22):14281–14347

Li Q, Zhu R, Li J, Wang X, Xu L, Li Y, Li P (2019) Highly specific chemiluminescence immunoassay for the determination of chloramphenicol in cosmetics. Int J Anal Chem 2019(1):7131907

Liu P, Hu J, Wang Q, Tan J, Wei J, Yang H et al (2023) Physicochemical characterization and cosmetic application of kelp blanching water polysaccharides. Int J Biol Macromol 248:125981

Lorenz C, Tiede K, Tear S, Boxall A, Von Goetz N, Hungerbühler K (2010) Imaging and characterization of engineered nanoparticles in sunscreens by electron microscopy, under wet and dry conditions. Int J Occup Environ Health 16(4):406–428

Lores M, Llompart M, Alvarez-Rivera G, Guerra E, Vila M, Celeiro M et al (2016) Positive lists of cosmetic ingredients: analytical methodology for regulatory and safety controls–a review. Anal Chim Acta 915:1–26

Maia AM, Baby AR, Yasaka WJ, Suenaga E, Kaneko TM, Velasco MVR (2007) Validation of HPLC stability-indicating method for vitamin C in semisolid pharmaceutical/cosmetic preparations with glutathione and sodium metabisulfite, as antioxidants. Talanta 71(2):639–643

Mildau G (2018) Chapter 4—General review of official methods of analysis of cosmetics. In: Salvador A, Chisvert ABT-A (eds) Analysis of cosmetic products, 2nd edn. Elsevier, Boston, MA, pp 67–83. ISBN 978-0-444-63508-2

Mishra L, Kurmi BD (2023) Cosmetics regulations and standardization guidelines. Pharmaspire 15:137–150

Mohamed HM (2015) Green, environment-friendly, analytical tools give insights in pharmaceuticals and cosmetics analysis. TrAC Trends Anal Chem 66:176–192

Monakhova YB, Diehl BW (2016) Rapid 1 H NMR determination of hydrogen peroxide in cosmetic products and chemical reagents. Anal Methods 8(23):4632–4639

Mukherjee PK, Bahadur S, Chaudhary SK, Kar A, Mukherjee K (2015) Quality related safety issue-evidence-based validation of herbal medicine farm to pharma. In: Evidence-based validation of herbal medicine. Elsevier, pp 1–28

Ngoc LTN, Moon JY, Lee YC (2023) Insights into bioactive peptides in cosmetics. Cosmetics 10(4):111

Organisation for Economic Co-Operation and Development (2022) Who we are. https://www.oecd.org/about/. Accessed 22 June 2022

Pasieczna-Patkowska S, Olejnik T (2013) Analysis of cosmetic products using different IR spectroscopy techniques. Annales Universitatis Mariae Curie-Skłodowska, sectio AA–Chemia 68(1–2)

Peddie CJ, Genoud C, Kreshuk A, Meechan K, Micheva KD, Narayan K et al (2022) Volume electron microscopy. Nat Rev Methods Primers 2(1):51

Pistollato F, Madia F, Corvi R, Munn S, Grignard E, Paini A et al (2021) Current EU regulatory requirements for the assessment of chemicals and cosmetic products: challenges and opportunities for introducing new approach methodologies. Arch Toxicol 95:1867–1897

Prado JM, Rostagno MA (eds) (2022) Natural product extraction: principles and applications, vol 71. Royal Society of Chemistry

Radoičić A, Ristivojević P, Radosavljević-Stevanović N (2024) Recent applications of high-performance thin-layer chromatography (HPTLC) in forensic drug analysis. JPC–J Planar Chromat Modern TLC:1–8

Rajasekar M, Mary J, Sivakumar M, Selvam M (2024) Recent developments in sunscreens based on chromophore compounds and nanoparticles. RSC Adv 14(4):2529–2563

Ribechini E, Modugno F, Pérez-Arantegui J, Colombini MP (2011) Discovering the composition of ancient cosmetics and remedies: analytical techniques and materials. Anal Bioanal Chem 401:1727–1738

Rico F, Mazabel A, Egurrola G, Pulido J, Barrios N, Marquez R, García J (2023a) Meta-analysis and analytical methods in cosmetics formulation: a review. Cosmetics 11(1):1

Rico L, Silva A, Gonzalez M (2023b) Chromatographic methods in cosmeceuticals: sustainability and precision. Chromatogr Today 12(5):78–85

Rizvi A, Mulvey JT, Carpenter BP, Talosig R, Patterson JP (2021) A close look at molecular self-assembly with the transmission electron microscope. Chem Rev 121(22):14232–14280

Rohman A, Windarsih A, Mustafa S (2020) The application of analytical methods for analysis of non-halal components in cosmetics products: a review. Int J Pharm Res 09752366

Roy A, Sahu RK, Matlam M, Deshmukh VK, Dwivedi J, Jha AK (2013) In vitro techniques to assess the proficiency of skin care cosmetic formulations. Pharmacogn Rev 7(14):97

Rujido-Santos I, Herbello-Hermelo P, Barciela-Alonso MC, Bermejo-Barrera P, Moreda-Piñeiro A (2022) Study design on the presence of metals in moisturisers, and compliance with regulation (EC) no. 1223/2009 of the European Parliament and of the Council of the European Union, on cosmetic products. Cosmetics 9(4):82

Sathyaseelan S, Rao BH, Anushmati S (2024) Cosmeceuticals: a transit state from synthetic to natural. Indian J Pharmacol 56(1):42–51

Shanbhag S, Nayak A, Narayan R, Nayak UY (2019) Anti-aging and sunscreens: paradigm shift in cosmetics. Adv Pharm Bull 9(3):348

Sharma A, Chauhan R, Kumar R, Mankotia P, Verma R, Sharma V (2021) A rapid and non-destructive ATR-FTIR spectroscopy method supported by chemometrics for discriminating between facial creams and the classification into herbal and non-herbal brands. Spectrochim Acta A Mol Biomol Spectrosc 258:119803

Sharma S, Ahmad U, Akhtar J, Islam A, Khan MM, Rizvi N, Sharma S, Ahmad U, Akhtar J, Islam A et al (2023) The art and science of cosmetics: understanding the ingredients. In: Cosmetic products and industry—new advances and applications. IntechOpen. ISBN 978-1-83768-623-0

Sharmeen JB, Mahomoodally FM, Zengin G, Maggi F (2021) Essential oils as natural sources of fragrance compounds for cosmetics and cosmeceuticals. Molecules 26(3):666

Siahaan EA, Agusman, Pangestuti R, Shin KH, Kim SK (2022) Potential cosmetic active ingredients derived from marine by-products. Mar Drugs 20(12):734
Singh P, Bhat SS, Singh N, Venkanna BU, Mohamed R, Rao RP (2022) Cell-based model systems for validation of various efficacy-based claims for cosmetic ingredients. Cosmetics 9(5):107
Sun Z, Shi J, Wang J, Jiang M, Wang Z, Bai X, Wang X (2022) A deep learning-based framework for automatic analysis of the nanoparticle morphology in SEM/TEM images. Nanoscale 14(30):10761–10772
Tranchida PQ, Franchina FA, Mondello L (2017) Analysis of essential oils through comprehensive two-dimensional gas chromatography: general utility. Flavour Fragr J 32(4):218–227
Tripathi S (2023) Decoding analytical excellence: navigating the pros and cons of HPLC technology. Int J Pharm Healthc Innov 1:1. https://doi.org/10.62752/mm8zjr64
Tuzimski T, Petruczynik A (2020) Review of new trends in the analysis of allergenic residues in foods and cosmetic products. J AOAC Int 103(4):997–1028
U.S. Food and Drug Administration. (2024) Cosmetics: safety and labeling guidelines. Retrieved from https://www.fda.gov
Union P (2009) Regulation (EC) no 1223/2009 of the european parliament and of the council. Off J Eur Union L 342:59
United Nations (2015). Transforming our world: the 2030 agenda for sustainable development. Retrieved from https://sdgs.un.org/2030agenda
Usman M, Cheng S, Boonyubol S, Cross JS (2023) Evaluating green solvents for bio-oil extraction: advancements, challenges, and future perspectives. Energies 16(15):5852
Wang X, Anton H, Vandamme T, Anton N (2023) Updated insight into the characterization of nano-emulsions. Expert Opin Drug Deliv 20(1):93–114
Yingling GL, Onel S (2013) Regulatory agencies, pharmaceuticals, food, and cosmetics. Kirk-Othmer Chemical Technology of Cosmetics, p 743
Zakaria Z (2013) Cosmetic safety regulations: a comparative study of Europe, the USA and Malaysia. The University of Manchester (United Kingdom)
Zangheri M, Calabretta MM, Calabria D, Fiori J, Guardigli M, Michelini E et al (2021) Immunological analytical techniques for cosmetics quality control and process monitoring. PRO 9(11):1982
Zeng J (2022, May) Analysis of analytical techniques for fragrances contained in perfumes and cosmetics. In: International conference on biometrics, microelectronic sensors, and artificial intelligence (BMSAI), vol 12252. SPIE, pp 79–84
Zorić M, Banožić M, Aladić K, Vladimir-Knežević S, Jokić S (2022) Supercritical CO2 extracts in cosmetic industry: current status and future perspectives. Sustain Chem Pharm 27:100688

Chapter 13
Stem Cell Treatment for Cosmetic Surgery

Bahzad Ahmad Farhan, Wooda Farooq, Ghulam Fatima, Gul Nabi Khan, Asmara Ahmad, and Sadia Javed

Contents

B. A. Farhan (✉)
Department of Kimia, FMIPA, Institut Teknologi Bandung, Bandung, Indonesia

Institute of Biological Sciences, Khwaja Fareed University of Engineering & Information Technology, Rahim Yar Khan, Pakistan

W. Farooq · G. Fatima
Institute of Biological Sciences, Khwaja Fareed University of Engineering & Information Technology, Rahim Yar Khan, Pakistan

G. N. Khan
Department of Zoology, Islamia College University, Peshawar, Pakistan

A. Ahmad · S. Javed
Department of Biochemistry, Government College University Faisalabad, Faisalabad, Pakistan

S. Javed et al. (eds.), *Sustainable Cosmeceuticals*,
https://doi.org/10.1007/978-3-031-86087-4_13

Abstract Stem cells (SCs) are special cells that have the unique power to renew themselves and transform into different types of specialized cells. This remarkable ability makes them essential for tissue growth, repair, and regeneration. There are four main types of stem cells: embryonic, adult, induced pluripotent, and tissue-resident stem cells, each coming from different sources and having unique capabilities. The balance of stem cell division is crucial for healthy tissues. If too few cells are produced, the body struggles to repair itself. On the other hand, if too many cells grow uncontrollably, it can lead to conditions like hyperplasia or even cancer. Stem cell therapies have brought new possibilities to regenerative medicine and aesthetics by offering natural ways to heal and repair tissues. Among these, mesenchymal stem cells (MSCs) are particularly popular, especially those derived from fat tissues, known as adipose-derived stem cells (ASCs). These cells are easy to obtain, plentiful, and have powerful properties like reducing inflammation and speeding up recovery. These therapies are used for various purposes, such as fixing bone and soft tissue defects, treating burns and stubborn wounds, rejuvenating the face, regrowing hair, reshaping the body, revising scars, and addressing signs of aging. Despite these advancements, challenges remain. Ethical debates surround the use of embryonic stem cells, and there are risks like immune rejection, tumor development, and inconsistent treatment practices. To overcome these hurdles, thorough research and clinical trials are needed to ensure that stem cell treatments are safe and effective by carefully controlling how these cells grow and specialize. As science progresses, stem cell therapies are set to transform regenerative medicine and aesthetics, offering innovative and lasting solutions for repairing tissues and treating conditions once thought to be incurable.

Keywords Stem cells · Tissue regeneration · Adipose-derived stem cells (ASCs) · Cosmetic surgery

13.1 Introduction of Stem Cells in Cosmetology

Stem cells (SCs) are unique cells in the body that possess the remarkable capacity for self-renewal through mitotic cell division and have the ability to differentiate into a diverse range of specialized cell types (Ilic and Polak 2011).

The concept of stem cells has its deepest roots in the study of hematopoiesis (formerly referred to as hemopoiesis). Notably, the German hematologist Alexander Pappenheim is credited as one of the first to describe this hypothetical cell type, which he termed *Stammzelle* (Maehle 2011); the idea that all blood cells arise from a single type of bone marrow cell is most often attributed to the Russian histologist Alexander Maximow (Gary and Alexander 2009). In later research on hematology and histology, the term "stem cell" was commonly used to describe blood precursor cells found in the bone marrow (Cowdry 1932; Dragutin et al. 2011). In the mid-twentieth century, research in hematopoiesis and radiation biology revived interest in stem cell theory. According to Maximow's unitary hypothesis, the "hemocytoblast"—later known as the "hematopoietic stem cell" (HSC)—was proposed as the common precursor for all blood cells, including both myeloid and lymphoid lineages. However, this view was not universally accepted, and for many years, it was believed that lymphoid cells had a distinct origin. At the time, it was difficult to distinguish between the unitary hypothesis and other theories, as the evidence relied primarily on histological analysis of normal bone marrow and observations of hematopoiesis under diseased conditions (Tan et al. 2024). The terms totipotency, pluripotency, and multipotency were used with considerable distinction, though their definitions have evolved over time, and cell differentiation is mentioned in Fig. 13.1 (Smith 2006).

13.2 Types of Stem Cells

13.2.1 Totipotent Cells

During the initial phases of development, the most primitive cells are referred to as totipotent (omnipotent) cells. These types of cells have the exceptional capability to divide into both the embryonic and extra-embryonic tissues. For instance, cells resulting from the first two divisions of a fertilized egg are totipotent, leading to the creation of both embryo and placenta (Rossant 2001).

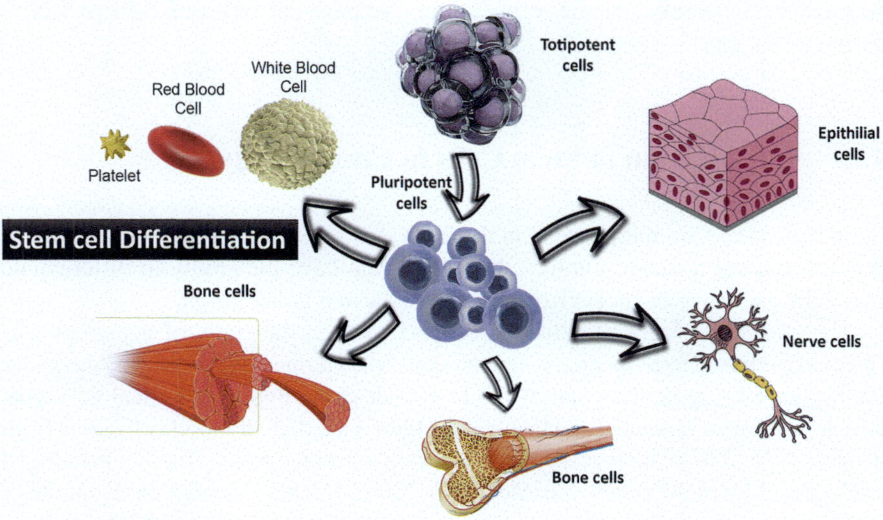

Fig. 13.1 Stem cell differentiation in totipotent cells

13.2.2 Pluripotent Cells

The three germ layers named as ectoderm, endoderm, and mesoderm, from which all tissues and organs emerge, can be differentiated into cells using pluripotent stem cells (De Miguel et al. 2010). Embryonic stem cells (ESCs), also known as pluripotent cells, were originally produced from the inner cell mass of blastocysts (Notarianni et al. 1991). Reprogramming somatic cells, as demonstrated by Takahashi and Yamanaka (Kolios and Moodley 2012), enabled the production of induced pluripotent stem cells (iPSCs), which exhibit traits comparable to embryonic stem cells (ESCs). Notably, a population of pluripotent cells specifically derived from the lungs has yet to be identified.

13.2.3 Multipotent Cells

Multipotent stem cells, present in various tissues, possess the ability to differentiate into multiple cell types within a single germ layer. A notable example is the mesenchymal stem cell (MSC), which can be derived from sources such as bone marrow, adipose tissue, bone, Wharton's jelly, umbilical cord blood, and peripheral blood. These cells have the capacity to generate mesodermal tissues, including adipose tissue, bone, cartilage, and muscle (Augello et al. 2010; Bruder et al. 1997; Azizi et al. 1998; Friedenstein et al. 1970).

13.2.4 Oligopotent Cells

Oligopotent stem cells are able to self-renew and form two or more lineages within a single tissue. Oligopotent stem cells, which may develop into both myeloid and lymphoid lineages, are frequently found in hematopoietic stem cells (Marone et al. 2002). The ocular surface of the pig, including the cornea, has been found to harbor oligopotent stem cells capable of forming distinct colonies of corneal and conjunctival cells (Majo et al. 2008).

13.2.5 Unipotent Cells

Unipotent stem cells exhibit the capacity for self-renewal and are specialized to differentiate into a specific type of cell, following a distinct lineage. An example of this is muscle stem cells, which exclusively give rise to adult muscle cells (Overturf et al. 1997; De Rooij and Grootegoed 1998; O'Neill et al. 2016). The spermatogonia of the testis possess a unipotent nature, being exclusively specialized for sperm production, which distinguishes them from most other adult stem cells that are typically multipotent and contribute to the maintenance of various tissue types (Slack 2018).

13.3 Characteristics of Stem Cell

Although originating from a hematologic tradition, it effectively integrates concepts from the embryonic stem cell (ESC) tradition, which emphasizes pluripotency, and the embryological tradition, characterized by repeated cell divisions:

- Stem cells possess the ability to self-replicate, maintaining their population over time.
- They produce progeny that eventually differentiate into specialized, functional cell types.
- Stem cells exhibit longevity, persisting in tissues for extended periods.
- Their activity and fate are significantly influenced by their microenvironment, known as the stem cell niche.

13.3.1 Stem Cell Environment

A dynamic and unique microenvironment known as a stem cell niche controls stem cell survival and function by secreting substances that have direct effects on the cells themselves (Gattazzo et al. 2014). Various components make up the diverse stem cell niches, including diffusible and cell surface-associated signaling molecules such as inflammatory cells, specific types of neurons, glial cells, vascular cells, and mesenchymal cells. Additionally, physical factors like oxygen levels, temperature, matrix rigidity, and shear stress also play crucial roles within these niches (Wang and Zou 2012). These niche-specific factors activate specific intracellular signaling systems that ultimately decide whether a cell will divide, develop, or go through apoptosis. The stem cell niche is a complex microenvironment that regulates stem cell behavior through various cellular and acellular components. While significant progress has been made in understanding these niches, many aspects remain unidentified and not fully characterized. For instance, the hematopoietic stem cell (HSC) niche in the bone marrow is among the best characterized, yet it may represent just a single example among many possible niches (Martinez et al. 2022). Additionally, tissue-specific stem cells are likely to reside in specialized niches that require further characterization in each tissue (Pennings et al. 2018). This underscores the need for continued research to fully elucidate the diverse components and mechanisms governing stem cell niches across different tissues. The majority of the stem cell niches remain unidentified and not yet fully characterized (Ceafalan et al. 2018). Even though the essential components of all the examined stem cell habitats are the same, depending on the organ, the stem cell niches use various signaling pathways to control the characteristics of stem cells, and they also contain differing degrees of biophysical conditions. The hematopoietic stem cell (HSC) niche in the bone marrow is the most extensively studied stem cell niche and was first proposed by Schofield in 1978. Numerous growth factors are secreted by HSC niche and support the HSCs' upkeep and hemopoiesis, including stem cell factor (SCF), CXC-chemokine ligand 12 (CXCL 12), and the thrombopoietin. The niche of stem cell is shown in Fig. 13.2.

13.4 Stem Cell Classification Based on Place of Origin

Stem cells can be categorized into four main groups based on their origin: Embryonic-stem cells (ESCs), fetal-stem cells, adult-stem cells, and induced pluripotent-stem cells (iPSCs) (Ilic and Polak 2011; Bongso and Richards 2004). Adult-stem cells are basically oligopotent (unipotent) cells, whereas ESCs and iPSCs are typically pluripotent cells.

13.4.1 Embryonic Stem Cells

Embryonic-stem cells (ESCs) are derived from the inner cell mass of the blastocyst, a pre-implantation embryo stage that forms approximately 5–6 days after conception (Notarianni et al. 1991). These ESCs are characterized by their pluripotency, enabling them to differentiate into tissues representing all three germ layers.

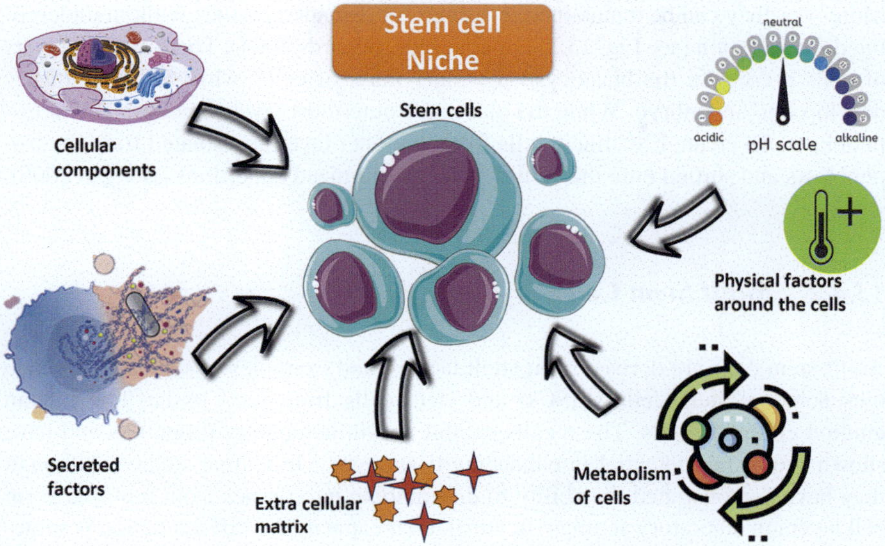

Fig. 13.2 Stem cell niche or factors that affect the cell survival of differentiation

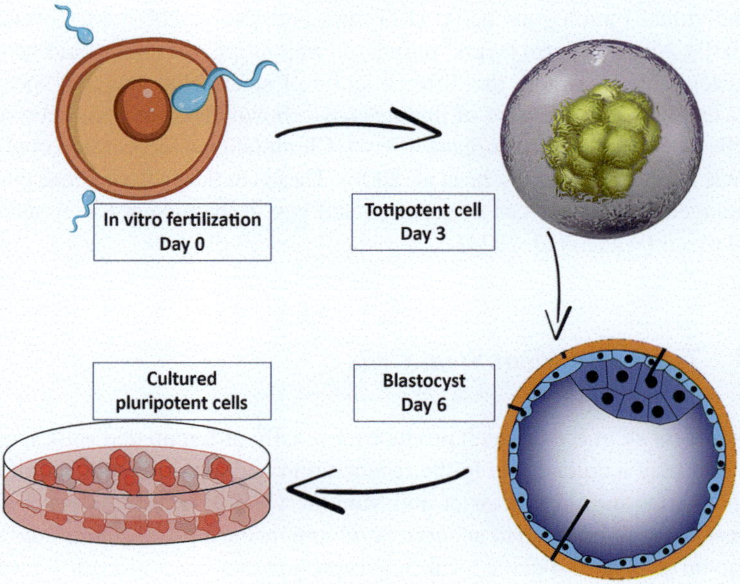

Fig. 13.3 Pluripotent stem cells can be cultured in vitro and are formed 3 days after fertilization as totipotent cells. On the sixth day after fertilization, blastocysts are formed, and these pluripotent stem cells can be effectively propagated on a dish

Moreover, they can be maintained in culture for extended periods without undergoing differentiation (see Fig. 13.3) (Bongso and Richards 2004). There are two layers of cells present in the blastocyst: the inner mass of cells, which will eventually develop into an embryo. While the placenta's outermost layer of cells is called trophoblasts. To create ESC lines, cells from the inner layer are isolated from the trophoblasts and shifted onto the culture plate at optimized conditions (Bongso 2006).

13.4.2 Adult Stem Cells

Adult-stem cells are derived from adult tissues, and examples of such cells include mesenchymal stem cells (MSCs) and stem cells from fetal tissue, like human amnion epithelial cells. These cells exhibit anti-inflammatory properties and have shown promising potential in enhancing tissue repair in animal studies. Although they have demonstrated the ability to differentiate into tissues from multiple germ cell layers in laboratory settings (in vitro), their capacity for differentiation is somewhat restricted (Moodley et al. 2010; Ilancheran et al. 2009). Adult stem cells offer significant advantages as they do not raise ethical issues or rejection-related concerns when used in therapies, particularly because they can be obtained from the same individual (autologous cells) (Körbling and Estrov 2003; McCormick and Huso 2010). All three germ layers' tissues, as well as the placenta, could be used to harvest stem cells. Through the differentiation of stem cells, new and specialized cells are created, and a number of findings have shown that transplantation of adult stem cells can restore injured organs in vivo (Chimutengwende-Gordon et al. 2012; Obradović et al. 2004; Menasché et al. 2008). These studies include those that show how damaged bone tissue can be repaired and how ischemic cardiac tissue can be revascularized (Wang et al. 2024).

13.4.3 Tissue-Resident Stem Cells

Tissue-resident stem cells, which produce terminally differentiated cells specific to each tissue, play a crucial role in the regeneration and repair of some adult tissues and organs after an injury (Passier and Mummery 2003). Based on research findings, these cells are believed to undergo development during ontogenesis and remain in a dormant state until specific local signals prompt their migration, differentiation, and proliferation (Smart and Riley 2008; Voog and Jones 2010). The "stem cell niche" serves as the habitat for tissue-resident stem cells (Kiefer 2011). The microenvironment commonly known as the stem cell niche regulates the self-renewal and differentiation of these cells (Yeung et al. 2011).

13.4.4 Induced Pluripotent Stem Cells

Induced pluripotent stem (iPS) cells are a form of pluripotent stem cell generated from adult somatic cells that have been genetically reprogrammed to resemble embryonic stem (ES) cells by inducing the expression of specific genes and factors critical for preserving the key characteristics of Embryonic stem cells (Rossant 2008). In laboratory conditions (in vitro), these cells demonstrated the ability to differentiate into cell types representing all three germ layers. Additionally, they exhibited various similarities with human ESCs, including their gene expression, morphology, surface antigen, proliferation state, the epigenetic profile of pluripotent cell-specific genes, and the telomerase activity (Ambasudhan et al. 2011).

13.5 Division of Stem Cells

The type of cell division a stem cell undergoes determines the types of cells it generates. When a stem cell divides symmetrically (see Fig. 13.4), it produces daughter cells that are identical to each other. This process is crucial for replenishing damaged cells following an injury (Knoblich 2001; Morrison and Kimble 2006). Maintaining stem cell homeostasis is vital, as a decrease in stem cell population could hinder organ repair, while uncontrolled stem cell proliferation could lead to stem cell hyperplasia and/or carcinogenesis. Striking a balance is essential to ensure the proper functioning and health of the body's tissues and organs (Yamashita 2010). When a stem cell produces two distinct daughter cells, one identical and the other differentiated, this is known as asymmetric division. This procedure keeps a

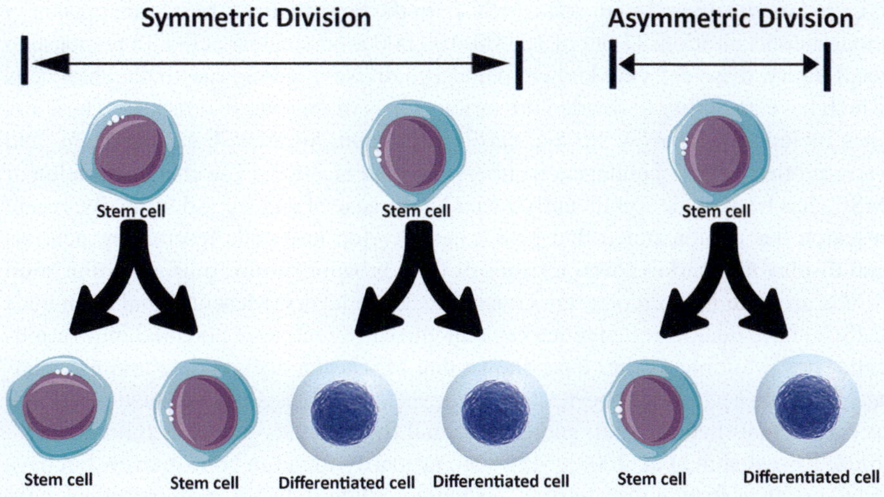

Fig. 13.4 Different types of the cell division

population of stem cells alive while permitting organ regeneration and repair (Neumüller and Knoblich 2009; Horvitz and Herskowitz 1992; Knoblich 2008).

13.6 Therapies Using Stem Cells

Regenerative medicine, often known as stem cell treatment, harnesses the power of stem cells or their derivatives to stimulate the healing and repair processes of damaged, dysfunctional, or injured tissues. Numerous medical problems, including autoimmune, inflammatory, and neurological disorders, can be treated with it (Wang et al. 2024).

There are various stem cell therapies available, such as amniotic fluid stem cell therapy and umbilical cord stem cell therapy. Hematopoietic stem cell transplantation is a very well-known FDA-approved stem cell-based treatment and is used to treat blood malignancies, including leukemia. In addition, stem cells can be used as regenerative therapy for corneal injuries and severe burns to the skin (Tan et al. 2024).

Somatic stem cells, also known as adult stem cells, are found in various tissues and organs throughout the body. Among them, hematopoietic stem cells (HSCs) are the most extensively studied. HSCs form the foundation of successful bone marrow transplants and have been widely utilized in clinical practice for over four decades. Similar to other types of adult stem cells, HSCs are relatively rare and difficult to isolate in large quantities from their in vivo environment. For example, only one HSC can typically be found among every 10,000 bone marrow cells (Spradling et al. 2001). Adult stem cells have also been found in a number of other organs, including the gut (intestinal crypt stem cells), eye (retinal stem cells), skin (epidermal stem cells), and brain (neuronal stem cells) (Spradling et al. 2001). Another group of adult-stem cells that have undergone extensive research and investigation is called mesenchymal stem cells (MSCs). In the bone marrow, MSCs are present in modest concentrations (1 out of 10,000–100,000 mononuclear cells). In response to local injury, these cells are known to undergo division, giving rise to daughter cells which have an ability to divide into various types of mesodermal tissues. These tissues include fat, marrow stroma, cartilage, tendon, muscles, bone, ligament, and various other forms of collagenous fibers (Short et al. 2003). The characterization of MSCs has been made significantly easier by the ease of culture. Additionally, recent research has demonstrated that MSCs can develop into cells resembling neurons and display the marker characteristics of fully grown neurons, indicating that adult MSCs are able to overcome the concept of germ layers. Mesenchymal stem cells (MSCs) have demonstrated remarkable in vitro differentiation potential into various cell types. Multiple studies have shown that MSCs can differentiate into mesodermal lineages such as adipocytes, chondrocytes, osteoblasts, and myocytes, as well as ectodermal (neural, glial) and endodermal (hepatocytes) lineages (Gimble et al. 2008; Rosenbaum et al. 2008). MSCs are primarily found in bone marrow but have been isolated from other tissues, including adipose tissue and umbilical cord (Rosenbaum et al. 2008). Their multipotency and ease of isolation make them attractive for tissue engineering applications (Rosenbaum et al. 2008; Jackson et al. 2007). Notably, MSCs have also shown the ability to differentiate into endothelial cells in vitro, forming capillary-like structures and expressing endothelial-specific markers when cultured with vascular endothelial growth factor (Oswald et al. 2004). This diverse differentiation potential highlights the promise of MSCs in regenerative medicine and tissue engineering. Although stem cells are present in many organs and tissues, they are not universally found throughout the body. For instance, the islet portion of the pancreas has been investigated through molecular tagging and lineage tracing of pancreatic cells, and no evidence of stem cells has been observed in this particular region (Dor et al. 2004).

Stem cell therapy offers a multitude of benefits, most notably in avoiding potentially risky surgical procedures and the associated donor-site complications. This innovative approach significantly reduces surgical risks, making it particularly advantageous for elderly patients. Indications for this treatment include a range of conditions such as traumatic skin defects, severe burns, scars, skin ulcers, and diabetic foot issues (You and Han 2014). Stem cell therapy shows great promise as a highly effective treatment option for addressing various bone and soft tissue defects, including injuries to the skin caused by trauma, burns, and persistent wounds that are exacerbated by conditions like ischemia, such as diabetic foot ulcers and sores. It also turns into a very alluring instrument for breast enhancement, scar healing, and skin renewal. Numerous plastic surgeons continue their tireless study employing diverse stem cells (Eun 2014).

13.6.1 Process of Stem Cell Treatment

Stem Cells are employed because of their capacity for regeneration. These pluripotent stem cells differentiate into diverse cell lineages as a result of specific inductive cues (Gimble et al. 2007). Stem cells can directly differentiate into target cells and modify their microenvironment under appropriate conditions. Additionally, endogenous stem cells can be recruited and stimulated by both autologous and allogeneic stem cells. Stem cells also exert significant effects through their paracrine actions, mediated by the secretome—a collection of bioactive substances such as growth factors and cytokines. The secretome plays a key role in processes like angiogenesis, immunomodulation, and wound healing (Kapur and Katz 2013).

In models of acute myocardial infection and limb ischemia, proangiogenic effects subsequent to stem cell transplantation have been demonstrated to be beneficial (Triana et al. 2024). It was believed that these effects were mediated by the upregulation of SCs because of their capacity for regeneration. Inductive stimuli, such as transforming growth factor (TGF), vascular endothelial growth factor, hepatocyte growth factor, granulocyte colony-stimulating factor, and basic fibroblast growth factor, prompt the differentiation of these pluripotent stem cells into distinct cell types (Rehman et al. 2004; Philpott et al. 1994).

Stem cells (SCs) have a demonstrated immunomodulatory effect that is mediated through the regulation of various cytokines. These cytokines can be broadly categorized into anti-inflammatory and pro-inflammatory factors. SCs upregulate several key anti-inflammatory cytokines, which contribute to their immunosuppressive properties. These anti-inflammatory cytokines are hepatocyte growth factor (HGF), prostaglandin E2 (PGE2), transforming growth factor-beta (TGF-β), indoleamine 2, 3-dioxygenase (IDO), nitric oxide (NO), interleukin-10 (IL-10). These cytokines work together to reduce the inflammatory response and promote tissue repair. For instance, IL-10 plays a crucial role in limiting inflammation, while TGF-β and PGE2 are involved in tissue remodeling and immune cell regulation. On the other hand, SCs have been shown to decrease the production of several pro-inflammatory cytokines, which further supports their anti-inflammatory effect. These pro-inflammatory cytokines include tumor necrosis factor-alpha (TNF-α), interleukin-6 (IL-6), interleukin-1 (IL-1), interleukin-12 (IL-12). The reduction of these pro-inflammatory cytokines contributes to a decrease in immune system activation, which is beneficial in conditions where excessive inflammation is harmful, such as autoimmune diseases and chronic inflammatory conditions (DelaRosa et al. 2009).

SCs' paracrine properties help start and speed up the healing process for wounds. Increased levels of IL-6, IL-8, and monocytes chemoattractant protein 1 were seen after ASC transplantation, which facilitates monocytes migration to the inflammatory areas (Lee et al. 2010). Subsequently, numerous secretory proteins, such as connective tissue growth factor, plasminogen activator inhibitor-1, hepatocyte growth factor, FGF-1, granulocyte colony-stimulating factor, granulocyte-macrophage colony-stimulating factor, vascular endothelial growth factor, and TGF-β, have been associated with the beneficial wound healing effects of adipose-derived stem cells (ASCs) (Moon et al. 2012).

13.6.2 Cosmetic Procedure

It might be challenging to distinguish reconstructive surgeries from cosmetic procedures when writing up a history of cosmetic surgery. The aim of facial reconstruction is to enhance and improve a distorted or abnormal facial appearance, with the goal of achieving beauty. Sushruta provides descriptions of various methods for addressing different types of defects, including skin release for small defects, flap rotation to compensate for partial loss, and the use of pedicle flaps to cover complete skin loss in a specific area (Wallace 1978). In Sushruta's writings, he discusses labioplasty and establishes the fundamental principles of plastic surgery, emphasizing the importance of appropriate physiotherapy before performing the procedure. Additionally, he describes several techniques, including pedicle grafts, sliding grafts, and rotating grafts. Sushruta's work on this subject highlights the significant attention given to essential surgical principles, including accuracy, precision, cost-effectiveness, hemostasis, and achieving optimal results (Wallace 1978; Shiffman and Di Giuseppe 2012).

Even though Tagliacozzi's 47 pages of illustrations and instructions for reconstructive surgery were published in 1597, some people still regarded him as the "Father of Plastic Surgery." Nonetheless, his accomplishments were lost when he passed away 2 years later (Wallace 1978; Shiffman and Di Giuseppe 2012).

In 1827, the first plastic surgeon in the United States, Dr. John Peter Mettauer, performed the nation's first cleft palate surgery using his own instruments. In 1869, Joseph performed an otoplasty at the University Polyclinic in Orthopedic Surgery under the direction of Professor Julius Wolff, which led to his dismissal from his position (Wallace 1978; Shiffman and Di Giuseppe 2012). In the early nineteenth century, Chopart used a flap from the neck to do an operation on the lip.

World War I (1914–1918) resulted in many soldiers suffering severe facial injuries, particularly from shrapnel and gunfire, which led to a need for reconstructive surgery. Gillies established a dedicated military hospital in London, known as the Queen's Hospital, where he developed and refined techniques for reconstructing the faces of these soldiers, often using flap surgery, skin grafts, and prosthetics. The "Reconstructive Surgery of the Face" paper that Blair published set the bar for craniofacial reconstruction techniques like rhinoplasticity for saddle noses (Blair 1921), nasal deformities related to the congenital cleft of lip (Blair 1925), Cleft palate with the orthodontic inclusion (Hopkins et al. 1999), correction of external nose abnormalities (Blair 1932), primary blockage of harelips (Blair and Robinson 1948), and the utilization of a switched lower lip flap to restore the upper lip (Blair and Letterman 1950).

13.6.3 Use of Stem Cells in Cosmetic Surgery

The uses of stem cells in cosmetic (aesthetic) surgery are lucrative. With $11 billion spent on aesthetic operations in the US in 2012 (Surgeons ASoP 2010). The demand for cosmetic operations has grown over the last 10 years. This is probably because of a mix of factors, including an aging population, a desire to seem young, a rise in the acceptance of cosmetic surgery in society, and the availability of minimally invasive alternatives (it highlights a shift in medical practices toward less traumatic, more efficient procedures that prioritize patient comfort, faster recovery, and fewer complications). The American Society of Plastic Surgeons said in 2012 that its members had conducted 14.6 million cosmetic treatments in the previous year, an increase of 5% over 2011 (Surgeons ASoP 2010).

It's interesting to note that there has been a decline in the number of surgical procedures performed and give rise to the number of minimally invasive treatments. In 2012, facial operations grew the fastest (Surgeons ASoP 2010). In cosmetic and plastic procedures, stem cell-based therapies are frequently used because of their capacity to regenerate and repair a variety of tissues and organs. Different stem cell types are used clinically in this process (Khandpur et al. 2021). Various stem cells have reportedly been used in cosmetic surgery procedures for breast augmentation, body contouring, skin rejuvenation, and lipoatrophy of the face. However, because they are simple to acquire and have the ability to differentiate into mesenchymal, ectodermal, and endodermal lineages, adipose-derived stem cells continue to be the most often used by aesthetic surgeons (Zarei and Abbaszadeh 2018).

Stem cell therapy will become more popular as long as there is a desire for cosmetic procedures. Therefore, it is crucial for clinicians to understand the present marketing practices and the ethical and professional responsibilities that underlie them.

13.6.4 Adult Stem Cells in Regenerative and Cosmetic Surgery

Embryonic-stem cells were initially identified and described by Dr. James at the University of Wisconsin in 1998 (Nguyen et al. 2016; Zuk et al. 2001). With their inherent pluripotent capabilities and ability to differentiate into various mature cell types, cosmetic surgeons can now utilize (embryonic stem cells) as regenerative solutions for tissue defect surgeries. The use of embryonic stem cells (ESCs) in research has been restricted due to both ethical and clinical concerns. Ethically, the isolation of ESCs involves the destruction of human embryos, which raises significant moral questions about the status of the embryo and the potential for exploitation. Clinically, ESCs pose safety risks due to their non-immunocompatibility, meaning they may be rejected by the recipient's immune system if transplanted, and their potential to form malignant tumors (teratomas), which raises concerns about their long-term safety in therapies (Ben-David and Benvenisty 2011). However, the successful isolation and cultivation of ESCs have facilitated this development, making it an efficient and viable option (Thomson et al. 1998).

Induced pluripotent cells or iPS cells are genetically altered cells that mimic embryonic stem cells. The original excitement surrounding the use of these cells centered on the possibility for researchers to make use of a cell with a similar level of pluripotency without having to worry about the ethical issues associated with embryonic stem cells (Jarajapu and Grant 2010). However, due to concerns regarding genetic change and tumorigenic potential, clinical indications should be used cautiously (Thesleff et al. 2011).

Adult stem cells, although not as strong as embryonic stem cells, have tremendous promise in regenerative medicine and are of great importance to plastic surgeons because of their abundance in fat (Rajashekhar et al. 2014; Liras 2010; Sándor et al. 2013). They can change into a small number of cellular phenotypes within a single family, and it has been suggested that they might use paracrine influences to trigger more local cell transformation in a microenvironment. In 2001, Zuk et al. discovered that mesenchymal stem cells are abundant in adipose tissue (Zuk et al. 2001). Due to their accessibility, lack of significant bioethical concerns, and abundant availability, mesenchymal stem cells (MSCs) are the most widely used stem cell source in aesthetic therapies (Wang et al. 2019). The most widely employed MSCs in dermatology are adipose-derived stem cells (ADSCs), which are generated following in vitro expansion or when combined with stromal vascular fraction

(SVF) (Ong and Sugii 2013). The initial approach is costly and time-consuming, yet it yields a cell fraction consisting of homogeneous cells with a well-defined phenotype. In contrast, SVF (stromal vascular fraction) is a more cost-effective alternative to other stem cell sources, such as embryonic stem cells (ESCs) or induced pluripotent stem cells (iPSCs), because it can be harvested directly from the patient's own tissue, typically through liposuction. This procedure is minimally invasive, can often be done in a single surgical session, and does not require the extensive cell culture and complex processing that other stem cell sources might need. Lipoaspirate cells, once isolated and centrifuged, can be immediately used for patient treatment. However, the exact cell type responsible for the regenerative effects remains unclear due to the heterogeneous nature of the cell fraction (Mehrabani et al. 2013). This terminology can be misleading, as the isolation process still yields a mixed population of cells, similar to those found in adipose tissue (Shiffman et al. 2014). It is possible to locate equipment for the automatic isolation of ADSCs on the market. In vitro cultivation after isolation is the only technique to get the right quantity of the homogeneous adipose-derived stem cells population. The stromal vascular fraction comprises fibroblasts, endothelial cells, smooth muscle cells, pericytes, immune cells, and preadipocytes. During SVF culture, many of these cell types are progressively eliminated, leading to a predominantly preadipocyte-rich population. The main challenge lies in the heterogeneity of the cell fraction, which accounts for its multipotent stem cell characteristics (Wankhade et al. 2016). For full-thickness skin defect healing, a cell source originating from adipose tissue was examined.

13.7 The Application of Stem Cells in Soft Tissue Augmentation

The regeneration and enhancement of soft tissue in cosmetic surgery face challenges due to limitations in current therapies. Biomaterials often result in complications like infection, fibrosis, and tissue contracture, while alternatives such as composite tissue flaps, autologous fat transplantation, and fat grafting show promise as potential solutions (Laochumroonvorapong et al. 2002).

13.8 Adipose-Derived Stem Cell-Derived Exosomes (ASC-Exos) in Fat Transfer

For tissue filling in cosmetic procedures like facial lipodystrophy, lower limb atrophy, and breast augmentation and reconstruction, autologous fat has been reported to contain a variety of cells (Gir et al. 2012), with ADSCs being the most significant (Brown et al. 2010). ADSCs are responsible for tissue regeneration through the secretion of vascular endothelial growth factor, which stimulates the neovascularization of new tissues (Sheng et al. 2011). ASCs and fat grafts can work together to inhibit collagen formation and reduce wrinkles (Kim et al. 2019). The survival rate of fat transplants remains low due to ischemia, devascularization, and absorption issues caused by injections and liposuction, often leading to partial necrosis and resorption (Lendeckel et al. 2004). However, technical advancements aim to enhance the survival and structural integrity of fat grafts for improved outcomes.

13.8.1 Cell-Assisted-Lipotransfer (CAL)

Cell-assisted lipotransfer (CAL) is an advanced technique that combines lipoaspirate with stromal vascular fraction (SVF), rich in adipose-derived stem cells (ASCs), to enhance adipogenesis and angiogenesis, improving fat graft reliability (Yoshimura et al. 2020). Research by Han et al. highlights that ASC-derived exosomes (ASC-exos) further support fat graft recovery and functional filling, making them a promising adjunct in enhancing fat transfer outcomes (Bai et al. 2018a). Recent studies reveal that hypoxia-treated ASC-derived exosomes (ASC-exos) significantly enhance neovascularization and reduce inflammation in fat grafts, as demonstrated in immunodeficient mouse models using subcutaneous fat grafting techniques. At the molecular level, hypoxia enhances ASC-exos' ability to promote angiogenesis in ischemic adipose tissues by regulating key signaling pathways, such as VEGF/VEGF-R, highlighting their potential to improve fat graft survival and functionality (Bai et al. 2018a). ASC-exosomes, the functional nanovesicles secreted by adipose-derived stem cells (ASCs), offer superior benefits over ASCs in improving fat graft volume retention. According to Chen B. et al. ASC-exos not only match ASCs in efficacy but also extend graft longevity by modulating inflammation and enhancing angiogenesis in fat grafting, as demonstrated in mouse models (Chen et al. 2019). ASC-exosomes demonstrate significant adipogenic effects and promote collagen synthesis during the mid-to-late stages of fat grafting, mirroring their parent ASCs. Zhu et al. reported that in a live mouse model of homologous fat grafting, ASC-exosomes notably prolonged graft viability, enhanced neovascularization, improved fat retention, and reduced fibrosis and necrosis compared to controls, showcasing their potential to optimize fat grafting outcomes (Zhu et al. 2020). The M2 macrophage-conditioned medium (M2 macrophage-CM) appears to play a role in promoting the transformation of adipose tissue into a type that resembles brown fat, leading to an increase in energy expenditure. This effect is attributed to the ASC-EVs being taken up by macrophages, which, in turn, stimulated M2-type polarization and the production of catecholamines (Zhu et al. 2020). These findings imply that, as a cell-free approach, ASC-exos might be a viable and desirable way to improve the survival of grafts in lipotransfer as shown in Fig. 13.5. However, the significant disadvantages of fat transplantation are mainly due to the unpredictable nature of analytical results. However, the significant rates of absorption by volume are typical, but the duration of this effect and whether AS cell-exos from healthy

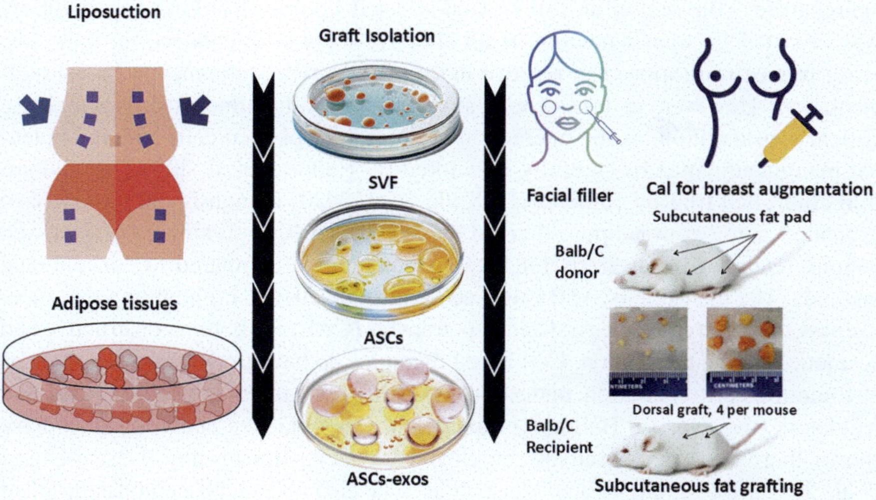

Fig. 13.5 Fat grafting process commences with the collection of sterile adipose tissue using liposuction. Subsequently, the collected tissue undergoes enzymatic digestion and centrifugation to obtain a heterogeneous mixture known as cell-assisted lipo-transfer (CAL). Within this mixture, the stromal vascular fraction (SVF) is isolated, which contains valuable adipose-derived stem cells (ASCs). Finally, the enriched SVF is combined with the fat and then injected into the targeted grafting site

individuals and sick individuals have equal efficacy remains unknown. In order to optimize fat grafting without causing adverse effects, the ideal concentration and sources of ASC-exos enrichment must be identified through more experimental and clinical investigations (see Fig. 13.5).

13.8.2 Breast Repair and Augmentation

Yoshimura and colleagues (Yoshimura et al. 2020) conducted clinical trials involving more than 40 patients to evaluate the efficacy of cell-assisted lipotransfer (CAL) for aesthetic breast augmentation. The findings revealed that 6 months post-surgery, all patients exhibited an increase in breast circumference without experiencing significant complications (Liu et al. 2024). Similarly, studies by Wang and Kamakura et al. confirmed that CAL effectively enhanced breast volume and improved contour. Notably, these studies also highlighted the absence of serious side effects associated with the procedure, further supporting its safety and efficacy (Wang et al. 2012). Breast reconstruction for cancer patients calls for the fusion of tumor safety and beauty. Fat grafting in mammoplasty can complicate breast imaging and cancer detection due to the formation of nodules, tissue calcifications, and other changes in breast tissue. These alterations may obscure mammographic findings, making it more challenging to differentiate between benign changes and potential malignancies (Bayram et al. 2019). Cell-assisted lipotransfer (CAL), particularly ASC-assisted fat transplantation, is an effective method to enhance fat graft survival, reduce resorption, and achieve desirable cosmetic outcomes in breast augmentation. However, in breast reconstruction, ensuring tumor safety remains a critical consideration, as the interaction between transplanted cells and any potential malignancy must be carefully monitored (O'Halloran et al. 2017). A randomized controlled trial by Kolle et al. (Kølle et al. 2020) demonstrated that ex-vivo expanded autologous adipose-derived stem cells (ASCs) significantly improved volume retention in fat grafts for breast augmentation compared to conventional methods. The inclusion of ASCs showed no adverse effects, highlighting the effectiveness and safety of Cell-Assisted Lipotransfer (CAL) for both reconstructive and cosmetic volume restoration. Cell-based therapies do have drawbacks, too, such as uncommitted differentiation, undesirable side effects, immunological rejection, and regulatory obstacles (O'Halloran et al. 2018). Castro-Govea and colleagues have shown that CAL was effectively employed for facial lipoatrophy (Castro-Govea et al. 2012). Facial augmentation, such as a facelift and facial contouring, all of which are cosmetic surgery specialties, and the subjective clinical improvements were all noted in these procedures (Lee et al. 2012). Koh and colleagues conducted a study to evaluate the efficacy of fat grafts enriched with concentrated adipose-derived stem cells (ADSCs) compared to fat grafts alone, focusing on fat volume reabsorption. This study involved patients undergoing cosmetic surgeries for hemifacial atrophy. The findings revealed that fat grafting enriched with concentrated ADSCs resulted in significantly less fat reabsorption compared to fat grafts alone (Koh et al. 2012). These results suggest that ADSCs may enhance the maintenance and volume-restoring capacity of transplanted fat.

13.9 The Role of Stem Cells in the Healing of Scars

The therapeutic potential of adipose-derived stem cells (ADSCs) in treating cosmetic scars has been explored extensively. Kim and colleagues demonstrated the effectiveness of scar therapy using differentiated adipocytes derived from ADSCs, showcasing their ability to improve scar appearance and promote skin regeneration. After 3 months of treatment with ADSCs, they found that the volume of scars had recovered by roughly 75% (Kim et al. 2011a). Other ADSC preparations have been described, such as Zhu et al.'s stem cell-enriched tissue injections, which led to long-term graft retention of fat-concentrated ADSCs (Zhu et al. 2010). The Kolle group's study revealed that using collagenase-processed adipose-derived stem cells (ADSCs) for cell-assisted lipotransfer (CAL) significantly improved fat graft survival in human studies compared to conventional fat grafting methods. Despite the promising outcomes of stem cell-based approaches in cosmetic procedures, significant health concerns remain, particularly regarding the potential risk of breast cancer associated with using stem cells in aesthetic applications like breast augmentation. When MSCs and ADSCs are combined and employed during cosmetic breast augmentation surgery, there has been some evidence that this may raise the metastatic risk of developing breast cancer (Duruisseaux et al. 2017; Rowan et al. 2014). Numerous exposed, unattractive, or persistent wounds, like diabetic ulcers, are challenging to cure and can have an impact on one's mental health in addition to their functional limitations. Poor wound healing eventually results in pigmentation, protracted healing, ulcerative skin abnormalities, hypertrophic scars, or keloid formation (Akita 2019). Skin transplantation, negative pressure wound therapy, advanced therapeutic dressings, low-intensity laser therapy, and hyperbaric oxygen therapy are only a few of the conventional procedures used to treat skin and soft tissue trauma (Bellei et al. 2018). However, because the healing of a wound is a complex procedure involving several types of cells, growth hormones, and the extracellular matrix, some of the conventional treatments merely serve as a supplemental measure followed by unfavorable healing. ASCs and SVF have a high secretome group production capacity, which promotes cell migration, proliferation, and differentiation, as well as a healing microenvironment. As a result of the PI3K-AKT pathways being activated by integrin 1 and enhancing the migration and activities of ASCs, chronic refractory wounds improved (Wang et al. 2020; Javed et al. 2022). Interestingly, Bi et al. (2019) discovered that both SVF and human ASCs enhanced the activity of endo-theliocytes and fibroblasts, controlled the expression of genes, and cooperatively accelerated skin healing in the wound mouse model (see Fig. 13.6). Due to the ease of SVF applications, this research demonstrated that SVF could replace ASCs for the purpose of wound healing. The trophic factors of ASCs and SVF, such as cytokines, chemokines, and growth factors, are often what enable this healing function. Notably, exosomes harboring the ASC secretome have made it possible for a recently developed cell-free therapy (Qiu et al. 2020).

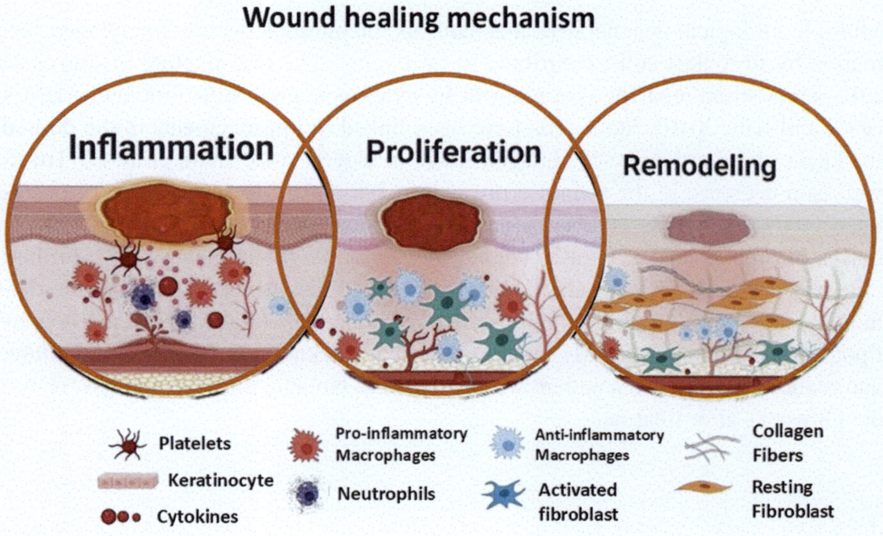

Fig. 13.6 Wound healing process

Furthermore, numerous findings have shown the function of ASC-exos miRNAs in skin regeneration. By raising the expression of matrix metalloproteinase 9 (MMP-9) via the PI3K-AKT pathways, Yang et al. (2020) discovered that miRNA-21, which is highly expressed and produced from ASC-exosomes, may promote the movement and proliferation of HaCaTs. The application of adipose-derived stem cells (ASCs) and their derivatives holds potential for advanced tissue repair methodologies characterized by minimally scarring regenerative healing. Exogenous adipose-derived stem cell-derived exosomes (ASC-exos) have been shown to accelerate the production of type I and type III collagen during the early stages of wound healing, thereby reducing healing time. In the later stages of healing, ASC-exos appeared to inhibit excessive collagen synthesis, contributing to reduce scarring (Yuan et al. 2016). This dual regulation aligns with the histological changes observed during the normal recovery process of soft tissue injuries. In the early phase, ASC-exos promoted significant collagen deposition essential for tissue repair, while in the late phase, they facilitated extracellular matrix (ECM) remodeling, emphasizing matrix reconstruction. By modulating the ratios of type III and type I collagen, balancing MMP-3/TIMP-1 (matrix metalloproteinase-3 and tissue inhibitor of metalloproteinase-1), and regulating TGF-β3/TGF-β1 signaling, ASC-exos optimized the differentiation of human dermal fibroblasts (HDFs). These actions collectively enhanced ECM rebuilding and minimized scar formation, suggesting a crucial role for ASC-exos in regenerative medicine and advanced wound healing therapies (Wang et al. 2017).

13.10 The Application of Stem Cells in Skin Rejuvenation

Multiple biological degenerative mechanisms, including a decline in collagen formation by fibroblast cells, contribute to skin aging. The manufacture of fibroblast collagen is stimulated for skin renewal by cytokines, growth factors, and ADSCs (Park and Kim 2010). Stem cells have been linked to improvements in the dermal thickness and the density of collagen fibers in elderly mice (Kim et al. 2011b) as well as a decrease in the wrinkles present on skin [119]. Researchers have postulated methods such as cutaneous angiogenesis and paracrine activation of fibroblasts, but the exact mechanism behind this action is yet unknown [118]. In their clinical pilot study (Kim et al. 2011b), Park and colleagues observed that intradermal injections of a mixture of adipose-derived stem cells (ADSCs) and autologous lipoaspirate cells significantly enhanced dermal thickness, improved skin texture, and reduced wrinkles in a patient with aging skin. Notable improvements were evident 8 weeks after treatment.

13.10.1 Skin Flap Procedures

Flap transplantation is a crucial technique for treating refractory injuries and organ restoration, such as deformities in the external ear, alar rim, and fingertip. Poor flap healing results are brought about by insufficient neovascularization and ischemia-reperfusion (I/R) damage (Sorkin et al. 2020). Despite the natural healing ability of a limited blood supply, it presents challenges, particularly around the facial region. This is due to the fact that during skin flap surgery, the distal part of the skin often receives insufficient blood flow, leading to potential issues. This issue has been addressed surgically using techniques including surgical delay prior to flap transfer (Atisha et al. 2009). According to Jeong and colleagues' research (Ho 2010). ASCs' angiogenic potential can effectively increase sufficient flow of blood to the distal part of the skin during aesthetic skin flap procedures because they can differentiate into flap tissue with new micro-vessels in ischemia circumstances. ASC-based therapy is now a practical way to stop I/R damage and support flap transplantation. Adipose-derived stem cells (ASCs) are effective in addressing defects in the nose, ear, and skin by stimulating angiogenesis, thereby enhancing the survivability of chondro-cutaneous composite grafts (Yucel et al. 2016). ASC-exosomes procedure gains importance by increasing the survival status of flaps produced by the skin. By using bioinformatics research, Xiong et al. discovered that the ASC-exos could be helpful in the regulation of expressions of the genes such as ITGA5 gene and HDAC5 gene, respectively, through upregulation of miRNA-760 and downregulation of miRNA-423-3p, which in turn aided in the vascularization of the skin flap. Microvascular angiography performed 28 days after the transplantation of skin flaps in a rat's leg wound model showed that the prefabricated artificial dermis flaps demonstrated higher levels of vascularization in the ASC-exos treated groups compared to control groups (Xie et al. 2020). Unquestionably, the vascularization of flaps is connected to a large number of differently expressed miRNAs in ASC-exos. In addition, Bai et al. (2018b) demonstrated that low-dose H_2O_2-stimulated ASC-exosomes could boost the flap's neovascularization and reduce inflammation and apoptosis following I/R injury, boosting the flap's in vivo survival rate. In conclusion, ASC-exos are crucial for the vascularization of skin flaps and help to tackle the issue of inadequate neovascularization, which broadens the scope of flap transplantation (see Fig. 13.7).

13.11 Benefits of Using Stem Cells in Cosmetic Surgery

In the fields of plastic and cosmetic surgery, a significant objective is to address congenital or acquired defects in morphology and function through diverse treatment methods. These approaches include microsurgery, nanotechnology, biomaterials, tissue allotransplantation, and medical imaging, all aimed at achieving effective repair and restoration (Yen et al. 2018). Important roles in immunological

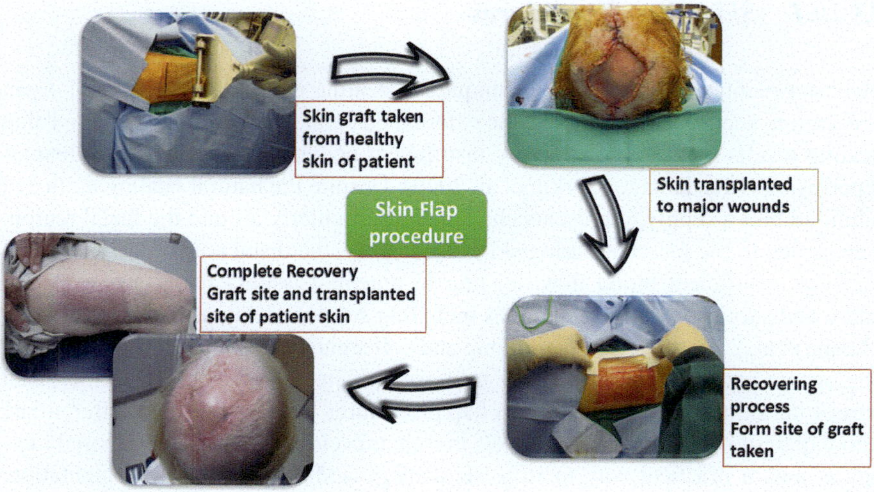

Fig. 13.7 The administration of ASC-derived exosomes (ASC-exos) can enhance neovascularization in the flap, reduce inflammation and apoptosis after ischemia/reperfusion (I/R) injury, and significantly improve the flap's in vivo survival rate

regulation, wound healing, and tissue regeneration are played by adipose tissue. Currently, the application of homologized adipose tissue is being researched for utilization in plastic and cosmetic surgery to improve contour abnormalities, regeneration of soft tissues, skin quality, and wound repair (Strong et al. 2016). Although its effectiveness and safety are well acknowledged, there aren't any broadly approved methods. However, because adipose tissue properties fluctuate greatly depending on donor status, the effect is unique to each person (Maki et al. 2009). Stem cell therapy has been used specifically in aesthetic surgery for scar reduction, skin renewal, and anti-aging (Nowacki et al. 2017). With its alleged anti-aging properties, its promotion to customers, particularly on social media, regularly alludes to the fountain of youth (McArdle et al. 2014).

13.12 Anti-Aging

The complex process of aging in the facial region affects the three-dimensional shape and textures. The skin is prone to both an exogenous factor-induced aging condition and an inherent, irreversible aging process. Premature skin aging, commonly referred to as extrinsic skin aging or photoaging, is caused mostly by UV radiation. The removal of suppleness and wrinkle development is attributable to the fact that skin cells and tissues undergo structural and functional alterations, which are the most common symptoms of skin aging (Li et al. 2019). Human dermal fibroblasts (HDFs) and keratinocytes (HaCaTs), which are the skin cells, are thought to act as defenses against UVB photoaging and skin aging over time. Antioxidants, peptides, homologous patient's fat, retinoids, growth factors, and collagen grafts are a few examples of improvement techniques that have been employed to combat skin aging (Kim et al. 2019). Adipose-derived stem cells (ASCs) have been shown to have a vital biological and therapeutic role in the reduction and prevention of photoaging dermal cells, in addition to the anti-aging properties of transforming growth factor-beta (TGF-b), which plays a critical role in maintaining skin integrity and promoting dermal rejuvenation (Zhou et al. 2013). They are typically used as seed cells in aesthetic plastic surgery. ASCs and ASC derivatives may be able to control HDF proliferation, expression, and migration of collagen and reduce skin aging. ASC-conditioned medium (ASC-CM) and ASC-derived exosomes (ASC-exos) contain essential cytokines and growth factors produced by ASCs, as shown in Fig. 13.8 (Li et al. 2019; Kim et al. 2008; Khan et al. 2012; Xu et al. 2020). These components play a vital role in the regeneration and repair of various tissues and organs. Moreover, they can enhance the skin's capacity to combat oxidation, reduce wrinkles, and address issues of discoloration (Zhou et al. 2013). It is proved that a well-prepared ASC-CM helped protect HDFs from some of the negative effects of intrinsic and extrinsic aging. The outcome also made clear that the PDGF-AA may help some other ASC-CM components provide better results. There is currently

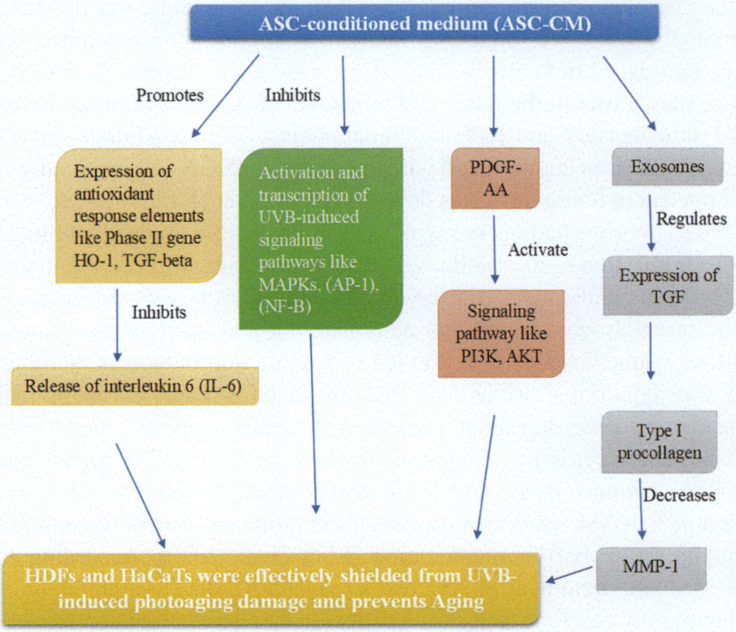

Fig. 13.8 Protective role of ASC-CM against UVB-induced damage. Transforming growth factor-beta (TGF-β), mitogen-activated protein kinases (MAPKs), activator protein 1 (AP-1), nuclear factor kappa B (NF-κB), platelet-derived growth factor AA (PDGF-AA), phosphatidylinositol-4,5-bisphosphate 3-kinase (PI3K), protein kinase B (AKT), interleukin-6 (IL-6), matrix metalloproteinase (MMP-1), Human dermal fibroblasts (HDFs) and keratinocytes (HaCaTs)

little ASC-exos skin aging research available. The human epidermis has a thickness of 2–5 mm and a layer of 50–120 m, making it possible for local exosome treatments to reach the epidermis and be absorbed (Xu et al. 2018). Exosomes generated from the stem cells of humans, such as ASCs, have a variety of bioactive properties for treating skin aging and merit further study (see Fig. 13.8).

13.12.1 Dermatitis Atopy

Dermatitis atopy is a long-lasting inflammatory skin condition that causes edema, pruritus, excoriation, thickening of the skin, and erythema. This reduces the skin's unkempt appearance (Yin et al. 2019). The development of dermatitis atopy or atopic dermatitis usually called AD is heavily influenced by both a compromised skin barrier and aberrant immunological responses. As a result, AD with immunologic abnormalities may be treated using various or combined strategies aimed at lowering the intensity and frequency of symptoms, including food management, medication, and UV-aided therapy (Park et al. 2019). Numerous studies have shown that bone marrow-derived stem cells (BMSCs) and ASCs can modulate numerous targets while suppressing allergy progression in AD (Shin et al. 2017). However, due to the chronic and recurrent nature of AD, the therapeutic use of MSCs has a number of disadvantages, including ineffective engraftment, unwanted immunological responses, a brief half-life, and issues with quality control (Lou et al. 2017). Exosomes play a role in the onset and prognosis of AD skin illnesses by mending damaged skin barriers and reducing inflammatory skin conditions (Alves et al. 2016). A model showing dermatitis atopy, using NC/NGA mice treated with mite antigens present in house dust, was developed by Cho et al. (2018). ASC-exos were discovered to improve pathogenic symptoms in this mouse model, including serum-IgE levels, infiltration of mast cells, the quantity of eosinophils present in the blood, and the dendritic cells present in the epidermis of skin lesions. In the subsequent study, the research team found that administering ASC-derived exosomes (ASC-exos) through subcutaneous injection led to notable improvements. It significantly reduced trans-epidermal water loss and enhanced stratum corneum hydration. Additionally, in a dose-dependent manner, it markedly decreased the levels of various inflammatory cytokines, including IL-4, IL-5, IL-13, TNF-alpha, interferon gamma (IFN-gamma), IL-17, and TSLP (Shi et al. 2020) (see Fig. 13.9). It's interesting to note that ASC-exos also stimulated ceramide and dihydroceramide synthesis to support skin barrier regeneration (Shin et al. 2020). According to these findings, systemic treatment of ASC-exos reduced AD-like symptoms by controlling inflammatory reactions and successfully reestablishing the functions of epidermal barrier present in AD, as shown in Fig. 13.9. ASC-derived exosomes might be a favorable cell-free choice for the few AD treatments available right now.

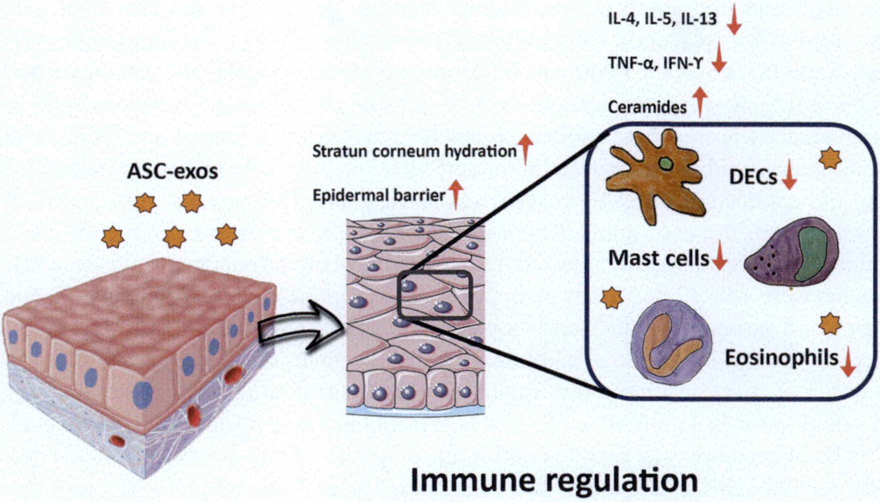

Fig. 13.9 ASC-derived exosomes (ASC-exos) are found to lead to considerable reductions in trans-epidermal water loss and significant enhancements in stratum corneum hydration. Furthermore, in a dose-dependent fashion, they demonstrated a remarkable ability to decrease the levels of various inflammatory cytokines, including IL-4, IL-5, IL-13, TNF-alpha, and interferon gamma (IFN-gamma). Importantly, ASC-exos also played a crucial role in stimulating the synthesis of ceramide and dihydroceramide, thereby supporting the regeneration of the skin barrier

13.12.2 The Application of Stem Cells in Hair Loss and Androgenic Alopecia

Loss of hair is primarily caused by a variety of factors, including genetics (trichodystrophy and androgenetic alopecia), coexisting medical conditions, hormonal interference (insulin resistance, thyroid disease), immune system interference (lupus erythematosus, patchy alopecia), nutritional interference, environmental interference (drugs, ultraviolet (UV) radiation), mental illness (stress, trichotillomania), and aging. Harmful factors affect the hair cycle (HC), decrease stem cell activity, and impair the ability of hair follicles to recover (Owczarczyk-Saczonek et al. 2018). The development of biotechnology is clinically required to promote hair growth and treat the hair loss problem, particularly in cases of androgenic alopecia (AGA). ASCCM (adipose-derived stem cell-conditioned media), which contained lyophilized proteins released by cultivated ASCs obtained from healthy adult women, has been studied for its effects (Fukuoka et al. 2017). Patients with AGA received four to six intradermal injections of the ASC-CM into the scalp every 3–5 weeks. Initial research found that patient satisfaction and the visual analog scale both significantly increased, along with subjective improvements in hair growth (Fukuoka et al. 2012). When compared to baseline, trichogram photos taken 1–3 months following the last treatment showed a considerable increase in the quantity of hairs. In a half-head comparative study, where one side of the scalp received injections of ASC-conditioned medium (ASC-CM) and the other side received saline (placebo), the number of hairs counted using a trichogram was significantly higher on the treatment side compared to the placebo side (Fukuoka and Suga 2015). Significant increases in hair density and thickness were observed in individuals who received another commercialized product named as ASC-CM in combination with the microneedle roller (Shin et al. 2015). A pilot study investigating the application of SVF in AGA was carried out. The pure fat graft and SVF created from the concentrated lipoaspirate following tumescent liposuction were mixed and subcutaneously injected into the scalp, where promising increases in hair counts were seen (Perez-Meza et al. 2017). The proportion of anagen hairs, telogen hairs, and variations in thickness, however, did not vary in a statistically significant way. In the case of hair loss, additional stem cell sources have been investigated. In a different study, mechanically fragmenting and centrifuging scalp punch biopsies resulted in the isolation of autologous SCs from human hair follicles (Gentile et al. 2017). Microscopy was used to confirm the existence of CD-44+ SCs from the dermal papilla and CD-200+ SCs from the bulge. After being administered intradermally, this suspension of SCs caused an increase in hair density and hair count in AGA patients.

13.13 Controversies, Challenges, and Safety Considerations in Stem Cell Therapy

Stem cell therapy holds significant promise for treating a wide range of previously incurable diseases, yet it faces several challenges and ethical concerns that must be addressed before it can be fully integrated into clinical practice. One of the major ethical dilemmas revolves around the use of embryonic stem cells (ESCs), which are the most well-known pluripotent stem cells. Since the discovery in 1998 that ESCs could be extracted from human embryos, their use has been a subject of ongoing controversy. The key issue lies in the fact that the process of isolating ESCs involves the destruction of the embryo, which has the potential to develop into a human being. This raises significant ethical questions and has led to opposition from various religious and societal groups. Recognizing the vast potential of stem cell therapy, scientists have worked to develop methods for obtaining stem cells without harming their source—the embryo. This approach seeks to make the process both scientifically viable and ethically acceptable while safeguarding the promising therapeutic possibilities that stem cell research offers. While human embryonic stem cells (hESCs) remain a controversial source due to ethical concerns, they hold significant potential for therapeutic applications in tissue regeneration.

Another critical challenge in stem cell therapy is ensuring good immunological tolerance between the stem cells and the patient's body. Both regenerative medicine and cell therapy face immunorejection issues. While mesenchymal stem cells (MSCs), placental tissue, and induced pluripotent stem cells (iPSCs) offer promising alternatives, immunorejection continues to be a major concern. One of the most promising strategies is to use the patient's own cells, reprogramming them to a pluripotent state. The goal is for these reprogrammed cells to fully replace damaged or lost natural cells, with seamless integration and functionality being crucial to the success of this approach.

The maintenance of genetic stability in stem cells, particularly induced pluripotent stem cells (iPSCs), is an area of ongoing concern. Genetic instability in these cells can lead to tumor formation, and their inherent properties—such as flexibility and self-renewal—may contribute to carcinogenesis within the host tissue (Filip et al. 2008). Additionally, the use of embryonic stem cells (ESCs) or iPSCs in therapeutic cell transplantation carries the risk of spontaneous teratoma formation and associated malignancies, further highlighting the need for rigorous safety measures in their clinical application (Cunningham et al. 2013).

Concerns regarding the safe administration of stem cells focus on ensuring the well-being of both the patient and the donor, without causing any morbidity or discomfort. Prior to applying stem cell therapy in regenerative procedures for living patients, thorough evaluations must be conducted to address potential risks, such as uncontrolled cell proliferation and improper differentiation (see Fig. 13.10). These evaluations are essential to ensure the safety and efficacy of stem cell-based treatments (Harris et al. 2004).

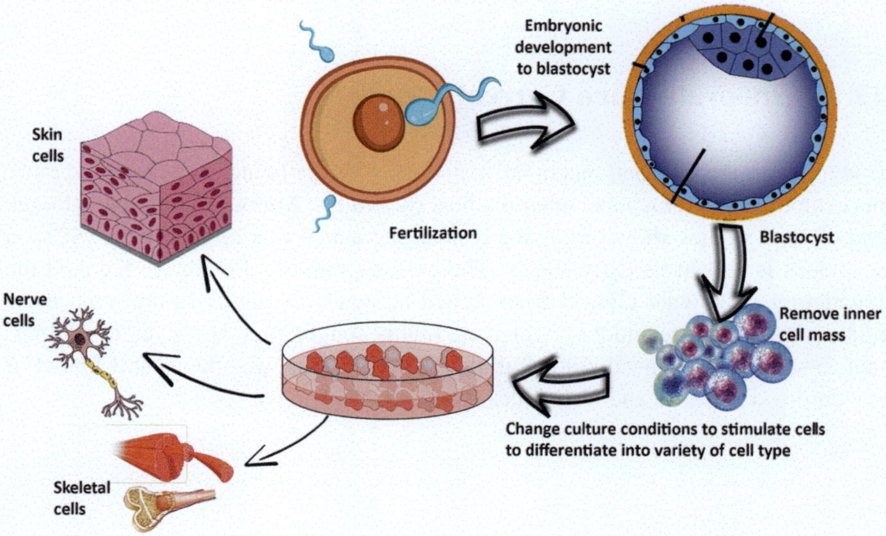

Fig. 13.10 Pluripotent stem cells from the inner cell mass are utilized and induced to differentiate into various cell types

13.14 The Role of Stem Cell Therapy in Advancing the Sustainable Development Goals (SDGs)

Stem cell therapy, with its immense potential in medicine and regenerative treatments, plays a crucial role in addressing several of the United Nations Sustainable Development Goals (SDGs) (Fatima et al. 2024). By enabling the regeneration of damaged tissues and organs, stem cells offer potential solutions to conditions that were once considered incurable. Their ability to regenerate and restore functionality aligns directly with SDG 3, which aims to ensure healthy lives and promote well-being for all at all ages. Stem cell therapy is not only a key player in healthcare but also a driving force in industrial innovation. The regenerative medicine sector, which utilizes stem cells for tissue engineering, biomanufacturing, and biofabrication, is expected to contribute to the growth of biotechnology industries globally. This aligns with SDG 9 s focus on fostering innovation, improving infrastructure, and increasing industrial capacity to support sustainable development. Stem cell therapy has the potential to contribute to responsible consumption and production by reducing waste in medical practices (SDG 12). In the case of organ transplantation, the need for donor organs could be reduced with advancements in tissue engineering. Stem cell research fosters global partnerships by bringing together research institutions, biotechnology companies, and health organizations from around the world (Farhan et al. 2023). This aligns with SDG 17, which promotes the strengthening of global partnerships for sustainable development. By contributing to healthcare innovation, industrial development, and environmental sustainability, stem cell therapy aligns with a variety of goals aimed at improving the quality of life for all people while promoting sustainable progress.

13.15 Innovation and Current Trends

Due to their paracrine and regenerative actions, SCs are being used more and more in cosmetic dermatology and other medical specialties. Although several preclinical and clinical studies show promising results, the actual role of stem cells (SCs) in aesthetics is still in its early stages. The existing published evidence has laid the foundation for crucial clinical trials aimed at standardizing and improving these minimally invasive techniques, which are highly sought after. However, it is important to note that advanced clinical trials are still lacking, and the establishment of standardized products and techniques is yet to be fully developed.

13.16 Future Perspective

In the last 10 years, the area of plastic surgery has successfully tested and used patient's own stem cells. The utilization of various stem cells in clinical research signifies a promising outlook for the advancement of innovative cell-based therapeutic approaches in the field of plastic surgery. However, to establish stem cell therapy as an accepted standard of care, further extensive clinical studies and research are necessary. The material that is now available supports the assertion that these methods are not yet ready for general clinical use by plastic surgeons. On the suitability of SCs for human usage, there is still inadequate information. Stem cells are currently not a first-choice therapy due to their short follow-up times and extremely expensive harvesting and culturing costs. However, the quick advancement of MSC research will undoubtedly enhance plastic surgeons' surgical methods with novel, ground-breaking therapeutic options.

Additional research is imperative to elucidate the fate of transplanted cells and determine the optimal quantity of cells essential for achieving clear therapeutic benefits. Validating the potential adverse effects of cell transplantation, such as tumor formation, along with comprehending the long-term consequences of these cells is of utmost importance. To ensure the safe collection, testing, and cryopreservation of donor cells, it is essential for facilities to adhere strictly to good manufacturing practices (GMPs).

13.17 Conclusion

According to the reviewed studies, stem cells from the adipose tissue are a breakthrough in plastic or cosmetic surgery, since they can differentiate into a variety of other tissues and are highly effective in cosmetic and reconstructive procedures. It was discovered that they favor angiogenesis when grafted, hastening the healing process and raising the procedure's success rate. In addition, compared to potential sources of external stem cells, it can reduce the risk of rejection and transmission of disease because it is an autologous transplant. It is currently unclear how most stem cells function exactly, how they interact biologically with extracellular matrix, how to isolate them, and what will happen to them in the long run. Stem cells have displayed significant promise in addressing tissue abnormalities, deformities, and illnesses. More recently, they have also proven to be a potent therapeutic tool for improving cosmetic and plastic surgery. While stem cells have demonstrated some successful outcomes in cosmetic breast augmentation using CAL (cell-assisted lipo-transfer), concerns persist regarding the long-term safety of this procedure with respect to potential breast cancer stimulation. Furthermore, because there is so little clinical and preclinical research looking into the use of stem cells in cosmetic surgery, it is still too early to draw any conclusive conclusions about their effectiveness. The combined efforts and studies of preclinical, clinical, and biomedical engineers will lead to any advancement in our understanding of the role of stem cells in cosmetic (aesthetic) surgery.

References

Akita S (2019) Wound repair and regeneration: mechanisms, signaling. Int J Mol Sci 20(24):6328

Alves NO, da Silva GT, Weber DM, Luchese C, Wilhelm EA, Fajardo AR (2016) Chitosan/poly (vinyl alcohol)/bovine bone powder biocomposites: a potential biomaterial for the treatment of atopic dermatitis-like skin lesions. Carbohydr Polym 148:115–124

Ambasudhan R, Talantova M, Coleman R, Yuan X, Zhu S, Lipton SA et al (2011) Direct reprogramming of adult human fibroblasts to functional neurons under defined conditions. Cell Stem Cell 9(2):113–118

Atisha D, Alderman AK, Janiga T, Singal B, Wilkins EG (2009) The efficacy of the surgical delay procedure in pedicle TRAM breast reconstruction. Ann Plast Surg 63(4):383–388

Augello A, Kurth TB, De Bari C (2010) Mesenchymal stem cells: a perspective from in vitro cultures to in vivo migration and niches. Eur Cell Mater 20(121):e33

Azizi SA, Stokes D, Augelli BJ, DiGirolamo C, Prockop DJ (1998) Engraftment and migration of human bone marrow stromal cells implanted in the brains of albino rats—similarities to astrocyte grafts. Proc Natl Acad Sci 95(7):3908–3913

Bai Y, Yan X-l, Ren J, Zeng Q, Li X-d, Pei X-t et al (2018a) Co-transplantation of exosomes derived from hypoxia-preconditioned adipose mesenchymal stem cells promotes neovascularization and graft survival in fat grafting. Biochem Biophys Res Commun 497(1):305–312

Bai Y, Yan X-l, Ren J, Zeng Q, Li X-d, Pei X-t et al (2018b) Adipose mesenchymal stem cell-derived exosomes stimulated by hydrogen peroxide enhanced skin flap recovery in ischemia-reperfusion injury. Biochem Biophys Res Commun 500(2):310–317

Bayram Y, Sezgic M, Karakol P, Bozkurt M, Filinte GT (2019) The use of autologous fat grafts in breast surgery: a literature review. Arch Plast Surg 46(06):498–510

Bellei B, Migliano E, Tedesco M, Caputo S, Papaccio F, Lopez G et al (2018) Adipose tissue-derived extracellular fraction characterization: biological and clinical considerations in regenerative medicine. Stem Cell Res Ther 9(1):1–18

Ben-David U, Benvenisty N (2011) The tumorigenicity of human embryonic and induced pluripotent stem cells. Nat Rev Cancer 11(4):268–277

Bi H, Li H, Zhang C, Mao Y, Nie F, Xing Y et al (2019) Stromal vascular fraction promotes migration of fibroblasts and angiogenesis through regulation of extracellular matrix in the skin wound healing process. Stem Cell Res Ther 10:1–21

Blair VP (1921) Rhinoplasty, with special reference to saddle nose. JAMA J Am Med Assoc 77(19):1479–1482

Blair P (1925) Nasal deformities associated with congenital cleft of the lip. JAMA J Am Med Assoc 84(3):185–187

Blair V (1932) Correction of losses and deformities of the external nose, including those associated with harelip. Cal West Med 36(5):308

Blair VP, Letterman GS (1950) The role of the switched lower lip flap in upper lip restorations. Plast Reconstr Surg (1946). 5(1):1–25

Blair V, Robinson R (1948) Primary closure of harelip. Surg Gynecol Obstet 86(4):502–504

Bongso A (2006) Blastocyst culture for deriving human embryonic stem cells. Methods Mol Biol 331:13–22

Bongso A, Richards M (2004) History and perspective of stem cell research. Best Pract Res Clin Obstet Gynaecol 18(6):827–842

Brown SA, Levi B, Lequex C, Wong VW, Mojallal A, Longaker MT (2010) Basic science review on adipose tissue for clinicians. Plast Reconstr Surg 126(6):1936–1946

Bruder SP, Jaiswal N, Haynesworth SE (1997) Growth kinetics, self-renewal, and the osteogenic potential of purified human mesenchymal stem cells during extensive subcultivation and following cryopreservation. J Cell Biochem 64(2):278–294

Castro-Govea Y, De La Garza-Pineda O, Lara-Arias J, Chacón-Martínez H, Mecott-Rivera G, Salazar-Lozano A et al (2012) Cell-assisted lipotransfer for the treatment of parry-romberg syndrome. Arch Plast Surg 39(06):659–662

Ceafalan LC, Enciu A-M, Fertig TE, Popescu BO, Gherghiceanu M, Hinescu ME et al (2018) Heterocellular molecular contacts in the mammalian stem cell niche. Eur J Cell Biol 97(6):442–461

Chen S, Tang Y, Liu Y, Zhang P, Lv L, Zhang X et al (2019) Exosomes derived from miR-375-overexpressing human adipose mesenchymal stem cells promote bone regeneration. Cell Prolif 52(5):e12669

Chimutengwende-Gordon M, Khan S, W. (2012) Advances in the use of stem cells and tissue engineering applications in bone repair. Curr Stem Cell Res Ther 7(2):122–126

Cho BS, Kim JO, Ha DH, Yi YW (2018) Exosomes derived from human adipose tissue-derived mesenchymal stem cells alleviate atopic dermatitis. Stem Cell Res Ther 9(1):1–5

Cowdry EV (1932) Special cytology: the form and functions of the cell in health and disease: a textbook for students of biology and medicine. (No Title)

Cunningham J, Ulbright T, Pera M, Looijenga L (2013) Lessons from human teratomas to guide development of safe stem cell therapies. Nat Biotechnol 31(6):849–857

De Miguel MP, Fuentes-Julián S, Alcaina Y (2010) Pluripotent stem cells: origin, maintenance and induction. Stem Cell Rev Rep 6:633–649

De Rooij DG, Grootegoed JA (1998) Spermatogonial stem cells. Curr Opin Cell Biol 10(6):694–701

DelaRosa O, Lombardo E, Beraza A, Mancheno-Corvo P, Ramirez C, Menta R et al (2009) Requirement of IFN-γ–mediated indoleamine 2, 3-dioxygenase expression in the modulation of lymphocyte proliferation by human adipose–derived stem cells. Tissue Eng Part A 15(10):2795–2806

Dor Y, Brown J, Martinez OI, Melton DA (2004) Adult pancreatic β-cells are formed by self-duplication rather than stem-cell differentiation. Nature 429(6987):41–46

Dragutin K, Maja P, Dragan K, Branislav M (2011) Alexander Brunschwig-110 years from birth September 11, 1901–August 7, 1969. Acta Chir Iugosl 58(3):21–23

Duruisseaux M, Besse B, Cadranel J, Pérol M, Mennecier B, Bigay-Game L et al (2017) Overall survival with crizotinib and next-generation ALK inhibitors in ALK-positive non-small-cell lung cancer (IFCT-1302 CLINALK): a French nationwide cohort retrospective study. Oncotarget 8(13):21903

Eun S-C (2014) Stem cell and research in plastic surgery. J Korean Med Sci 29(Suppl 3):S167–S1S9

Farhan BA, Zhihe L, Ali S, Shah TA, Zhiyu L, Zhang A et al (2023) Multiple strategies for the development of multienzyme complex for one-pot reactions. Environ Sci Pollut Res 30(24):64904–64931

Fatima G, Magomedova A, Parvez S (2024) Biotechnology and sustainable development. Shineeks Publishers

Filip S, Mokry J, Horacek J, English D (2008) Stem cells and the phenomena of plasticity and diversity: a limiting property of carcinogenesis. Stem Cells Dev 17(6):1031–1038

Friedenstein A, Chailakhjan R, Lalykina K (1970) The development of fibroblast colonies in monolayer cultures of guinea-pig bone marrow and spleen cells. Cell Prolif 3(4):393–403

Fukuoka H, Suga H (2015) Hair regeneration treatment using adipose-derived stem cell conditioned medium: follow-up with trichograms. Eplasty 15:e10

Fukuoka H, Suga H, Narita K, Watanabe R, Shintani S (2012) The latest advance in hair regeneration therapy using proteins secreted by adipose-derived stem cells. Am J Cos Surg 29(4):273–282

Fukuoka H, Narita K, Suga H (2017) Hair regeneration therapy: application of adipose-derived stem cells. Curr Stem Cell Res Ther 12(7):531

Gary G, Alexander S (2009) The maximow 1909 centenary: a reappraisal. Cell Therap Transplant 1(3):31–34

Gattazzo F, Urciuolo A, Bonaldo P (2014) Extracellular matrix: a dynamic microenvironment for stem cell niche. Biochim Biophys Acta 1840:2506–2519

Gentile P, Scioli MG, Bielli A, Orlandi A, Cervelli V (2017) Stem cells from human hair follicles: first mechanical isolation for immediate autologous clinical use in androgenetic alopecia and hair loss. Stem Cell Investig 4:58

Gimble JM, Katz AJ, Bunnell BA (2007) Adipose-derived stem cells for regenerative medicine. Circ Res 100(9):1249–1260

Gimble JM, Guilak F, Nuttall ME, Sathishkumar S, Vidal M, Bunnell BA (2008) In vitro differentiation potential of mesenchymal stem cells. Transfus Med Hemother 35(3):228–238

Gir P, Brown SA, Oni G, Kashefi N, Mojallal A, Rohrich RJ (2012) Fat grafting: evidence-based review on autologous fat harvesting, processing, reinjection, and storage. Plast Reconstr Surg 130(1):249–258

Harris M, Butler D, Boivin G, Florer J, Schantz E, Wenstrup R (2004) Mesenchymal stem cells used for rabbit tendon repair can form ectopic bone and express alkaline phosphatase activity in constructs. J Orthop Res 22(5):998–1003

Ho JJ (2010) Adipose stem cells and skin repair. Curr Stem Cell Res Ther 5(2):137–140

Hopkins J, Jackson I, Smith W, LeVasseur F (1999) Large tongue flaps to close massive palatal defects. Eur J Plast Surg 22:387–393

Horvitz HR, Herskowitz I (1992) Mechanisms of asymmetric cell division: two Bs or not two Bs, that is the question. Cell 68(2):237–255

Ilancheran S, Moodley Y, Manuelpillai U (2009) Human fetal membranes: a source of stem cells for tissue regeneration and repair? Placenta 30(1):2–10

Ilic D, Polak JM (2011) Stem cells in regenerative medicine: introduction. Br Med Bull 98(1):117–126

Jackson L, Jones D, Scotting P, Sottile V (2007) Adult mesenchymal stem cells: differentiation potential and therapeutic applications. J Postgrad Med 53(2):121–127

Jarajapu YP, Grant MB (2010) The promise of cell-based therapies for diabetic complications: challenges and solutions. Circ Res 106(5):854–869

Javed S, Farhan BA, Shabbir M, Tahseen A, Ahmedah HT, Moga M (2022) Morphology, pathogenesis, genome organization, and replication of coronavirus (COVID-19). The COVID-19 pandemic. Apple Academic Press, pp 3–43

Kapur SK, Katz AJ (2013) Review of the adipose derived stem cell secretome. Biochimie 95(12):2222–2228

Khan A, Vaibhav K, Javed H, Moshahid Khan M, Tabassum R, Ahmed ME et al (2012) Attenuation of Aβ-induced neurotoxicity by thymoquinone via inhibition of mitochondrial dysfunction and oxidative stress. Mol Cell Biochem 369:55–65

Khandpur S, Gupta S, Gunaabalaji D (2021) Stem cell therapy in dermatology. Indian J Dermatol Venereol Leprol 87(6):753–767

Kiefer JC (2011) Primer and interviews: the dynamic stem cell niche. Dev Dyn 240(3):737–743

Kim W-S, Park B-S, Kim H-K, Park J-S, Kim K-J, Choi J-S et al (2008) Evidence supporting antioxidant action of adipose-derived stem cells: protection of human dermal fibroblasts from oxidative stress. J Dermatol Sci 49(2):133–142

Kim M, Kim I, Lee SK, Bang SI, Lim SY (2011a) Clinical trial of autologous differentiated adipocytes from stem cells derived from human adipose tissue. Dermatologic Surg 37(6):750–759

Kim JH, Jung M, Kim HS, Kim YM, Choi EH (2011b) Adipose-derived stem cells as a new therapeutic modality for ageing skin. Exp Dermatol 20(5):383–387

Kim K, Fan Y, Lin G, Park YK, Pak CS, Jeong JH et al (2019) Synergistic effect of adipose-derived stem cells and fat graft on wrinkles in aged mice. Plast Reconstr Surg 143(6):1637–1646

Knoblich JA (2001) Asymmetric cell division during animal development. Nat Rev Mol Cell Biol 2(1):11–20

Knoblich JA (2008) Mechanisms of asymmetric stem cell division. Cell 132(4):583–597

Koh KS, Oh TS, Kim H, Chung IW, Lee KW, Lee HB et al (2012) Clinical application of human adipose tissue–derived mesenchymal stem cells in progressive hemifacial atrophy

(Parry-Romberg disease) with microfat grafting techniques using 3-dimensional computed tomography and 3-dimensional camera. Ann Plast Surg 69(3):331–337
Kolios G, Moodley Y (2012) Introduction to stem cells and regenerative medicine. Respiration 85(1):3–10
Kølle S-FT, Duscher D, Taudorf M, Fischer-Nielsen A, Svalgaard JD, Munthe-Fog L et al (2020) Ex vivo-expanded autologous adipose tissue-derived stromal cells ensure enhanced fat graft retention in breast augmentation: a randomized controlled clinical trial. Stem Cells Transl Med 9(11):1277–1286
Körbling M, Estrov Z (2003) Adult stem cells for tissue repair—a new therapeutic concept? N Engl J Med 349(6):570–582
Laochumroonvorapong P, Kokta V, Quan MB (2002) Trichilemmal carcinoma in an African American. Dermatologic Surg 28(3):284–286
Lee MJ, Kim J, Kim MY, Bae Y-S, Ryu SH, Lee TG et al (2010) Proteomic analysis of tumor necrosis factor-α-induced secretome of human adipose tissue-derived mesenchymal stem cells. J Proteome Res 9(4):1754–1762
Lee SK, Kim D-W, Dhong E-S, Park S-H, Yoon E-S (2012) Facial soft tissue augmentation using autologous fat mixed with stromal vascular fraction. Arch Plast Surg 39(05):534–539
Lendeckel S, Jödicke A, Christophis P, Heidinger K, Wolff J, Fraser JK et al (2004) Autologous stem cells (adipose) and fibrin glue used to treat widespread traumatic calvarial defects: case report. J Craniofac Surg 32(6):370–373
Li L, Ngo HT, Hwang E, Wei X, Liu Y, Liu J et al (2019) Conditioned medium from human adipose-derived mesenchymal stem cell culture prevents UVB-induced skin aging in human keratinocytes and dermal fibroblasts. Int J Mol Sci 21(1):49
Liras A (2010) Future research and therapeutic applications of human stem cells: general, regulatory, and bioethical aspects. J Transl Med 8:1–15
Liu L, Kim L, Teotia SS, Haddock NT (2024) Long-term Implications of Cosmetic Breast Surgeries on Subsequent Breast Reconstruction. Aesthet Surg J 44(12):1300–1308
Lou G, Chen Z, Zheng M, Liu Y (2017) Mesenchymal stem cell-derived exosomes as a new therapeutic strategy for liver diseases. Exp Mol Med 49(6):e346-e
Maehle A-H (2011) Ambiguous cells: the emergence of the stem cell concept in the nineteenth and twentieth centuries. Notes Rec R Soc Lond 65(4):359–378
Majo F, Rochat A, Nicolas M, Jaoudé GA, Barrandon Y (2008) Oligopotent stem cells are distributed throughout the mammalian ocular surface. Nature 456(7219):250–254
Maki M, Leung KT, Qin W (2009) The prospects of cellulase-producing bacteria for the bioconversion of lignocellulosic biomass. Int J Biol Sci 5(5):500
Marone M, Ritis D, Bonanno G, Mozzetti S, Rutella S, Scambia G et al (2002) Cell cycle regulation in human hematopoietic stem cells: from isolation to activation. Leuk Lymphoma 43(3):493–501
Martinez P, Ballarin L, Ereskovsky A, Gazave E, Hobmayer B, Manni L et al (2022) Articulating the "stem cell niche" paradigm through the lens of non-model aquatic invertebrates. BMC Biol 20(1):23
McArdle A, Senarath-Yapa K, Walmsley GG, Hu M, Atashroo DA, Tevlin R et al (2014) The role of stem cells in aesthetic surgery: fact or fiction? Plast Reconstr Surg 134(2):193
McCormick JB, Huso HA (2010) Stem cells and ethics: current issues. J Cardiovasc Transl Res 3:122–127
Mehrabani D, Mehrabani G, Zare S, Manafi A (2013) Adipose-derived stem cells (ADSC) and aesthetic surgery: a mini review. World J Plast Surg 2(2):65
Menasché P, Alfieri O, Janssens S, McKenna W, Reichenspurner H, Trinquart L et al (2008) The Myoblast Autologous Grafting in Ischemic Cardiomyopathy (MAGIC) trial: first randomized placebo-controlled study of myoblast transplantation. Circulation 117(9):1189–1200
Moodley Y, Ilancheran S, Samuel C, Vaghjiani V, Atienza D, Williams ED et al (2010) Human amnion epithelial cell transplantation abrogates lung fibrosis and augments repair. Am J Respir Crit Care Med 182(5):643–651

Moon KM, Park Y-H, Lee JS, Chae Y-B, Kim M-M, Kim D-S et al (2012) The effect of secretory factors of adipose-derived stem cells on human keratinocytes. Int J Mol Sci 13(1):1239–1257

Morrison SJ, Kimble J (2006) Asymmetric and symmetric stem-cell divisions in development and cancer. Nature 441(7097):1068–1074

Neumüller RA, Knoblich JA (2009) Dividing cellular asymmetry: asymmetric cell division and its implications for stem cells and cancer. Genes Dev 23(23):2675–2699

Nguyen A, Guo J, Banyard DA, Fadavi D, Toranto JD, Wirth GA et al (2016) Stromal vascular fraction: a regenerative reality? Part 1: Current concepts and review of the literature. J Plast Reconstr Aesthet Surg 69(2):170–179

Notarianni E, Galli C, Laurie S, Moor R, Evans M (1991) Derivation of pluripotent, embryonic cell lines from the pig and sheep. Methods 10:250

Nowacki M, Kloskowski T, Pietkun K, Zegarski M, Pokrywczyńska M, Habib SL et al (2017) The use of stem cells in aesthetic dermatology and plastic surgery procedures. A compact review of experimental and clinical applications. Postepy Dermatol Alergol 34(6):526–534

O'Halloran N, Courtney D, Kerin MJ, Lowery AJ (2017) Adipose-derived stem cells in novel approaches to breast reconstruction: their suitability for tissue engineering and oncological safety. Breast Cancer (Auckl) 11:1178223417726777

O'Halloran N, Potter S, Kerin M, Lowery A (2018) Recent advances and future directions in post-mastectomy breast reconstruction. Clin Breast Cancer 18(4):e571–ee85

O'Neill CA, Monteleone G, McLaughlin JT, Paus R (2016) The gut-skin axis in health and disease: a paradigm with therapeutic implications. Bioessays 38(11):1167–1176

Obradović S, Rusović S, Balint B, Ristić-Anđelkov A, Romanović R, Baškot B et al (2004) Autologous bone marrow-derived progenitor cell transplantation for myocardial regeneration after acute infarction. Vojnosanit Pregl 61(5):519–529

Ong WK, Sugii S (2013) Adipose-derived stem cells: fatty potentials for therapy. Int J Biochem Cell Biol 45(6):1083–1086

Oswald J, Boxberger S, Jørgensen B, Feldmann S, Ehninger G, Bornhäuser M et al (2004) Mesenchymal stem cells can be differentiated into endothelial cells in vitro. Stem Cells 22(3):377–384

Overturf K, Al-Dhalimy M, Ou C-N, Finegold M, Grompe M (1997) Serial transplantation reveals the stem-cell-like regenerative potential of adult mouse hepatocytes. Am J Pathol 151(5):1273

Owczarczyk-Saczonek A, Krajewska-Włodarczyk M, Kruszewska A, Banasiak Ł, Placek W, Maksymowicz W et al (2018) Therapeutic potential of stem cells in follicle regeneration. Stem Cells Int 2018:1049641

Park B-S, Kim W-S (2010) Adipose-derived stem cells and their secretory factors for skin aging. In: Textbook of aging skin. Springer, Berlin, pp 201–212

Park A, Park H, Yoon J, Kang D, Kang M-H, Park Y et al (2019) Priming with Toll-like receptor 3 agonist or interferon-gamma enhances the therapeutic effects of human mesenchymal stem cells in a murine model of atopic dermatitis. Stem Cell Res Ther 10:1–11

Passier R, Mummery C (2003) Origin and use of embryonic and adult stem cells in differentiation and tissue repair. Cardiovasc Res 58(2):324–335

Pennings S, Liu KJ, Qian H (2018) The stem cell niche: interactions between stem cells and their environment. Stem Cells Int 2018:4879379

Perez-Meza D, Ziering C, Sforza M, Krishnan G, Ball E, Daniels E (2017) Hair follicle growth by stromal vascular fraction-enhanced adipose transplantation in baldness. Stem Cells Cloning 10:1–10

Philpott MP, Sanders D, Westgate GE, Kealey T (1994) Human hair growth in vitro: a model for the study of hair follicle biology. J Dermatol Sci 7:S55–S72

Qiu H, Liu S, Wu K, Zhao R, Cao L, Wang H (2020) Prospective application of exosomes derived from adipose-derived stem cells in skin wound healing: a review. J Cosmet Dermatol 19(3):574–581

Rajashekhar G, Ramadan A, Abburi C, Callaghan B, Traktuev DO, Evans-Molina C et al (2014) Regenerative therapeutic potential of adipose stromal cells in early stage diabetic retinopathy. PloS one. 9(1):e84671

Rehman J, Traktuev D, Li J, Merfeld-Clauss S, Temm-Grove CJ, Bovenkerk JE et al (2004) Secretion of angiogenic and antiapoptotic factors by human adipose stromal cells. Circulation 109(10):1292–1298

Rosenbaum AJ, Grande DA, Dines JS (2008) The use of mesenchymal stem cells in tissue engineering: a global assessment. Organogenesis 4(1):23–27

Rossant J (2001) Stem cells from the mammalian blastocyst. Stem Cells 19(6):477–482

Rossant J (2008) Stem cells and early lineage development. Cell 132(4):527–531

Rowan BG, Gimble JM, Sheng M, Anbalagan M, Jones RK, Frazier TP et al (2014) Human adipose tissue-derived stromal/stem cells promote migration and early metastasis of triple negative breast cancer xenografts. PloS one 9(2):e89595

Sándor GK, Tuovinen VJ, Wolff J, Patrikoski M, Jokinen J, Nieminen E et al (2013) Adipose stem cell tissue–engineered construct used to treat large anterior mandibular defect: a case report and review of the clinical application of good manufacturing practice–level adipose stem cells for bone regeneration. J Oral Maxillofac Surg 71(5):938–950

Sheng L, Yang M, Li H, Du Z, Yang Y, Li Q (2011) Transplantation of adipose stromal cells promotes neovascularization of random skin flaps. Tohoku J Exp Med 224(3):229–234

Shi R, Jin Y, Hu W, Lian W, Cao C, Han S et al (2020) Exosomes derived from mmu_circ_0000250-modified adipose-derived mesenchymal stem cells promote wound healing in diabetic mice by inducing miR-128-3p/SIRT1-mediated autophagy. Am J Phys Cell Phys 318(5):C848–CC56

Shiffman MA, Di Giuseppe A (2012) Cosmetic surgery: art and techniques. Springer

Shiffman MA, Di Giuseppe A, Bassetto F (2014) Stem cells in aesthetic procedures: art, science, and clinical techniques. Springer

Shin H, Ryu HH, Kwon O, Park BS, Jo SJ (2015) Clinical use of conditioned media of adipose tissue-derived stem cells in female pattern hair loss: a retrospective case series study. Int J Dermatol 54(6):730–735

Shin T-H, Kim H-S, Choi SW, Kang K-S (2017) Mesenchymal stem cell therapy for inflammatory skin diseases: clinical potential and mode of action. Int J Mol Sci 18(2):244

Shin K-O, Ha DH, Kim JO, Crumrine DA, Meyer JM, Wakefield JS et al (2020) Exosomes from human adipose tissue-derived mesenchymal stem cells promote epidermal barrier repair by inducing de novo synthesis of ceramides in atopic dermatitis. Cells 9(3):680

Short B, Brouard N, Occhiodoro-Scott T, Ramakrishnan A, Simmons PJ (2003) Mesenchymal stem cells. Arch Med Res 34(6):565–571

Slack JM (2018) What is a stem cell? Wiley Interdiscip Rev Dev Biol 7(5):e323

Smart N, Riley PR (2008) The stem cell movement. Circ Res 102(10):1155–1168

Smith A (2006) A glossary for stem-cell biology. Nature 441(7097):1060

Sorkin A, Heller L, Landau G, Sherf M, Hartstein ME, Hadad E (2020) Inferiorly based preauricular flap for anterior ear reconstruction. Ann Plast Surg 84(4):394–396

Spradling A, Drummond-Barbosa D, Kai T (2001) Stem cells find their niche. Nature 414(6859):98–104

Strong AL, Bowles AC, Wise RM, Morand JP, Dutreil MF, Gimble JM et al (2016) Human adipose stromal/stem cells from obese donors show reduced efficacy in halting disease progression in the experimental autoimmune encephalomyelitis model of multiple sclerosis. Stem Cells 34(3):614–626

Surgeons ASoP (2010) National clearinghouse of plastic surgery statistics. American Society of Plastic Surgeons Arlinghton, Heights, IL

Tan F, Li X, Wang Z, Li J, Shahzad K, Zheng J (2024) Clinical applications of stem cell-derived exosomes. Signal Transduct Target Ther 9(1):17

Thesleff T, Lehtimäki K, Niskakangas T, Mannerström B, Miettinen S, Suuronen R et al (2011) Cranioplasty with adipose-derived stem cells and biomaterial: a novel method for cranial reconstruction. Neurosurgery 68(6):1535–1540

Thomson JA, Itskovitz-Eldor J, Shapiro SS, Waknitz MA, Swiergiel JJ, Marshall VS et al (1998) Embryonic stem cell lines derived from human blastocysts. Science 282(5391):1145–1147

Triana L, Palacios Huatuco RM, Campilgio G, Liscano E (2024) Trends in surgical and nonsurgical aesthetic procedures: a 14-year analysis of the International Society of Aesthetic Plastic Surgery—ISAPS. Aesth Plast Surg 48(20):4217–4227

Voog J, Jones DL (2010) Stem cells and the niche: a dynamic duo. Cell Stem Cell 6(2):103–115

Wallace AF (1978) History of plastic surgery: the early development of pedicle flaps. SAGE, London

Wang S, Zou D (2012) Viral dynamics in a distributed time delayed HCV pathogenesis model. Int J Biomath 5(06):1250056

Wang L, Lu Y, Luo X, Fu M-G, Hu X, Dong H et al (2012) Cell-assisted lipotransfer for breast augmentation: a report of 18 patients. Zhonghua Zheng Xing Wai Ke Za Zhi 28(1):1–6

Wang L, Hu L, Zhou X, Xiong Z, Zhang C, Shehada HM et al (2017) Exosomes secreted by human adipose mesenchymal stem cells promote scarless cutaneous repair by regulating extracellular matrix remodelling. Sci Rep 7(1):13321

Wang JV, Schoenberg E, Rohrer T, Zachary CB, Saedi N (2019) Stem cells in aesthetic dermatology: bioethical and professional obligations. Arch Dermatol Res 311:833–835

Wang J, Guo X, Kang Z, Qi L, Yang Y, Wang J et al (2020) Roles of exosomes from mesenchymal stem cells in treating osteoarthritis. Cell Reprogram 22(3):107–117

Wang S, Han J, Huang J, Islam K, Shi Y, Zhou Y et al (2024) Deep learning-based predictive classification of functional subpopulations of hematopoietic stem cells and multipotent progenitors. Stem Cell Res Ther 15(1):74

Wankhade UD, Shen M, Kolhe R, Fulzele S (2016) Advances in adipose-derived stem cells isolation, characterization, and application in regenerative tissue engineering. Stem Cells Int 2016:3206807

Xie X, Hui T, Luo Y, Li H, Li G, Wang Z (2020) Research on the properties of low temperature and anti-UV of asphalt with nano-ZnO/nano-TiO2/copolymer SBS composite modified in high-altitude areas. Adv Mater Sci Eng 2020:1–15

Xu P, Yu Q, Huang H, Zhang WJ, Li W (2018) Nanofat increases dermis thickness and neovascularization in photoaged nude mouse skin. Aesth Plast Surg 42:343–351

Xu T, Ma Y, Yuan Q, Hu H, Hu X, Qian Z et al (2020) Enhanced ferroptosis by oxygen-boosted phototherapy based on a 2-in-1 nanoplatform of ferrous hemoglobin for tumor synergistic therapy. ACS Nano 14(3):3414–3425

Yamashita YM (2010) Cell adhesion in regulation of asymmetric stem cell division. Curr Opin Cell Biol 22(5):605–610

Yang C, Luo L, Bai X, Shen K, Liu K, Wang J et al (2020) Highly-expressed micoRNA-21 in adipose derived stem cell exosomes can enhance the migration and proliferation of the HaCaT cells by increasing the MMP-9 expression through the PI3K/AKT pathway. Arch Biochem Biophys 681:108259

Yen YH, Pu CM, Liu CW, Chen YC, Chen YC, Liang CJ et al (2018) Curcumin accelerates cutaneous wound healing via multiple biological actions: the involvement of TNF-α, MMP-9, α-SMA, and collagen. Int Wound J 15(4):605–617

Yeung TM, Chia LA, Kosinski CM, Kuo CJ (2011) Regulation of self-renewal and differentiation by the intestinal stem cell niche. Cell Mol Life Sci 68:2513–2523

Yin F, Lin L, Zhan S (2019) Preparation and properties of cellulose nanocrystals, gelatin, hyaluronic acid composite hydrogel as wound dressing. J Biomater Sci Polym Ed 30(3):190–201

Yoshimura K, Sato K, Aoi N, Kurita M, Hirohi T, Harii K (2020) Cell-assisted lipotransfer for cosmetic breast augmentation: supportive use of adipose-derived stem/stromal cells. Aesth Plast Surg 44:1258–1265

You H-J, Han S-K (2014) Cell therapy for wound healing. J Korean Med Sci 29(3):311–319

Yuan Z, Druzhinina IS, Labbé J, Redman R, Qin Y, Rodriguez R et al (2016) Specialized microbiome of a halophyte and its role in helping non-host plants to withstand salinity. Sci Rep 6(1):32467

Yucel E, Alagoz MS, Eren GG, Yasar EK, Izmirli HH, Duruksu G et al (2016) Use of adipose-derived mesenchymal stem cells to increase viability of composite grafts. J Craniofac Surg 27(5):1354–1360

Zarei F, Abbaszadeh A (2018) Stem cell and skin rejuvenation. J Cosmet Laser Ther 20(3):193–197

Zhou B-R, Xu Y, Guo S-L, Xu Y, Wang Y, Zhu F et al (2013) The effect of conditioned media of adipose-derived stem cells on wound healing after ablative fractional carbon dioxide laser resurfacing. Biomed Res Int 2013:519126

Zhu M, Zhou Z, Chen Y, Schreiber R, Ransom JT, Fraser JK et al (2010) Supplementation of fat grafts with adipose-derived regenerative cells improves long-term graft retention. Ann Plast Surg 64(2):222–228

Zhu Y-z, Zhang J, Hu X, Wang Z-h, Wu S, Yi Y-y (2020) Supplementation with extracellular vesicles derived from adipose-derived stem cells increases fat graft survival and browning in mice: a cell-free approach to construct beige fat from white fat grafting. Plast Reconstr Surg 145(5):1183–1195

Zuk PA, Zhu M, Mizuno H, Huang J, Futrell JW, Katz AJ et al (2001) Multilineage cells from human adipose tissue: implications for cell-based therapies. Tissue Eng 7(2):211–228

Chapter 14
Gene Array Technology and the Search for Cosmeceutical Actives

Naveed Sultan, Faisal Mehmood, Muhammad Zulqurnain Haider, Qasim Ali, Amtul Sami, Sadia Javed, Shahid Adeel, Irum, Asia Shaheen, Muhammad Afzaal, and Muhammad Azeem

Contents

N. Sultan · M. Z. Haider (✉) · Q. Ali · Irum · A. Shaheen
Department of Botany, Government College University Faisalabad, Faisalabad, Pakistan
e-mail: drmzhaider@gcuf.edu.pk

F. Mehmood
Department of Environment Sciences, Government College University Faisalabad, Faisalabad, Pakistan

A. Sami
Department of Health Biotechnology, Women University, Swabi, Pakistan

S. Javed
Department of Biochemistry, Government College University Faisalabad, Faisalabad, Pakistan

S. Adeel
Department of Applied Chemistry, Government College University Faisalabad, Faisalabad, Pakistan

M. Afzaal
School of Environment and Life Sciences, Salford, UK

M. Azeem
Department of Biology, College of Science, University of Bahrain, Zallaq, Bahrain

S. Javed et al. (eds.), *Sustainable Cosmeceuticals*,
https://doi.org/10.1007/978-3-031-86087-4_14

Abstract Gene array technology enhances our understanding of gene identification and expression in various bacteria and higher species. Various techniques have been utilized to identify and express microorganism genes, but gene array methods excel in their sophistication. This chapter thoroughly explores gene array innovations, emphasizing their significance and potential applications in identifying and controlling genes within microorganisms that produce bioactive compounds with cosmeceutical and anti-aging properties. Bacteria, filamentous fungi, and higher fungi are recognized as prolific producers of metabolites with potential cosmeceutical actives. This technology aligns with SDGs by promoting the sustainable production of bioactive compounds for cosmeceuticals by reducing the reliance on synthetic chemicals, fostering innovation in eco-friendly manufacturing, supporting healthier skincare formulation, and minimizing environmental impact. The primary objective of this chapter is to elaborate on the impact of these bioactive chemicals' relevant genes, which can be used for skin health in a sustainable and natural manner without potential adverse effects.

Keywords Gene array technology · Microbes · Bioactive compounds · Cosmeceutical · Natural safe products · Microbial biotechnology

14.1 Introduction

The skin, recognized as the largest and most intricate organ in the human body, serves as a vital protective barrier against toxic compounds and disease-causing agents. Additionally, it plays a pivotal role in osmotic regulation and body temperature maintenance (Grice and Segre 2011; Yousef et al. 2017). Human skin comprises three layers—the outermost epidermis, the middle dermis, and the innermost hypodermis, as per histological classification. The collaborative functioning of these layers is instrumental (Agarwal and Krishnamurthy 2019; Lopez-Ojeda et al. 2017).

The epidermis hosts a diverse array of microorganisms, establishing both commensal and mutualistic relationships with the epidermal cells in a complex network of interactions (Simmering and Breves 2009; Bay et al. 2020). Many dermatological microbes residing on the skin provide substantial benefits to the host. Some activate innate immunity, while others produce antimicrobial substances inhibiting pathogen growth (Schommer and Gallo 2013). Maintaining a healthy skin microflora is crucial for human health, considering that the skin faces external challenges from natural environmental contaminants and optional cosmetics and cleansing products. These external factors are integral considerations in preserving the skin's well-being (Arshad et al. 2022; Adu et al. 2020).

Cosmeceutical products typically provide protection, nourishment, enhanced barrier functions, prevent pathogen formation, and significantly contribute to the overall health of the skin (Heinrich et al. 2014; Rodan et al. 2016; Gupta et al. 2019). Chemical surfactants play a central role in formulating these skin health products, serving as emulsifiers and foaming agents. However, research indicates that approximately half of the chemical surfactants on the market are derived from non-sustainable petrochemical sources (Arshad et al. 2018; Akbari et al. 2018).

While these petrochemical-based surfactants are successful in formulations, they may pose risks to the skin's flora (Holland and Bojar 2002; Lee et al. 2018). Some products have been reported to alter the natural skin flora by binding to lipids and proteins in the epidermis, leading to allergic reactions and skin irritability. Prolonged use and high concentrations can compromise the skin's barrier functions and structural integrity (Bujak et al. 2015). Consequently, there has been a shift toward replacing chemical surfactants with natural bioactives sourced from inexpensive and sustainable materials. These bioactive alternatives are characterized by low harmfulness, biodegradability, and skin compatibility, ensuring no adverse effects on consumers or the environment (Marchant and Banat 2012; Banat et al. 2014; Seweryn 2018).

Biological surfactants, originating from bacteria, yeasts, or filamentous fungi, are surface-active chemicals produced as secondary metabolic products. Obtaining biosurfactants involves extraction, precipitation, and distillation from these microorganisms, without the need for organic synthesis before, during, or after production. Consequently, they are termed naturally derived surfactants (Banat et al. 2014; Santos et al. 2016; Varvaresou and Iakovou 2015). The molecular structures of biosurfactants from microbes vary widely, influenced by environmental conditions, culture, and substrate type (Santos et al. 2016).

In addition to their role in moistening, emulsification, surface tension reduction, and detergency, biosurfactants offer several advantages. They are less toxic, biodegradable, compatible with human skin, and exhibit stability under extreme conditions such as pH, temperature, and salinity. Moreover, they are renewable (Banat et al. 2014; Varvaresou and Iakovou 2015; Sahnoun et al. 2014; Vijayakumar and Saravanan 2015). Noteworthy biologically derived biosurfactants, including rhamnolipids, sophorolipids, mannosyl erythritol lipids (MELs), and surfactin, are gaining increased attention globally and show potential applications in various industries, including the cosmeceutical sector (Lourith and Kanlayavattanakul 2009; Naughton et al. 2019).

The incorporation of biosurfactants into cosmetic formulations highlights the critical role in advancing sustainability in the cosmetic industry by addressing the goals within the framework of sustainability. Biosurfactants derived from renewable sources through innovative bioprocesses offer a cost-effective, biodegradable, and nontoxic alternative to synthetic surfactants (Mgbechidinma et al. 2022). Their utilization in cosmetic formulation, such as shower gel, enhances product efficacy as well as ensures adherence to eco-friendly and socially responsible practices (Manga et al. 2021). Their utilization of cosmetic products supports sustainable developmental goals 12 (SDG) 12 by fostering sustainable production, reducing waste, and optimizing the use of natural ingredients. This strategy guarantees the development of better environmentally friendly products that satisfy the growing demand from customers for eco-friendly solutions (Seweryn et al. 2023).

Retinol possesses anti-aging properties, including the stimulation of collagen synthesis in photoaged skin and inhibition of UV-induced matrix metalloproteinase induction (Fisher et al. 1997; Blaner and Olson 1994). Despite the significant commercial value of retinol, attempts to enhance its yield through genetic modification of *E. coli* and *S. cerevisiae* pathways have proven challenging and inaccessible to researchers (Jang et al. 2015; Sun et al. 2019). The susceptibility of retinol and retinoic acid to break down in the presence of oxygen, UV light, high temperatures, and other environmental factors due to their chemical makeup poses challenges (Lian et al. 2017).

The chemical production of retinyl palmitate presents disadvantages, such as non-site-specific esterification, the need for multiple refinements, and the use of corrosive catalysts during esterification (Xu et al. 2018). In contrast, the microbial synthesis of retinyl palmitate involves mild chemical processes, efficient catalytic reactions, and naturally high selectivity, making it a promising alternative for biological compound formation (Muthusamy et al. 2016). However, information regarding the natural production of retinyl palmitate within the microbial body is currently unavailable.

Enzymes play a crucial role in various biological pathways during retinoid biosynthesis. The production of retinaldehyde, also known as retinal, commences with the breakdown of β-carotene by β-carotene 15,15′-oxygenase (Bcox) (Kim and Oh 2009). Subsequent steps involve retinal dehydrogenase or retinal reductase oxidizing or reducing retinal for the formation of retinoic acid or retinol, respectively (Hoog and Ostberg 2011). Cytoplasmic retinol-binding protein (CRBP) binds retinol in vivo for storage and homeostasis before recombining with ester to form retinyl esters for further use (Menozzi et al. 2017). Retinyl ester, considered a stored form of retinol, is synthesized when a fatty acyl group combines with the hydroxyl terminal of retinol by lecithin retinol acyltransferase (LRAT) (O-Byrne and Blaner 2013). The fatty acid moiety in retinyl esters in the human body can be palmitic acid, oleic acid, or linoleic acid (Blaner and Olson 1994).

The phenotypic characteristics of organisms result from a combination of their genetic makeup and environmental influences (Lester and Bonner 1952; Jacob and Monod 1961). In simpler terms, genes must be "expressed" for their phenotypes to manifest (Gann 2010). Gene expression regulation occurs at multiple levels, as emphasized by Koch and Genomics. While the term "gene expression" is often used interchangeably with mRNA measurements, it's crucial to recognize that individual genes can produce a range of transcripts and proteoforms, contrary to the assumption of single-gene, single-product relationships in previous studies (Aebersold et al. 2018; Salovska et al. 2020).

The blueprint for biochemical reactions leading to the production of bioactive substances is stored in the biosynthetic genes of the fungal genome. These genes are typically physically linked and are often associated with regulators and transporters, forming extensive biological genomes (Robey et al. 2021; Keller 2019). Biosynthetic gene clusters (BGCs) control the production of specific compounds in response to specific signals. Initially silenced by heterochromatic structures at the chromosomal level, BGCs act as orchestrated factories, producing the compound for the cluster once the signal is activated (Schuller et al. 2022).

The CRISPR-Cas technology is employed for editing or transcriptional control of individual genetic loci based on the complementarity of their sequence with designed guiding RNAs (gRNAs). While CRISPR-Cas technologies are undoubtedly valuable, the use of individual gRNAs has limitations, impacting the technology's efficiency and potential applications in biotechnology (McCarty et al. 2020).

The application of a GeoChip enables the analysis of microbial communities in various environments, providing insights into their function, composition, and structure based on the examination of thousands of different gene families influencing biogeochemical processes. Developed multiple times over the last decade, the GeoChip microarray has proven to be a potent tool for high-throughput assessment of microbial communities (Wu et al. 2001; Rhee et al. 2004; He et al. 2007, 2010). Utilizing the GeoChip, researchers can identify diverse microbial communities and assess their compositions, structural properties, metabolic potential, and dynamics. These parameters can then be correlated with ecosystem interactions and processes.

This chapter delves into various bioactive compounds derived from microorganisms, encompassing bacteria, fungi, and microalgae, and their potential applications in the cosmetics and cosmeceutical industries. Emphasizing the significance of gene array technology, the chapter explores the identification and enhancement of responsible genes and their expression in microbes involved in the biological cycle for the biosynthesis of cosmeceutically important compounds. Notably, existing literature underscores a gap in researchers' attention to the utilization of gene array technology and its applications in exploring gene expression in microbes crucial for synthesizing bioactive secondary metabolites. This neglect extends to their potential roles in anti-aging, photo-anti-aging, and shielding the skin from UV damage.

14.2 Understanding Microbial Genes and Bioactives

To evaluate the value of secondary metabolites produced by microorganisms, understanding the molecular framework of these compounds is crucial for defining their effectiveness against harmful substances. Numerous bacterial strains have been documented for producing diverse secondary metabolites through specific biosynthetic pathways. Genetic studies on these bacterial strains have revealed the biological production of secondary metabolites, including non-ribosomal peptides (NRP), polyketides (PK), terpenes (TP), and hybrid PK/NRP (Kim 2011a, b; Song et al. 2013; Phonghanpot and Jarintanan 2022).

Several procedures must be followed for the isolation, identification, and maintenance of new microbial strains from the desired site, establishing an appropriate method for their study. After extracting 16S ribosomal RNA gene fragments from bacterial genomes, these fragments are amplified, sequenced, and phylogenetically compared to those of closely related strains using phylogenetic analysis (Tshikhudo et al. 2013). The use of degenerate primers allows the amplification of genes involved in secondary metabolite biosynthesis, reflecting the bacteria's ability to synthesize these compounds (Nicholson et al. 2001).

Even though a large number of genes have been identified from genome sequence data, only a small portion have been described for their roles in biosynthesis. Degenerate PCR amplification and genome sequencing data mining represent a new age of discovery, involving the evaluation of candidate genes potentially participating in the synthesis of new bioactive metabolites from bacterial species (Donadio et al. 2007; Omura et al. 2001; Weber et al. 2015; Phonghanpot and Jarintanan 2022). Newly identified microbial species may contain genes directly correlated with significant biosynthesis of secondary metabolites, as observed in several research studies.

Under standard laboratory conditions, the genes responsible for secondary metabolism in these microbes may remain inactive, preventing the synthesis of bioactive compounds. Techniques such as culture condition optimization or genetic engineering are required to activate the biosynthetic process (Bergmann et al. 2007).

In recent years, various high-throughput technological techniques, including next-generation sequencing and methods like single-cell genomics (Woyke et al. 2010; Huang and Zhou 2012), PhyloChip (Brodie et al. 2006, 2007; Schatz et al. 2010), and functional gene arrays (FGAs) such as GeoChip (He et al. 2007, 2010, 2012a), have been developed. These techniques have proven successful in the analysis of microbial communities and the identification of genes playing crucial roles in the biosynthetic pathways of secondary metabolites (Ansorge 2009; MacLean et al. 2009; Metzker 2010).

Primers compatible with both Roche 454 and Illumina platforms are currently available as part of next-generation sequencing technologies (Brisson et al. 2012; Deng et al. 2012; Díez et al. 2012; Sintes et al. 2013). GeoChip, an array of functional genetic markers targeting thousands of genes related to biogeochemical processes, enables comprehensive examination of microbial communities in different environments. Several GeoChip-like FGAs developed in the last decade have proven to be valuable tools for high-throughput microbial community investigation (Wu et al. 2001; Rhee et al. 2004; He et al. 2007, 2010). GeoChip 2.0 and GeoChip 3.0, earlier versions of GeoChip, have been utilized to explore microbial populations in diverse habitats (He et al. 2012a, b).

Extensive literature attests to the power of GeoChip as an FGA-based technology, offering insights into microbial community function, composition, structure, metabolism, dynamics, and their correlation with ecosystem processes (Tu et al. 2014).

Ferulic acid (FA) and other hydroxycinnamic acids are released by feruloyl esterase (FAE), a subclass of carboxylic acid esterases. FAEs are widely biosynthesized in bacteria, fungi, and plants, as observed in various biotechnological applications (Furuya et al. 2017; Gopalan et al. 2015; Oliveira et al. 2019). The antimicrobial and antioxidant properties of ferulic acid make it valuable as a food preservative, and it is also utilized as a medicinal agent due to its anti-inflammatory, antibacterial, antidiabetic, neuroprotective qualities, and sun protection factor (Paiva et al. 2013; Oliveira et al. 2019).

Carbohydrate Esterase Family 1 (CE1) is currently identified as FAEs, existing in diverse forms labeled A, B, C, D, and a probable type E. These forms share similarities in their amino acid sequence (Oliveira et al. 2019). Phylogenetic studies indicate that some microbial FAEs don't align with Crepin's classification, suggesting a putative type E subclass, particularly prevalent among bacterial enzymes. Members of this subclass, including *Ruminococcus* sp. RspXyn1, *Orpinomyces* sp. OspFaeA, *Clostridium thermocellum* XynZ, *R. flavefaciens* RfXynE, and *R. albus* RaXynB, are distinguished primarily by their amino acid sequences (Crepin et al. 2004).

Despite the functional classification based on its fungal origin, the probiotic FAE from *L. acidophilus* F46 belongs to the reputed type E, emphasizing the need for more comprehensive characterization (Kim and Baik 2015).

The literature review provides a clear illustration of the biosynthesis of small molecules by microbes, as evidenced by the identification of biosynthetic pathways through whole-genome sequencing. While the environmental conditions under which microbes produce these metabolites remain unclear, the products can be revealed through the heterologous expression of gene clusters (Zerikly and Challis 2009).

A landmark study sequencing the genome of *Streptomyces coelicolor* A3 (Wasinger et al. 1995) revealed a significant biosynthetic capability beyond what was expected from fermentation studies based on the genome sequence (Bentley et al. 2002). The genome analysis identified routes for natural compounds such as peptides, terpenoids, and polyketides, accurately predicting about 30 biosynthetic gene clusters. Laboratory purification had been achieved for only five natural products at that time—actinorhodin, prodiginine, calcium-dependent antibiotic, dodecaketide spore pigment, and hopenes. Molecular analysis of S. coelicolor's culture extract provided clear evidence of an additional 11 gene clusters involved in the biosynthesis of natural products or secondary metabolites (Nett et al. 2009).

Genome efforts of soil-dwelling bacteria, including *Streptomyces avermitilis* MA-4680 (Ikeda et al. 2003), *Streptomyces griseus* IFO 13350 (Ohnishi et al. 2008), and *Saccharopolyspora erythraea* NRL2338 (Oliynyk et al. 2007), known for therapeutically treasured natural compound production, increased the discovery of previously unseen compounds. Genome sequencing of *Myxococcus xanthus*, a species recognized for swarming and fruiting body-forming behaviors, revealed approximately 18 gene clusters implicated in the synthesis of bioactive metabolites (Goldman et al. 2006). Metabolic profiling of 98 *Myxococcus xanthus* strains led to the discovery of 37 new natural compounds (Krug et al. 2008). Additionally, three sequenced Frankia strains, nitrogen-fixing actinorhizal symbionts (Normand et al. 2007; Udwary et al. 2011), and *Rhodococcus* sp. RHA1, with 18 putative natural product gene clusters (Nett et al. 2009; McLeod et al. 2006), revealed a wealth of secondary metabolite gene clusters.

It is evident that many bacterial strains, especially those belonging to the phylum Actinobacteria, have the ability to produce natural products. However, genome sequencing alone cannot provide information on the conditions or confirm whether encoded natural products are produced in natural conditions (Davies and Ryan 2012).

It is very interesting that fungi inherently lack the ability to produce carotenoid pigments through biosynthetic pathways. However, through biotechnological means, the insertion of bacterial genes into yeast enables the production of carotenoid pigments (Kanamoto et al. 2021). The establishment of the biosynthetic pathway for astaxanthin began with the introduction of different bacterial genes into *Candida utilis*. The crtB gene, in conjunction with other genes, facilitates the synthesis of phytoene, its desaturation to lycopene, cyclization to β-carotene, and further ketolation and hydroxylation to astaxanthin (Miura et al. 1998).

To modify key metabolic pathways, next-generation *C. utilis* recombinant strains overexpressed HMG-CoA reductase, also known as the terpenoid enzyme, a crucial enzyme in the terpenoid system (Shimada et al. 1998). Overexpression of the geranylgeranyl pyrophosphate synthase gene crtE and a single allele of the squalene synthase gene in conjunction with lycopene gene insertion in *Lipomyces starkeyi* and *Pichia pastoris*, through genetic engineering, increased lycopene production in both strains, respectively (Kanamoto et al. 2021; Bhataya et al. 2009; Araya-Garay et al. 2012).

Moreover, metabolic modifications of *Yarrowia lipolytica*, using CRISPR-Cas9, involved integrating the phytoene synthase gene crtB and lycopene synthase gene crtL. Additionally, increased expression of these integrated genes for HMG, phosphomevalonate kinate ERG8, and crtE further enhanced the terpenoid pathway (Schwartz et al. 2017). Substituting crtYB for crtB and ERG8, as a bipotent phytoene synthase/lycopene cyclase, improved the biosynthetic pathway up to β-carotene (Liu et al. 2021). Similarly, introducing the bacterial β-carotene hydroxylase gene crtZ enhanced the metabolic pathway to zeaxanthin, another form of carotenoid (Xie et al. 2021).

Following the metabolic engineering of zeaxanthin, the final step for the biological production of astaxanthin from yeast involved replacing the bacterial gene crtZ with the algal β-carotene ketolase gene bkt, which enabled the production of astaxanthin from carotenoids (Tramontin et al. 2019). A *Rhodotorula glutinis* strain, with an increased pathway achieved through the overexpression of the limiting gene crtYB, demonstrated an elevated product yield of 27.1 mg/g dry weight in the form of β-carotene (Pi et al. 2018). The crtYB gene plays a crucial role in both β-carotene formation and an alternate pathway leading to torularhodin in R. glutinis (Herz et al. 2007).

A systematic genetic strategy was employed to interfere with the astaxanthin production pathway in *Xanthophyllomyces dendrorhous* (Sandmann et al. 2021). Overcoming limiting stages by increasing the expression of HMG, CrtI, and CrtYB genes in a high-yield astaxanthin mutant revealed enhanced efficiency. Additional copies of the astaxanthin synthase gene further improved the conversion of carotene to astaxanthin (Gassel et al. 2014). Carotenoid concentrations in transformants were measured under regular laboratory conditions. Optimized media and fed-batch fermentation techniques, when applied to the transformation of *Y. lipolytica*, substantially increased carotenoid contents (Larroude et al. 2018; Matthäus et al. 2014).

Yeast can easily grow on agricultural waste; for example, modified *X. dendrorhous* grown on wheat straw hemicellulose hydrolysate demonstrated increased zeaxanthin production (Breitenbach et al. 2019). Metabolic engineering techniques were also employed to enhance *Y. lipolytica*'s ability to utilize a variety of substrates (Spagnuolo et al. 2018).

14.3 Importance of Identifying and Regulating Genes for Cosmeceutical Applications

14.3.1 Overview of Gene Array Technology

The terms "biochip," "DNA chip," and "GeneChip" (a registered brand of Affymetrix, Inc.) are just a few examples used to describe gene arrays. "Microarray" and "macroarray" may be employed to distinguish between spot size or the quantity of spots on the support, with the former typically referring to arrays on glass supports.

A gene array is a highly efficient method for rapidly assessing the differential expression of hundreds to thousands of genes in a single experiment, widely regarded as the preferred approach (Epstein and Butow 2000; Khan et al. 1999; Schena et al. 1998). In gene array analysis, target sequences encoding individual genes are applied to a substrate, which is then studied alongside a labeled sample, enabling immediate observation of the results after hybridization. This fundamental concept underlies gene array experiments. The variations in gene array techniques primarily involve the substrate to which target samples are attached, the samples employed on the gene array, and the procedure used to prepare the probe for subsequent hybridization to the gene array (Shackel et al. 2002).

14.3.2 Gene Array Technology

Gene arrays serve as substantial platforms where nucleic acids of numerous genes are either spotted or directly synthesized at predetermined locations. In array analysis, a sample containing nucleic acids is labeled and then allowed to hybridize with the gene-specific targets on the array. In the literature, the term "target" may refer to the labeled nucleic acid in the sample or the nucleic acids attached to the array. In this context, the labeled nucleic acids comprising the sample are referred to as "probes," while the nucleic acids connected to arrays are referred to as targets. The specific nucleic acid content of the sample is determined based on the amount of probe hybridized to each target location.

The primary advantage of gene arrays lies in their capability to provide data on hundreds of targets in a single experiment (Thermofisher). Gene arrays are employed for various purposes, including the analysis of differential expression in two or more RNA samples or the identification of sequences through mutation analysis. This discussion will predominantly focus on the use of arrays for expression analysis.

The procedure involves extracting RNA from the samples to be compared. Subsequently, reverse transcription, coupled with aRNA amplification, is utilized to produce labeled cDNA from the RNA samples. This labeled cDNA is then subjected to hybridization on identical membrane or glass slide arrays. The analysis processes encompass locating and measuring the hybridized cDNA, extracting the unhybridized cDNA, and comparing quantitative data from different specimens.

14.4 Applications of Gene Array Technology

14.4.1 Advanced Nature and Capabilities of Gene Array Methods

There are three main platforms for gene arrays: oligonucleotide Gene Chip arrays, microarrays, and membranes (usually made of nylon or plastic). Gene Chip arrays have the highest density of DNA targets on each chip, while membrane-based gene arrays have the lowest density of target genes. The term "Gene Chip" was initially used to describe high-density oligonucleotide gene arrays, but it is also applied to all microarrays on glass. Affymetrix Corporation holds the exclusive trademark "Gene Chip" (Shackel et al. 2002).

Membrane-based gene arrays represent the simplest and most straightforward array experiment, allowing for a semi-quantitative estimate of gene abundance. In membrane-based gene arrays, a radioisotope-labeled (P32 or P33) probe is used, and signals are detected and quantified using phosphoimage. P33 labeled probes, starting from 5 micrograms of total RNA, can achieve a sensitivity of 1/1000 (mRNA abundance), requiring a minimum of 25 × 10^6 molecules for detection (Bertucci et al. 1999). P32 is considered similar to fluorometric detection and is about four times more sensitive than P33. P33 has become more popular due to its superior signal saturation and reduced blooming characteristics compared to P32, which suffers from excessive blooming (Cox 2001). P33 also shows a linear response over five orders of magnitude, enabling the quantification of genes with both high and low abundances using phosphoimager quantitation.

Fluorescent microarray detection typically uses fluorophores cyanine-3 (green) and cyanine-5 (red) to highlight probes for the relative abundance of the target sequence. The minimal sample size for fluorescent microarray detection is 20 × 10^6 molecules, with a detection limit for mRNA abundance of 1/100,000 (mRNA abundance). Fluorometric detection provides a linear signal response over a wide range of 3–4 orders of magnitude. The lowest projected sensitivity for any microarray element is 2 pg mRNA. Signal variability from the same elements reproduced on different microarrays ranges from 12% to 14%, and a four-fold difference is the smallest reported detectable difference on microarrays (Yue et al. 2001). However, triple hybridizations are necessary for the repeatable detection of more than 99% of gene expression (Lee et al. 2000).

High-density oligonucleotide gene arrays are offered by several companies, and these arrays use cutting-edge methods (Lipshutz 2000; Lipshutz et al. 1999; McGall et al. 1996). Affymetrix Corporation pioneered this technology, producing Gene Chip gene arrays with a density greater than 250,000 oligonucleotides per cm^2. In contrast to microarrays, the oligonucleotides on the Gene Chip are typically shorter, ranging between 15 and 30 nucleotides. They are manufactured using a photolithographic method on a silica substrate. Each GeneChip contains 16–21 oligonucleotides that play a crucial role in the functional representation of each gene.

For each oligonucleotide on the array, there is a corresponding partner oligonucleotide that contains a single mismatch out of all the bases. This design allows the regulation of nonspecific binding by comparing it to the partner oligonucleotide. To determine the likelihood of a gene being expressed, algorithms are constructed based on the signals from each oligonucleotide on the array. These algorithms estimate the probability of expression, contributing to the analysis of gene expression levels (Lipshutz 2000).

In the comparison of several gene array platforms, membrane-based arrays are often considered the most appropriate technique for the best quantification of gene expression, and they are also reusable. However, each platform has its own unique advantages and disadvantages. Glass slides and Gene Chips have higher density and cover entire gene targets, but the Gene Chip substrate's homogeneity offers the best optical performance when analyzing gene signals on the array. Fluorophores used on Gene Chips do not experience blooming, a concern with radioisotopes frequently utilized on membranes. Additionally, Gene Chip experiments require a uniform reference for comparing gene signals, in contrast to stand-alone membrane-based array expression data. Each platform has its strengths and considerations, and the choice depends on the specific needs of the experiment or analysis.

14.4.2 Microorganisms as Bioactive Compound Producers

Microbial natural products, in addition to their structural novelty, are biologically highly active and hold potential applications against various diseases, including diabetes, immunosuppression, antibiotics, anticancer medications, and hormone antagonists. Despite the significant achievements in natural product research, there are challenges. One major obstacle is the limited number of easily cultivable microorganisms in laboratory settings, considering that a gram of fertile soil is estimated to contain a vast number of different bacterial and fungal species. Only a small percentage of these microorganisms can be cultivated (Hug et al. 2016).

The complex mixtures found in microbial extracts can slow down the process of isolating and characterizing microbial natural products. Another challenge lies in determining the concentrations of targeted metabolites in laboratory cultures. Modern technologies, such as OSMAC methodologies, genome mining, and metabolomic methods, are being developed to address these issues. These technologies, along with advancements in ionization and culturing techniques, enable researchers to explore novel bacteria and study the genomes of a significant portion of the microbiota in soil samples (Harvey et al. 2015).

Despite the extensive research in the field, bacteria and marine environments continue to produce novel structures with potential applications in drug discovery. Over the past three decades, a substantial percentage of new antibiotics and anticancer medicines have been linked to natural products or their compounds. The ongoing exploration of microbial natural products and the development of innovative technologies are expected to contribute to further breakthroughs in cosmeceutical bioactive compound discovery (Baltz 2019).

14.4.3 Polyketides Bioactive Compounds

Several strains of bacteria biosynthetically produce polyketides, which are classified into two major classes based on their structure and biosynthetic pathways. There are two groups of polyketides: Iterative polyketide synthases produce polyketides with condensed aromatic structures, whereas modular polyketide synthases generate linear polyketides with copious alcohols, methyl branches, etc. (Khosla 2009). The bacterial strain *Bacillus amyloliquefaciens*, isolated from a marine environment, has been found to produce several known polyketides, including macrolactin V, a 24-membered macrolide, along with macrolactin S (Gao et al. 2010). Similarly, various species of *Streptomyces* potentially produce polyketides, and a new compound, 6-deoxyglycosidic elaiophylin derivatives, namely, 11′,12′-dehydroelaiophylin and 11,11′-O-dimethyl-14′-deethyl-14′-methylelaiophylin, have been extracted from the liquid culture of *Streptomyces* (Wu et al. 2013a, b). However, *Streptomyces* sp. HK-2006-1 potentially synthesizes dihydrochalcomycin, chalcomycin, and a new additional compound, chalcomycine E (Jiang et al. 2017). Meanwhile, another strain, *Streptomyces koyangensis* SCSIO 5802, produces the bioactive compound neoabyssomicins A–C and some familiar compounds. Notably, neoabyssomicins A and C demonstrate potential activity against HIV-1 viral replication in white blood cells (Song et al. 2017). A detailed study on *Streptomyces koyangensis* reveals the production of diverse neoabyssomicins bioactive compounds, including neoabyssomicins D, E, F, A2, and G (Huang et al. 2018).

Another significant polyketide, formed through a biosynthetic pathway, is found, particularly in marine-derived fungal strains. Investigations revealed that various strains of the *Dendrodochium* genera produce diverse types of polyketides in the culture broth. A detailed study of the culture broth elucidated 13 new bioactive polyketide compounds, including 12-membered macrolides, known as dendrodolides A–M (Sun et al. 2013). Wu et al. (2017) demonstrated that *Penicillium sumatrense* synthesizes sumalactones A–D, along with known bioactive metabolites such as curvularin and dehydrocurvularin. Dehydrocurvularin plays a significant role in inflammatory disorders caused by the formation of NO. Polyester polyketides, such as hansforesters A–M, were successfully isolated from the marine-derived fungal strain *Hansfordia sinuosae*. Further investigation into these metabolites revealed that hansforester A bioactive compound potentially inhibits the growth of diverse bacterial strains (Wu et al. 2018).

Aromatic polyketides derived from marine fungi are predominantly found in fungi of marine origin. Zhang et al. (2018a, b) identified an isolate of *Cladosporium cladosporioides* HDN14-342 from the deep sea as a new source of four tetralone derivatives. These included homodimeric cladosporols F and G, along with heterodimeric clindanones A and B. Moreover, *Penicillium* sp. SCSGAF 0023 yielded eight antifouling and antibacterial polyketides, including 6,8,5,6-tetrahydroxy-30-methylflavone and paecilin C. These polyketides, also present in gorgonian coral cells, help prevent fouling (Bao et al. 2013). Furthermore, *Aspergillus* sp. F40's fermentation broth produced two new polyketides (Tian et al. 2018), and sponge-derived fungi extracted novel anti-inflammatory polyketides, including ascomfurans C and ascomarugosin A with a unique 2,3-diaryl indole moiety (Liu et al. 2018). Various fungal aromatic polyketides, such as benzophenone derivatives like peniphenone and methyl peniphenone, were extracted from the endophytic fungal strain *Penicillium* sp. ZJ-SY2 (Liu et al. 2016a, b).

Similarly, in *Penicillium citrinum*, a marine sediment-derived fungus, Chen et al. (2017a, b) discovered the novel antimitotic polyketide dicitrinone D, a citrinin derivative. In a recent study of *Aspergillus* sp. 16-02-1 in deep-sea water, nine new C9 polyketides were identified, including aspiketolactonol, aspilactonols A–F, and epiaspinonediol, alongside five familiar antibiotics (Chen et al. 2014). *Aspergillus ochraceus* MA-15, another marine mangrove-derived fungus, produced three asperochrins named asperochrins A to C (Liu et al. 2015). Fang et al. (2017) obtained the novel cyclopentenone 5-hydroxycyclopenicillone from *Trichoderma* sp. HPQJ-34, a sponge fungus isolated from a sponge.

14.4.4 Peptides

Marine microorganisms have the capacity to produce various peptides containing amino acids, showcasing significant potential against bacterial strains and other diseases. The marine-derived bacterium *Bacillus* sp. 176, for instance, was shown to synthesize two pumilacidin-like cyclic lipopeptides (Xing et al. 2017; Wang et al. 2017). Another marine-derived organism, *Streptomyces* sp. IMB094, produced neoactinomycins A and B, novel chromopeptides (Wang et al. 2017). *Micromonospora chalcea* FIM 02-523, also marine-derived, generated three novel cyclic depsipeptides, rakicidins G–I (Chen et al. 2018).

Fungi represent another critical source of polypeptide-like bioactive compounds. The *Aspergillus versicolor* strain LZD-14-1, for example, yielded versicotides D, E, and F, cyclopeptide-like substances. Versicotides E and F exhibit unique structures, combining two anthranilic acid residues with a proline to form a singular cyclic peptide system (Chen et al. 2017a, b). A detailed study of the culture broth of *Sarocladium kiliense* HDN11-112 revealed two new bioactive compounds containing cyclic depsipeptides, saroclides A and B. Further investigation highlighted the significant value of saroclide A in minimizing lipid activities (Guo et al. 2018). Additionally, the deep-sea sediment-isolated marine-derived fungus *Simplicillium obclavatum* EIODSF 020 developed several novel linear peptides named simplicilliumtides A–H (Liang et al. 2016).

14.4.5 Alkaloids

Alkaloids, a broad class of substances with diverse structural and biological properties, are synthesized by various groups of bacteria and fungi. In bacteria, alkaloids are predominantly discovered in the *Actinomycete* genus, including a new species. For instance, the alkaloid 3-((6-methylpyrazin-2-yl) methyl)-1H-indole was found in a culture extract of *Serinicoccus profundi* sp., a deep-sea *actinomycete* of the family *Intrasporangiaceae*, subordinate *Micrococcineae*, marking a previously unreported alkaloid (Yang et al. 2012). *Streptomyces* sp., a marine-derived organism, produced streptopertusacin A, a distinct indolizinium alkaloid with potential antibacterial properties (Zhang et al. 2017). Wang et al. identified 12 known chemicals and a novel alkaloid, 2-(furan-2-yl)-6-(2S,3S,4-trihydroxybutyl) pyrazine, from the *actinomycete Jishengella endophytica* 161111 found in mangroves (Wang et al. 2014a, b). Deep-sea bacteria *Shewanella piezotolerans* yielded three indole-based alkaloids, shewanellines A–C, along with 12 other familiar bioactive metabolites (Wang et al. 2014a, b).

Marine-isolated fungi, over the study period, emerged as the most active producers of variant alkaloids, especially those formed from quinazolines. Eleven new unknown fumiquinazoline-type alkaloids, such as versiquinazolines A–K from the fungus *Aspergillus versicolor* LZD-14-1, derived from gorgonian-derived materials, were discovered (Cheng et al. 2016). Further research led to the identification of six new bioactive secondary metabolites or alkaloids, including versiquinazolines L, M, N, O, P, and Q (Cheng et al. 2018). Additionally, three novel tryptophan-derived quinazolinone alkaloids, cottoquinazolines B-D, were reported to be produced by another marine coral fungal strain *A. versicolor* LCJ-5-4 (Zhuang et al. 2011). Similarly, the mangrove endophytic fungus *Campylocarpon* sp. produced four new pyridone alkaloids known as campyridones A–D, along with the recognized compound ilicicolin (Zhu et al. 2016).

14.4.6 Terpenoids, Terpenes, and Meroterpenoids

Marine fungi have been found to produce a variety of terpenoid compounds with intriguing biological properties, including eremophilane and carotene sesquiterpenoids, bisabolane sesquiterpenoids, eudesmane sesquiterpenoids, pimarane diterpenes, and others, as reported by Chinese researchers. However, sesquiterpenoids or sesquiterpene compounds are predominantly observed among numerous kinds of terpenoids. For instance, the marine-derived fungus *A. ochraceus* Jcma1F17 was the source of the exceedingly rare nitrobenzoyl sesquiterpenoid 6b,9a-dihydroxy-14-p-nitrobenzoylcinnamolide and its known counterpart, insulicolide A (Fang et al. 2014). Moreover, *Penicillium* sp. J–54 obtained from mangrove areas yielded four novel eudesmane-type sesquiterpenoids, penicieudesmol A–D (Qiu et al. 2018). From the fungus associated with soft corals called *Chondrostereum* sp., five novel triquinane-type sesquiterpenoids called chondrosterins A, B, C, D, and E were extracted from the nutrient growth medium (Li et al. 2012). Additionally, *Aspergillus* species, known for being excellent sources of terpenoid compounds, including sesterterpenoids, demonstrated the discovery of a novel pentacyclic sesterterpenoid, Asperterpenoid A, in the *Aspergillus* sp. 16-5c endophytic fungal strain (Huang et al. 2013). Moreover, Asperterpenols A and B, two other comparable sesterterpenoids with a unique 5,8,6,6 tetracyclic ring skeleton, were produced by the endophytic *Aspergillus* strain 085242 (Xiao et al. 2013).

14.4.7 Other Bioactive Compounds

In the literature, researchers discovered three new itaconic acid derivatives designated as asperitaconic acids A–C from the culture extract of the marine isolate *Aspergillus niger* (Ding et al. 2018). Furthermore, the marine-derived strain *Streptomyces xinghaiensis* SCSIO S15077 was found to produce the novel nucleoside antibiotic tunicamycin E along with six analogs (Zhang et al. 2018a, b).

14.4.8 Significance of Bioactives in the Cosmeceutical Industry

A study revealed adverse effects on the skin and general health of individuals exposed to UVA radiation (320–400 nm) and UVB radiation (280–320 nm) over prolonged periods (Rittié and Fisher 2002; Young et al. 2017). In response to the growing awareness of the dangers of UV radiation, there has been an unforeseen surge in the manufacturing and usage of solar products in recent decades (Sánchez-Quiles and Tovar-Sánchez 2015). To protect themselves from the harmful effects of UV radiation, various marine microbes have developed a range of strategies (Carreto

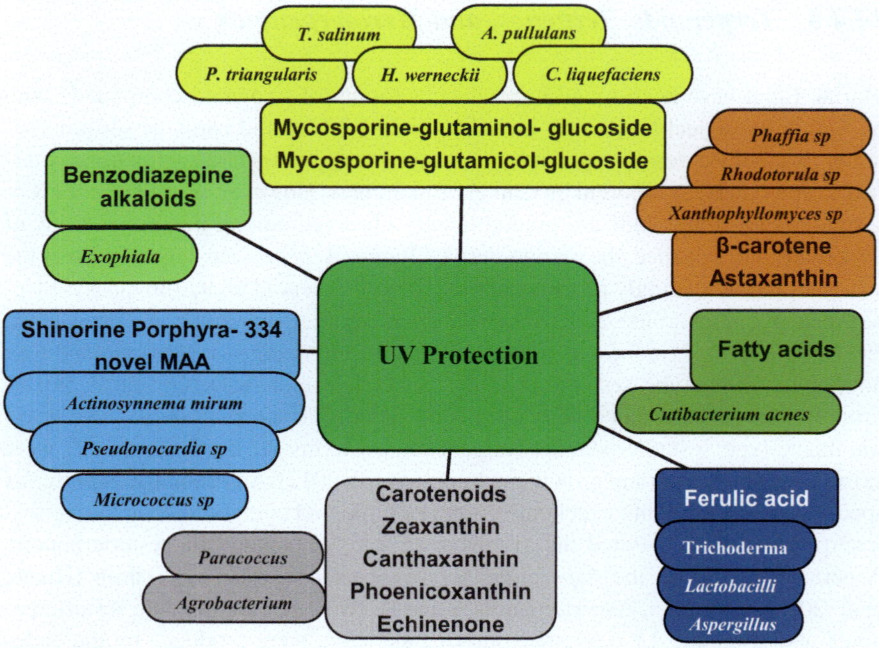

Fig. 14.1 Bioactive compounds obtained from diverse microbes potentially used in cosmeceutical industries to make cosmetics products useful for UV protection

and Carignan 2011; Gao and Garcia-Pichel 2011). These organisms produce UV-absorbing substances such as scytonemins (found only in cyanobacteria), mycosporines, mycosporine-like amino acids (MAAs), carotenoids, and melanin (Fig. 14.1). In the future, these materials could serve as UV filters in sunblock formulations. Common sunscreen ingredients include inorganic or organic filters (Table 14.1) (Morabito et al. 2011). Previous literature has provided a clear roadmap of the biological synthesis of various bioactive compounds from diverse groups of marine-derived and terrestrial microorganisms (Corinaldesi et al. 2017).

14.4.8.1 Bacteria

Among microbes, marine bacteria, such as *Pseudonocardia* sp. strain P1, *Micrococcus* sp. AK-334, and *Actinosynnema mirum*, have been utilized for the production of mycosporine-like amino acids and derivatives like shinorine, porphyra-334, and mycosporine-glycine-alanine (MAA) (Fig. 14.1). The chemical investigation of these extracted compounds plays a significant role in the cosmetics and cosmeceutical industries (Table 14.1) (Miyamoto et al. 2014).

Carotenoids and their derivatives are commonly produced across various organisms, from simple unicellular to complex multicellular organisms. These pigments are well-recognized for their extensive use in food, textiles as colorants, cosmetics,

Table 14.1 Microbes and their bioactive products for cosmeceuticals

Microbes	Bioactive compounds	Main Function	Reference
Phaeotheca triangularis	Mycosporine (Mycosporine-glutaminol- glucoside Mycosporine-glutamicol-glucoside)	UV protection	Rastogi et al. (2010)
Trimmatostroma salinum			
Hortaea werneckii			
Aureobasidium pullulans			
Cryptococcus liquefaciens			
Rhodotorula sp.	Carotenoids β-carotene astaxanthin	Skin photo-protection by solar UV radiation	Vílchez et al. (2011), Mata-Gomez et al. (2014), Sy et al. (2015), Ambati et al. (2014)
Phaffia sp.			
Xanthophyllomyces sp.			
Pseudozyma	Mannosylerythritol lipids (4-O-β-D-mannopyranosyl erythritol or 1-O-β-D-mannopyranosyl erythritol)	Protect from UV-damaged	Morita et al. (2015), Bae et al. (2019)
Exophiala	Benzodiazepine alkaloids (circumdatins I, C, G)	UV protection	Zhang et al. (2008)
Pseudonocardia sp	Amino acids similar to Mycosporine (Shinorine, Porphyra- 334, and novel MAA (mycosporine-glycine-alanine))	UV protection	Miyamoto et al. (2014)
Micrococcus p.			
Actinosynnema mirum			
Paracoccus	Carotenoids Zeaxanthin Cantaxanthin Phoenicoxanthin and Echinenone	Skin photo-protection by solar UV radiation	Vílchez et al. (2011), Mata-Gomez et al. (2014), Sy et al. (2015), Ambati et al. (2014)
Agrobacterium			
Cutibacterium acnes	Fatty acids	Prevent skin from UV light radical	Smith et al. (2020)
Trichoderma	Ferulic acid	UV absorption	Gottschalk et al. (2010), Minami et al. (1998)
Aspergillus			
Lactobacilli			Tomaro-Duchesneau et al. (2012)
Streptococcus zooepidemicus	Hyaluronan (Hyaluronic acid)	Anti-aging	Ferreira et al. (2021)
Re-*Corynebacterium glutamicum*			
Re-*Bacillus subtilis*			Sajna et al. (2015)
Pichia pastoris	Resveratrol		
Saccharomyces cerevisiae			

(continued)

Table 14.1 (continued)

Microbes	Bioactive compounds	Main Function	Reference
Aureobasidium pullulans	EPS	Anti-aging	Martins et al. (2014); Poli et al. (2010); Nwodo et al. (2012); Freitas et al. (2011); Courtois et al. (2014); Donot et al. (2012)
Eduarsiella tarda	EPS	Anti-aging Anti-wrinkles	
Alcaligenes faecalis			
Xanthomonas campestris			
Bacillus sp.			
Zymonas mobilis			
Agrobacterium sp.			
Alteromonas macleodii			
Pseudoalteromonas sp.			
E coli	Retinoic acid	Anti-aging	Choi et al. (2020)
Vibrio diabolicus	HE 800 (Analogous to hyaluronic acid)	Anti-aging	Courtois et al. (2014)
Trichoderma sp.	Polyunsaturated fatty acids (DHA, EPA, and omega-3 fatty acids)	Anti-aging	Aasen et al. (2016)
Rhodotorula mucilaginosa			
Moritella dasanensis			Hutchings and Reynolds (2004)
Vibrio sp.			Adarme-Vega et al. (2014)
Pseudomonas sp.			Armenta and Valentine (2013)
Shewanella sp.			Xie and Wang (2015)
Colwellia sp.			Cohen and Ratledge (2005)
Engineered microbes (*E. coli*)	Polyphenol (Resveratrol)	Photoaging	Baxter (2008); Lim et al. (2011)
Micro algae *Haematococcus pluvialis* and *Xanthophyllomyces dendrorhous*	Astaxanthin	Photoaging and antioxidant	Yuan et al. (2011); Olaizola (2000); Soudant and Koulbanis (1998)
Symbiotic culture of acetic acid bacteria and yeast	Keratinocytes and fibroblasts	Anti-aging	Zagórska-Dziok et al. (2023)
		Antioxidant	
Xanthophyllomyces dendrorhous	Astaxanthin	Antioxidant Terpene	

(continued)

Table 14.1 (continued)

Microbes	Bioactive compounds	Main Function	Reference
Aspergillus niger Candida utilis coculture	Ellagic acid	Antioxidant	Shi et al. (2005)
Aspergillus niger	Tannin-rich substrate to Gallic acid Whiting cream		Umakumar et al. (2012)
Penicillium			
Rhizopus oryzae			Kim (2007)
Trichoderma	Lignocellulosic biomass to ferulic acid skin lightening, and improved uptake of other active ingredients by skin		Gottschalk et al. (2010)
Aspergillus			
Lactobacilli			Tomaro-Duchesneau et al. (2012)
Aspergillus niger	Citric acid		Kubicek et al. (2011)
Agaricus bisporus	Phenolics contents (p-hydroxybenzoic and gallic acid)		Machado-Carvalho et al. (2023)
Portobello			
Lentinula edodes			
Pleurotus ostreatus			
Lactobacillus pentosus	Glycolipopeptide		Vecino et al. (2015)
Lactobacillus paracasei	BS		Ferreira et al. (2017)
Streptomyces coelicolor	Carotenoids (Anthraquinones) (Kalafungin)	Antioxidant	Rao et al. (2017)
Actinomycetales	Mycosporine-like amino acids		Miyamoto et al. (2014); Wada et al. (2015); Suh et al. (2014); Coba et al. (2007)
Flavobacteriaceae	Saproxanthin and myxol		Shindo et al. (2007)
Acremonium sp.	Phenols (Hydroquinone derivatives (wentiquinone ethyl 4-(3,4-dihydroxybenzamido)-butanoate))		Abdel-Lateff et al. (2002)
Aspergillus wentii			Li et al. (2014)
Epicoccum sp.	Isobenzofuranone (4,5,6-trihydroxy-7-methylphthalide)	Antioxidant (Radical scavenging)	Abdel-Lateff et al. (2003)
Aspergillus sp.	Kojic acid (2-hydroxymethyl-5-hydroxy-γ-pyrone)	Radical scavenger	Pandey (1992)
Micro algae *Porphyra umbilicalis*	MAAs	Natural sunscreens against harmful rays	Oren and Gunde-Cimerman (2007)

(continued)

Table 14.1 (continued)

Microbes	Bioactive compounds	Main Function	Reference
Algae *Porphyra umbilicalis*	MAA (palythine, porphyra, shinorine)	Prevention of skin cells damaged by UV radiation	Biosiltech (2014)
Microalga *Aphanizomenon flos aquae*	MAA	Wrinkle reduction	Scoglio et al. (2012)
Keissleriella sp.	Exopolysaccharides	Antioxidant (Radical scavenging)	Sun et al. (2004)
Aspergillus sp.	Diketopiperazine alkaloids (Diketopiperazine alkaloids)	Radical scavenging UV-A screening function	Li et al. (2004a, b)
Aspergillus sp.	Dioxopiperazine alkaloids (Dihydroxyisoechinulin A and related echinulin)		Li et al. (2004a, b)
Pseudomonas sp., e.g., *P. Aeruginosa*	Glycolipids (mono, di-rhamnolipids)	Biosurfactants	Peyrat et al. (2019), Vecino et al. (2017), Santos et al. (2016)
Candida sp., e.g., *C. Bombicola*	Sophorolipids (Lactonic and acidic)		
	MELs		
Mycobacterium	Trehalose lipid		
Rhodococcus sp., e.g., *R. Erythopolis*			
Pseudozyma yeasts	MELs		Sajna et al. (2013)
Lactobacillus plantarum	Glycoprotein		Sharma et al. (2014)
Escherichia coli	Glycine betaine from choline	Biosurfactants Cleansing agents	Kiene (1998), Nyyssölä et al. (2000), Arkin et al. (2011)
Staphylococcus aureus			
Bacillus subtilis			
Rhizobium meliloti			
Rhodobacter sphaeroides			
Vibrio costicola			
Alcaligenes			
Pseudomonas fluorescens	Palmitic acid	Biosurfactants (Lipopeptide)	Janek et al. (2010)
Bacillus licheniformis	b-hydroxy		Coronel-León et al. (2015)
Bacillus amyloliquefaciens	b-hydroxy-C17		Lee et al. (2010)
Bacillus mojavensis	Saturated b-amino fatty acid		Ma and Hu (2014)
Enterobacter	n-C12, iso-C13, anteiso-C13, iso-C14, n-C14, iso-C15, anteiso-C15, iso-C16, n-C16 and iso-C17 b-hydroxy		You et al. (2015)

(continued)

Table 14.1 (continued)

Microbes	Bioactive compounds	Main Function	Reference
Corynebacterium xerosis	C10 3OH acid	Biosurfactants (Mannosylerythritol lipids)	Dalili et al. (2015)
Pseudozyma	MEL-C acid		Sajna et al. (2013)
Pseudozyma churashimaensis	C16 acid		Morita et al. (2013)
Pseudozyma aphidis	MEL-A and MEL-B		Onghena et al. (2011)
Pseudomonas aeruginosa	Mono-rhamnolipids	Biosurfactants (Rhamnolipids)	Hazra et al. (2011)
Pseudomonas aeruginosa	Saturated or unsaturated alkyl chains between C8 and C12		Cui et al. (2010)
Burkholderia kururiensis	Mono, di-rhamnolipids		Tavares et al. (2013)
Lactobacillus pentosus	Palmitic acid (C16)	Biosurfactants (Glycolipopeptid)	Vecino et al. (2015)
	Stearic acid (C18)		
	Oleic or elaidic acid (C18:1)		
	Linoelaidic acid (C18:2)		
Pseudomonas strains	Lipids	Biosurfactants	Ben Belgacem et al. (2015)
Rhodococcus pyridinivorans	Trehalose lipids		Kundu et al. (2013)
Enterobacter	Glycolipids		Jadhav et al. (2011)
Aspergillus sp.	Pyrone (kojic acid, kojic acid dimethyl ether, and kojic acid monomethyl ether)	Inhibition of tyrosinase	Lee et al. (2003)
Penicillium			
Alternaria species			Li et al. (2003)
Botrytis sp.	α-Pyrone derivate (6-[(E)-Hept-1-enyl]-α-pyrone)	Inhibition of tyrosinase	Zhang et al. (2007)
Alternaria sp.	Phomaligol A		Li et al. (2003)
Genus Myrothecium	6-n-pentyl-α-pyrone and myrothenone A		Li et al. (2005)
T. viridae strain H1–7	Homothallin II		Tsuchiya et al. (2008)
	Seven different compounds	Inhibition of melanin	

(continued)

Table 14.1 (continued)

Microbes	Bioactive compounds	Main Function	Reference
Malasseziales sp.	Dicarboxylic acid 1,7-heptanedicarboxylic acid (azelaic acid)	Inhibition of tyrosinase	Rendon et al. (2006); Amend et al. (2014)
Pestalotiopsis sp. Z233.	Sesquiterpenes		Wu et al. (2013)
Microsporum sp.	Anthraquinones (Chrysophanol)		US patent 20140056834A1
Gram-negative bacterium, *Thalassotalea* sp. PP2-459	Thalassotalic acids A, B and C		Deering et al. (2016)
Bacteria and Fungi	Carotenoids (Astaxanthin)		Ambati et al. (2014)
Pseudomonas sp.	Alkyl halides Methylene chloride		Kang et al. (2011)

and nutraceuticals on an industrial scale (Kirti et al. 2014). Although approximately 750 different types of carotenoid pigments and their derivatives have been discussed, only a few, including lycopene, β-carotene, astaxanthin, zeaxanthin, and lutein, hold significant commercial importance (Vílchez et al. 2011). Marine-isolated bacterial genera, such as *Paracoccus* sp. and *Agrobacterium* sp., produce carotenoid bioactive compounds like astaxanthin, which are potentially used in cosmetic products for their multifunctional properties (Fig. 14.1 and Table 14.1) (Vílchez et al. 2011; Kim 2011a, b; Ambati et al. 2014).

Exopolysaccharides (EPS), a series of high molecular weight carbohydrate polymers, play a crucial role in various natural mechanisms, including attachment, communication, and competition between species (Poli et al. 2010). While EPSs are produced by various microbes, bacteria, especially *Agrobacterium* sp., *Alcaligenes faecalis*, *Xanthomonas campestris*, and *Bacillus* sp., are considered the main EPS-producing microorganisms, offering characteristics such as emulsifying, thickening, absorption, and gel formation with potential applications in cosmeceutical industries (Donot et al. 2012). Similarly, the marine bacterium *Alteromonas macleodii* has found commercial use in cosmetics, and EPSs from *Pseudoalteromonas* sp. have been utilized to create anti-aging compounds, enhancing collagen production and improving skin structural qualities through fermentation (Fig. 14.2) (Martins et al. 2014).

Fats, besides being dietary supplements, are effective treatments for various skin conditions, promoting soft tissue repair and skin nourishment. Fatty acids, in addition to stimulating collagen production, offer anti-inflammatory and wound-healing properties, making them vital components of cosmetics and cosmeceuticals (Ziboh et al. 2000). Polyunsaturated fatty acids (PUFA), such as eicosapentaenoic acid (EPA) and docosahexaenoic acid (DHA), particularly the omega-3 fatty acids EPA and DHA, provide a range of health benefits (Fig. 14.2 and Table 14.1).

Antioxidants (Fig. 14.3 and Table 14.1) play a crucial role in protecting the ingredients in cosmetic formulations from oxidation. Reactive oxygen species (ROS) generated by UV radiation can adversely affect skin cells' metabolic processes, but antioxidants help mitigate the oxidative impact. Various petroleum-based antioxidants, such as propyl gallate, butylated hydroxyanisole, butylated hydroxytoluene, and tertiary butyl hydroquinone, have historically been employed in cosmetics and cosmeceutical products (Kim 2011a, b; Kang et al. 2011).

In addition to traditional antioxidants, the marine bacterium *Thalassotalea* sp. strain PP2-459 has biologically synthesized thalassotalic acid derivatives (A, B, and C), which exhibit tyrosinase inhibitory properties. Thalassotalic acid derivatives, particularly thalassotalic acid A, demonstrate comparable inhibitory functions to arbutin and may find applications as food preservatives or whiteners (Deering et al. 2016).

Surfactants and emulsifiers (being used in cosmetic products), characterized by amphipathic substances with both hydrophilic and hydrophobic components, are derived from various marine bacteria and fungi (McClements and Gumus 2016). These compounds encompass glycolipids, lipopeptides, and protein-polysaccharide complexes. Researchers have explored marine microorganisms such as *Alteromonas* sp., *Pseudomonas*, *Halomonas*, *Myroides*, *Corynebacteria*, and *Arthrobacter* for the production of biosurfactants and bio-emulsifiers (Satpute et al. 2010).

Chitosan, in fiber or film form, serves various purposes, including strengthening the epithelium and increasing the granular density of the skin (Table 14.1) (Ito et al. 2014). It promotes the development of fibroblasts, known for their remarkable healing qualities, along with moisturizing and antimicrobial properties (Kumirska et al.

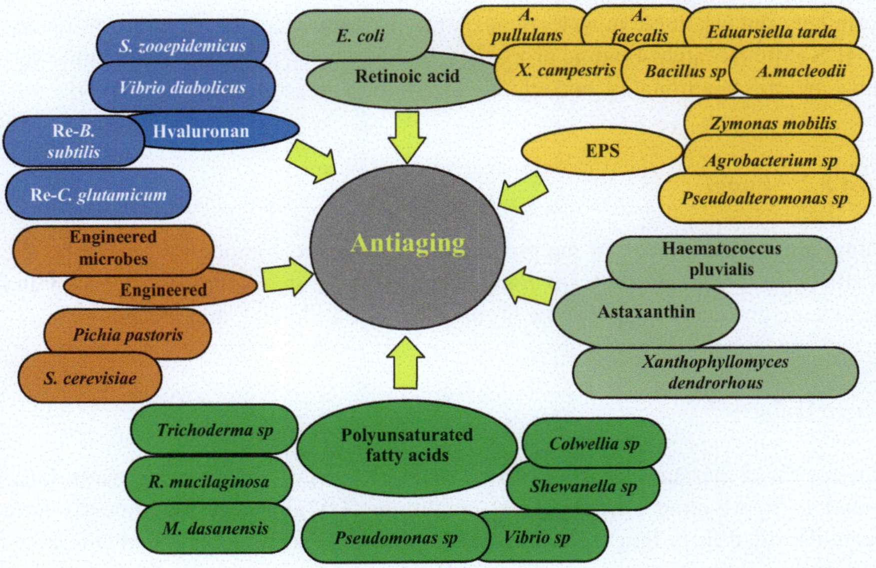

Fig. 14.2 The microbial bioactive secondary metabolites containing anti-aging properties

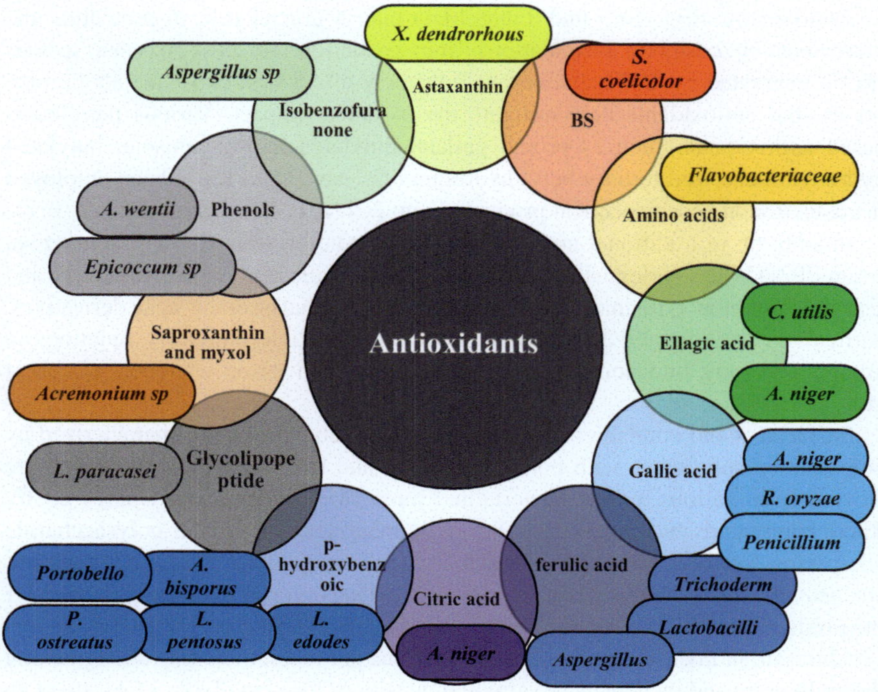

Fig. 14.3 The diverse group of biologically active compounds produced by bacteria and fungi is significantly used in antioxidants and radical scavenging

2011). Carboxymethyl chitosan (CMCS), a chitosan derivative, has the ability to form gels and is soluble in water at a neutral pH (Mourya et al. 2010). Additionally, CMCS exhibits cation-binding properties, high viscosity, a large hydrodynamic volume, anionic functionality, and significant osmotic pressure (Ito et al. 2014).

14.4.8.2 Fungi

Fungi are indeed known for the production of bioactive compounds crucial for the cosmetic industry, and they are considered both safe and sustainable for skin and healthcare products. Fungi can be grouped into the following major categories.

Micro Fungi

The reviewed literature clearly describes the synthesis of mycosporines from fungi found in hypersaline environmental conditions. UV-absorbing compounds have been identified from fungal strains, including black yeasts such as *Phaeotheca triangularis*, *Trimmatostroma salinum*, *Hortaea werneckii*, and *Aureobasidium pullulans*. These isolated compounds are specifically known as mycosporine–glutamicol–glucoside and mycosporine–glutaminol–glucoside (Fig. 14.1) (Kogej et al. 2006).

Microbes, microalgae, and fungi, including protists, are renowned for producing carotenoid pigments, with fungi playing a significant role in carotenoid synthesis. Diverse yeast strains from the genera *Rhodotorula*, *Phaffia*, and *Xanthophyllomyces* also produce astaxanthin, although their production rates are relatively low (Fig. 14.2 and Table 14.1) (Ambati et al. 2014).

Both terrestrial and marine fungi are known to produce benzodiazepine alkaloids, including circumdatins A–H. The mycelium of a marine fungus belonging to the genus *Exophiala* yields these bioactive chemicals. Detailed investigations have shown that circumdatin I, C, and G demonstrate strong UV-A screening abilities (Fig. 14.1 and Table 14.1) (Zhang et al. 2008). Other antioxidants, including diketopiperazine alkaloid golmaenone, related alkaloids, and dihydroxy isoechinulin A, an echinulin identified using chromatographic approaches, have been isolated from a marine fungus, *Aspergillus* sp., and have proven to be powerful antioxidants (Li et al. 2004a, b).

Fungi found in the ocean serve as an excellent source of antioxidant compounds beneficial to health (Fig. 14.3 and Table 14.1). The marine fungus *Acremonium* sp. has been identified as a producer of four novel hydroquinone derivatives with antioxidant properties (Abdel-Lateff et al. 2002). A derivative of *Epicoccum* sp., isolated from *Fucus vesiculosus*, exhibits strong DPPH radical scavenging properties (4,5,6-trihydroxy-7-methylphthalide), similar to those found in *Epicoccum* sp. (Abdel-Lateff et al. 2003). Moreover, *Aspergillus wentii* strain EN-48, isolated from brown algae, produces eight novel secondary metabolites with potent free radical scavenging properties, establishing itself as a rich source of antioxidants. These naturally occurring antioxidants show significant potential compared to commonly used synthetic antioxidants such as butylated hydroxytoluene (Li et al. 2004a, b). Another noteworthy antioxidant is the exopolysaccharide EPS2, isolated from the marine filamentous fungus *Keissleriella* sp. YS 4108, which demonstrates notable superoxide radical scavenging capabilities (Sun et al. 2004).

Aspergillus and *Penicillium* species have been extensively studied for their biosynthesis of kojic acid (5-hydroxy-2-(hydroxymethyl)-4H-pyran-4-one), known for its water-soluble characteristics and unique properties as an antioxidant, antimicrobial, anti-inflammatory, and tyrosinase inhibitor (Lee et al. 2003; Rana et al. 2013; Chang 2009). Similarly, broth culture medium of *Alternaria* sp. has been found to contain phomaligol A along with two novel kojic acid derivatives, kojic acid monomethyl ether and kojic acid dimethyl ether. Further studies on these compounds revealed their ability to inhibit tyrosinase activity (Li et al. 2003).

Moreover, *Malassezia* sp., on the other hand, produces azelaic acid (1,7-heptane dicarboxylic acid), a reliable inhibitor of tyrosinase (Baliña and Graupe 1991; Rendon et al. 2006; Balboa et al. 2015). Various strains of the genus *Trichoderma* have been shown to produce bioactive compounds that efficiently suppress tyrosinase activity (Hong et al. 2015). From the marine sediment-derived *T. viridae* strain H1-7, a competitive inhibitor of tyrosinase with structural similarities to homothallin II was identified. It blocks the enzyme by attaching to its copper active site. Similarly, *T. viridae* also synthesizes melanogenesis suppressor compounds that potentially inhibit tyrosinase activity (Tsuchiya et al. 2008).

Different species of the genus *Botrytis* potentially produce the α-Pyrone derivative (6-[(E)-Hept-1-enyl]-α-pyrone) with the capability to inhibit tyrosinase action (Zhang et al. 2007). Myrothenone A and 6-n-pentyl-α-pyrone from *Myrothecium* strain MFA 58 were found to perform better against tyrosinase than kojic acid (Li et al. 2005). Sesquiterpenes extracted from the culture broth medium of *Pestalotiopsis* sp. Z233 were illustrated by Wu et al. (2013a, b), showing their potential to inhibit tyrosinase activity. Furthermore, *Aspergillus unguis* SPMD-EGY mycelia were found to contain various bioactive compounds, including 2,4-dihydroxybenzoic acid, caffeic acid, benzene-acetic acid-2-hydroxy, and benzene-acetic acid-2-hydroxy (Fitton and Goa 1991).

Macro Fungi

The diversity of beneficial bioactives offered by mushrooms in addition to their role as food and medicine. Environmental stressors drive the development of secondary metabolites in mushrooms, leading to the production of various compounds with potential health benefits (Qaderi et al. 2023).

Phenolic compounds, commonly found in microorganisms, plants, and mushrooms, are known for their antioxidant properties and diverse biological activities (Abdelshafy et al. 2022; de la Rosa et al. 2019). Some wild mushrooms, such as *Butyriboletus regius* (*Boletus regius*), are rich sources of phenolic compounds, tocopherols, and citric acids, exhibiting high antioxidant properties (Leal et al. 2013).

Similarly, *Ganoderma lucidum*, a well-researched wild mushroom, has been found to possess various medicinal properties, including cholesterol prevention, immunomodulation, liver protection, and anti-HIV, anti-metastatic, anti-tumor, and anti-viral activities (Kumar et al. 2021; Walton 2014; Xu et al. 2010).

Moreover, the medicinal mushroom *Inonotus hispidus*, known for its immunomodulatory and anticancer effects, synthesizes a high concentration of phenolic compounds, which have potent antioxidant properties against free radicals (Wang et al. 2022; Yang et al. 2022; Zhang et al. 2022). Studies have also reported that various edible mushrooms, including *Lanmaoa fragrans*, *Pisolithus tinctorius*, and *Suillellus luridus*, produce polar and other types of phenolic compounds with antioxidant properties. These compounds could be potential natural ingredients in cosmeceuticals (Grangeia et al. 2011; Pringle et al. 2021; Liu et al. 2016a, b).

Comparatively, the polysaccharide extract from *Butyriboletus regius* exhibits the strongest antioxidant potential among other boletus mushrooms (Zhang et al. 2018a, b). However, it's important to note that some mushroom species, such as *Suillellus mendax* and *Xerocomus subtomentosus*, may have adverse health effects, and further studies are needed to understand their potential implications (Martins et al. 2023). Moreover, among the eight mushroom strains investigated (*Butyriboletus regius*, *Ganoderma lucidum*, *Inonotus hispidus*, *Lanmaoa fragrans*, *Pisolithus tinctorius*, *Suillellus luridus*, *Suillellus mendax*, and *Xerocomus subtomentosus*), I. hispidus and P. tinctorius were found to synthesize high concentrations of phenolic compounds with significant antioxidant, antibacterial (against gram-positive bacteria), and anti-hyaluronidase potential (Martins et al. 2023).

Various mushrooms, including *Agaricus bisporus* (commonly known as the button mushroom), *Lentinula edodes* (shiitake mushroom), and *Pleurotus ostreatus*, have been reported to possess medicinal properties with applications in the treatment of tumors, antibiotics, immunomodulation, anti-inflammatory effects, and antioxidant activity. For instance, *Agaricus bisporus* has been explored for its potential medicinal benefits, including tumor treatment, antibiotic properties, immunomodulation, anti-inflammatory effects, and antioxidant activity (Stojkovic et al. 2014). Similarly, *Lentinula edodes* (shiitake mushroom) has been associated with the inhibition of tumor cell proliferation and immune system boosters in humans (Baba et al. 2015). The mushroom is reported to produce bioactive compounds such as L-ergothioneine and lentinan, which exhibit antioxidant and anti-inflammatory properties (Cheah and Halliwell 2012; Zembron-Lacny et al. 2013; Mizuno et al. 2009). Moreover, *Pleurotus ostreatus*, commonly known as the oyster mushroom, has been studied for its anti-inflammatory, antioxidant, immunomodulatory, antibacterial, and anticancer properties (Taofiq et al. 2015; Stefan et al. 2015; Facchini et al. 2014; Fernandes et al. 2015). Rich in polysaccharides, proteins, organic acids, vitamins, and minerals, *Pleurotus ostreatus* represents a versatile mushroom with potential health benefits.

The diverse bioactive compounds present in mushrooms not only enhance their medicinal properties but also render them invaluable in diverse applications for human health and well-being. This extends to skincare, showcasing their potential in anti-aging formulations within the cosmeceutical industry. These bioactive compounds, with their multifaceted nature, have the capacity to provide numerous advantages for skin health and overall skincare. Noteworthy contributions encompass antioxidant properties, anti-inflammatory effects, immunomodulation, hydration, moisturization, antibacterial characteristics, stimulation of collagen production, and the promotion of skin-brightening effects.

14.4.8.3 Other Microorganisms

Mycosporines and MAAs, particularly those derived from microalgae, hold significant promise for cosmetics, as evidenced by numerous patents and well-documented research (Rastogi et al. 2010; Shick and Dunlap 2002; Whitehead and Hedges 2005; Oren and Gunde-Cimerman 2007). Although cyclosporine and certain MAAs, such as those produced by the red alga *Porphyra umbilicalis*, are present in a relatively limited number of commercially available UV-screening and anti-aging products (Pallela et al. 2010; Rastogi et al. 2010; Cardozo et al. 2007).

Thraustochytrids, both planktonic and benthonic, exhibit a global presence, ranging from polar to tropical regions. These species thrive in diverse habitats, including mangrove sediments, estuaries, and deep-sea ecosystems (Raghukumar

2002). The *Thraustochytrium* strains ONC-T18 and CHN-1, *Thraustochytriidae* sp. AS4-A1 (Ulkenia sp.), and *Aurantiochytrium* sp. KH105 are capable of producing various carotenoid derivatives, such as β-carotene, astaxanthin, zeaxanthin, cantaxanthin, phoenicoxanthin, and echinenone (Aasen et al. 2016).

Certain studies suggest that chitosan in fiber or film forms is more effective in enhancing skin granular density and improving the epithelial layer (Ito et al. 2014). Its stimulation of fibroblast production contributes to the creation of moisturizing and antimicrobial properties, known for their remarkable healing capabilities (Kumirska et al. 2011). Carboxymethyl chitosan (CMCS), a chitosan derivative with active hydroxyl, carboxyl, and amine groups, exhibits high viscosity, significant hydrodynamic volumes, cation-binding properties, large osmotic pressures, and gel-forming capabilities (Ito et al. 2014). Chitosan and its derivatives, such as CMCS, are highly desirable candidates for various applications, including delivery systems, stabilizers, antimicrobials, antioxidants, absorption boosters, and hydration agents (Gautier et al. 2008).

14.4.9 Advantages of Natural and Sustainable Cosmeceuticals

Personal care products and cosmetics are formulated to nourish, protect, and supply essential nutrients to the skin, its microbiome, and associated cells. These products play a crucial role in maintaining skin health by enhancing the skin's barrier function, inhibiting pathogen growth, and effectively removing dirt and impurities. Researchers and cosmetic manufacturers are consistently exploring new and promising ingredients to incorporate into formulations, aiming to deliver healthier and more efficient skincare products in line with the specified objectives (Heinrich et al. 2014; Rodan et al. 2016; Gupta et al. 2019).

Approximately 50% of the chemical surfactants available in the market are derived from non-sustainable petrochemical sources, posing a challenge to the sustainability of personal skincare formulations. Some of these chemicals, utilized for foaming and emulsification purposes, have raised concerns (Akbari et al. 2018). Despite their acknowledged efficacy in formulations, scientific research indicates that these chemical surfactants can have detrimental effects on both the skin and its microbiome (Holland and Bojar 2002; Lee et al. 2018). Recent investigations highlight that certain surfactants alter the skin's flora, leading to allergies and irritations by binding to the lipids and proteins in the epidermis and hair. Prolonged use and high concentrations may render the skin's intracellular and epidermal lipids soluble, compromising the skin's foundational strength and protective functions (Bujak et al. 2015).

The growing awareness of these concerns has prompted a shift toward replacing chemical surfactants with compounds that are cost-effective, environmentally friendly, and sustainable. These alternatives are highly tolerant, biodegradable, and skin-compatible, presenting no environmental or health risks to consumers (Marchant and Banat 2012; Banat et al. 2014; Seweryn 2018).

14.4.10 Gene Array Technology Reshaping Cosmeceuticals

Gene array technology has made significant strides in the cosmeceuticals sector by uncovering active chemicals with desirable biological activity, laying the foundation for more effective and targeted skincare solutions. Employing a meticulous screening program based on cell and molecular biology, this method identifies active chemicals that stimulate specific genes, such as collagen I, III, or VII. The "active" components thus discovered undergo thorough testing to ensure they have no adverse effects on the skin (Arshad et al. 2023).

An instrumental aspect of gene array technology's impact on the cosmeceuticals business lies in its ability to facilitate the development of topical formulations that efficiently deliver these "active" substances to their target cells. This involves ensuring the compounds can traverse the stratum corneum, the outermost layer of the skin, and reach the deeper layers where they can exert their intended biological effects. This precise distribution strategy is pivotal for the performance of cosmeceutical products.

Moreover, double-blind, placebo-controlled clinical trials have become integral to cosmeceutical product development. These trials, involving a substantial number of patients, generate statistically significant data on product efficacy, providing empirical evidence to support the claimed benefits. Gene array technology contributes to these studies by offering a solid scientific basis for selecting "active" components and creating solutions with a higher likelihood of success.

Looking forward, predicting the exact impact of gene array technology on the future of the cosmeceuticals business is challenging, given its current early phases and continuous study and development. However, it is reasonable to anticipate its continued significant role in the sector's growth. One potential avenue is the creation of more accurate and customized cosmeceutical solutions. Gene array technology may pave the way for skincare treatments targeting specific skin issues, potentially leading to personalized cosmeceuticals tailored to each individual's unique genetic composition.

In summary, gene array technology has already exerted a substantial influence on the cosmetics sector by identifying active chemicals and optimizing their transport to target cells. The prospect of more personalized and effective cosmeceuticals is on the horizon as research in this sector advances. While the precise nature of its future impact remains uncertain, one certainty is that gene array technology will persist as a driving force in the growth of cosmeceuticals.

14.5 Future Prospects and Trends

The researcher should focus on discovering and identifying new bioactive compounds from diverse microbes for potential application in the cosmeceutical industry. The literature review underscores the importance of enhancing bioactive pigments from a variety of microorganisms, including bacteria, fungi, and microalgae. Additionally, it highlights a knowledge gap in the gene expression of key genes involved in the biosynthetic pathways of bioactive compounds.

Emphasizing genetic engineering and other biotechnological tools, the researcher should aim to empower microbial strains to produce essential cosmeceutical compounds. This involves a comprehensive exploration of the regulatory mechanisms governing the biosynthesis of vitamins, amino acids, exopolysaccharides, and other metabolites in microbes, such as algae, fungi, and bacteria. To advance the application of genetic engineering, gaining a thorough understanding of these intricate regulatory mechanisms is crucial.

Utilizing high-throughput "omics" technologies and gene array techniques, researchers can identify potential bottlenecks and redirect carbon fluxes toward the desired metabolic pathways for specific compounds. Achieving a high level of production is essential for the commercialization of microbial metabolites, necessitating the development of large-scale growth systems capable of producing both biomass and bioactive compounds at economies of scale (Arshad et al. 2022).

14.6 Conclusions

Gene array technology emerges as a pivotal force in securing the future sustainability of cosmeceutical industries. Advanced gene array techniques, such as GeneChip and Cas9, offer avenues to explore unidentified genes responsible for the synthesis of potential and multifunctional bioactive compounds. With the aid of gene array technologies and genetic engineering, significant genes have been effectively integrated into diverse bacterial and fungal genomes.

In addition, various bacterial, fungal, and microalgal groups produce a diverse array of compounds, encompassing glycosides, isoprenoids, polyketides, alkaloids, peptides, proteins, lipids, mycosporins, and amino acids resembling mycosporins. These compounds, known for their anti-aging, antimicrobial, antioxidant, and moisturizing properties, can find application in cosmeceutical and cosmetic products (Imhoff et al. 2011; Jin et al. 2016). Notably, the chemical structures and biological activity of compounds from these groups surpass those of species inhabiting terrestrial habitats by several orders of magnitude.

References

Aasen IM, Ertesvåg H, Heggeset TMB, Liu B, Brautaset T, Vadstein O, Ellingsen TE (2016) Thraustochytrids as production organisms for docosahexaenoic acid (DHA), squalene, and carotenoids. Appl Microbiol Biotechnol 100:4309–4321

Abdel-Lateff A, König GM, Fisch KM, Höller U, Jones PG, Wright AD (2002) New antioxidant hydroquinone derivatives from the algicolous marine fungus Acremonium sp. J Nat Prod 65(11):1605–1611

Abdel-Lateff A, Fisch KM, Wright AD, König GM (2003) A new antioxidant isobenzofuranone derivative from the algicolous marine fungus Epicoccum sp. Planta Med 69(09):831–834
Abdelshafy AM, Belwal T, Liang Z, Wang L, Li D, Luo Z, Li L (2022) A comprehensive review on phenolic compounds from edible mushrooms: occurrence, biological activity, application and future prospective. Crit Rev Food Sci Nutr 62(22):6204–6224
Adarme-Vega TC, Thomas-Hall SR, Schenk PM (2014) Towards sustainable sources for omega-3 fatty acids production. Curr Opin Biotechnol 26:14–18
Adu SA, Naughton PJ, Marchant R, Banat IM (2020) Microbial biosurfactants in cosmetic and personal skincare pharmaceutical formulations. Pharmaceutics 12(11):1099
Aebersold R, Agar JN, Amster IJ, Baker MS, Bertozzi CR, Boja ES et al (2018) How many human proteoforms are there? Nat Chem Biol 14(3):206–214
Agarwal S, Krishnamurthy K (2019) Histology, skin. StatPearls Publishing, Treasure Island
Akbari S, Abdurahman NH, Yunus RM, Fayaz F, Alara OR (2018) Biosurfactants—a new frontier for social and environmental safety: a mini review. Biotechnol Res Innov 2(1):81–90
Ambati RR, Phang SM, Ravi S, Aswathanarayana RG (2014) Astaxanthin: sources, extraction, stability, biological activities and its commercial applications—a review. Mar Drugs 12(1):128–152
Amend A (2014) From dandruff to deep-sea vents: Malassezia-like fungi are ecologically hyper-diverse. PLoS pathogens 10(8):e1004277
Ansorge WJ (2009) Next-generation DNA sequencing techniques. New Biotechnol 25(4):195–203
Araya-Garay JM, Ageitos JM, Vallejo JA, Veiga-Crespo P, Sánchez-Pérez A, Villa TG (2012) Construction of a novel Pichia pastoris strain for production of xanthophylls. AMB Express 2(1):1–8
Arkin M, Zeevi A, Avram N, Uzan R, Shilo-Volin H, Hollander E, Buriakovsky O (2011) U.S. patent application no. 12/887,021.
Armenta RE, Valentine MC (2013) Single-cell oils as a source of omega-3 fatty acids: an overview of recent advances. J Am Oil Chem Soc 90:167–182
Arshad M, Bano I, Younus M, Khan A, Rahman A (2018) Health concerns associated with biofuel production. In: Arshad M (ed) Perspectives on water usage for biofuels production: aquatic contamination and climate change. Springer, Cham, pp 97–105
Arshad M, Ansari AR, Qadir R, Tahir MH, Nadeem A, Mehmood T, Alhumade H, Khan N (2022) Green electricity generation from biogas of cattle manure: an assessment of potential and feasibility in Pakistan. Front Ener Res 10:911485
Arshad M, Syed-Hassan SSA, Masood R, Ansari AR, Mumtaz A, Rahman A, Batool S, Atif FA, Muhammad A, Rehman AU (2023) Utilization of waste animal fat for sustainable biodiesel production. In: Climate changes mitigation and sustainable bioenergy harvest through animal waste: sustainable environmental implications of animal waste. Springer, pp 149–169
Baba E, Uluköy G, Öntaş C (2015) Effects of feed supplemented with Lentinula edodes mushroom extract on the immune response of rainbow trout, Oncorhynchus mykiss, and disease resistance against Lactococcus garvieae. Aquaculture 448:476–482
Bae IH, Lee SH, Oh S, Choi H, Marinho PA, Yoo JW et al (2019) Mannosylerythritol lipids ameliorate ultraviolet A-induced aquaporin-3 downregulation by suppressing c-Jun N-terminal kinase phosphorylation in cultured human keratinocytes. Korean J Physiol Pharmacol 23(2):113–120
Balboa EM, Conde E, Soto ML, Pérez-Armada L, Domínguez H (2015) Cosmetics from marine sources. In: Springer handbook of marine biotechnology. Springer, Berlin, Heidelberg, pp 1015–1042
Baliña LM, Graupe K (1991) The treatment of melasma 20% azelaic acid versus 4% hydroquinone cream. Int J Dermatol 30(12):893–895
Baltz RH (2019) Natural product drug discovery in the genomic era: realities, conjectures, misconceptions, and opportunities. J Ind Microbiol Biotechnol 46(3–4):281–299
Banat IM, De Rienzo MAD, Quinn GA (2014) Microbial biofilms: biosurfactants as antibiofilm agents. Appl Microbiol Biotechnol 98:9915–9929

Bao J, Sun YL, Zhang XY, Han Z, Gao HC, He F et al (2013) Antifouling and antibacterial polyketides from marine gorgonian coral-associated fungus Penicillium sp. SCSGAF 0023. J Antibiot 66(4):219–223

Baxter RA (2008) Anti-aging properties of resveratrol: review and report of a potent new antioxidant skin care formulation. J Cosmet Dermatol 7(1):2–7

Bay L, Barnes CJ, Fritz BG, Thorsen J, Restrup MEM, Rasmussen L et al (2020) Universal dermal microbiome in human skin. MBio 11(1):e02945

Ben Belgacem Z, Bijttebier S, Verreth C, Voorspoels S, Van de Voorde I, Aerts G et al (2015) Biosurfactant production by pseudomonas strains isolated from floral nectar. J Appl Microbiol 118(6):1370–1384

Bentley SD, Chater KF, Cerdeño-Tárraga AM, Challis GL, Thomson NR, James KD et al (2002) Complete genome sequence of the model actinomycete Streptomyces coelicolor A3 (2). Nature 417(6885):141–147

Bergmann S, Schümann J, Scherlach K, Lange C, Brakhage AA, Hertweck C (2007) Genomics-driven discovery of PKS-NRPS hybrid metabolites from Aspergillus nidulans. Nat Chem Biol 3(4):213–217

Bertucci F, Bernard K, Loriod B, Chang YC, Granjeaud S, Birnbaum D et al (1999) Sensitivity issues in DNA array-based expression measurements and performance of nylon microarrays for small samples. Hum Mol Genet 8(9):1715–1722

Bhataya A, Schmidt-Dannert C, Lee PC (2009) Metabolic engineering of Pichia pastoris X-33 for lycopene production. Process Biochem 44(10):1095–1102

Biosiltech (2014) www.biosiltech.com/sites/default/files/Helionori.pdf [accessed 05.02.14]

Blaner WS, Olson JA (1994) Retinol and retinoic acid metabolism. In: Sporn MB, Roberts AB, Goodman DS (eds) The retinoids: biology, chemistry and medicine. Raven Press, Ltd., New York

Breitenbach J, Pollmann H, Sandmann G (2019) Genetic modification of the carotenoid pathway in the red yeast Xanthophyllomyces dendrorhous: engineering of a high-yield zeaxanthin strain. J Biotechnol 289:112–117

Brisson VL, West KA, Lee PK, Tringe SG, Brodie EL, Alvarez-Cohen L (2012) Metagenomic analysis of a stable trichloroethene-degrading microbial community. ISME J 6(9):1702–1714

Brodie EL, DeSantis TZ, Joyner DC, Baek SM, Larsen JT, Andersen GL et al (2006) Application of a high-density oligonucleotide microarray approach to study bacterial population dynamics during uranium reduction and reoxidation. Appl Environ Microbiol 72(9):6288–6298

Brodie EL, DeSantis TZ, Parker JPM, Zubietta IX, Piceno YM, Andersen GL (2007) Urban aerosols harbor diverse and dynamic bacterial populations. Proc Natl Acad Sci 104(1):299–304

Bujak T, Wasilewski T, Nizioł-Łukaszewska Z (2015) Role of macromolecules in the safety of use of body wash cosmetics. Colloids Surf B: Biointerfaces 135:497–503

Cardozo KH, Guaratini T, Barros MP, Falcão VR, Tonon AP, Lopes NP et al (2007) Metabolites from algae with economical impact. Comp Biochem Physiol C Toxicol Pharmacol 146(1–2):60–78

Carreto JI, Carignan MO (2011) Mycosporine-like amino acids: relevant secondary metabolites. Chemical and ecological aspects. Mar Drugs 9(3):387–446

Chang TS (2009) An updated review of tyrosinase inhibitors. Int J Mol Sci 10(6):2440–2475

Cheah IK, Halliwell B (2012) Ergothioneine; antioxidant potential, physiological function and role in disease. Biochim Biophys Acta 1822(5):784–793

Chen XW, Li CW, Cui CB, Hua W, Zhu TJ, Gu QQ (2014) Nine new and five known polyketides derived from a deep sea-sourced Aspergillus sp. 16-02-1. Mar Drugs 12(6):3116–3137

Chen L, Zhao YY, Lan RF, Du L, Wang BS, Zhou T et al (2017a) Dicitrinone D, an antimitotic polyketide isolated from the marine-derived fungus Penicillium citrinum. Tetrahedron 73(40):5900–5911

Chen R, Cheng Z, Huang J, Liu D, Wu C, Guo P, Lin W (2017b) Versicotides D–F, new cyclopeptides with lipid-lowering activities. RSC Adv 7(78):49235–49243

Chen L, Zhao W, Jiang HL, Zhou J, Chen XM, Lian YY et al (2018) Rakicidins G-I, cyclic depsipeptides from marine Micromonospora chalcea FIM 02-523. Tetrahedron 74(30):4151–4154

Cheng Z, Lou L, Liu D, Li X, Proksch P, Yin S, Lin W (2016) Versiquinazolines A–K, fumiquinazoline-type alkaloids from the gorgonian-derived fungus Aspergillus versicolor LZD-14-1. J Nat Prod 79(11):2941–2952

Cheng Z, Liu D, Cheng W, Proksch P, Lin W (2018) Versiquinazolines L–Q, new polycyclic alkaloids from the marine-derived fungus Aspergillus versicolor. RSC Adv 8(55):31427–31439

Choi BH, Hwang HJ, Lee JE, Oh SH, Hwang JS, Lee BY, Lee PC (2020) Microbial production of retinyl palmitate and its application as a cosmeceutical. Antioxidants 9(11):1130

Cohen Z, Ratledge C (2005) Single cell oils: microbial and algal oils. AOCS Publishing

Corinaldesi C, Barone G, Marcellini F, Dell'Anno A, Danovaro R (2017) Marine microbial-derived molecules and their potential use in cosmeceutical and cosmetic products. Mar Drugs 15(4):118

Coronel-León J, de Grau G, Grau-Campistany A, Farfan M, Rabanal F, Manresa A, Marqués AM (2015) Biosurfactant production by AL 1.1, a Bacillus licheniformis strain isolated from Antarctica: production, chemical characterization and properties. Ann Microbiol 65(4):2065–2078

Courtois A, Berthou C, Guézennec J et al (2014) Exopolysaccharides isolated from hydrothermal vent bacteria can modulate the complement system. PLoS One 9:e94965

Cox JM (2001) Applications of nylon membrane arrays to gene expression analysis. J Immunol Methods 250(1–2):3–13

Crepin, V.F., Faulds, C.B., Connerton, I.F. (2004). Functional classification of the microbial feruloyl esterases. Appl. Microbiol. Biotechnol., 63(6), 647–52.

Cui C, Shen P, Zhang J (2010) Structural and physicochemical characterization of rhamnolipids produced by Pseudomonas aeruginosa W3. Acta Sci Circum 30:2030–2034

Dalili D, Amini M, Faramarzi MA, Fazeli MR, Khoshayand MR, Samadi N (2015) Isolation and structural characterization of Coryxin, a novel cyclic lipopeptide from Corynebacterium xerosis NS5 having emulsifying and anti-biofilm activity. Colloids Surf B: Biointerfaces 135:425–432

Davies J, Ryan KS (2012) Introducing the parvome: bioactive compounds in the microbial world. ACS Chem Biol 7(2):252–259

De la Coba F (2007) Use of mycosporine-type aminoacid shinorine as an antioxidant. Intl Patent WO2007/026038A2

de la Rosa LA, Moreno-Escamilla JO, Rodrigo-García J, Alvarez-Parrilla E (2019) Phenolic compounds. In: Postharvest physiology and biochemistry of fruits and vegetables. Woodhead Publishing, pp 253–271

Deering RW, Chen J, Sun J, Ma H, Dubert J, Barja JL et al (2016) N-acyl dehydrotyrosines, tyrosinase inhibitors from the marine bacterium Thalassotalea sp. PP2-459. J Nat Prod 79(2):447–450

Deng Y, He Z, Xu M, Qin Y, Van Nostrand JD, Wu L et al (2012) Elevated carbon dioxide alters the structure of soil microbial communities. Appl Environ Microbiol 78(8):2991–2995

Díez B, Bergman B, Pedrós-Alió C, Antó M, Snoeijs P (2012) High cyanobacterial nifH gene diversity in Arctic seawater and sea ice brine. Environ Microbiol Rep 4(3):360–366

Ding L, Li T, Liao X, He S, Xu S (2018) Asperitaconic acids A–C, antibacterial itaconic acid derivatives produced by a marine-derived fungus of the genus Aspergillus. J Antibiot 71(10):902–904

Donadio S, Monciardini P, Sosio M (2007) Polyketide synthases and nonribosomal peptide synthetases: the emerging view from bacterial genomics. Nat Prod Rep 24(5):1073–1109

Donot F, Fontana A, Baccou JC, Schorr-Galindo S (2012) Microbial exopolysaccharides: main examples of synthesis, excretion, genetics and extraction. Carbohydr Polym 87(2):951–962

Epstein CB, Butow RA (2000) Microarray technology—enhanced versatility, persistent challenge. Curr Opin Biotechnol 11(1):36–41

Facchini JM, Alves EP, Aguilera C, Gern RMM, Silveira MLL, Wisbeck E, Furlan SA (2014) Antitumor activity of Pleurotus ostreatus polysaccharide fractions on Ehrlich tumor and Sarcoma 180. Int J Biol Macromol 68:72–77

Fang W, Lin X, Zhou X, Wan J, Lu X, Yang B et al (2014) Cytotoxic and antiviral nitrobenzoyl sesquiterpenoids from the marine-derived fungus aspergillus ochraceus Jcma1F17. Med Chem Comm 5(6):701–705

Fang F, Zhao J, Ding L, Huang C, Naman CB, He S et al (2017) 5-Hydroxycyclopenicillone, a new β-amyloid fibrillization inhibitor from a sponge-derived fungus Trichoderma sp. HPQJ-34. Mar Drugs 15(8):260

Fernandes Â, Barros L, Martins A, Herbert P, Ferreira IC (2015) Nutritional characterisation of Pleurotus ostreatus (Jacq. ex Fr.) P. Kumm. produced using paper scraps as substrate. Food Chem 169:396–400

Ferreira A, Vecino X, Ferreira D, Cruz JM, Moldes AB, Rodrigues LR (2017) Novel cosmetic formulations containing a biosurfactant from lactobacillus paracasei. Colloids Surf B: Biointerfaces 155:522–529

Ferreira RG, Azzoni AR, Santana MHA, Petrides D (2021) Techno-economic analysis of a hyaluronic acid production process utilizing streptococcal fermentation. PRO 9(2):241

Fisher GJ, Wang Z, Datta SC, Varani J, Kang S, Voorhees JJ (1997) Pathophysiology of premature skin aging induced by ultraviolet light. N Engl J Med 337(20):1419–1429

Fitton A, Goa KL (1991) Azelaic acid: a review of its pharmacological properties and therapeutic efficacy in acne and hyperpigmentary skin disorders. Drugs 41:780–798

Freitas F, Alves VD, Reis MA (2011) Advances in bacterial exopolysaccharides: from production to biotechnological applications. Trends Biotechnol 29(8):388–398

Furuya T, Kuroiwa M, Kino K (2017) Biotechnological production of vanillin using immobilized enzymes. J Biotechnol 243:25–28

Gann A (2010) Jacob and monod: from operons to evodevo. Curr Biol 20(17):R718–R723

Gao Q, Garcia-Pichel F (2011) Microbial ultraviolet sunscreens. Nat Rev Microbiol 9(11):791–802

Gao CH, Tian XP, Qi SH, Luo XM, Wang P, Zhang S (2010) Antibacterial and antilarval compounds from marine gorgonian-associated bacterium Bacillus amyloliquefaciens SCSIO 00856. J Antibiot 63(4):191–193

Gassel S, Breitenbach J, Sandmann G (2014) Genetic engineering of the complete carotenoid pathway towards enhanced astaxanthin formation in Xanthophyllomyces dendrorhous starting from a high-yield mutant. Appl Microbiol Biotechnol 98:345–350

Gautier S, Xhauflaire-Uhoda E, Gonry P, Piérard GE (2008) Chitin–glucan, a natural cell scaffold for skin moisturization and rejuvenation. Int J Cosmet Sci 30(6):459–469

Goldman BS, Nierman WC, Kaiser D, Slater SC, Durkin AS, Eisen JA et al (2006) Evolution of sensory complexity recorded in a myxobacterial genome. Proc Natl Acad Sci 103(41):15200–15205

Gopalan N, Rodriguez-Duran LV, Saucedo-Castaneda G, Nampoothiri KM (2015) Review on technological and scientific aspects of feruloyl esterases: a versatile enzyme for biorefining of biomass. Bioresour Technol 193:534–544

Gottschalk LMF, Oliveira RA, da Silva Bon EP (2010) Cellulases, xylanases, β-glucosidase and ferulic acid esterase produced by Trichoderma and aspergillus act synergistically in the hydrolysis of sugarcane bagasse. Biochem Eng J 51(1–2):72–78

Grangeia C, Heleno SA, Barros L, Martins A, Ferreira IC (2011) Effects of trophism on nutritional and nutraceutical potential of wild edible mushrooms. Food Res Int 44(4):1029–1035

Grice EA, Segre JA (2011) The skin microbiome. Nat Rev Microbiol 9(4):244–253

Guo W, Wang S, Li N, Li F, Zhu T, Gu Q et al (2018) Saroclides A and B, cyclic depsipeptides from the mangrove-derived fungus Sarocladium kiliense HDN11-112. J Nat Prod 81(4):1050–1054

Gupta PL, Rajput M, Oza T, Trivedi U, Sanghvi G (2019) Eminence of microbial products in cosmetic industry. Nat Prod Bioprospect 9:267–278

Harvey AL, Edrada-Ebel R, Quinn RJ (2015) The re-emergence of natural products for drug discovery in the genomics era. Nat Rev Drug Discov 14(2):111–129

Hazra C, Kundu D, Ghosh P, Joshi S, Dandi N, Chaudhari A (2011) Screening and identification of Pseudomonas aeruginosa AB4 for improved production, characterization and application of a glycolipid biosurfactant using low-cost agro-based raw materials. J Chem Technol Biotechnol 86(2):185–198

He Z, Gentry TJ, Schadt CW, Wu L, Liebich J, Chong SC et al (2007) GeoChip: a comprehensive microarray for investigating biogeochemical, ecological and environmental processes. ISME J 1(1):67–77

He Z, Deng Y, Van Nostrand JD, Tu Q, Xu M, Hemme CL et al (2010) GeoChip 3.0 as a high-throughput tool for analyzing microbial community composition, structure and functional activity. ISME J 4(9):1167–1179
He Z, Deng Y, Zhou J (2012a) Development of functional gene microarrays for microbial community analysis. Curr Opin Biotechnol 23(1):49–55
He Z, Van Nostrand JD, Zhou J (2012b) Applications of functional gene microarrays for profiling microbial communities. Curr Opin Biotechnol 23(3):460–466
Heinrich K, Heinrich U, Tronnier H (2014) Influence of different cosmetic formulations on the human skin barrier. Skin Pharmacol Physiol 27(3):141–147
Herz S, Weber RW, Anke H, Mucci A, Davoli P (2007) Intermediates in the oxidative pathway from torulene to torularhodin in the red yeasts Cystofilobasidium infirmominiatum and C. capitatum (Heterobasidiomycetes, Fungi). Phytochemistry 68(20):2503–2511
Holland KT, Bojar RA (2002) Cosmetics: what is their influence on the skin microflora? Am J Clin Dermatol 3:445–449
Hong JH, Jang S, Heo YM, Min M, Lee H, Lee YM et al (2015) Investigation of marine-derived fungal diversity and their exploitable biological activities. Mar Drugs 13(7):4137–4155
Hoog JO, Ostberg LJ (2011) Mammalian alcohol dehydrogenases–a comparative investigation at gene and protein levels. Chem Biol Interact 191(1–3):2–7
Huang WE, Zhou J (2012) When single cell technology meets omics, the new toolbox of analytical biotechnology is emerging. Curr Opin Biotechnol 23(1):1
Huang X, Huang H, Li H, Sun X, Huang H, Lu Y et al (2013) Asperterpenoid A, a new sesterterpenoid as an inhibitor of mycobacterium tuberculosi s protein tyrosine phosphatase B from the culture of Aspergillus sp. 16-5c. Org Lett 15(4):721–723
Huang H, Song Y, Li X, Wang X, Ling C, Qin X et al (2018) Abyssomicin monomers and dimers from the marine-derived Streptomyces koyangensis SCSIO 5802. J Nat Prod 81(8):1892–1898
Hug LA, Baker BJ, Anantharaman K, Brown CT, Probst AJ, Castelle CJ, Butterfield CN, Hernsdorf AW, Amano Y, Ise K (2016) A new view of the tree of life. Nat Microbiol 1(5):1–6
Hutchings JA, Reynolds JD (2004) Marine fish population collapses: consequences for recovery and extinction risk. BioScience 54(4):297–309
Ikeda H, Ishikawa J, Hanamoto A, Shinose M, Kikuchi H, Shiba T et al (2003) Complete genome sequence and comparative analysis of the industrial microorganism Streptomyces avermitilis. Nat Biotechnol 21(5):526–531
Imhoff JF, Labes A, Wiese J (2011) Bio-mining the microbial treasures of the ocean: new natural products. Biotechnol Adv 29(5):468–482
Ito I, Osaki T, Ifuku S, Saimoto H, Takamori Y, Kurozumi S et al (2014) Evaluation of the effects of chitin nanofibrils on skin function using skin models. Carbohydr Polym 101:464–470
Jacob F, Monod J (1961) Genetic regulatory mechanisms in the synthesis of proteins. J Mol Biol 3(3):318–356
Jadhav M, Kagalkar A, Jadhav S, Govindwar S (2011) Isolation, characterization, and antifungal application of a biosurfactant produced by Enterobacter sp. MS16. Eur J Lipid Sci Technol 113:1347–1356
Janek T, Łukaszewicz M, Rezanka T, Krasowska A (2010) Isolation and characterization of two new lipopeptide biosurfactants produced by Pseudomonas fluorescens BD5 isolated from water from the Arctic Archipelago of Svalbard. Bioresour Technol 101(15):6118–6123
Jang HJ, Ha BK, Zhou J, Ahn J, Yoon SH, Kim SW (2015) Selective retinol production by modulating the composition of retinoids from metabolically engineered E. coli. Biotechnol Bioeng 112(8):1604–1612
Jiang S, Zhang L, Pei X, Deng F, Hu D, Chen G et al (2017) Chalcomycins from marine-derived Streptomyces sp. and their antimicrobial activities. Mar Drugs 15(6):153
Jin L, Quan C, Hou X, Fan S (2016) Potential pharmacological resources: natural bioactive compounds from marine-derived fungi. Mar Drugs 14(4):76
Kanamoto H, Nakamura K, Misawa N (2021) Carotenoid production in oleaginous yeasts. In: Carotenoids: biosynthetic and biofunctional approaches. Springer, Singapore, pp 153–163

Kang HY, Yoon TJ, Lee GJ (2011) Whitening effects of marine pseudomonas extract. Ann Dermatol 23(2):144–149

Keller NP (2019) Fungal secondary metabolism: regulation, function and drug discovery. Nat Rev Microbiol 17(3):167–180

Khan J, Bittner ML, Chen Y, Meltzer PS, Trent JM (1999) DNA microarray technology: the anticipated impact on the study of human disease. Biochim Biophys Acta 1423(2):M17–M28

Khosla C (2009) Structures and mechanisms of polyketide synthases. J Org Chem 74(17):6416–6420

Kiene RP (1998) Uptake of choline and its conversion to glycine betaine by bacteria in estuarine waters. Appl Environ Microbiol 64(3):1045–1051

Kim YJ (2007) Antimelanogenic and antioxidant properties of gallic acid. Biol Pharm Bull 30(6):1052–1055

Kim J (2011a) *Natural product biosynthesis in uncultured bacteria.*

Kim SK (ed) (2011b) Marine cosmeceuticals: trends and prospects. CRC Press

Kim J-H, Baik S-H (2015) Properties of recombinant novel cinnamoyl esterase from Lactobacillus acidophilus F46 isolated from human intestinal bacterium. J Mol Catal B Enzym 116:9–15

Kim YS, Oh DK (2009) Substrate specificity of a recombinant chicken β-carotene 15, 15′-monooxygenase that converts β-carotene into retinal. Biotechnol Lett 31:403–408

Kirti K, Amita S, Priti S, Mukesh Kumar A, Jyoti S (2014) Colorful world of microbes: carotenoids and their applications. Adv Biol 2014:1–13

Kogej T, Gostinčar C, Volkmann M, Gorbushina AA, Gunde-Cimerman N (2006) Mycosporines in extremophilic fungi—novel complementary osmolytes? Environ Chem 3(2):105–110

Krug D, Zurek G, Revermann O, Vos M, Velicer GJ, Müller R (2008) Discovering the hidden secondary metabolome of Myxococcus xanthus: a study of intraspecific diversity. Appl Environ Microbiol 74(10):3058–3068

Kubicek CP, Punt P, Visser J (2011) Production of organic acids by filamentous fungi. Springer, Berlin Heidelberg, pp 215–234

Kumar K, Mehra R, Guiné RP, Lima MJ, Kumar N, Kaushik R et al (2021) Edible mushrooms: a comprehensive review on bioactive compounds with health benefits and processing aspects. Food Secur 10(12):2996

Kumirska J, Weinhold MX, Thöming J, Stepnowski P (2011) Biomedical activity of chitin/chitosan based materials—influence of physicochemical properties apart from molecular weight and degree of N-acetylation. Polymers 3(4):1875–1901

Kundu D, Hazra C, Dandi N, Chaudhari A (2013) Biodegradation of 4-nitrotoluene with biosurfactant production by Rhodococcus pyridinivorans NT2: metabolic pathway, cell surface properties and toxicological characterization. Biodegradation 24:775–793

Larroude M, Celinska E, Back A, Thomas S, Nicaud JM, Ledesma-Amaro R (2018) A synthetic biology approach to transform Yarrowia lipolytica into a competitive biotechnological producer of β-carotene. Biotechnol Bioeng 115(2):464–472

Leal AR, Barros L, Barreira JC, Sousa MJ, Martins A, Santos-Buelga C, Ferreira IC (2013) Portuguese wild mushrooms at the "pharma–nutrition" interface: nutritional characterization and antioxidant properties. Food Res Int 50(1):1–9

Lee MLT, Kuo FC, Whitmore GA, Sklar J (2000) Importance of replication in microarray gene expression studies: statistical methods and evidence from repetitive cDNA hybridizations. Proc Natl Acad Sci 97(18):9834–9839

Lee GT, Lee SY, Jeong JH, Jo BK, Li XF, Son BW (2003) PP-35 screening of tyrosinase inhibiting activity from the marine-derived fungus. Pigment Cell Res 16(5):604–604

Lee SC, Kim SH, Park IH, Chung SY, Subhosh Chandra M and Choi YL (2010) Isolation,purification, and characterization of novel fengycin S from Bacillus amyloliquefaciens LSC04degrading-crude oil. Biotechnol. Bioproc. Eng. 15:246–253

Lee HJ, Jeong SE, Lee S, Kim S, Han H, Jeon CO (2018) Effects of cosmetics on the skin microbiome of facial cheeks with different hydration levels. Microbiology 7(2):e00557

Lester G, Bonner DM (1952) The occurrence of betagalactosidase in Escherichia coli. J Bacteriol 63(6):759–769

Li X, Jeong JH, Lee KT, Rho JR, Choi HD, Kang JS, Son BW (2003) γ-Pyrone derivatives, kojic acid methyl ethers from a marine-derived fungus Altenaria sp. Arch Pharm Res 26:532–534

Li Y, Li X, Kang JS, Choi HD, Son BW (2004a) New radical scavenging and ultraviolet-a protecting prenylated dioxopiperazine alkaloid related to isoechinulin A from a marine isolate of the fungus Aspergillus. J Antibiot 57(5):337–340

Li Y, Li X, Kim SK, Kang JS, Choi HD, Rho JR, Son BW (2004b) Golmaenone, a new diketopiperazine alkaloid from the marine-derived fungus Aspergillus sp. Chem Pharm Bull 52(3):375–376

Li X, Kim MK, Lee U, Kim SK, Kang JS, Choi HD, Son BW (2005) Myrothenones A and B, cyclopentenone derivatives with tyrosinase inhibitory activity from the marine-derived fungus Myrothecium sp. Chem Pharm Bull 53(4):453–455

Li HJ, Xie YL, Xie ZL, Chen Y, Lam CK, Lan WJ (2012) Chondrosterins A–E, triquinane-type sesquiterpenoids from soft coral-associated fungus Chondrostereum sp. Mar Drugs 10(3):627–638

Li X, Li XM, Xu GM, Li CS, Wang BG (2014) Antioxidant metabolites from marine alga-derived fungus Aspergillus wentii EN-48. Phytochem Lett 7:120–123

Lian X, Fang Y, Joseph E, Wang Q, Li J, Banerjee S et al (2017) Enzyme–MOF (metal–organic framework) composites. Chem Soc Rev 46(11):3386–3401

Liang X, Zhang XY, Nong XH, Wang J, Huang ZH, Qi SH (2016) Eight linear peptides from the deep-sea-derived fungus Simplicillium obclavatum EIODSF 020. Tetrahedron 72(22):3092–3097

Lim CG, Fowler ZL, Hueller T, Schaffer S, Koffas MA (2011) High-yield resveratrol production in engineered Escherichia coli. Appl Environ Microbiol 77(10):3451–3460

Lipshutz RJ (2000) Applications of high-density oligonucleotide arrays. In: Novartis foundation symposium. Wiley, Chichester/New York, pp 84–93

Lipshutz RJ, Fodor S, Gingeras TR, Lockhart DJ (1999) High density synthetic oligonucleotide arrays. Nat Genet 21(1):20–24

Liu Y, Li XM, Meng LH, Wang BG (2015) Polyketides from the marine mangrove-derived fungus Aspergillus ochraceus MA-15 and their activity against aquatic pathogenic bacteria. Phytochem Lett 12:232–236

Liu H, Chen S, Liu W, Liu Y, Huang X, She Z (2016a) Polyketides with immunosuppressive activities from mangrove endophytic fungus Penicillium sp. ZJ-SY2. Mar Drugs 14(12):217

Liu Y, Chen D, You Y, Zeng S, Li Y, Tang Q et al (2016b) Nutritional composition of boletus mushrooms from Southwest China and their antihyperglycemic and antioxidant activities. Food Chem 211:83–91

Liu Z, Qiu P, Li J, Chen G, Chen Y, Liu H, She Z (2018) Anti-inflammatory polyketides from the mangrove-derived fungus Ascomycota sp. SK2YWS-L. Tetrahedron 74(7):746–751

Liu L, Qu YL, Dong GR, Wang J, Hu CY, Meng YH (2021) Elevated β-carotene production using codon-adapted CarRA&B and metabolic balance in engineered Yarrowia lipolytica. Front Microbiol 12:627150

Lopez-Ojeda W, Pandey A, Alhajj M, Oakley AM (2017) Anatomy, skin (integument). StatPearls Publishing, Treasure Island

Lourith N, Kanlayavattanakul M (2009) Natural surfactants used in cosmetics: glycolipids. Int J Cosmet Sci 31(4):255–261

Ma Z, Hu J (2014) Production and characterization of iturinic lipopeptides as antifungal agents and biosurfactants produced by a marine Pinctada martensii-derived Bacillus mojavensis B0621A. Appl Biochem Biotechnol 173:705–715

Machado-Carvalho L, Martins T, Aires A, Saavedra MJ, Marques G (2023) Antioxidant, antibacterial, and cosmeceutical potential of four common edible mushrooms. Appl Sci 13(13):7357

MacLean D, Jones JD, Studholme DJ (2009) Application of 'next-generation' sequencing technologies to microbial genetics. Nat Rev Microbiol 7(4):96–97

Manga EB, Celik PA, Cabuk A, Banat IM (2021) Biosurfactants: opportunities for the development of a sustainable future. Curr Opin Colloid Interface Sci 56:101514

Marchant R, Banat IM (2012) Biosurfactants: a sustainable replacement for chemical surfactants? Biotechnol Lett 34:1597–1605

Martins A, Vieira H, Gaspar H, Santos S (2014) Marketed marine natural products in the pharmaceutical and cosmeceutical industries: tips for success. Mar Drugs 12(2):1066–1101

Martins T, Machado-Carvalho L, Aires A, Saavedra MJ, Marques G (2023) Antioxidant, antimicrobial and cosmeceutical potential of wild mushroom extracts. Appl Microbiol 3(2):562–579

Mata-Gómez LC, Montañez JC, Méndez-Zavala A, Aguilar CN (2014) Biotechnological production of carotenoids by yeasts: an overview. Microb Cell Fact 13:1–11

Matthäus F, Ketelhot M, Gatter M, Barth G (2014) Production of lycopene in the non-carotenoid-producing yeast Yarrowia lipolytica. Appl Environ Microbiol 80(5):1660–1669

McCarty NS, Graham AE, Studená L, Ledesma-Amaro R (2020) Multiplexed CRISPR technologies for gene editing and transcriptional regulation. Nat Commun 11(1):1281

McClements DJ, Gumus CE (2016) Natural emulsifiers—Biosurfactants, phospholipids, biopolymers, and colloidal particles: molecular and physicochemical basis of functional performance. Adv Colloid Interf Sci 234:3–26

McGall G, Labadie J, Brock P, Wallraff G, Nguyen T, Hinsberg W (1996) Light-directed synthesis of high-density oligonucleotide arrays using semiconductor photoresists. Proc Natl Acad Sci 93(24):13555–13560

McLeod MP, Warren RL, Hsiao WW, Araki N, Myhre M, Fernandes C et al (2006) The complete genome of Rhodococcus sp. RHA1 provides insights into a catabolic powerhouse. Proc Natl Acad Sci 103(42):15582–15587

Menozzi I, Vallese F, Polverini E, Folli C, Berni R, Zanotti G (2017) Structural and molecular determinants affecting the interaction of retinol with human CRBP1. J Struct Biol 197(3):330–339

Metzker ML (2010) Sequencing technologies—the next generation. Nat Rev Genet 11(1):31–46

Mgbechidinma CL, Akan OD, Zhang C, Huang M, Linus N, Zhu H, Wakil SM (2022) Integration of green economy concepts for sustainable biosurfactant production–a review. Bioresour Technol 364:128021

Minami R, Uchiyama K, Murakami T, Kawai J, Mikami K, Yamada T, ... Honma M (1998) Properties, sequence, and synthesis in Escherichia coli of 1-aminocyclopropane-l-carboxylate deaminase from Hansenula saturnus. The Journal of Biochemistry 123(6):1112–1118

Miura Y, Kondo K, Saito T, Shimada H, Fraser PD, Misawa N (1998) Production of the carotenoids lycopene, β-carotene, and astaxanthin in the food yeast Candida utilis. Appl Environ Microbiol 64(4):1226–1229

Miyamoto KT, Komatsu M, Ikeda H (2014) Discovery of gene cluster for mycosporine-like amino acid biosynthesis from Actinomycetales microorganisms and production of a novel mycosporine-like amino acid by heterologous expression. Appl Environ Microbiol 80(16):5028–5036

Mizuno M, Nishitani Y, Hashimoto T, Kanazawa K (2009) Different suppressive effects of fucoidan and lentinan on IL-8 mRNA expression in in vitro gut inflammation. Biosci Biotechnol Biochem 73(10):2324–2325

Morabito K, Shapley NC, Steeley KG, Tripathi A (2011) Review of sunscreen and the emergence of non-conventional absorbers and their applications in ultraviolet protection. Int J Cosmet Sci 33(5):385–390

Morita T, Kawamura D, Morita N, Fukuoka T, Imura T, Sakai H et al (2013) Characterization of mannosylerythritol lipids containing hexadecatetraenoic acid produced from cuttlefish oil by Pseudozyma churashimaensis OK96. J Oleo Sci 62(5):319–327

Morita T, Fukuoka T, Imura T, Kitamoto D (2015) Mannosylerythritol lipids: production and applications. J Oleo Sci 64(2):133–141

Mourya VK, Inamdara N, Ashutosh Tiwari N (2010) Carboxymethyl chitosan and its applications. Adv Mater Lett 1(1):11–33

Muthusamy K, Sridharan V, Maheswari CU, Nagarajan S (2016) Lipase catalyzed synthesis of fluorescent glycolipids: gelation studies and graphene incorporated self-assembled sheet formation for semiconductor applications. Green Chem 18(13):3722–3731

Naughton PJ, Marchant R, Naughton V, Banat IM (2019) Microbial biosurfactants: current trends and applications in agricultural and biomedical industries. J Appl Microbiol 127(1):12–28
Nett M, Ikeda H, Moore BS (2009) Genomic basis for natural product biosynthetic diversity in the actinomycetes. Nat Prod Rep 26(11):1362–1384
Nicholson TP, Rudd BA, Dawson M, Lazarus CM, Simpson TJ, Cox RJ (2001) Design and utility of oligonucleotide gene probes for fungal polyketide synthases. Chem Biol 8(2):157–178
Normand P, Lapierre P, Tisa LS, Gogarten JP, Alloisio N, Bagnarol E et al (2007) Genome characteristics of facultatively symbiotic Frankia sp. strains reflect host range and host plant biogeography. Genome Res 17(1):7–15
Nwodo UU, Green E, Okoh AI (2012) Bacterial exopolysaccharides: functionality and prospects. Int J Mol Sci 13(11):14002–14015
Nyyssölä A, Kerovuo J, Kaukinen P, von Weymarn N, Reinikainen T (2000) Extreme halophiles synthesize betaine from glycine by methylation. J Biol Chem 275(29):22196–22201
O-Byrne SM, Blaner WS (2013) Retinol and retinyl esters: biochemistry and physiology: thematic review series: fat-soluble vitamins: vitamin A. J Lipid Res 54:1731
Ohnishi Y, Ishikawa J, Hara H, Suzuki H, Ikenoya M, Ikeda H et al (2008) Genome sequence of the streptomycin-producing microorganism Streptomyces griseus IFO 13350. J Bacteriol 190(11):4050–4060
Olaizola M (2000) Commercial production of astaxanthin from Haematococcus pluvialis using 25,000-liter outdoor photobioreactors. J Appl Phycol 12(3–5):499–506
Oliveira DM, Mota TR, Oliva B, Segato F, Marchiosi R, Ferrarese-Filho O et al (2019) Feruloyl esterases: biocatalysts to overcome biomass recalcitrance and for the production of bioactive compounds. Bioresour Technol 278:408–423
Oliynyk M, Samborskyy M, Lester JB, Mironenko T, Scott N, Dickens S et al (2007) Complete genome sequence of the erythromycin-producing bacterium Saccharopolyspora erythraea NRRL23338. Nat Biotechnol 25(4):447–453
Omura S, Ikeda H, Ishikawa J, Hanamoto A, Takahashi C, Shinose M et al (2001) Genome sequence of an industrial microorganism Streptomyces avermitilis: deducing the ability of producing secondary metabolites. Proc Natl Acad Sci 98(21):12215–12220
Onghena M, Geens T, Goossens E, Wijnants M, Pico Y, Neels H et al (2011) Analytical characterization of mannosylerythritol lipid biosurfactants produced by biosynthesis based on feedstock sources from the agrofood industry. Anal Bioanal Chem 400:1263–1275
Oren A, Gunde-Cimerman N (2007) Mycosporines and mycosporine-like amino acids: UV protectants or multipurpose secondary metabolites? FEMS Microbiol Lett 269(1):1–10
Paiva LB, Goldbeck R, dos Santos WD, Squina FM (2013) Ferulic acid and derivatives: molecules with potential application in the pharmaceutical field. Braz J Pharm Sci 49(3):395–411
Pallela R, Na-Young Y, Kim SK (2010) Anti-photoaging and photoprotective compounds derived from marine organisms. Mar Drugs 8(4):1189–1202
Pandey A (1992) Recent process developments in solid-state fermentation. Process Biochem 27(2):109–117
Peyrat LA, Tsafantakis N, Georgousaki K, Ouazzani J, Genilloud O, Trougakos IP, Fokialakis N (2019) Terrestrial microorganisms: cell factories of bioactive molecules with skin protecting applications. Molecules 24(9):1836
Phonghanpot S, Jarintanan F (2022) Secondary metabolism gene diversity and cocultivation toward isolation and identification of potent bioactive compounds producing bacterial strains from Thailand's natural resources. Scientifica 2022:2827831
Pi HW, Anandharaj M, Kao YY, Lin YJ, Chang JJ, Li WH (2018) Engineering the oleaginous red yeast Rhodotorula glutinis for simultaneous β-carotene and cellulase production. Sci Rep 8(1):10850
Poli A, Anzelmo G, Nicolaus B (2010) Bacterial exopolysaccharides from extreme marine habitats: production, characterization and biological activities. Mar Drugs 8(6):1779–1802

Pringle NA, van de Venter M, Boukes GJ, Koekemoer TC (2021) Therapeutic potential of selected south African macrofungi in diabetic wound healing: an in vitro evaluation. S Afr J Bot 138:337–347

Qaderi MM, Martel AB, Strugnell CA (2023) Environmental factors regulate plant secondary metabolites. Plan Theory 12(3):447

Qiu L, Wang P, Liao G, Zeng Y, Cai C, Kong F et al (2018) New eudesmane-type sesquiterpenoids from the mangrove-derived endophytic fungus Penicillium sp. J-54. Mar Drugs 16(4):108

Raghukumar S (2002) Ecology of the marine protists, the Labyrinthulomycetes (Thraustochytrids and Labyrinthulids). Eur J Protistol 38(2):127–145

Rana J, Diwakar G, Qiang H, Li T, Scholten J (2013) U.S. Patent Application No. 13/887,729

Rao MPN, Xiao M, Li WJ (2017) Fungal and bacterial pigments: secondary metabolites with wide applications. Front Microbiol 8:1113

Rastogi RP, Richa, Sinha RP, Singh SP, Häder DP (2010) Photoprotective compounds from marine organisms. J Ind Microbiol Biotechnol 37(6):537–558

Rendon M, Berneburg M, Arellano I, Picardo M (2006) Treatment of melasma. J Am Acad Dermatol 54(5):S272–S281

Rhee SK, Liu X, Wu L, Chong SC, Wan X, Zhou J (2004) Detection of genes involved in biodegradation and biotransformation in microbial communities by using 50-mer oligonucleotide microarrays. Appl Environ Microbiol 70(7):4303–4317

Rittié L, Fisher GJ (2002) UV-light-induced signal cascades and skin aging. Ageing Res Rev 1(4):705–720

Robey MT, Caesar LK, Drott MT, Keller NP, Kelleher NL (2021) An interpreted atlas of biosynthetic gene clusters from 1,000 fungal genomes. Proc Natl Acad Sci 118(19):e2020230118

Rodan K, Fields K, Majewski G, Falla T (2016) Skincare bootcamp: the evolving role of skincare. Plast Reconstr Surg Glob Open 4(12 Suppl):e1152

Sahnoun R, Mnif I, Fetoui H, Gdoura R, Chaabouni K, Makni-Ayadi F et al (2014) Evaluation of Bacillus subtilis SPB1 lipopeptide biosurfactant toxicity towards mice. Int J Pept Res Ther 20:333–340

Sajna KV, Sukumaran RK, Jayamurthy H, Reddy KK, Kanjilal S, Prasad RB, Pandey A (2013) Studies on biosurfactants from Pseudozyma sp. NII 08165 and their potential application as laundry detergent additives. Biochem Eng J 78:85–92

Sajna KV, Gottumukkala LD, Sukumaran RK, Pandey A (2015) White biotechnology in cosmetics. In: Industrial biorefineries & white biotechnology. Elsevier, pp 607–652

Salovska B, Zhu H, Gandhi T, Frank M, Li W, Rosenberger G et al (2020) Isoform-resolved correlation analysis between mRNA abundance regulation and protein level degradation. Mol Syst Biol 16(3):e9170

Sánchez-Quiles D, Tovar-Sánchez A (2015) Are sunscreens a new environmental risk associated with coastal tourism? Environ Int 83:158–170

Sandmann G, Pollmann H, Gassel S, Breitenbach J (2021) Xanthophyllomyces dendrorhous, a versatile platform for the production of carotenoids and other acetyl-CoA-derived compounds. In: Carotenoids: biosynthetic and biofunctional approaches. Springer, Singapore, pp 137–151

Santos DKF, Rufino RD, Luna JM, Santos VA, Sarubbo LA (2016) Biosurfactants: multifunctional biomolecules of the 21st century. Int J Mol Sci 17(3):401

Satpute SK, Banat IM, Dhakephalkar PK, Banpurkar AG, Chopade BA (2010) Biosurfactants, bioemulsifiers and exopolysaccharides from marine microorganisms. Biotechnol Adv 28(4):436–450

Schatz MC, Phillippy AM, Gajer P, DeSantis TZ, Andersen GL, Ravel J (2010) Integrated microbial survey analysis of prokaryotic communities for the PhyloChip microarray. Appl Environ Microbiol 76(16):5636–5638

Schena M, Heller RA, Theriault TP, Konrad K, Lachenmeier E, Davis RW (1998) Microarrays: biotechnology's discovery platform for functional genomics. Trends Biotechnol 16(7):301–306

Schommer NN, Gallo RL (2013) Structure and function of the human skin microbiome. Trends Microbiol 21(12):660–668

Schuller A, Studt-Reinhold L, Strauss J (2022) How to completely squeeze a fungus—advanced genome mining tools for novel bioactive substances. Pharmaceutics 14(9):1837

Schwartz C, Shabbir-Hussain M, Frogue K, Blenner M, Wheeldon I (2017) Standardized markerless gene integration for pathway engineering in Yarrowia lipolytica. ACS Synth Biol 6(3):402–409

Scoglio S, Canestrari F, Benedetti S, Zolla L (2012) U.S. Patent No. 8,337,858. Washington, DC: U.S. Patent and Trademark Office

Seweryn A (2018) Interactions between surfactants and the skin–theory and practice. Adv Colloid Interf Sci 256:242–255

Seweryn A, Wasilewski T, Hordyjewicz-Baran Z, Bochynek M, Pannert D, Łukaszewicz M, Lewińska A (2023) Implementation of sustainable development goals in the cosmetics industry based on the example of cleansing cosmetics containing a surfactin-rich digestate extract. Clean Techn Environ Policy 25(9):3111–3125

Shackel NA, Gorrell MD, McCaughan GW (2002) Gene array analysis and the liver. Hepatology 36(6):1313–1325

Sharma D, Saharan BS, Chauhan N, Bansal A, Procha S (2014) Production and structural characterization of Lactobacillus helveticus derived biosurfactant. Sci World J 2014:493548

Shi B, He Q, Yao K, Huang W, Li Q (2005) Production of ellagic acid from degradation of valonea tannins by Aspergillus niger and Candida utilis. J Chem Technol Biotechnol Int Res Process Environ Clean Technol 80(10):1154–1159

Shick JM, Dunlap WC (2002) Mycosporine-like amino acids and related gadusols: biosynthesis, accumulation, and UV-protective functions in aquatic organisms. Annu Rev Physiol 64(1):223–262

Shimada H, Kondo K, Fraser PD, Miura Y, Saito T, Misawa N (1998) Increased carotenoid production by the food yeast Candida utilis through metabolic engineering of the isoprenoid pathway. Appl Environ Microbiol 64(7):2676–2680

Shindo K, Kikuta K, Suzuki A et al (2007) Rare carotenoids, (3 R)-saproxanthin and (3 R, 2′ S)-myxol, isolated from novel marine bacteria (Flavobacteriaceae) and their antioxidative activities. Appl Microbiol Biotechnol 74:1350–1357

Simmering R, Breves R (2009) Pre-and probiotic cosmetics. Hautarzt 60:809–814

Sintes E, Bergauer K, De Corte D, Yokokawa T, Herndl GJ (2013) Archaeal amoA gene diversity points to distinct biogeography of ammonia-oxidizing Crenarchaeota in the ocean. Environ Microbiol 15(5):1647–1658

Smith R, Russo J, Fiegel J, Brogden N (2020) Antibiotic delivery strategies to treat skin infections when innate antimicrobial defense fails. Antibiotics 9(2):56

Song JW, Dong XY, Jiao BH, Wang LH (2013) Directly accessing the diversity of bacterial type I polyketide synthase gene in Chinese soil and seawater. Afr J Microbiol Res 7(31):4065–4072

Song Y, Li Q, Qin F, Sun C, Liang H, Wei X et al (2017) Neoabyssomicins A–C, polycyclic macrolactones from the deep-sea derived Streptomyces koyangensis SCSIO 5802. Tetrahedron 73(36):5366–5372

Soudant E, Koulbanis C (1998) U.S. Patent No. 5,712,311. Washington, DC: U.S. Patent and Trademark Office

Spagnuolo M, Shabbir Hussain M, Gambill L, Blenner M (2018) Alternative substrate metabolism in Yarrowia lipolytica. Front Microbiol 9:1077

Stefan RI, Vamanu E, Angelescu GC (2015) Antioxidant activity of crude methanolic extracts from Pleurotus ostreatus. Res J Phytochem 9(1):25–32

Stojkovic D, Reis FS, Glamočlija J, Ćirić A, Barros L, Van Griensven LJ et al (2014) Cultivated strains of Agaricus bisporus and A. brasiliensis: chemical characterization and evaluation of antioxidant and antimicrobial properties for the final healthy product–natural preservatives in yoghurt. Food Funct 5(7):1602–1612

Suh S-S, Hwang J, Park M et al (2014) Anti-inflammation activities of mycosporine-like amino acids (MAAs) in response to UV radiation suggest potential anti-skin aging activity. Mar Drugs 12:5174–5187

Sun C, Wang JW, Fang L, Gao XD, Tan RX (2004) Free radical scavenging and antioxidant activities of EPS2, an exopolysaccharide produced by a marine filamentous fungus Keissleriella sp. YS 4108. Life Sci 75(9):1063–1073

Sun P, Xu DX, Mandi A, Kurtan T, Li TJ, Schulz B, Zhang W (2013) Structure, absolute configuration, and conformational study of 12-membered macrolides from the fungus Dendrodochium sp. associated with the sea cucumber Holothuria nobilis Selenka. J Org Chem 78(14):7030–7047

Sun L, Kwak S, Jin YS (2019) Vitamin A production by engineered Saccharomyces cerevisiae from xylose via two-phase in situ extraction. ACS Synth Biol 8(9):2131–2140

Sy C, Dangles O, Borel P, Caris-Veyrat C (2015) Interactions between carotenoids from marine bacteria and other micronutrients: impact on stability and antioxidant activity. Mar Drugs 13(11):7020–7039

Taofiq O, Calhelha RC, Heleno S, Barros L, Martins A, Santos-Buelga C et al (2015) The contribution of phenolic acids to the anti-inflammatory activity of mushrooms: screening in phenolic extracts, individual parent molecules and synthesized glucuronated and methylated derivatives. Food Res Int 76:821–827

Tavares LF, Silva PM, Junqueira M, Mariano DC, Nogueira FC, Domont GB et al (2013) Characterization of rhamnolipids produced by wild-type and engineered Burkholderia kururiensis. Appl Microbiol Biotechnol 97:1909–1921

Tian YQ, Lin ST, Kumaravel K, Zhou H, Wang SY, Liu YH (2018) Polyketide-derived metabolites from the sponge-derived fungus Aspergillus sp. F40. Phytochem Lett 27:74–77

Tomaro-Duchesneau C, Saha S, Malhotra M, Coussa-Charley M, Al-Salami H, Jones M et al (2012) Lactobacillus fermentum NCIMB 5221 has a greater ferulic acid production compared to other ferulic acid esterase producing lactobacilli. Int J Prob Preb 7(1):23–32

Tramontin LRR, Kildegaard KR, Sudarsan S, Borodina I (2019) Enhancement of astaxanthin biosynthesis in oleaginous yeast Yarrowia lipolytica via microalgal pathway. Microorganisms 7(10):472

Tshikhudo P, Nnzeru R, Ntushelo K, Mudau F (2013) Bacterial species identification getting easier. Afr J Biotechnol 12(41):5975

Tsuchiya T, Yamada K, Minoura K, Miyamoto K, Usami Y, Kobayashi T et al (2008) Purification and determination of the chemical structure of the tyrosinase inhibitor produced by Trichoderma viride strain H1-7 from a marine environment. Biol Pharm Bull 31(8):1618–1620

Tu Q, Yu H, He Z, Deng YE, Wu L, Van Nostrand JD et al (2014) GeoChip 4: a functional gene-array-based high-throughput environmental technology for microbial community analysis. Mol Ecol Resour 14(5):914–928

Udwary DW, Gontang EA, Jones AC, Jones CS, Schultz AW, Winter JM et al (2011) Significant natural product biosynthetic potential of actinorhizal symbionts of the genus Frankia, as revealed by comparative genomic and proteomic analyses. Appl Environ Microbiol 77(11):3617–3625

Umakumar G, Naiman AM, Samuel John K, Hajera T (2012) Bio-conversion studies on gallic acid production from Chebulic myrobalan and Emblic myrobalan by Aspergillus niger mtcc 281 and Rhizopus oryzae mtcc 1987. Int J Pharma Biosci 3(2):146–155

Varvaresou A, Iakovou K (2015) Biosurfactants in cosmetics and biopharmaceuticals. Lett Appl Microbiol 61(3):214–223

Vecino X, Barbosa-Pereira L, Devesa-Rey R, Cruz JM, Moldes AB (2015) Optimization of extraction conditions and fatty acid characterization of lactobacillus pentosus cell-bound biosurfactant/bioemulsifier. J Sci Food Agric 95(2):313–320

Vecino X, Cruz JM, Moldes AB, Rodrigues LR (2017) Biosurfactants in cosmetic formulations: trends and challenges. Crit Rev Biotechnol 37(7):911–923

Vijayakumar S, Saravanan V (2015) Biosurfactants-types, sources and applications. Res J Microbiol 10(5):181–192

Vílchez C, Forján E, Cuaresma M, Bédmar F, Garbayo I, Vega JM (2011) Marine carotenoids: biological functions and commercial applications. Mar Drugs 9(3):319–333

Wada N, Sakamoto T, Matsugo S (2015) Mycosporine-like amino acids and their derivatives as natural antioxidants. Antioxidants 4(3):603–646

Walton E (2014) Buried treasure: unlocking the secrets of medicinal mushrooms. Biom J 37(6):339

Wang P, Kong F, Wei J, Wang Y, Wang W, Hong K, Zhu W (2014a) Alkaloids from the mangrove-derived actinomycete Jishengella endophytica 161111. Mar Drugs 12(1):477–490

Wang Y, Tang X, Shao Z, Ren J, Liu D, Proksch P, Lin W (2014b) Indole-based alkaloids from deep-sea bacterium Shewanella piezotolerans with antitumor activities. J Antibiot 67(5):395–399

Wang Q, Zhang Y, Wang M, Tan Y, Hu X, He H et al (2017) Neo-actinomycins a and B, natural actinomycins bearing the 5 H-oxazolo [4, 5-b] phenoxazine chromophore, from the marine-derived Streptomyces sp. IMB094 Sci Rep 7(1):3591

Wang ZX, Feng XL, Liu C, Gao JM, Qi J (2022) Diverse metabolites and pharmacological effects from the basidiomycetes Inonotus hispidus. Antibiotics 11(8):1097

Wasinger VC, Cordwell SJ, Cerpa-Poljak A, Yan JX, Gooley AA, Wilkins MR et al (1995) Progress with gene-product mapping of the Mollicutes: Mycoplasma genitalium. Electrophoresis 16(1):1090–1094

Weber T, Blin K, Duddela S, Krug D, Kim HU, Bruccoleri R et al (2015) antiSMASH 3.0—a comprehensive resource for the genome mining of biosynthetic gene clusters. Nucleic Acids Res 43(W1):W237–W243

Whitehead K, Hedges JI (2005) Photodegradation and photosensitization of mycosporine-like amino acids. J Photochem Photobiol B Biol 80(2):115–121

Woyke T, Tighe D, Mavromatis K, Clum A, Copeland A, Schackwitz W et al (2010) One bacterial cell, one complete genome. PLoS One 5(4):e10314

Wu L, Thompson DK, Li G, Hurt RA, Tiedje JM, Zhou J (2001) Development and evaluation of functional gene arrays for detection of selected genes in the environment. Appl Environ Microbiol 67(12):5780–5790

Wu B, Wu X, Sun M, Li M (2013a) Two novel tyrosinase inhibitory sesquiterpenes induced by CuCl2 from a marine-derived fungus Pestalotiopsis sp. Z233. Mar Drugs 11(8):2713–2721

Wu C, Tan Y, Gan M, Wang Y, Guan Y, Hu X et al (2013b) Identification of elaiophylin derivatives from the marine-derived actinomycete Streptomyces sp. 7-145 using PCR-based screening. J Nat Prod 76(11):2153–2157

Wu YH, Zhang ZH, Zhong Y, Huang JJ, Li XX, Jiang JY et al (2017) Sumalactones A–D, four new curvularin-type macrolides from a marine deep sea fungus Penicillium Sumatrense. RSC Adv 7(63):40015–40019

Wu Z, Liu D, Huang J, Proksch P, Zhu K, Lin W (2018) Hansforesters A–M, polyesters from the sponge-associated fungus Hansfordia sinuosae with antibacterial activities. RSC Adv 8(69):39756–39768

Xiao ZE, Huang H, Shao C, Xia X, Ma L, Huang X et al (2013) Asperterpenols A and B, new sesterterpenoids isolated from a mangrove endophytic fungus Aspergillus sp. 085242. Org Lett 15(10):2522–2525

Xie Y, Wang G (2015) Mechanisms of fatty acid synthesis in marine fungus-like protists. Appl Microbiol Biotechnol 99:8363–8375

Xie Y, Chen S, Xiong X (2021) Metabolic engineering of non-carotenoid-producing yeast Yarrowia lipolytica for the biosynthesis of zeaxanthin. Front Microbiol 12:699235

Xing PY, Liu R, Zhang DC, Sun CM (2017) Pumilacidin-like lipopeptides derived from marine bacterium Bacillus sp. strain 176 suppress the motility of Vibrio alginolyticus. Appl Environ Microbiol 83:e00450-17

Xu JW, Zhao W, Zhong JJ (2010) Biotechnological production and application of ganoderic acids. Appl Microbiol Biotechnol 87:457–466

Xu L, Yu J, Wang A, Zuo C, Li H, Chen X et al (2018) Efficient synthesis of vitamin A palmitate in nonaqueous medium using self-assembled lipase TLL@ apatite hybrid nanoflowers by mimetic biomineralization. Green Chem Lett Rev 11(4):476–483

Yang XW, Zhang GY, Ying JX, Yang B, Zhou XF, Steinmetz A et al (2012) Isolation, characterization, and bioactivity evaluation of 3-((6-methylpyrazin-2-yl) methyl)-1 H-indole, a new alkaloid from a deep-sea-derived actinomycete serinicoccus profundi sp. nov. Mar Drugs 11(1):33–39

Yang H, Li S, Qu Y, Li L, Li Y, Wang D (2022) Anti-colorectal cancer effects of inonotus hispidus (Bull.: Fr.) p. karst. spore powder through regulation of gut microbiota-mediated JAK/STAT signaling. Nutrients 14(16):3299

You J, Yang SZ, Mu BZ (2015) Structural characterization of lipopeptides from Enterobacter sp. strain N18 reveals production of surfactin homologues. Eur J Lipid Sci Technol 117(6):890–898

Young AR, Claveau J, Rossi AB (2017) Ultraviolet radiation and the skin: photobiology and sunscreen photoprotection. J Am Acad Dermatol 76(3):S100–S109

Yousef H, Alhajj M, Sharma S (2017) Anatomy, skin (integument), epidermis. StatPearls Publishing, Treasure Island

Yuan JP, Peng J, Yin K, Wang JH (2011) Potential health-promoting effects of astaxanthin: a high-value carotenoid mostly from microalgae. Mol Nutr Food Res 55(1):150–165

Yue H, Eastman PS, Wang BB, Minor J, Doctolero MH, Nuttall RL et al (2001) An evaluation of the performance of cDNA microarrays for detecting changes in global mRNA expression. Nucleic Acids Res 29(8):e41–e41

Zagórska-Dziok M, Ziemlewska A, Mokrzyńska A, Nizioł-Łukaszewska Z, Sowa I, Szczepanek D, Wójciak M (2023) Comparative study of cytotoxicity and antioxidant, anti-aging and antibacterial properties of unfermented and fermented extract of cornus mas L. Int J Mol Sci 24(17):13232

Zembron-Lacny A, Gajewski M, Naczk M, Siatkowski I (2013) Effect of shiitake (Lentinus edodes) extract on antioxidant and inflammatory response to prolonged eccentric exercise. J Physiol Pharmacol 64(2):249–254

Zerikly M, Challis GL (2009) Strategies for the discovery of new natural products by genome mining. Chembiochem 10(4):625–633

Zhang D, Li X, Kang JS, Choi HD, Son BW (2007) A new α-pyrone derivative, 6-[(*E*)-Hept-1-enyl]-α-pyrone, with tyrosinase inhibitory activity from a marine isolate of the fungus *Botrytis*. Notes 28(5):887

Zhang D, Yang X, Kang JS, Choi HD, Son BW (2008) Circumdatin I, a new ultraviolet-a protecting benzodiazepine alkaloid from a marine isolate of the fungus Exophiala. J Antibiot 61(1):40–42

Zhang X, Chen L, Chai W, Lian XY, Zhang Z (2017) A unique indolizinium alkaloid streptopertusacin A and bioactive bafilomycins from marine-derived Streptomyces sp. HZP-2216E. Phytochemistry 144:119–126

Zhang L, Hu Y, Duan X, Tang T, Shen Y, Hu B et al (2018a) Characterization and antioxidant activities of polysaccharides from thirteen boletus mushrooms. Int J Biol Macromol 113:1–7

Zhang S, Gui C, Shao M, Kumar PS, Huang H, Ju J (2018b) Antimicrobial tunicamycin derivatives from the deep sea-derived Streptomyces xinghaiensis SCSIO S15077. Nat Prod Res 34(11):1499–1504

Zhang Y, Hao J, Liu Z, Li Z, Teng L, Wang D (2022) Inonotus hispidus protects against hyperlipidemia by inhibiting oxidative stress and inflammation through Nrf2/NF-κB signaling in high fat diet fed mice. Nutrients 14(17):3477

Zhu M, Zhang X, Feng H, Che Q, Zhu T, Gu Q, Li D (2016) Campyridones A–D, pyridone alkaloids from a mangrove endophytic fungus Campylocarpon sp. HDN13-307. Tetrahedron 72(37):5679–5683

Zhuang Y, Teng X, Wang Y, Liu P, Li G, Zhu W (2011) New quinazolinone alkaloids within rare amino acid residue from coral-associated fungus, Aspergillus versicolor LCJ-5-4. Org Lett 13(5):1130–1133

Ziboh VA, Miller CC, Cho Y (2000) Metabolism of polyunsaturated fatty acids by skin epidermal enzymes: generation of antiinflammatory and antiproliferative metabolites. Am J Clin Nutr 71(1):361s–366s

Chapter 15
Nanotechnology in Skincare: Unlocking the Future

Sadia Javed, Ayesha Khan, Andleeb Shahzadi, Muhammad Arshad, and Shahzad Ali Shahid Chatha

Contents

The original version of the chapter has been revised. A correction to this chapter can be found at https://doi.org/10.1007/978-3-031-86087-4_19

S. Javed (✉) · A. Khan
Department of Biochemistry, Government College University Faisalabad, Faisalabad, Pakistan
e-mail: sadiajaved@gcuf.edu.pk

A. Shahzadi
Department of Medical Pharmacology, Istanbul University, Istanbul, Turkey

M. Arshad
Biochemistry Section, Jhang-campus, University of Veterinary and Animal Sciences Lahore, Lahore, Pakistan

S. A. S. Chatha
Department of Chemistry, Government College University, Faisalabad, Pakistan

S. Javed et al. (eds.), *Sustainable Cosmeceuticals*,
https://doi.org/10.1007/978-3-031-86087-4_15

Abstract Nanotechnology has the potential to generate advancements and innovations in formulations and delivery systems. This fast-developing technology has been widely exploited for diagnostic and therapeutic purposes. Cosmetic manufacturers employ nano-sized ingredients to improve UV protection, skin penetration, color, the release of fragrance, finish quality, anti-aging effect, and a variety of other properties. Consumers all over the world are looking for more natural and less harmful products to use on their skin, as it is one of the most important features. Drawbacks of traditional cosmetics can be overcome by using nanocosmeceuticals. Nanocosmetics and nanocosmeceuticals have been extensively explored for skin, hair, nails, lips, and teeth, and the inclusion of nanomaterials has been found to improve product efficacy and consumer satisfaction. Nanocosmeceuticals are a combination of cosmetics and pharmaceuticals; they just don't conceal and make the skin fine but also cure and heal it now. The use of nanomaterials (NMs) by the cosmetic industry aims for long-lasting effects and increased stability. The high surface area of nanomaterials allows for more efficient transport of the ingredients through the skin. Also, many other features of nanoparticles help improve the usage of cosmetics, like controlled drug release, enhanced bioavailability, moisturizing power, entrapment and penetration efficiency, enhanced stability, and improved results. In this chapter, the emerging role of nanotechnology in skincare will be discussed.

Keywords Nanotechnology · Cosmetics · Skincare · Nano-delivery · Nanomaterials

15.1 Introduction

Nanotechnology is revolutionizing the skincare sector by facilitating the creation of highly effective, targeted therapies. Active compounds can penetrate the skin more deeply with the help of nanoparticles, which improves absorption and improves the efficacy of products like sunscreens, moisturizers, and serums. As nanotechnology develops further, it has the potential to revolutionize skincare practices by providing individualized and innovative ways to attain younger, healthier skin. The development of lighter, more stable formulations that provide long-lasting advantages including anti-aging, skin regeneration, and defense against environmental stresses is made possible by this revolutionary technology. In the cosmetics sector, nanomaterials are of tremendous interest in developing innovative formulations with improved functionality. The cosmeceutical company has numerous opportunities for growth due to the rising use of nanomaterials in the cosmetics industry (Santos et al. 2019).

The ability of nanotechnology to modify and control materials at the atomic and molecular levels is one of its key advantages in this industry. This makes it possible to create novel formulations with better qualities like targeted distribution, higher efficacy, and improved stability. Because of their unique features, silver and gold nanoparticles are widely employed in cosmetics. For example, gold nanoparticles can give the skin a bright glow and offer exceptional stability. Additionally, they have stronger skin penetration properties, which improve the absorption of a variety of other beneficial ingredients. However, due to their antibacterial qualities, silver nanoparticles are useful in healthcare treatments that aim to treat skin infections and acne (Barkat et al. 2018). Skincare products are in high demand all over the world as a result of emerging trends in the cosmetics industry and developments in global marketing.

The largest organ in the human body, the skin is made up of lipids, amino acids, and carbohydrates. It serves as a barrier against pathogens and guards against cutaneous infection. Sufficient dietary intake is crucial for maintaining healthy, glowing skin. Herbs and other natural substances have antibacterial, nutritious, and antioxidant properties that make them useful for treating skin conditions. Antibiotic resistance and adverse consequences in cutaneous infections are caused by conventional therapy, such as antibiotics. By omitting disease-causing microorganisms, natural products, and plant extracts, the regeneration of the skin accelerates. They also contain antibacterial and anti-inflammatory properties (Bittner et al. 2021). Skin is easily injured, and thermal or physical trauma interrupts the epithelium mucosa, necessitating the use of therapeutics to promote wound healing. In order to restore injured skin, wound healing and repair include inflammation, hemostasis, proliferation, and tissue remodeling. Dermatologists still find it difficult to treat wounds because of intricate processes, even with advances in medicine. By integrating inexpensive herbs to avoid infection, the alternative medicine system promotes health care and treats wounds. Globally, the use of acupuncture, herbal remedies, spiritual therapy, and conventional medicine in skincare is becoming more popular as a means of improving illness administration (Mesraoua et al. 2021).

15.2 Sustainable Development Goals

The United Nations established the 17 Sustainable Development Goals (SDGs) in 2015 as a component of the 2030 Agenda for Sustainable Development. A wide range of global issues, including poverty, inequality, environmental degradation, peace, and justice, are intended to be addressed by these objectives. The economic, social, and environmental facets of sustainable development are all intended to be balanced by their connectedness (Bilderback 2024). Environmental sustainability and the Sustainable Development Goals (SDGs) are closely related because several of the SDGs directly address conservation, environmental protection, and the prudent use of natural resources. The following explains how several SDGs relate to environmental concerns:

With the goal of reducing the adverse impacts on the environment and human health, green nanotechnology (GNT) incorporates the concepts of sustainability and nanotechnology. This method assures that nanomaterials are produced and employed in ways that are safer for the environment and people by utilizing green chemistry, renewable resources, and energy-efficient techniques. The use of non-toxic, biodegradable materials, green solvents, and sustainable synthesis techniques that lower energy consumption and hazardous waste are important components. By ensuring that nanomaterials are created with low toxicity and recyclability in mind, a life cycle assessment (LCA) method lowers pollution and promotes recycling, waste reduction, environmental conservation, and process optimization.

GNT's potential is multifaceted, tackling global issues including water treatment, eco-friendly agriculture, and renewable energy. It is essential to the advancement of the Sustainable Development Goals (SDGs) of the UN. Nanotechnology, for instance, improves the effectiveness of insect repellents and fertilizers in agriculture, supporting SDG 2 (Zero Hunger).

As the field develops, it is turning into an important tool for striking a balance between social well-being, environmental preservation, and economic growth. GNT is considered to be an essential component in accomplishing the SDGs and advancing a more ecologically conscious and just future because of its emphasis on resource efficiency and pollution reduction (Aithal and Aithal 2018).

Sustainability in Cosmetics The demand for cosmetic firms to embrace more sustainable practices is emphasized by the growing global market for sustainable cosmetics, which is expected to reach USD 937.13 million by 2030 at a compound annual growth rate of 7.7% from 2024 to 2030 (Nhani 2024). The development is a result of increasing demand from customers for items that are energy-efficient, ethically sourced, and environmentally conscientious. Goals including climate action (SDG 13), life on land (SDG 15), and responsible consumption and production (SDG 12) are all highlighted in the 2030 Agenda for Sustainable Development and have a direct bearing on the cosmetics sector. By supporting these objectives and increasing its commercial appeal, the sector can help promote global sustainability (Schröder 2024).

One essential tool for assessing the environmental impact of cosmetic items from cradle until their disposal is the lifecycle assessment (LCA). It assists in pinpointing areas that require improvement, including the procurement of raw materials, manufacturing procedures, packaging, and disposal of end-of-life products. By using life cycle assessment (LCA), organizations can make sensible choices that lessen their environmental impact. Ethical and sustainable sourcing of raw materials, respect for local communities and biodiversity, the use of environmentally friendly chemical processes and the reduction of hazardous substances, the adoption of recyclable, biodegradable, or reusable packaging materials, and the investment in energy-efficient technologies and renewable energy sources to power manufacturing processes are all examples of good practices in the production of sustainable cosmetics (Hopewell et al. 2009).

Sustainability has been successfully incorporated into the operations and product lines of certain brands. By taking steps to lessen their influence on the environment, such as using minimum packaging, sourcing raw ingredients responsibly, and actively opposing animal testing, Lush, for example, has set a noteworthy example. Furthermore, a thorough approach to sustainability is demonstrated by Lush's investment in technologies to lower energy use and its carbon footprint (Nhani 2024).

Deforestation, destruction of habitat, biodiversity loss from the extraction of natural components, and pollution from synthetic chemicals are only a few of the environmental issues associated with traditional cosmetic production. Waste problems are made worse by the industry's heavy reliance on packaging (Santos et al. 2019). The cosmetics industry is under pressure to change its methods as a result of these effects. Reusing waste materials, for example, reduces environmental effects and satisfies consumer aspirations for sustainable products, meeting both market and ecological criteria (Fröhlich and Steinbiß 2021). Niosomes, silver nanoparticles, and nanocapsules are a few instances of how the firm applies nanotechnology in its products (Hameed et al. 2019).

15.3 Development of Cosmetics

Cosmetics are defined as skincare items that enhance skin appearance in order to improve its aesthetic appeal. As mentioned, Egyptians are said to have used makeup as early as 4000 BC, while Greeks, Chinese, Romans, Americans, and Japanese people later started making cosmetics (González et al. 2023). Moreover, during the twenty-first century, technology has advanced, and innovative cosmetic formulae that are widely utilized incorporate the most recent inventions (Sathyaseelan et al. 2024). As seen by the rise in the Asia-Pacific market, which is now worth over US 70 billion dollars, Asian cosmetics are among the best-performing and most sought-after in the world.

Furthermore, statistics from the Malaysia External Trade Development Corporation (MATRADE) indicate that the value of Malaysia's cosmetics sector was five billion dollars in 2019. From 2012 to 2017, the nation's export of Halal cosmetics items increased by 68% from $433 million to $725 million (Kim et al. 2020a, b). Nowadays, a lot of cosmetics make the claim that they protect the environment from pollution. Air pollution is caused by extremely reactive nanoparticles produced by combustion. Burning fuel in the engine of a car, coal in power plants, garbage in an incinerator, or trees in a forest fire are all examples of combustion.

These nanoparticles invade dermal collagen and elastin fibers to release reactive oxygen species that damage the skin when they come into contact with it. If the nanoparticles are not allowed to come into touch with the skin, then no damage will occur because the reactive oxygen species won't develop. The majority of moisturizers and foundations use dimethicone, mineral oil, and shea butter to apply a thin layer of an oily occlusive material to the skin's surface. The skin remains protected from pollution-related nanoparticle damage by this cosmetic layer. Therefore, with the right cosmetic application, one can shield their skin from environmental damage.

15.4 Key Benefits of Nanocosmetics

In cosmeceuticals, nanomaterials are used to treat a range of issues, such as wrinkles, photoaging, hyperpigmentation, hair damage, and skin damage. Because of their unique qualities and use of nanoscale, they can penetrate the skin more effectively, improving the active compounds' absorption and bioavailability. This makes it easier for substances that support skin regeneration, healing, and rejuvenation to be delivered effectively. The ability of nanoparticles to encapsulate and protect active substances is an essential feature of their use in cosmeceuticals (Souto et al. 2020). Nanocarriers that can encapsulate fragile bioactive molecules and improve their stability include liposomes, solid lipid nanoparticles, and nanoemulsions. Active substances, such as antioxidants or anti-inflammatory drugs, can be delivered to specific skin layers or cells by functionalizing nanoparticles with certain ligands or antibodies (Ramanunny et al. 2021). This targeted strategy minimizes any negative effects on tissues nearby while increasing therapy efficacy. The use of nanoparticles in cosmeceuticals has several benefits, but it's important to recognize the disadvantages as well (Gupta et al. 2022).

15.4.1 Advantages and Disadvantages of Imaging Techniques for Nanoparticles Associated with Cosmetic and Skin Research

Imaging Techniques	Advantages	Disadvantages	References
Transmission electron microscopy (TEM)	– High-resolution imaging at the atomic level – Detailed visualization of the size, shape, and distribution of nanoparticles	– Requires complex sample preparation – Potential for sample damage – Limited field of view	Tang and Yang (2017)
Scanning electron microscopy (SEM)	– Detailed surface morphology – Provides a broader view of nanoparticle dispersion – High depth of field	– Sample preparation can alter or damage samples – Lower resolution compared to TEM – Limited to surface imaging	Mohammed and Abdullah (2018)
Confocal laser scanning microscopy (CLSM)	– 3D imaging and depth profiling – Can visualize nanoparticles within skin layers – Non-invasive imaging	– Limited penetration depth – Requires specific sample preparation – Can be expensive	Tata and Raj (1998)
Two-photon microscopy	– Deeper tissue imaging with minimal damage – Allows for dynamic studies of nanoparticle interactions	– Limited spatial resolution compared to TEM and SEM – More complex and costly equipment	Helmchen and Denk (2005)
Fluorescence imaging	– Allows for quantitative assessment of distribution and concentration – Can use fluorescently labeled nanoparticles	– May require nanoparticle labeling – Limited penetration depth in tissues – Potential for fluorescence quenching or background noise	Rao et al. (2007)
Mass spectrometry imaging	– Detailed spatial distribution and chemical characterization – Can provide quantitative information	– Complx sample preparation – Expensive and requires specialized equipment – Potential for ionization interference	Buchberger et al. (2017)
X-ray computed tomography (CT)	– Non-invasive, 3D imaging – Useful for in vivo studies – Provides information on internal structure	– Lower resolution compared to electron microscopy – Limited detail on nanoparticles' size and shape	Withers et al. (2021)

15.5 Smart Delivery Systems

Because they can distribute active bioactive compounds in a targeted way, nanocarriers have attracted a lot of interest in the cosmetics industry. Numerous benefits are associated with these nanoscale delivery methods, including enhanced stability, increased bioavailability, regulated release, and focused distribution of active substances to certain skin cells or layers.

The following list of frequently used nanocarriers for targeted distribution in cosmetics includes:

15.5.1 Liposomes

Lipid bilayers form the spherical vesicles known as liposomes. Both hydrophilic and hydrophobic active substances can be encapsulated in them. Liposomes have the ability to target particular layers of the skin, such as the dermis or epidermis, and improve the penetration of active chemicals into the skin. These nanocarriers' lipid bilayers imitate the natural structure of the skin, which improves the encapsulated chemicals' absorption and bioavailability (Yang et al. 2020). In their pursuit of novel and potent cosmetic components, researchers have turned their attention to natural sources. *Asterias pectinifera*, a starfish found in East Asian coastal waters, is one such source. The possibility of collagen peptides produced from *Asterias pectinifera* in cosmetic formulations has been explored in recent studies.

To improve the stability and efficacy of collagen peptides in cosmetic formulations (Powar et al. 2024). Their main goal was to create elastic nanoliposomes that could contain *Asterias pectinifera* collagen peptides. The main goal was to overcome challenges pertaining to the stability and bioavailability of collagen peptides in cosmetic compositions. The stability of the collagen peptides, the duration of their release, and their penetration into the skin may all be increased by encasing them in nanoliposomes. Additionally, the collagen peptides isolated from *Asterias pectinifera* showed high encapsulation effectiveness when encapsulated in elastic nanoliposomes.

This indicates that a significant amount of collagen peptides can be effectively contained in the nanoliposomes, guaranteeing their stability during application and storage. Collagen peptides are shielded from oxidation and breakdown by their encapsulation in the nanoliposomes, which increases their shelf life. This discovery is significant because it ensures the integrity of the active ingredient over time and answers questions about the stability of collagen peptides in cosmetic compositions. As of right now, liposomes are acknowledged as one of the most successful cosmetic delivery technologies (Nikam et al. 2020). Because liposomes improve skin penetration and provide protection against dangerous radiation (Ultraviolet (UV) Radiation and Ionizing Radiation) that may cause cancer, they are widely used in cosmetic formulations.

Phospholipids, which are generally acknowledged as safe (GRAS) substances with few side effects, make up the majority of these lipid bilayers (Magri et al. 2020). Drugs encapsulated in liposomes are shielded from metabolic breakdown and have the ability to release their therapeutic ingredients gradually. Furthermore, the structure of liposomes can be unilamellar or multilamellar (Nsairat et al. 2022). To improve the stability of liposomes in water, antioxidants like lycopene or carotenoids can be added. Because of its emollient and conditioning qualities, phosphatidylcholine, another crucial component of liposomes, is frequently used in skincare and hair care products (shampoos, conditioners, and moisturizers) (Ahmad and Ahsan 2020).

15.5.2 Nanosomic

Smaller than liposomes, lipid-based nanocarriers are known as nanosomic. Their development aimed to improve the distribution of active substances via the stratum corneum, the skin's outermost layer (Kim et al. 2020a, b). Numerous substances, such as vitamins, antioxidants, and moisturizers, can be encapsulated in nanosomes. In contrast to liposomes, which are a wider variety of vesicles that can reach sizes of several micrometers. Greater surface area and a size range of 20–100 nm for small liposomes and above 100 nm for large liposomes are characteristics of nanoliposomes. The main components of these vesicles are lipids and phospholipids, while they may also include additional substances including proteins, carbohydrates, antioxidants, or sterols.

According to their distinct makeup, hydrophilic and hydrophobic compounds can be encapsulated and released by nanoliposomes, offering a combination of advantages. They can improve the distribution of bioactive chemicals to target sites since their bilayer structure is bio-compatible (Kouassi et al. 2022). In contrast to previous nanoscale drug delivery technologies, natural and affordable materials can be used to produce nanoliposomes on a large scale (Alavi et al. 2022). Nanoliposomes are highly desirable as "smart" drug delivery systems because of these qualities as well as their environmental perspective and sustainability. Research is ongoing to improve the properties of nanoliposomes by investigating novel synthesis techniques and surface modifications to boost their potential, stability, and durability (Dabiri and Hasanpouri Notash 2020). Investigated the potential of vesicular nanostructures, particularly liposomes, to enhance the transdermal delivery of tetracycline. In the present study, they prepared liposomes encapsulating tetracycline and examined their physicochemical properties, such as size, stability, and drug release behavior. They also used an animal model to assess these liposomes' capability for delivering tetracycline into the skin.

Tetracycline is an antibiotic that is frequently used to treat various skin infections, and improving its delivery can enhance therapeutic outcomes. The results of the investigation demonstrated that the liposomes exhibited favorable properties such as small size and significant stability, and they were also successful in encapsulating tetracycline. Additionally, the liposomes containing the encapsulated tetracycline demonstrated a prolonged release, suggesting that the drug's skin-active properties may be extended. In addition, they discovered that, in comparison to a traditional tetracycline solution, the liposomal for tetracycline imitation showed improved skin penetration. This suggests that the liposomes served as carriers, enabling the tetracycline to be transported across the layers of the skin. This strategy may improve the effectiveness of topical tetracycline-based therapies for a range of skin disorders by improving medication penetration and supplying continuous release.

The disorder known as melasma, which is characterized by excessive skin pigmentation and poses a considerable issue for dermatology, was the subject of another study carried out in 2012 by (Powar et al. 2024). The researchers looked into topical nanosome vitamin C iontophoresis and a 70% glycolic acid peel as two potential therapy modalities for this prevalent skin condition. The use of glycolic acid peels as a successful melasma treatment has grown in popularity. This technique stimulates skin cell turnover and produces regulated exfoliation by applying a 70% glycolic acid solution to the afflicted areas. The general texture of the skin is improved and hyperpigmentation is lessened with this exfoliation. The comparative study covered in this article demonstrated that applying the glycolic acid peel could improve melasma and reduce pigmentation. Topical nanosome vitamin C iontophoresis was another therapy method that was examined in this investigation. This technique uses iontophoresis to apply a specially prepared vitamin C in nanosomes to the melasma-affected areas. When iontophoresis and nanosomes are combined, vitamin C is better absorbed and penetrates the skin, reaching the melanocytes and minimizing pigmentation. The comparison study found that although this approach is marginally less effective than a glycolic acid peel, it still had positive effects when treating melasma. A group of volunteers with melasma were randomly randomized to either the glycolic acid peel group (70%) or the topical nanosome vitamin C iontophoresis group in this single-blinded comparative trial.

It was discovered that both therapy modalities were successful in treating melasma. On the other hand, the group that experienced the glycolic acid peel exhibited a marginally more noticeable improvement in the severity of their melasma and a decrease in pigmentation. In both groups, patient satisfaction scores were comparable. Both treatments had little side effects, the most often reported ones being temporary redness, slight burning, and dryness. The 70% glycolic acid peel and topical nanosome vitamin C iontophoresis are both effective treatments for melasma, according to the findings of this comparative study. When it came to lowering pigmentation and improving melasma, the glycolic acid peel performed better.

However, both strategies produced acceptable outcomes and were well-tolerated. In-depth investigations and additional studies are required to verify these findings and assess the treatments' long-term consequences. In order to properly cure melasma and enhance their patients' quality of life, dermatologists may take these treatment options into account.

15.5.3 Nanoemulsions

Transparent or translucent, stable dispersions of oil and water, known as nanoemulsions, are stabilized by surfactants. Because of their nanometer-sized droplets, these emulsions are better able to act and penetrate in particular regions. They are efficient vehicles for a wide range of active components, including vitamins, antioxidants, and essential oils. Because of the numerous benefits that come with microscopic droplet sizes, cosmetic goods employ nanoemulsions. These include enhanced penetration into small spaces like hair follicles and scales, faster absorption of active ingredients by the skin, higher fluidity without segregation, a glossy coating on the skin, a desirable visual appearance (clear to slightly cloudy and pearlescent), and use in cosmetic wet wipes and spray able products (Syperek and Wade 2023). Oil-in-water nanoemulsions are the most common kind utilized in cosmetic compositions. Many benefits come with them, including preservation of the encapsulated active substances, huge surface area, high stability, low viscosity, clear or translucent appearance, high solubility capacity, and little requirement for surfactants. Because of these characteristics, they are useful nanocarriers in cosmetics that enhance skin penetration and regulate the release of active components. One can use low-energy or high-energy techniques to create nanoemulsions. The versatile plant *Agave sisalana*, which is well known for producing fiber, may find use in the cosmetics sector. Safety investigations both in vivo and in vitro were included in the study.

Studies conducted in vitro on cultured cells and tissues revealed that the *Agave sisalana* by-product does not significantly indicate the potential for cytotoxicity or irritation. These findings suggest that the raw ingredient is safe to use in formulations for cosmetic products. They also used the by-product of *Agave sisalana* to create a nanoemulsion. Small droplet sizes in stable oil-water solutions, known as nanoemulsions, enhance the penetration and distribution of active chemicals. The goal of creating the nanoemulsion was to increase skin moisture. The efficiency of the nanoemulsion in enhancing skin moisture was evaluated clinically. Comparing the results to a control group, it was evident that the nanoemulsion was successful in hydrating and moisturizing the skin. These findings demonstrate the potential uses of the by-product *Agave sisalana* in the cosmetic industry, particularly in the creation of skincare products that increase skin hydration.

A thorough investigation into the variations in transdermal transport resulting from the delivery of hydrophilic and hydrophobic medicines via water-in-oil (W/O) nanoemulsion systems (NE) was presented by (Roy et al. 2022). Increased lipophilicity is necessary for the skin to permeabilize hydrophilic medicines. Furthermore, it was emphasized how crucial water mobility is for the NE vehicle and for transdermal delivery via the skin barrier. In order to make hydrophilic medications available for transdermal delivery, this water flow is essential. Furthermore, a number of variables affecting hydrophilic drug transdermal transit in NE systems were examined. These include the drug's thermodynamic activity, variations in the ionic drugs' surface electrical charge, the NE's constituents' ability to solubilize sebum, the NE's pores' ability to hold large water-soluble molecules, and the NE's ability to carry small water-soluble molecules for follicular delivery. The complex mechanisms behind transdermal medication delivery are clarified by these mechanistic hypotheses. It is important to remember that a number of variables affect these pathways, including drug dispersion, diffusion, solubilization, and thermodynamic activity. These variables can improve transdermal medication delivery by increasing the permeability and penetration of hydrophilic medicines when they are correctly handled. Nevertheless, additional investigation is required to verify and enhance these methods for actual implementations. Lastly, by making agricultural chemicals more effective and reducing their negative effects on soil health and terrestrial biodiversity, nanoemulsions pursue sustainable agriculture in connection with SDG 15: life on land. Additionally, they provide solutions for soil remediation, providing substances that assist in treating contamination and lessen the harm that pollutants cause to ecosystems. Essentially, nanoemulsions help the development of environmental sustainability and ecosystem preservation by offering effective, environmentally friendly solutions in a variety of regions.

Nanoemulsions can play an integral part in accomplishing SDGs 13, 14, and 15 because of their unique characteristics. By permitting more accurate distribution of fertilizers, insecticides, and herbicides, nanoemulsions can improve agricultural techniques' efficiency and reduce chemical waste and farming's carbon footprint, which is important for SDG 13: Climate Action. Nanoemulsions can also enhance biofuel formulations, increasing their sustainability and efficiency and lowering dependency on fossil fuels. By improving the regulated release of active components in medications and personal care products and lowering hazardous chemical discharge into the oceans, nanoemulsions could help in the mitigation of marine pollution in relation to SDG 14: Life Below Water. Additionally, in the event of an oil spill, nanoemulsions can help clean up marine environments by dissolving oil into smaller, biodegradable droplets. Because nanoemulsions enhance nutrition delivery and medicinal efficacy while lowering the need for hazardous chemicals, they also hold potential for sustainable aquaculture.

15.5.3.1 Formation of Nanoemulsions

Water-in-oil or oil-in-water colloidal dispersions with droplets ranging in diameter from a few to 200 nm are known as nanoemulsions. A growing interest in nanoemulsions by industry is shown by an improvement in patent protection actions connected to nanoemulsions (Mohd et al. 2022). The droplets can offer better stability and more advantageous optical, rheological, and ingredient distribution qualities than traditional emulsions because of their small size (Rachman et al. 2023). Two properties of nanoemulsions are stability and transparency.

Moreover, their production raises the stratum corneum's water content, demonstrating their effectiveness as a moisturizer. Because of their physicochemical properties, nanoemulsions may significantly alter the permeation profiles of molecules. In particular, when compared to traditional emulsions, oil/water nanoemulsions significantly improve the permeation profiles of polar components. Nanoemulsions function as efficient carriers for many bioactive chemicals with improved qualities, such as increased bioavailability, physical stability, and high optical consistency, in comparison to traditional emulsions (Choi and McClements 2020). However, because of their lipophilic core, nanoemulsions are especially well-suited for the water transportation of hydrophobic materials (Barradas and de Holanda e Silva 2021). Nanoemulsions are able to permeate and distribute throughout the skin than nanocapsules. Their lack of flexibility and polymers may be the cause of this; it interacts with the stratum corneum. Water-immiscible active substances, such as lipids, antioxidants, and retinol, are encouraged by nanoemulsions to enter target tissues more deeply and to reach higher effective concentrations.

15.5.3.2 Penetration of Nanoemulsion Systems into the Skin

Penetration of Nanoemulsion Systems Into the Skin: The skin barrier, which keeps the medication from entering the systemic circulation, is the main obstacle to topical drug delivery. Topical medication delivery based on nanoemulsions can greatly overcome this barrier. The three most common routes for medications to enter the skin are through hair follicles, sweat ducts, and the stratum corneum itself. Figure 15.4 shows how easily small-sized nanoparticles can pass through the pores of nanoemulsions. Furthermore, penetration of the hydrophilic sweat ducts and the hydrophobic stratum corneum is facilitated by the presence of both hydrophilic and hydrophobic units. Psoriasis and dermatitis can be efficiently treated with nano-emulsion (Algahtani et al. 2020). It has been demonstrated that positively charged nanoemulsions function better and pierce through negatively charged tissue more than negatively charged nanoemulsions (Vaz et al. 2022). This is because positively charged nanoemulsions interact more when they travel inside the tissue (Fig. 15.1).

15.5.4 Formation of Micelle Nanoparticles

Amphibians frequently form the basic structure known as a micelle in aquatic solutions. Micelles typically have a particle size between 5 and 50 nm. They are a class of associations or amphiphilic colloids whose molecules comprise two clearly defined areas with opposite affinities to a given solvent and which, at specific temperatures and concentrations, instantly form amphiphilic or surface-active agents. In an aqueous medium, these amphiphilic molecules exist independently at low concentrations. But when their concentration is increased to a particular range, aggregation occurs. The concentration at which the monomeric amphiphile forms micelles is known as the critical micelle concentration, or CMC (Perinelli et al. 2020).

The total number of individual molecules that make up a micelle is known as its aggregation number. The reduction of free energy in the system causes the creation of micelles, which clear out hydrophobic segments from the aqueous solution and

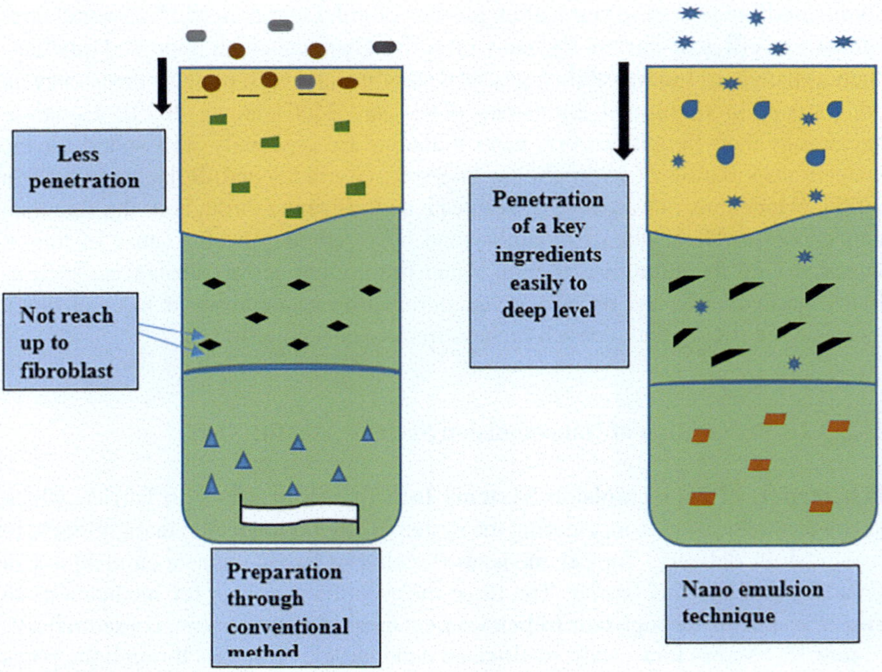

Fig. 15.1 Comparing nanoemulsions and classical transdermal formulations for skin barrier crossing (Rai et al. 2018)

restore the hydrogen bond network in water. An increase in energy gain results from the creation of Van der Waals attraction between hydrophobic blocks in the middle of the produced micelles (Lu et al. 2020). The micelle's shell is created by the hydrophilic parts of amphiphilic molecules, whereas its core is created by its hydrophobic sections (Mohd et al. 2022). When employed as drug carriers in aquatic environments, polar molecules are adsorbed in the micelle shell, whereas molecules of weakly soluble nonpolar medicines are dissolved within the micelle core. Drug carrier micelles have a number of advantages, including (Kotta et al. 2022).

1. Drug solubilization using micelle-forming amphiphiles increases the water solubility and bioavailability of sparingly soluble medicines.
2. The use of particular amphiphilic substances may also contribute to the property of micelles' blood having an extended half-life.
3. Permeability across physiological barriers is enhanced.
4. There are positive improvements in drug biodistribution.
5. The reduction of toxicity and other adverse effects.
6. Because of their small size, micelles can spontaneously enter bodily compartments and cause leaky vasculature in the interstitium.

7. The medication in micellar form is sufficiently protected against biological surround inactivation and has no negative side effects on tissues or organs that are not the intended target.

15.6 Application of Nanotechnology in Cosmetics

15.6.1 Cosmetics Nanoemulsion

In recent years, NEs have gained significance as possible delivery systems for cosmetics with precise dosing and for the best possible distribution of active substances in specific skin layers. NEs are better suited than liposomes for the transportation of lipophilic substances because of their lipophilic content. The interest in NEs continues to rise as a result of their unique bioactive characteristics. This may reduce trans-epidermal water loss (TEWL), a symptom of better skin barrier function. There are numerous cosmetic products on the market that contain nanoemulsions, such as the Red Vine Hair sunscreen from Kores (Dhawan et al. 2020).

Emulsifiers that are ready to use are available from several businesses, such as Sinerga's Nanocream® and Kemira's nanogel, for the creation of stable nanoemulsions intended for cosmetic uses. Because of the microscopic size of the droplets, nanoemulsions are transparent and hold their stability for a longer amount of time. The majority of products that contain them are deodorants, sunscreens, shampoos, and skin and hair care items. The excellent sensory qualities of nanoemulsions, such as their quick penetration, ability to blend textures, and biophysical qualities, particularly their capacity to hydrate, make them highly recognized in the skincare industry. Using a cationic nanoemulsion results in a substantial and long-lasting improvement in the dry hair aspect (after multiple shampooing) (Santos et al. 2022).

Nanogel one such instance is the recent introduction of a novel gel with nanotechnology by New Jersey-based TRI-K Industries and its patent firm Kemira, which aims to improve the effectiveness of various skincare products. According to reports, Kemira nanogel is a special nanoemulsion carrier system that combines the advantages of nanotechnology with an easy-to-form formulation. With the use of nanogel technology, one may easily produce submicron emulsions from an oil-in-water concentrate using a straightforward technique and equipment. The formulation is especially well-suited to reducing trans-epidermal water loss, increasing skin production, and improving active component penetration. These qualities imply that it would be especially helpful for sun care products, as well as moisturizing and anti-aging creams—specific areas where nanotechnology is already being used in a wide range of goods that are presently available for purchase. It is also emphasized that it contributes to the skin-feeling quality of skincare formulations, which is becoming a more crucial attribute for formulators (Meftahi et al. 2022).

15.6.2 PEG-Free Cosmetic Nanoemulsions

Making and processing low-viscosity oil-in-water nanoemulsions devoid of emulsifiers based on polyethylene glycol (PEG) is another example. In the expanding market for impregnating emulsions for moistened tissue, these mixtures are quite appealing. Emulsion-based wet wipes are a relatively new but rapidly expanding field of use for things like makeup removal and newborn care. Low-viscosity and well-storable O/W emulsions are the primary constituents of these products. The characteristic white appearance of classical emulsions is caused by their usual particle radii of 0.5 to 10 μm and viscosities of over 1000 m per second Pas. They can be produced using a homogenizer and are kinetically stable. Comparable viscosity systems, however, cream up and become unstable because of the relatively big particles they contain. On the other hand, O/W microemulsions' thermodynamic stability makes them simple to make. Their usual particle radii are between 10 and 40 nm, and they are translucent (Akinribide et al. 2022).

15.6.3 A Topical Application of Vitamin and Gold-Loaded Nanofiber Face Masks

Cotton masks that have been pre-moistened with skin nutrients are the standard beauty face masks found on the market. The pre-moistened mask's aqueous phase has the potential to accelerate the breakdown of unstable components like ascorbic acid. A unique polymeric face mask that can incorporate collagen, ascorbic acid, retinoic acid, and gold has been created as a solution to this issue. Antioxidants and other skin nutrients are frequently incorporated into cosmetic products as part of marketing strategies. Collagen, which is crucial for skin renewal and the ability to reverse the signs of aging, determines the strength and function of the skin. Collagen is widely employed as a moisturizer in cosmetic creams and cosmetics because as we age, our skin produces less of it. Because of its capacity to neutralize oxidizing agents, free radicals, and photoprotective properties, vitamin C (L-ascorbic acid) is typically utilized in dermatological and cosmetic treatments.

Additionally, it can inhibit skin pigmentation and promote the manufacture of collagen. Since vitamin C is easily oxidized due to its chemical instability, the pharmaceutical industry makes extensive use of its more stable derivatives, which can transform into ascorbic acid after ingestion. Examples of these derivatives are ascorbyl palmitate, ascorbyl tetra isopalmitate, and magnesium ascorbyl phosphate, which are all formulated as an emulsion (Akinribide et al. 2022). In addition to assisting in healing sun-damaged skin and reducing wrinkles from photoaging, retinoic acid can be used to treat acne. Gold nanoparticles have been investigated for transdermal distribution and as possible vaccination carriers. These days, salons and beauty centers utilize gold face masks. It revitalizes the skin and lessens the appearance of wrinkles by enhancing blood circulation and skin elasticity. Studies on skin penetration show that spherical gold nanoparticles are not intrinsically harmful to human skin cells (Chen and Feng 2022).

15.6.4 Cosmetics Nanoparticles Hair Disease Therapy with a Nanogold Facial Mask

Particulate medication delivery technologies are becoming more important in the treatment of hair disorders such as alopecia than aqueous alcohol solutions. They can serve as a depot for a prolonged medication release within the hair follicle and increase drug penetration into the follicle openings. Solid lipid nanoparticles, neutral liposomes, poly(lactic-co-glycolic) acid, and poly(e-caprolactone)-block polyethylene glycol are a few examples of nanoparticles used to treat hair diseases. Gene therapy of hair and new particle-based drug delivery technologies for a prospective active follicular targeting of disease-related cell populations in the hair follicle are becoming more and more important due to the absence of alternative treatment options (Sharma et al. 2023).

15.7 Non-Invasive Imaging Techniques

15.7.1 Single-Photon Confocal Microscopy

Laser scanning confocal microscopy (LSCM), also referred to as single-photon confocal microscopy, is a frequently used non-invasive real-time imaging technology. By focusing coherent light through a pinhole filter and eliminating out-of-focus light from above or below the object before it reaches the detector, the fluorescence-based LSCM approach produces high-resolution images (Ayardulabi et al. 2021). Researchers can study and comprehend nanoparticle interactions with skin and characterize nanoparticle and biomolecule distribution, penetration, and overall delivery pathway across the cutaneous barrier, thanks to LSCM's ability to optically section biological samples without compromising their structure. However, the microscope's configuration—specifically, the kind of laser and optics—determines the LSCM's capacity to visualize the morphological structure of biological materials. A laser with a visible or near-infrared wavelength is usually needed to visualize the samples' cellular morphology.

A type of LSCM called reflectance confocal microscopy (RCM) provides face optical slices of the skin at a quasi-histological resolution by using the visible to near-infrared spectrum (Navarrete et al. 2021). RCM has been applied clinically to a variety of dermatological disorders, including the differential diagnosis of melanoma (Longo et al. 2024) and basal cell carcinoma (Saginala et al. 2021). We are the first group to use RCM to evaluate the topical nanoparticles' penetration profile in human skin (Darvin 2023). We noticed that highly reflecting aggregates were formed when topically placed nanoparticles were dissolved. We looked at the penetration profiles of silver (Ag-NP) and gold (Au-NP) nanoparticles in 2011 using human skin models both in vivo and ex vivo. It makes sense that if the stratum corneum, the skin's outermost layer of protection, were to become weakened, there would be a greater chance of nanoparticle penetration. Protocols for tape stripping were used to simulate this kind of damage. Even in cases where the outer skin barrier was compromised, Au-NP and Ag-NP accumulated on the skin's surface and in its furrows without penetrating the viable epidermis (Rabiee et al. 2022). Evaluating the natural rate of nanoparticle removal following topical application was our secondary objective in the silver nanoparticle investigation. In distinct regions of skin injury, visible aggregates were seen for up to 10 days, but not in intact skin (Fig. 15.2) (Fuchs et al. 2021).

The US Food and Drug Administration (FDA) also classifies permanent makeup and tattoo inks as cosmetic items. It has been observed that carbon nanoparticles are present in pigments and ink (Negi et al. 2022). Typically, these medications are injected into the superficial dermis layer; however, the FDA has not approved these application methods (Parhi and Mandru 2021), between 2003 and 2004, more than 150 permanent makeup adverse events were reported to the FDA. The first group to employ RCM to examine a 40-year-old woman's lip skin modification following a permanent lip lining treatment at a cosmetic institute (Fiuza et al. 2024). The patient continued to have significant soreness, swelling, and scaling even after receiving topical corticosteroid treatment. Using high-definition optical coherence tomography (HD-OCT) and RCM, pigment is seen as bright clusters of particles distributed across the epidermis and superficial dermis. Both HD-OCT (Figs. 15.3 and 15.4) and RCM images revealed a non-infectious granuloma composed of highly reflecting structures with macrophages laden with pigment; this was subsequently validated by a skin biopsy (Fiuza et al. 2024).

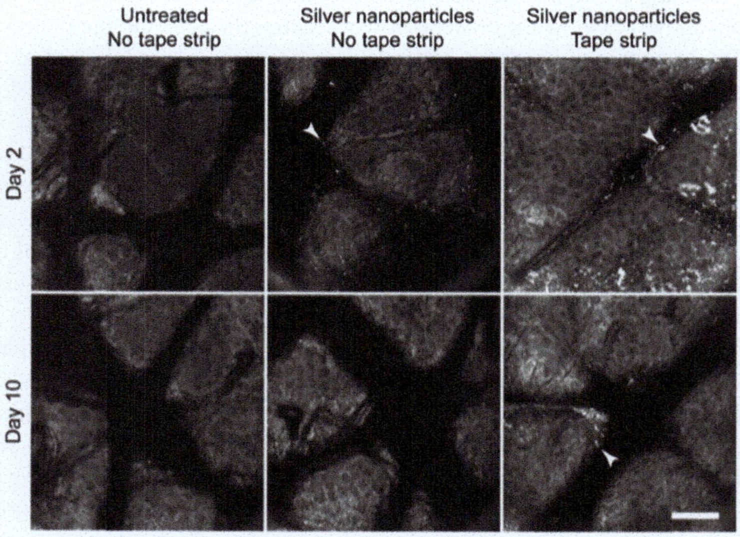

Fig. 15.2 ZnO-NP applied in vivo to reflectance confocal slices of human skin. The white arrowheads on the skin surface represent the silver nanoparticles, which were seen to aggregate for a maximum of 6 days (data not provided) in the skin that was intact and up to 10 days in the skin that had a damaged barrier. 200 μm is represented by the white bar, published in 2011 by Elsevier B.V. with permission

For sunscreen products, dispersion and particle size are essential performance parameters (Catalano et al. 2020). It's unclear if adding raw TiO_2-NP to sunscreen compositions affects these qualities. To effectively evaluate these modifications, the characterization techniques used must be carried out without requiring samples to be diluted or altered. Wokovich and colleagues utilized three distinct methodologies to ascertain the influence of the formulation procedure on the dimensions or configuration of TiO_2-NP (Bevacqua et al. 2023). In intact swine epidermis, four distinct sunscreen formulations were evaluated: an uncoated 21 nm TiO_2-NP sunscreen, a coated 50 nm TiO_2-NP sunscreen, and a submicron-sized 250 nm TiO_2-NP sunscreen vehicle. The resolution limit of LCSM is 200 nm, beyond which single nanoparticle imaging and identification are not achievable. This implies that although a single nanoparticle signal may be caught in a single pixel in theory, the

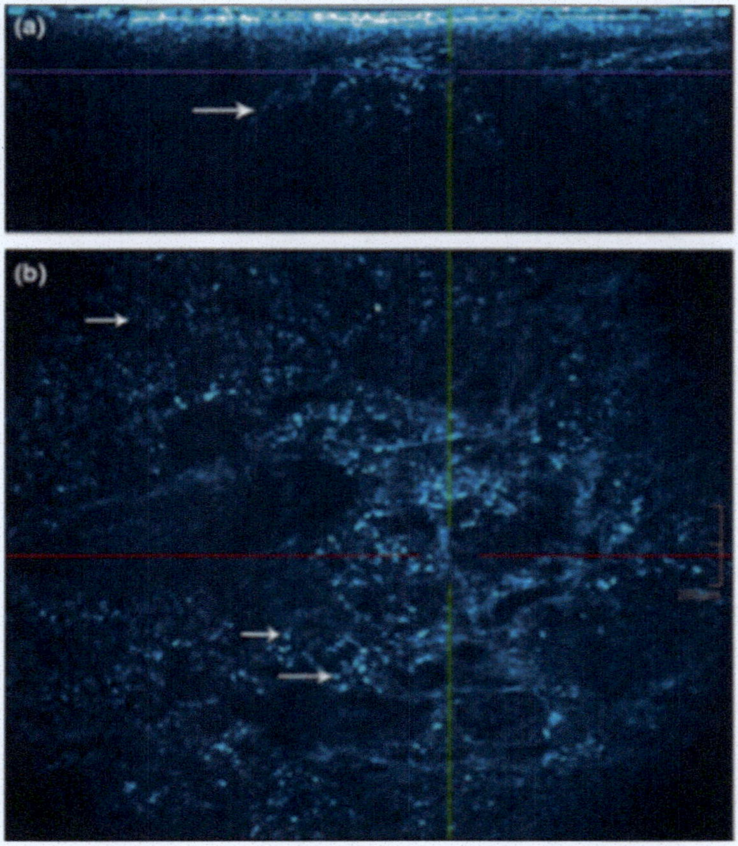

Fig. 15.3 Optical coherence tomography high-definition image of a lip covered with permanent makeup. From a cross-sectional image (**a**), aggregated pigmentation and the development of granuloma-like formations (white arrow) were observed beneath the epidermis. An en-face high-definition optical coherence tomography (HD-OCT) image (**b**) revealed scattered pigment that was composed of brilliant structures with a dark center that corresponded to a pigment-loaded macrophage (white arrow). Used by permission from (Lin et al. 2015), which was released in 2014 by the European Academy of Dermatology and Venereology

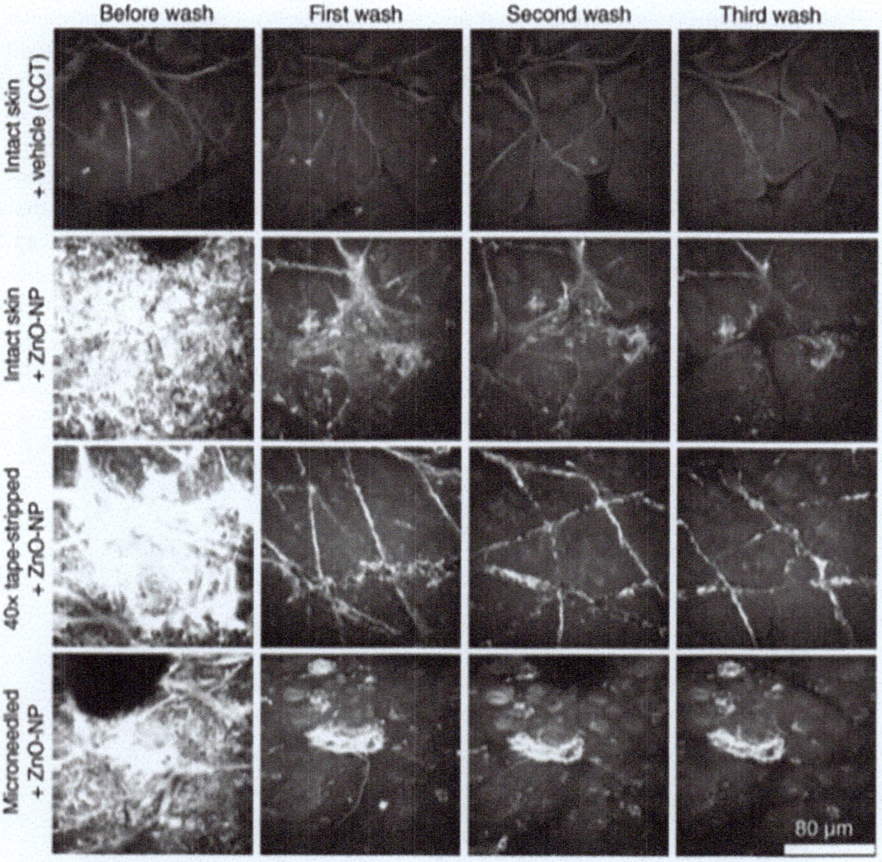

Fig. 15.4 ZnO-NP applied to human skin multiphoton tomography. The rows show the epidermal layers' typical z-projection to the superficial dermal-epidermal junction both before and after washing. Triglyceride buffer, or CCT for short. Reproduced with permission from (Lin et al. 2015), which was released in 2013 by Future Medicine Ltd.

apparatus is unable to differentiate between the signal's spatial origins from one or many nanoparticles.

The dispersion and depth of penetration of nanoparticles into the skin were assessed by Wokovich et al. X-ray diffraction (elemental analysis) and scanning electron microscopy (SEM) were employed in conjunction with invasive techniques at higher resolutions to examine the delivery and skin penetration of nanoparticles. Nevertheless, we have discovered that low signal intensity and processing artifacts from contamination are common in electron microscopy techniques, which complicates the interpretation of results. An equal distribution was seen in the LSCM of sunscreens comprising uncoated and submicron TiO_2-NP, which was verified by SEM. SEM confirmed that the particles in the coated TiO_2-NP sunscreen were unevenly distributed and appeared as aggregates on the skin's surface (Bevacqua et al. 2023).

Apart from the limitations of the technique's resolution, LSCM analysis of skin and its cellular constituents is frequently challenging. In vivo research in particular runs the danger of irradiation damage to living tissues when doing repetitive imaging utilizing laser light sources. One of the most difficult challenges with LSCM imaging is the production of autofluorescence. By separating signals from the fluorophore and sample using a dual-channel arrangement, this can be prevented (Li et al. 2020). Further more expensive to buy and maintain are laser confocal microscopes.

15.7.2 Multiphoton Microscopy and Time-Correlated Single-Photon Counting

Due to its high picture resolution, contrast, and ability to exploit skin autofluorescence, multiphoton microscopy (MPM), also known as multiphoton tomography (MPT), provides real-time in vivo imaging and has emerged as a useful method for the study of nanoparticles in living cells and tissue (Wizenty et al. 2020). MPM uses femtosecond pulsed lasers to stimulate unlabelled nanoparticles as well as endogenous and exogenous fluorophores (Graf 2012). In order to differentiate signals between endogenous fluorophores and nanoparticles, MPM is frequently used in conjunction with additional detectors. Fluorescence lifetime imaging is made possible by time-correlated single-photon counting (TCSPC) (FLIM). Data generated from simultaneous evaluations can be quantitated, and a dye-free assessment is produced by coupling TCSPC detectors (Graf 2012). To find out if ZnO permeates beyond the stratum corneum, numerous investigations using human volunteers and skin have been conducted (Filipe et al. 2009). Our lab investigated several methods, including MPM, to perform ZnO-NP in vivo monitoring on both healthy and sick skin. The study involved two volunteer cohorts: individuals without a history of skin conditions and those with psoriasis or atopic dermatitis. Subjects received dosages of 2 and 14 mg/cm^2 of ZinClear-S 60 capric/caprylic triglyceride (CCT) nanoparticles, an active component in sunscreens, applied to their forearms. Subjects with intact, healthy skin were tape-stripped and treated for 4 and 24 h; volunteers with sick skin received treatment for 2 h.

Using multiphoton tomography in conjunction with TCSPC-FLIM, ZnO-excited photoluminescence (MEP) signals were measured concurrently with nicotinamide adenine dinucleotide phosphate (NAD(P)H), a metabolic intracellular molecule. At 4 h, a ZnO-NP signal was seen in the furrows of both intact and tape-stripped skin, as well as in the stratum corneum, according to MPM pictures. There were more ZnO-NP signals in the stratum corneum of the skin that had been tape-stripped. After 24 h, the ZnO-NP signal in tape-stripped skin dropped to a minute level, indicating that ZnO-NP was dissolved for the whole 24-h treatment period (Lin et al.

2011). At the 2-h time point, a strong ZnO-NP signal was found in the thick stratum corneum layer. Prior to the stratum granulosum, this signal deteriorated and vanished from later, deeper levels. In both lesional and non-lesional ZnO-NP treated skin, there was no discernible ZnO-NP signal in the superficial viable epidermis (Lin et al. 2011).

These findings provide credence to the theory that topically applied ZnO-NP does not penetrate living human skin. This was further supported by a different investigation in which we demonstrated that washing with soap and water and then re-imaging eliminated the topical nanoparticle signal from sunscreens (Fig. 15.5) (Lin et al. 2015). In this investigation, ZnO-NP was administered topically at a dose of 2 mg/cm^2, rubbed for 30 s, exposed for 2 h, and then MPT imaging was performed. After 10 s of hospital-grade soap washing, the skin was re-imaged. Stepwise optical biopsy with 40 sections measuring 5.1¬µm was used to analyze the epidermis. The proportion of the ZnO-NP signal strength was reported. After the third wash, the signal in intact skin decreased from 36% ± 17% to 2% ± 1%. Before washing, 36% ± 19% ZnO-NP was detected on tape-stripped skin; after three washes, this percentage dropped to 3% ± 3%. The ZnO-NP signal before washing did not significantly differ between the skin that was tape-stripped and the skin that was not (Lin et al. 2015). All things considered, this work demonstrated that ZnO penetration could be assessed using MPM and that these nanoparticles could be located in the skin using MPT through semi-quantification of signal processing. ZnO-NP could not pass through healthy human skin unless the barrier had already been compromised. This study also showed that cleaning the area can eliminate ZnO-NP applied topically.

The advancement of cosmetics and raising consumer awareness both benefit greatly from these studies. In vivo nanoparticle interactions in human skin are rarely evaluated by multiphoton microscopy because of expensive purchase, operation,

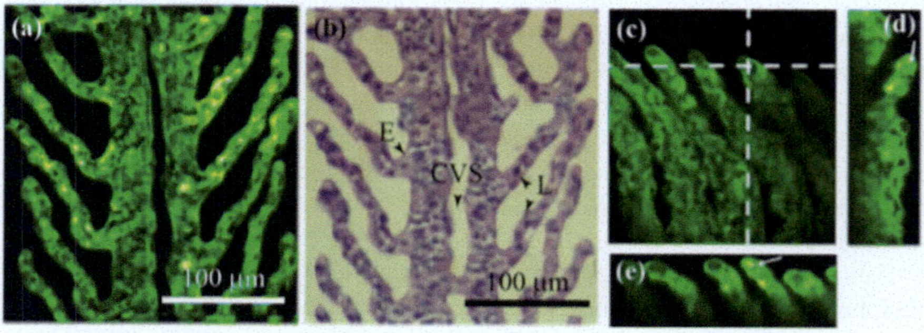

Fig. 15.5 Gill lamella in coherent anti-Stokes Raman scattering (CARS) picture. 8 µm slices of a tomographic CARS picture were obtained with pump and Stokes wavelengths of 125 nm and 924 nm, respectively (**a**). Schiff-stained slice image captured in white light at a magnification of 60× (**b**); L is lamella, E is the epithelium, and CVS is the central venous space. An illustration of a trans-axial slice of gill lamella treated with TiO_2 for a duration of 2 weeks (**c**). Cross-sectional images of the vertical and horizontal dotted lines are displayed in panels (**d**, **e**), respectively. The white arrows in panels (**d**, **e**) indicated the locations of the nanoparticle clusters (Roshchina 2020)

and maintenance costs as well as the need for substantial user training. This finding was intriguing because it demonstrated how quickly artifacts from high nanoparticle concentrations could skew the interpretation of imaging data.

15.7.3 Optical Coherence Tomography

Real-time internal cross-sectional imaging of the skin's microstructural morphology is made possible by the non-invasive method known as optical coherence tomography, or OCT. One new clinical imaging technique for inflammatory skin conditions, skin malignancies, and wound healing is OCT. It is currently growing into the fields of cosmetic testing and cosmetic dermatology. Using in vivo technology, OCT imaging produces high-resolution, 2- or 3-dimensional images, albeit at a lower resolution than confocal methods. The testing of cosmetic products can then be done in real-time using both qualitative and quantitative evaluations. Because this imaging method is non-invasive, it can be used to frequently photograph treatment areas and track the skin's progressive changes after applying cosmetics.

Using low-coherence interferometry, OCT imaging calculates the echo time delay of reflected light in the near-infrared range (Arshad and Abbas 2018; Li and Fujimoto 2004). Low-coherence near-infrared light is directed at a designated skin spot during OCT imaging. Next, backscattered light from the interior microstructures is observed. A high-resolution image is produced by these structures' backscattered light's location and magnitude (Schwarzer et al. 2009). OCT is frequently used to evaluate cosmetic goods qualitatively. In order to investigate the penetration and accumulation of four cosmetic creams on human hand skin as well as their effects on collagen synthesis levels, Han et al. (2009) conducted a study using OCT.

While one cream claimed to increase the formation of collagen, another included soluble collagen. There were no active components in the other two tested creams. Male adult participants underwent swept-source optical coherence tomography in vivo for a duration ranging from 5 to 120 min. Only the cream containing soluble collagen exhibited an increase in tissue depth in the epidermis, according to single-intensity OCT pictures of hand skin. 30 and 120 min were peak times (Geng et al. 2009). Additionally, it was determined from OCT pictures that the collagen-soluble cream had no effect on tissue during the short term (5–15 min). Researchers were able to use this data to provide people with timetables for getting the optimum cosmetic results.

15.7.4 Advanced Raman Spectroscopy

Clinical research labs are now able to use cutting-edge label-free imaging technologies with remarkable sensitivity and fast time resolution for developing biological applications and therapeutics. In several scientific domains, such as chemistry

(Moreira et al. 2008), physics (Kakani 2004), material science (Kakani 2004), and medicine, Raman spectroscopy has been extensively employed as an analytical and diagnostic tool since the 1930s. Due to its label-free optical imaging capabilities and ability to produce native chemical fingerprints of cells, tissues, or biofluids based on the inelastic scattering of light, this approach has lately been thoroughly investigated for biomedical applications (Arshad et al. 2018; Mabrouk et al. 2021). The Raman effect is the name for this inelastic scattering of a photon from a specific molecule.

The distinctive vibrational signature of a molecule linked to particular chemical structural bonds is the basis of the molecular chemical fingerprinting theory. Unlike other intrinsic imaging methods, Raman spectroscopy is not restricted by the quantity of endogenous fluorophores and is very slightly impacted by the surrounding environment. The extremely weak scattering impact resulting from the poor photon conversion efficiency of 1:1018 and the lengthy data collecting time is a significant disadvantages of this approach. Studies of living biological specimens are severely limited, in addition to the requirement of high laser power consumption and extensive integration periods. Over time, variations of Raman spectroscopy with increased sensitivity have been created to address the present issues with spontaneous Raman scattering. Coherent anti-Stokes Raman Scattering is one example of enhanced vibrational signals combined with video-rate imaging speed Raman spectroscopy (CARS).

Evan and Xie (2008) presented a thorough review that addresses the setup and theory behind CARS microscopy (Goodhead et al. 2015). This review section's focus is on the imaging of sunscreen nanoparticles using sophisticated Raman spectroscopy technology. To appropriately analyze the biological interactions of metal nanoparticles inside a structurally complex biological system like skin, current imaging methods must overcome current limitations in localizing the particles. The majority of optical imaging methods are deficient in-depth penetration and/or spatial resolution, which are necessary to identify the particles at the cellular level in intact tissues. The inability of current label-free, non-invasive imaging methods to separate metal nanoparticles from biological specimens without intrusive preprocessing is a significant constraint. Except for ZnO-NP, which was covered in the preceding sections and has inherent multiphoton fluorescence features. Over the past 10 years, CARS microscopy has drawn a lot of attention, especially for biomedical applications.

To date, there are no reported data on the use of this approach for analyzing sunscreen-linked nanoparticles in human volunteers. The application of CARS microscopy for ex vivo detection of ZnO-NP and TiO_2-NP up till now was limited to fish models (Arshad 2018; Fadoju 2018). Using this non-invasive method, Moger et al. carried out an exploratory investigation to find metal oxide nanoparticles in the gills of rainbow trout, *Oncorhynchus mykiss*, including ZnO-NP, cerium dioxide (CeO_2), and TiO_2-NP (Fujihara and Nishimoto 2024). They were going after the lamella, which are leaf-like structures found inside fish gills. First, the lamella's structure was determined by contrasting the CARS picture with a portion of the

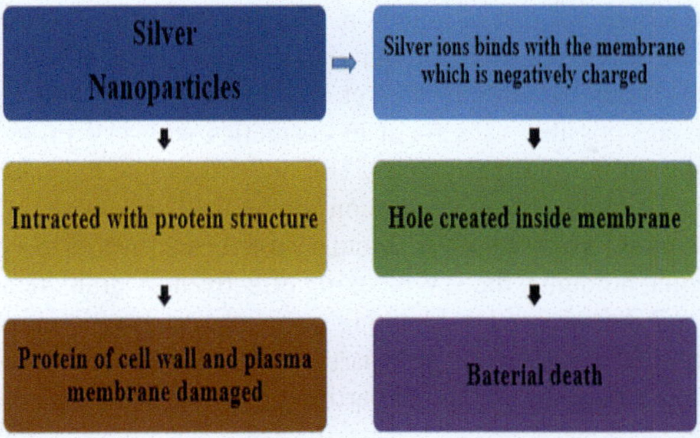

Fig. 15.6 Silver nanoparticles in clinical research for wound healing

same area that had been dyed histologically (Fig. 15.6a, b). Carbon-hydrogen (CH)-rich structures were the source of the contrast seen in CARS microscopy, which coincided with the areas of strong dye uptake.

The CARS images show the different sizes of the nanoparticles suspended in agarose; bigger particle aggregations were only visible when the Forward-CARS (F-CARS) detection scheme was applied. The absence of smaller particles in the F-CARS images could be attributed to either the signal dropping below the background level or the reduction in the forward-detected signal caused by the particle-agarose interface. To locate nanoparticle aggregations within the tissue, three-dimensional pictures of fish gills subjected to 5000 μg/L TiO_2 for either 24 h or 14 days were taken. After 2 weeks of treatment, they noted tiny particle aggregates in the lamella's periphery (Fig. 15.5c, e). Using CARS imaging, large clusters of particles up to 100 μm in diameter were also reported to be visible inside the secondary lamella close to the capillary following a prolonged duration of exposure.

15.8 Inorganic Materials Nanomaterials in the Cosmetic Metal Oxides

Because of the incorporation of inorganic nanoparticles, the cosmetics sector has seen a significant upheaval in recent years. These nanoparticles are particularly sought for cosmetic applications due to their amazing qualities, which include a high surface-to-volume ratio, customized optical and catalytic capabilities, and outstanding stability. A noteworthy advancement has been made in the area of sun protection since nanoparticles like zinc oxide (ZnO) and titanium dioxide (TiO_2) demonstrate exceptional abilities to block harmful ultraviolet (UV) radiation and protect skin (Dunaway et al. 2018). These nanoparticles vigorously attack microbes that cause acne and stop harmful microorganisms from growing, which improves skin health and lowers the risk of infection. The subsequent discussion offers an in-depth analysis of certain research that demonstrates the importance of nanoparticles in the cosmetics sector. A study by Lee et al. looked into the production of planar (ZnO) powder and possible uses for it in cosmetics with UV protection (Arshad et al. 2023; Raha and Ahmaruzzaman 2022). They discovered that each type of ZnO has comparatively poor light transmission values in terms of light transmission properties. However, when the LAR acicular and vertical wall ZnO powders were combined in equal amounts, the UV-blocking effectiveness dramatically increased. Both UVB and UVA bands of UV blocking were enhanced by the combination. The uniform distribution of the various ZnO particles, along with their synergistic effects to maximize light scattering and avoid particle agglomeration, was linked with this improvement.

Furthermore, the mixture demonstrated effective functional group adsorption on the ZnO surface, improving dispersibility and avoiding particle agglomeration. The 1:1 powder mixture's strong UV-blocking efficacy was validated by an in vivo experiment. These discoveries enhance our knowledge of light transmission qualities and optimize ZnO particle dispersion and distribution, which helps the cosmetics industry create sunscreen remedies that are more effective. These discoveries can be used by producers and scientists to generate sophisticated sunscreens with enhanced UV protection. Nickels et al. (2016) conducted a recent study to examine the detrimental effects of ultraviolet (UV) radiation on human skin (Gallagher and Lee 2006).

They emphasized issues including sunburn, photoaging, and the chance of getting skin cancer. It is well known that sunscreens, which contain both organic and inorganic UV compounds, are crucial for offering complete UV protection. But inorganic medications, especially (TiO_2) and (ZnO), come with special difficulties. Typically, these medications are applied as nanoparticles to avoid the unwanted whitening effects on the skin. These inorganic medications have the ability to function as photocatalysts in the presence of UV radiation, which can cause the organic components in cosmetic formulas to break down. Furthermore, UV radiation exposure can harm cells and DNA. In earlier research, a number of tactics have been investigated to lessen these deficiencies. TiO_2 and ZnO particle surfaces have been altered, covered in silicon dioxide, or substituted with other materials like cerium oxide.

They suggested a novel method in this work for creating inorganic UV drugs for use in cosmetic formulations. Their study's primary objective was to enhance semiconductor nanoparticles' absorption capabilities in the 320–400 nm region while shielding the active phases from outside effects. To do this, they used the mesoporous silica nanoparticles (MSN) pores as a carrier material to disseminate the semiconductor, specifically TiO_2. After that, a heat treatment followed by the addition of bismuth ions as a reactive element sealed the MSNs' surfaces. Bismuth titanate phases were created when the bismuth ions interacted with the TiO_2 nanoparticles that were imprisoned in the silica pores. These substances presented intriguing features and were thoroughly investigated in a number of domains, including visible light photocatalysis, microelectronics, electro-optics, and dielectrics.

The reason behind selecting a bismuth salt originated from its capacity to react with TiO_2 to generate novel active phases that offer extensive UV shielding. As the temperature of calcination rose, the sealing mechanism was supported by the diffusion of bismuth ions into the MSN. A glassy silica coating that successfully sealed the nanoparticles' exterior was formed as a result of this technique. As a result, the UV drug's active components were given a protected environment, which improved UV protection. The drawbacks of traditional inorganic medications are removed by this creative self-sealing system. Thus, the goal of the study was to include bismuth titanate phases into mesoporous silica nanoparticles in order to provide a better inorganic UV medication for cosmetic formulations. This method successfully addressed the issues with conventional inorganic medicines while improving UV protection. The findings of this study support continued attempts to create sophisticated sunscreens that are more effective and safe.

15.8.1 Quantum Dots

Due to their amazing potential in the cosmetics industry, particularly in skincare and beauty products, quantum dots (QDs) are attracting a lot of interest. The optical characteristics of these semiconductor nanoparticles are amazing, especially their bright and adjustable luminescence. QDs can produce different forms of light depending on their size, shape, and composition. This makes them an attractive addition to cosmetic formulations that aim to achieve different visual effects including skin illumination and color refinement (Aslam et al. 2021). Furthermore, these QDs can be precisely designed to release particular colors, like captivating blues and greens, which are very useful for balancing out dull or uneven skin tones (Starke 2017).

By incorporating QDs with certain emission wavelengths into cosmetics, imperfections appear less noticeable, and the complexion appears more even and stunning. By deftly adjusting their size and arrangement, amazing optical phenomena are produced, like eye-catching color shifts that vary according to the viewing angle. This distinctive feature has been used in intriguing ways in cosmetics like lipsticks and nail varnishes, producing visually striking results that entice the senses. Additionally, these dots can be fitted with particular molecules or compounds that allow them to distribute active chemicals to the targeted skin areas. This improves the effectiveness of skincare products by guaranteeing accurate distribution and regulated release of their active ingredients (Reddy et al. 2016).

Thus, in order to considerably lessen UVA band edge emission, Adersh et al. looked into employing atomic flaws in ZnO QDs (Unger et al. 2015). Defect centers in ZnO QDs are known to create a luminescence that is broadly observable and may be adjusted in relation to particle size. Correlating this visible emission caused by the defect with the skin's emission profile may have potential uses in hiding imperfections in the complexion. Different layers of human skin produce autofluorescence; the dermis, which contains collagen and elastin, exhibits strong emissions in the green range. Age and UV exposure can lessen the components that cause green fluorescence. Creating a sunscreen that contains ZnO QDs with broad green fluorescence could be a multipurpose tool for skin brightening and camouflaging, giving the illusion of younger skin.

15.8.2 Nanoclay

A type of clay known as nanoclay is made of tiny particles—typically less than 100 nm—through processing. These special nanoparticles are being utilized more and more in a variety of sectors, including the cosmetics industry (Dubey et al. 2022). Because of its beneficial properties, nanoclay is frequently utilized as a component in skincare and personal care products in the cosmetics industry. Because of their enormous surface area, nanoclay particles are useful in the production of oil-controlling products like face masks, foundations, and mattifying powders because they can absorb excess oil and sebum from the skin. Furthermore, nanoclay functions as a rheological modifier to enhance the texture and viscosity of cosmetic compositions while also having thickening and stabilizing qualities. It keeps separation from happening, stabilizes emulsions, and enhances the consistency of gels, creams, and lotions.

Because of their small size, nanoclay particles can better distribute active substances to the skin and allow for easier penetration. Better stability and control over the release of active chemicals can be achieved in cosmetic compositions by encasing them in nanoclay particles. Furthermore, research has been done on nanoclay's capacity to control the skin's production of sebum. A more balanced complexion and control over excessive oiliness can be achieved by using nanoclay in cosmetic products (Arshad et al. 2014; Dubey et al. 2022). Especially in dermatology, there has been a growing interest in creating innovative delivery methods to increase the effectiveness of topical medication delivery in recent years. Using active chemicals with carriers made of nanoclay is one such fascinating method.

In order to improve the distribution of this important active ingredient to the skin (Viseras et al. 2019), carried out a thorough investigation on the development and application of nanoclay hybrids loaded with melatonin (Borrego et al. 2022). The pineal gland naturally produces melatonin, a hormone that has attained interest for its several biological benefits, such as its capacity to reduce inflammation, protect the skin, and act as an antioxidant. Effective melatonin transfer over the skin barrier is still problematic. According to their findings, melatonin's stability and regulated release are enhanced when it is incorporated into nanoclay-based carriers. Because of their high surface area and remarkable adsorption powers, nanoclays are able to effectively encapsulate melatonin molecules, shielding them from deterioration and extending the duration of their release. Furthermore, the skin penetration of these nanoclay carriers is excellent, allowing for the effective transportation of melatonin into the skin's deeper layers. Melatonin can more easily reach its target areas by penetrating the stratum corneum, the outermost layer of skin, thanks to the interstitial voids in the nanoclays. Furthermore, there is a synergistic effect when melatonin and nanoclay are combined. Melatonin's therapeutic effects are enhanced by the natural anti-inflammatory and antibacterial capabilities of nanoclays. This interaction creates new opportunities for the treatment of inflammatory conditions and microbial infections, among other skin ailments. As such, melatonin/nanoton hybrids show promise for a number of dermatological uses, including wound healing, inflammatory skin conditions, and skin regeneration.

15.8.3 Silver Nanoparticles

In cosmetic items, nanoparticles serve as transporters, active substances, and texture enhancers to improve the overall performance of the product. Ag nanoparticles are useful ingredients in cosmetic formulations because of their extensive antibacterial range. Ag nanoparticles are useful for disinfecting skin and wounds because they have potent antibacterial properties in addition to acting as preservatives. Ag nanoparticles have been shown to be very helpful when added to creams since they strengthen the product's antibacterial activity and aid in wound healing and skin hygiene. Ag nanoparticles are also utilized in toothpaste and other oral hygiene products because of their antibacterial qualities, which aid with conditions like dandruff and greasy, itching scalps. Ag nanoparticles' special qualities have made them popular in dental materials (Arshad 2024).

For the purpose of maintaining oral hygiene and preventing infections, they provide enhanced antibacterial action. Furthermore, the use of Ag nanoparticles in a range of cosmetic products, such as peel-off face masks, has been made possible by advancements in nanotechnology. These Ag nanoparticle-infused masks make use of the antibacterial qualities of Ag nanoparticles produced by microorganisms, including Bacillus Nakamura. The latest development strengthens the peel-off masks' antibacterial qualities, increases their efficacy, and encourages healthier, cleaner skin. Ag nanoparticles are extremely useful in cosmetic goods due to their multipurpose characteristics. In oral hygiene, skincare, and other areas, they provide enhanced efficacy and antibacterial advantages (Choo et al. 2001).

15.9 Commercialized Nanotechnology-Based Micellar Cosmetics

Nanoparticle-based micellar products are becoming more and more common in the cosmetics industry across a broad spectrum of product categories. Additionally, a lot of local and international businesses have adopted this groundbreaking technique of nanotechnology to give their cosmetic products exceptional quality and efficacy. The production of bioactive component-enriched micellar nanoparticles, which serve as a stable delivery vehicle for cosmeceuticals, is greatly influenced by the nanoemulsion system.

15.9.1 L'Oreal

This section presents the current line of commercial cosmetic products that make use of micellar nanoparticle technology. Numerous personal care products that were created using nanotechnology and touted as the ideal skincare solution have been produced. L'Oreal Paris actually seemed to be the most successful skincare brand in the world. Statistics between 2010 and 2021 demonstrate that the L'Oreal Group raised its consolidated sales by roughly 10 billion US dollars globally, according to Statista Research Department (Pantelimon et al. 2020) study. Following years of steady expansion, the business was able to produce about 29.8 billion US dollars in worldwide sales in 2019.

15.9.2 Garnier

Micellar water from the Garnier brand is used to cleanse, tone, and refresh skin. Micellar technology allows the micelles to absorb and extract oil, grime, and makeup without the need for vigorous rubbing, leaving the skin clean but not overly dried out. All skin types can benefit from Garnier Micellar Water's oil-free, alcohol-free formulation. In addition, Micellar Cleaning Water has won 23 beauty awards and over 10,000 five-star evaluations. According to Garnier, one bottle of Micellar Cleaning Water is bought every 4 s (Rocha 2020).

15.10 Safety Assessment and Risk Management of Novel Nanomaterials in Cosmetics and Their Role in SDG

Managing the safety and risks related to novel nanomaterials in cosmetics (Fig. 15.7) entails identifying possible dangers and putting precautions in place to reduce risks to the environment and public health. A detailed characterization of nanomaterials, including their physicochemical characteristics such as size, shape, surface charge, and composition, is necessary to accomplish this. These characteristics may affect their toxicity, behavior, and possible hazards. For nanoparticles to be used in cosmetics, thorough toxicity analyses are necessary. It could be necessary to modify or create specialized toxicity tests for nanomaterials in order to take into account their special characteristics.

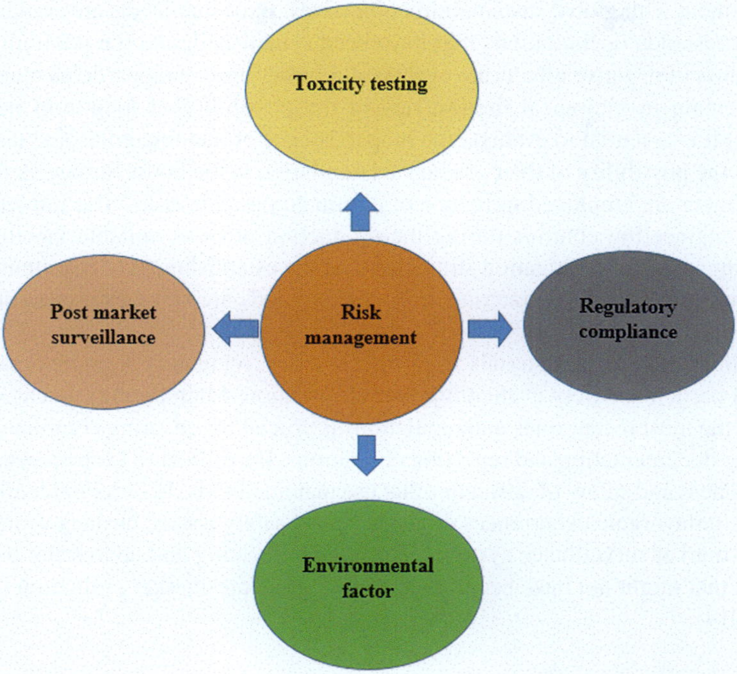

Fig. 15.7 Risk management consideration when utilizing nanoparticles in cosmetic products

Potential side effects. Skin irritation, sensitization, and genotoxicity are assessed by these tests. The ways in which workers and consumers might be exposed to nanoparticles in cosmetics are identified. The evaluation of potential exposure routes, such as oral, inhaled, and facial exposure, is part of this (Enyoh et al. 2020). Considerations must be made for elements including product concentration, frequency and duration of usage, and potential release during use. To evaluate the possible risks of utilizing nanoparticles in cosmetics, risk characterization is based on toxicity and exposure data.

Although the use of innovative nanomaterials in cosmetics has many advantages, including improved product performance and innovation, it also necessitates thorough safety evaluations and risk management to protect the environment and consumer health. This involves complying with regulations while evaluating potential toxicity, skin penetration, and environmental impact. The industry can reduce risks by implementing techniques such as transparency, controlled formulation, and post-market monitoring into practice. Additionally, by enhancing product efficacy, encouraging sustainable production practices, and lowering environmental impact, nanomaterials in cosmetics assist with achieving a number of Sustainable Development Goals (SDGs), such as good health (SDG 3), responsible consumption (SDG 12), and climate action (SDG 13). This corresponds to technological advancement with global sustainability objectives. Exposure levels are compared to safety thresholds or limitations that have been established, and the possibility and seriousness of negative effects are evaluated. Furthermore, nanoparticles may eventually remain in biological tissues, raising the possibility of long-term negative impacts. It is essential to evaluate the biopersistence of nanomaterials and take into account the possibility of their gradual accumulation in the body. Strategies for risk management are employed to reduce or eliminate possible risks. The implementation of engineering controls during the production process, suitable labeling and packaging, exposure mitigation strategies, and the establishment of maximum concentration limits for nanomaterials in cosmetic goods are some examples of these methods.

For innovative nanomaterials used in cosmetics, regulatory agencies may also demand particular safety evaluations and information. Maintaining current knowledge of the most recent rules and regulations is crucial for ensuring compliance and fulfilling documentation and reporting obligations. It's critical to keep an eye on the effectiveness and safety of cosmetics that use nanomaterials. In order to identify and evaluate unfavorable occurrences or newly raised safety issues, businesses must set up post-market surveillance systems. This aids in locating and addressing possible hazards that might not have been obvious during the pre-market evaluation (Zhang et al. 2014).

15.11 Future Prospects and Outlook

The potential for revolutionizing the cosmetics industry through the integration of nanomaterials with emerging technology is immense. In combination with 3D printing and personalized medicine, nanomaterials hold great promise for improving beauty items and therapies. It is possible to create more effective delivery methods for cosmetic components by utilizing nanoparticles. Nanomaterials and 3D printing work together to create customized cosmetics with distinctive surfaces, colors, and textures (Dubey et al. 2022). Active compounds can be encapsulated in nanoparticles to help them penetrate the skin more effectively and be delivered to certain cells or locations for better results. These developments have the potential to improve hydration, accelerate the absorption of anti-aging components, and raise sunscreen efficacy. By improving product efficacy, enhancing skin penetration, and providing targeted treatment solutions for a range of skin issues, nanotechnology's growing role in skincare is revolutionizing the sector. Active substances can reach deeper layers of the skin through the use of nanomaterials like nanoparticles and nanostructures, which improves moisturization, anti-aging, and UV protection. A number of Sustainable Development Goals (SDGs) are supported by this innovation, especially SDG 3 (good health and well-being), which encourages healthier skin and better cosmetic procedures, and SDG 12 (responsible consumption and production), requiring for the creation of more environmentally friendly, efficient, and sustainable formulations. Furthermore, by minimizing the requirement for greater amounts of ingredients, nanotechnology can help create more resource-efficient production processes, supporting low-impact, sustainable manufacturing methods that support SDGs 9 (industry, innovation, and infrastructure) and 13 (climate action). Long-term health and preservation of the environment are ensured by the incorporation of nanotechnology into skincare products, which offers both increased customer advantages and a route toward a more ethical and sustainable industry.

Incorporating nanomaterials into cosmetics can result in "smart" products that react to particular stimuli or situations (Gottardo et al. 2021). For example, nanoparticles can be manufactured to release active substances in response to variations in the pH, moisture content, or temperature of the skin. This method maximizes the effectiveness of active compounds by enabling their precise and regulated release. Virtual product testing is made possible by the integration of augmented reality technology with nanomaterials. Realistic virtual makeup can be produced with nanoparticles, enabling consumers to see how various surfaces or hues will look on their skin before making a purchase. Furthermore, by combining sensors and diagnostics, nanomaterials can facilitate customized skincare. Cosmetic goods with nanosensors included in them can track current skin conditions, such as moisture content or UV exposure, and offer customized treatment advice. Nanotechnology and diagnostics work together to create targeted and efficient skincare products. Multidisciplinary research and collaboration are essential to the advancement of multifunctional nanomaterials for intricate cosmetic applications. Experts in a variety of disciplines, including materials science, chemistry, biology, dermatology, and engineering, can work together to provide innovative solutions and comprehensive approaches to problem-solving.

In conclusion, the targeted distribution of bioactive substances has advanced significantly with the use of nanocarriers in the cosmetics industry. These nanocarriers, which give active ingredients superior stability, controlled release, and improved penetration capabilities, are mainly liposomes, nanosomes, and nanoemulsions. The ability of these nanocarriers to transport both hydrophilic and hydrophobic active substances makes them dynamic platforms that can be used to tailor formulations to the unique needs of dermatologists. Inorganic nanomaterials like quantum dots and metal oxides have also made a difference. While quantum dots provide modularly luminosity for the visual attractiveness and enhancement of skincare products, zinc oxide and titanium dioxide, as metal oxides, offer sufficient protection against ultraviolet light. The paradigms governing cosmetic formulations and their interactions with skin substrates may shift as a result of these innovative materials. Working together makes it easier to share viewpoints, knowledge, and experience, which generates fresh ideas and a greater understanding of the potential and problems involved in creating multifunctional nanomaterials for cosmetics.

References

Ahmad A, Ahsan H (2020) Lipid-based formulations in cosmeceuticals and biopharmaceuticals. Biomed Dermatol 4:1–10

Aithal PS, Aithal S (2018) Study of various general-purpose technologies and their comparison towards developing sustainable society. Int J Manag Technol Soc Sci 3(2):16–33

Akinribide OJ, Mekgwe GN, Akinwamide SO, Gamaoun F, Abeykoon C, Johnson OT, Olubambi PA (2022) A review on optical properties and application of transparent ceramics. J Mater Res Technol 21:712–738

Alavi M, Mozafari MR, Hamblin MR, Hamidi M, Hajimolaali M, Katouzian I (2022) Industrial-scale methods for the manufacture of liposomes and nanoliposomes: pharmaceutical, cosmetic, and nutraceutical aspects. Micro Nano Bio Aspects 1(2):26–35

Algahtani MS, Ahmad MZ, Ahmad J (2020) Nanoemulsion loaded polymeric hydrogel for topical delivery of curcumin in psoriasis. J Drug Deliv Sci Technol 59:101847

Arshad M (2018) Cogeneration policy to contest climate changes and the future of sustainable energy. In: IOP conference series: earth and environmental science, vol 200. IOP Publishing

Arshad M (2024) Sustainability of renewable energy options as compared to coal-fired power plants in Pakistan. Green Low-Carbon Economy 2(3):162–173

Arshad M, Abbas M (2018) Water sustainability issues in biofuel production. In: Perspectives on water usage for biofuels production: aquatic contamination and climate change. Springer, Cham, pp 55–76

Arshad M, Adil M, Sikandar A, Hussain T (2014) Exploitation of meat industry by-products for biodiesel production: Pakistan's perspective. Pakistan J Life Soc Sci 12:120–125

Arshad M, Bano I, Younus M, Khan A, Rahman A (2018) Health concerns associated with biofuel production. In: Perspectives on water usage for biofuels production: aquatic contamination and climate change. Springer, pp 97–105

Arshad M, Syed-Hassan SSA, Masood R, Ansari AR, Mumtaz A, Rahman A, Batool S, Atif FA, Muhammad A, Rehman AU (2023) Utilization of waste animal fat for sustainable biodiesel production. In: Climate changes mitigation and sustainable bioenergy harvest through animal waste: sustainable environmental implications of animal waste. Springer, Cham, pp 149–169

Aslam A, Bahadar A, Liaquat R, Saleem M, Waqas A, Zwawi M (2021) Algae as an attractive source for cosmetics to counter environmental stress. Sci Total Environ 772:144905

Ayardulabi R, Khamespanah E, Abbasinia S, Ehtesabi H (2021) Point-of-care applications of smartphone-based microscopy. Sensors Actuators A Phys 331:113048

Barkat MA, Harshita FN, Beg S, Pottoo FH, Garg A, Singh SP, Ahmad FJ (2018) Silver nanoparticles and their antimicrobial applications. Curr Nanomed (Formerly: Recent Patents on Nanomedicine) 8(3):215–224

Barradas TN, de Holanda e Silva KG (2021) Nanoemulsions of essential oils to improve solubility, stability and permeability: a review. Environ Chem Lett 19(2):1153–1171

Bevacqua E, Occhiuzzi MA, Grande F, Tucci P (2023) TiO_2-NPs toxicity and safety: an update of the findings published over the last six years. Mini Rev Med Chem 23(9):1050–1057

Bilderback S (2024) Integrating training for organizational sustainability: the application of sustainable development goals globally. Euro J Training Dev 48(7/8):730–748

Bittner Fialová S, Rendeková K, Mučaji P, Nagy M, Slobodníková L (2021) Antibacterial activity of medicinal plants and their constituents in the context of skin and wound infections, considering European legislation and folk medicine—a review. Int J Mol Sci 22(19):10746

Borrego-Sánchez A, Muñoz-Santiburcio D, Viseras C, Hernández-Laguna A, Sainz-Díaz CI (2022) Melatonin/nanoclay hybrids for skin delivery. Appl Clay Sci 218:106417

Buchberger AR, DeLaney K, Johnson J, Li L (2017) Mass spectrometry imaging: a review of emerging advancements and future insights. Anal Chem 90(1):240

Catalano R, Masion A, Ziarelli F, Slomberg D, Laisney J, Unrine JM, Labille J (2020) Optimizing the dispersion of nanoparticulate TiO_2-based UV filters in a non-polar medium used in sunscreen formulations–the roles of surfactants and particle coatings. Colloids Surf A Physicochem Eng Asp 599:124792

Chen Y, Feng X (2022) Gold nanoparticles for skin drug delivery. Int J Pharm 625:122122

Choi SJ, McClements DJ (2020) Nanoemulsions as delivery systems for lipophilic nutraceuticals: strategies for improving their formulation, stability, functionality and bioavailability. Food Sci Biotechnol 29:149–168

Choo A, Delac DM, Messer LB (2001) Oral hygiene measures and promotion: review and considerations. Aust Dent J 46(3):166–173

Dabiri R, Hasanpouri Notash N (2020) Evaluation of geofoam effects on seismic response in cantilever retaining wall. Geotech Geol Eng 38:2097–2116

Darvin ME (2023) Optical methods for non-invasive determination of skin penetration: current trends, advances, possibilities, prospects, and translation into in vivo human studies. Pharmaceutics 15(9):2272

Dhawan S, Sharma P, Nanda S (2020) Cosmetic nanoformulations and their intended use. In: Nanocosmetics. Elsevier, pp 141–169

Dubey SK, Dey A, Singhvi G, Pandey MM, Singh V, Kesharwani P (2022) Emerging trends of nanotechnology in advanced cosmetics. Colloids Surf B: Biointerfaces 214:112440

Dunaway S, Odin R, Zhou L, Ji L, Zhang Y, Kadekaro AL (2018) Natural antioxidants: multiple mechanisms to protect skin from solar radiation. Front Pharmacol 9:392

Enyoh CE, Shafea L, Verla AW, Verla EN, Qingyue W, Chowdhury T, Paredes M (2020) Microplastics exposure routes and toxicity studies to ecosystems: an overview. Environ Anal Health Toxicol 35(1):e2020004

Fadoju O (2018) Genetic and systemic toxicity induced by titanium dioxide and zinc oxide nanoparticles and their mixture in somatic and germ cells of mice

Filipe P, Silva J, Silva R, Cirne de Castro J, Marques Gomes M, Alves L, Santus R, Pinheiro T (2009) Stratum corneum is an effective barrier to TiO_2 and ZnO nanoparticle percutaneous absorption. Skin Pharmacol Physiol 22(5):266–275

Fiuza S, Marques T, Padin I, Carvalho MT, Veiga N, Fernandes JCH, Fernandes GVO, Couto P (2024) Histological, clinical assessment, and treatment of a permanent filler complication in the upper lip: a case report with 16-year follow-up. Cosmetics 11(2):50

Fröhlich E, Steinbiß K (2021) Circular economy applies to beauty industry. In: Responsible consumption and production. Springer, Berlin/Heidelberg, pp 1–11. https://doi.org/10.1007/978-3-319-71062-4_119-1

Fuchs C, Pham L, Henderson J, Stalnaker KJ, Anderson RR, Tam J (2021) Multi-faceted enhancement of full-thickness skin wound healing by treatment with autologous micro skin tissue columns. Sci Rep 11(1):1688

Fujihara J, Nishimoto N (2024) Review of zinc oxide nanoparticles: toxicokinetics, tissue distribution for various exposure routes, toxicological effects, toxicity mechanism in mammals, and an approach for toxicity reduction. Biol Trace Elem Res 202(1):9–23

Gallagher RP, Lee TK (2006) Adverse effects of ultraviolet radiation: a brief review. Prog Biophys Mol Biol 92(1):119–131

Geng W, Nakajima T, Takanashi H, Ohki A (2009) Analysis of carboxyl group in coal and coal aromaticity by Fourier transform infrared (FT-IR) spectrometry. Fuel 88(1):139–144

González-Minero FJ, Bravo-Díaz L, Moreno-Toral E (2023) Pharmacy and fragrances: traditional and current use of plants and their extracts. Cosmetics 10(6):157

Goodhead RM, Moger J, Galloway TS, Tyler CR (2015) Tracing engineered nanomaterials in biological tissues using coherent anti-stokes Raman scattering (CARS) microscopy–a critical review. Nanotoxicology 9(7):928–939

Gottardo S, Mech A, Drbohlavová J, Małyska A, Bøwadt S, Sintes JR, Rauscher H (2021) Towards safe and sustainable innovation in nanotechnology: state-of-play for smart nanomaterials. NanoImpact 21:100297

Graf BW (2012) Multimodal intravital imaging of tissue structure and cell dynamics in skin using integrated optical coherence and multiphoton microscopy. University of Illinois at Urbana-Champaign

Gupta V, Mohapatra S, Mishra H, Farooq U, Kumar K, Ansari MJ, Iqbal Z (2022) Nanotechnology in cosmetics and cosmeceuticals—a review of latest advancements. Gels 8(3):173

Hameed A, Fatima GR, Malik K, Muqadas A, Fazal-ur-Rehman M (2019) Scope of nanotechnology in cosmetics: dermatology and skin care products. J Med Chem Sci 2:9–16. https://doi.org/10.26655/JMCHEMSCI.2019.6.2

Han J-W, Zheng H-F, Cui Y, Sun L-D, Ye D-Q, Hu Z, Xu J-H, Cai Z-M, Huang W, Zhao G-P (2009) Genome-wide association study in a Chinese Han population identifies nine new susceptibility loci for systemic lupus erythematosus. Nat Genet 41(11):1234–1237

Helmchen F, Denk W (2005) Deep tissue two-photon microscopy. Nat Methods 2(12):932–940

Hopewell J, Dvorak R, Kosior E (2009) Plastics recycling: challenges and opportunities. Philos Trans Royal Soc B Biol Sci 364:2115–2126. https://doi.org/10.1098/rstb.2008.0311

Kakani S (2004) Material science. New Age International (P) Ltd., Publishers

Kim B, Cho HE, Moon SH, Ahn HJ, Bae S, Cho HD, An S (2020a) Transdermal delivery systems in cosmetics. Biomed Dermatol 4:1–12

Kim S, Cho DH, Kweon DK, Jang EH, Hong JY, Lim ST (2020b) Improvement of mechanical properties of orodispersible hyaluronic acid film by carboxymethyl cellulose addition. Food Sci Biotechnol 29:1233–1239

Kotta S, Aldawsari HM, Badr-Eldin SM, Nair AB, Yt K (2022) Progress in polymeric micelles for drug delivery applications. Pharmaceutics 14(8):1636

Kouassi MC, Grisel M, Gore E (2022) Multifunctional active ingredient-based delivery systems for skincare formulations: a review. Colloids Surf B: Biointerfaces 217:112676

Li X, Fujimoto JG (2004) Optical coherence tomography. In: Encyclopedia of biomaterials and biomedical engineering, vol 1. Informa Healthcare, pp 1120–1132

Li Z, Li J, Zhang D, Zhu X, Ye Y, Zhao Y (2020) A novel dual-channel fluorescent probe for nitroxyl detection and its application in HepG-2 cells. Sensors Actuators B Chem 312:127944

Lin LL, Grice JE, Butler MK, Zvyagin AV, Becker W, Robertson TA, Soyer HP, Roberts MS, Prow TW (2011) Time-correlated single photon counting for simultaneous monitoring of zinc oxide nanoparticles and NAD (P) H in intact and barrier-disrupted volunteer skin. Pharm Res 28:2920–2930

Lin LL, Nufer KL, Tomihara S, Prow TW (2015) Non-invasive nanoparticle imaging technologies for cosmetic and skin care products. Cosmetics 2(3):196–210

Longo C, Guida S, Mirra M, Pampena R, Ciardo S, Bassoli S, Casari A, Rongioletti F, Spadafora M, Chester J (2024) Dermatoscopy and reflectance confocal microscopy for basal cell carcinoma diagnosis and diagnosis prediction score: a prospective and multicenter study on 1005 lesions. J Am Acad Dermatol 90(5):994–1001

Lu Y, Lin J, Wang L, Zhang L, Cai C (2020) Self-assembly of copolymer micelles: higher-level assembly for constructing hierarchical structure. Chem Rev 120(9):4111–4140

Mabrouk M, Das DB, Salem ZA, Beherei HH (2021) Nanomaterials for biomedical applications: production, characterisations, recent trends and difficulties. Molecules 26(4):1077

Magri A, Petriccione M, Cerqueira MA, Gutierrez TJ (2020) Self-assembled lipids for food applications: a review. Adv Colloid Interf Sci 285:102279

Meftahi A, Samyn P, Geravand SA, Khajavi R, Alibkhshi S, Bechelany M, Barhoum A (2022) Nanocelluloses as skin biocompatible materials for skincare, cosmetics, and healthcare: formulations, regulations, and emerging applications. Carbohydr Polym 278:118956

Mesraoua B, Kissani N, Deleu D, Elsheikh L, Ali M, Melikyan G, Asadi-Pooya AA (2021) Complementary and alternative medicine (CAM) for epilepsy treatment in the Middle East and North Africa (MENA) region. Epilepsy Res 170:106538

Mohammed A, Abdullah A (2018) Scanning electron microscopy (SEM): a review. In: Proceedings of the 2018 international conference on hydraulics and pneumatics—HERVEX, Băile Govora, Romania 2018, pp 7–9

Mohd-Setapar SH, John CP, Mohd-Nasir H, Azim MM, Ahmad A, Alshammari MB (2022) Application of nanotechnology incorporated with natural ingredients in natural cosmetics. Cosmetics 9(6):110

Moreira LM, Silveira L Jr, Santos FV, Lyon JP, Rocha R, Zângaro RA, Villaverde AB, Pacheco MT (2008) Raman spectroscopy: a powerful technique for biochemical analysis and diagnosis. Spectroscopy 22(1):1–19

Navarrete-Dechent C, Cordova M, Sahu A, Liopyris K, Rishpon A, Chen C, Chen CSJ (2021) Optical imaging guided-'precision'biopsy of skin tumors: a novel approach for targeted sampling and histopathologic correlation. Arch Dermatol Res 313:517–529

Negi S, Bala L, Shukla S, Chopra D (2022) Tattoo inks are toxicological risks to human health: a systematic review of their ingredients, fate inside skin, toxicity due to polycyclic aromatic hydrocarbons, primary aromatic amines, metals, and overview of regulatory frameworks. Toxicol Ind Health 38(7):417–434

Nickels KC, Zaccariello MJ, Hamiwka LD, Wirrell EC (2016) Cognitive and neurodevelopmental comorbidities in paediatric epilepsy. Nat Rev Neurol 12(8):465–476

Nikam NR, Patil PR, Vakhariya RR, Magdum CS (2020) Liposomes: a novel drug delivery system—an overview. Asian J Pharm Res 10(1):23–28

Nsairat H, Khater D, Sayed U, Odeh F, Al Bawab A, Alshaer W (2022) Liposomes: structure, composition, types, and clinical applications. Heliyon 8(5):e09394

Pantelimon FV, Georgescu TM, Posedaru BŞ (2020) The impact of mobile e-commerce on gdp: a comparative analysis between Romania and Germany and how covid-19 influences the e-commerce activity worldwide. Inform Econom 24(2):27–41

Parhi R, Mandru A (2021) Enhancement of skin permeability with thermal ablation techniques: concept to commercial products. Drug Deliv Transl Res 11:817–841

Perinelli DR, Cespi M, Lorusso N, Palmieri GF, Bonacucina G, Blasi P (2020) Surfactant self-assembling and critical micelle concentration: one approach fits all? Langmuir 36(21):5745–5753

Powar NS, Chaeeun A, In SI (2024) A comprehensive exploration of nanomaterials in cosmetics. IEEE Nanotechnol Mag 18:15

Rabiee N, Ahmadi S, Akhavan O, Luque R (2022) Silver and gold nanoparticles for antimicrobial purposes against multi-drug resistance bacteria. Materials 15(5):1799

Rachman ES, Soeratri W, Erawati T (2023) Characteristics and physical stability of Nanoemulsion as a vehicle for anti-aging cosmetics: a systematic review. J Farm Ilmu Kefarmasian Indones 10(1):62–85

Raha S, Ahmaruzzaman M (2022) ZnO nanostructured materials and their potential applications: progress, challenges and perspectives. Nanoscale Adv 4(8):1868–1925

Rai VK, Mishra N, Yadav KS, Yadav NP (2018) Nanoemulsion as pharmaceutical carrier for dermal and transdermal drug delivery: formulation development, stability issues, basic considerations and applications. J Control Release 270:203–225

Ramanunny AK, Wadhwa S, Gulati M, Singh SK, Kapoor B, Dureja H, Pandey NK (2021) Nanocarriers for treatment of dermatological diseases: principle, perspective and practices. Eur J Pharmacol 890:173691

Rao J, Dragulescu-Andrasi A, Yao H (2007) Fluorescence imaging in vivo: recent advances. Curr Opin Biotechnol 18(1):17–25

Reddy KR, Paul MJ, Basha SKM (2016) Ethnobotanical study of Durgamkonda of Veligonda Hill range, eastern Ghats, Andhra Pradesh. Int J Latest Trends Eng Technol 8(3):258–262
Rocha ASV (2020) Formulation of sustainable concentrated rinse-off products (doctoral dissertation)
Roshchina VV (2020) Fluorescence of living plant cells for Phytomedicine preparations. CRC Press
Roy A, Nishchaya K, Rai VK (2022) Nanoemulsion-based dosage forms for the transdermal drug delivery applications: a review of recent advances. Expert Opin Drug Deliv 19(3):303–319
Saginala K, Barsouk A, Aluru JS, Rawla P, Barsouk A (2021) Epidemiology of melanoma. Med Sci 9(4):63
Santos AC, Morais F, Simões A, Pereira I, Sequeira JA, Pereira-Silva M, Veiga F, Ribeiro A (2019) Nanotechnology for the development of new cosmetic formulations. Expert Opin Drug Deliv 16(4):313–330
Santos JS, Barradas TN, Tavares GD (2022) Advances in nanotechnology-based hair care products applied to hair shaft and hair scalp disorders. Int J Cosmet Sci 44(3):320–332
Sathyaseelan S, Rao BH, Anushmati S (2024) Cosmeceuticals: a transit state from synthetic to natural. Indian J Pharmacol 56(1):42–51
Schwarzer RA, Field DP, Adams BL, Kumar M, Schwartz AJ (2009) Present state of electron backscatter diffraction and prospective developments. In: Electron backscatter diffraction in materials science. Springer, Boston, pp 1–20
Sharma A, Mohapatra H, Arora K, Babbar R, Arora R, Arora P, Sindhu RK (2023) Bioactive compound-loaded nanocarriers for hair growth promotion: current status and future perspectives. Plan Theory 12(21):3739
Souto EB, Fernandes AR, Martins-Gomes C, Coutinho TE, Durazzo A, Lucarini M, Santini A (2020) Nanomaterials for skin delivery of cosmeceuticals and pharmaceuticals. Appl Sci 10(5):1594
Starke M (2017) The visual appearance of skin in motion picture. Bachelor Thesis, Media University Stuttgart
Syperek P, Wade S (eds) (2023) Oceans. MIT Press
Tang CY, Yang Z (2017) Transmission electron microscopy (TEM). In: Membrane characterization. Elsevier, pp 145–159
Tata BVR, Raj B (1998) Confocal laser scanning microscopy: applications in material science and technology. Bull Mater Sci 21:263–278
Unger N, Dechand S, Bonneau J, Fahl S, Perl H, Goldberg I, Smith M (2015) SoK: secure messaging. In: 2015 IEEE symposium on security and privacy
Vaz G, Clementino A, Mitsou E, Ferrari E, Buttini F, Sissa C, Dora CL (2022) In vitro evaluation of curcumin-and quercetin-loaded nanoemulsions for intranasal administration: effect of surface charge and viscosity. Pharmaceutics 14(1):194
Viseras C, Carazo E, Borrego-Sánchez A, García-Villén F, Sánchez-Espejo R, Cerezo P, Aguzzi C (2019) Clay minerals in skin drug delivery. Clay Clay Miner 67(1):59–71
Withers PJ, Bouman C, Carmignato S, Cnudde V, Grimaldi D, Hagen CK, Stock SR (2021) X-ray computed tomography. Nat Rev Methods Primer 1(1):18
Wizenty J, Schumann T, Theil D, Stockmann M, Pratschke J, Tacke F, Aigner F, Wuensch T (2020) Recent advances and the potential for clinical use of autofluorescence detection of extra-ophthalmic tissues. Molecules 25(9):2095
Yang S, Liu L, Han J, Tang Y (2020) Encapsulating plant ingredients for dermocosmetic application: An updated review of delivery systems and characterization techniques. Int J Cosmet Sci 42(1):16–28

Chapter 16
Microbiological Safety and Cosmetics

Ayesha Nazir, Sadia Javed, Asmara Ahmad, Sibtain Ahmed, and Ameer Fawad Zahoor

Contents

A. Nazir · S. Javed (✉) · A. Ahmad
Department of Biochemistry, Government College University Faisalabad, Faisalabad, Pakistan
e-mail: sadiajaved@gcuf.edu.pk

S. Ahmed
Department of Biochemistry, Bahauddin Zakariya University, Multan, Pakistan

A. F. Zahoor
Department of Chemistry, Government College University, Faisalabad, Pakistan

S. Javed et al. (eds.), *Sustainable Cosmeceuticals*,
https://doi.org/10.1007/978-3-031-86087-4_16

Abstract Cosmetics are items used to improve or enhance the appearance of the face and body. They also include personal hygiene items. Cosmetics cover a broad spectrum of products, including cosmetics, skincare, hair care, and perfumes. The main purposes of cosmetics are to protect, cleanse, and beautify the skin and hair. Because of the possible hazards connected with microbial contamination, ensuring their microbiological safety is critical. These products, which come in a variety of forms and are extensively used across the world, seek to enhance skin function and appearance. However, questions have been raised about product quality, safety, and the presence of pathogens, especially antibiotic-resistant bacteria. To ensure the microbiological safety of cosmetic goods, strict monitoring and restrictions for manufacture and sale are required. Cosmetic products can be subjected to microbial contamination from a variety of sources throughout the production process and everyday usage, including raw ingredients, processing equipment, the environment, water, and manufacturing staff. Manufacturers must prioritize quality control, regulatory compliance, research, education, and timely reporting of adverse product incidents to ensure microbiological safety. Efforts to combat microbiological contamination will increase customer trust in cosmetic goods and assure their sustained safety. Green packaging can have a strong impact in this because it uses eco-friendly designs and biodegradable, sustainable materials to decrease waste and lessen environmental effects, which is in line with the Sustainable Development Goals (SDGs). In addition to guaranteeing product safety and microbial prevention, it favors SDG objectives for climate action, responsible consumerism, and environmental conservation by substituting traditional plastics with recyclable and biodegradable alternatives.

Keywords Microbial contamination · Cosmetic preservation · Pathogenic microorganisms · Microbiological testing · Cosmetic spoilage · Green packaging

Abbreviations

CFU	Colony-forming unit
FDA	Food and Drug Administration
GMP	Good manufacturing practice
IATA	Integrated Approaches to Testing and Assessment
LATA	Approaches to Testing and Assessment
MQM	Microbiological Quality Management
PCPs	Personal Care Products
QA	Quality Assurance
RAPEX	Rapid Exchange of Information System
SDGs	Sustainable Development Goals

16.1 Introduction

Cosmetics are defined by the Federal Food, Drug, and Cosmetic Act as products that are meant to be applied to the human body or to its various parts, such as the skin, hair, nails, and eyes. This can involve rubbing, pouring, sprinkling, or spraying the product. The Food and Drug Administration (FDA) defines cosmetics as "a product (excluding pure soap) intended to be applied to the human body for cleansing, beautifying, promoting attractiveness, or altering the appearance" (Alharbi and Alhashim 2021b). The primary purposes of cosmetics are to cleanse, beautify, promote attractiveness, or alter one's appearance. Some cosmetics can be used as components in the formulation of other cosmetic products. And it is worth mentioning that the definition of cosmetics excludes soap (Arshad 2018; Bashir and Lambert 2020). Cosmetics, originating from the Greek word "adorn," are products made from synthetic and natural sources that aim to enhance the texture and functioning of the skin. Throughout history, various substances derived from natural sources, including olive oil, peppermint, lavender, rosemary, cedar, almond oil, rose petals, aloe, henna, and sesame oil, have been utilized as cosmetics. In modern times, cosmetic products have become an integral part of daily skincare routines (Fig. 16.1).

Fig. 16.1 Most commonly used items as cosmetics

Cosmetics can be consumed orally, inhaled, or applied topically. Every technique has a unique function in boosting appearance. Because of their efficiency and practicality, some of these applications are utilized more frequently than others. These widely used applications address a range of skincare, haircare, and general aesthetic requirements. These items come in different forms such as suspensions, gels, foams, ointments, sprays, lotions, pastes, creams, and aerosols. Personal care cosmetic products are designed to be applied by rubbing, sprinkling, pouring, or spraying on various body parts, including the hair, skin, and nails, for the purposes of cleansing, beautification, enhancing attractiveness, or altering physical appearance. Cosmetics and equipment provide the ideal conditions for the proliferation of viral, parasite, and bacterial organisms that contribute to and transmit illness (Alharbi and Alhashim 2021b). Therefore, it is crucial for these preparations to be stable, effective, and suitable for topical use (Almukainzi et al. 2022).

While the majority of cosmetic and personal care products are generally safe, there have been growing concerns regarding product quality and marketing practices. It is essential to implement stringent monitoring and regulations for the manufacturing and sale of these products. Collecting and evaluating data on any adverse effects caused by these products is crucial to ensure their safety for consumers. Contamination with microorganisms is a significant concern that can lead to product spoilage and pose health risks to users. Therefore, it is critical to address these concerns and take necessary measures to maintain the quality and safety of cosmetic products (Jairoun et al. 2020). Disease-causing organisms can get into cosmetic items from a wide range of sources, but antibiotic-resistant bacteria are the most harmful (Arshad et al. 2022; Elzoughby 2022).

The quality of pharmaceutical and cosmetic products can be significantly influenced by the presence of microorganisms. Throughout the manufacturing process and daily use, these products can be exposed to various sources of microbial contamination (Gilmore and Denyer 2023). Such instability may result in changes to the physicochemical qualities of the product, which may compromise its shelf life or, in certain situations, pose life-threatening risks (Smart and Spooner 1972). The contamination sources include raw materials, processing equipment, the environment, water used in manufacturing, and manufacturing personnel. Controlling the water activity of a product is one of the unique and different ways for minimizing microbial instability. In this overview, we summarize all elements of product stability, from contamination sources to processing conditions, preventative techniques, how microorganisms acquire resistance to preservatives, and novel approaches to this issue (Dao et al. 2018).To ensure the quality and safety of personal care products and cosmetics, manufacturers must focus on applying quality control and assurance best practices in collaboration with government inspectors. It is important to emphasize good manufacturing practices, regulatory compliance, education, research, and the notification of any adverse situations associated with these products (Dao et al. 2018).

16.1.1 Use of Sustainable and Biodegradable Constituents in Cosmetics

Cosmetics sustainability and biodegradability are essential for lowering pollution levels in the environment and promoting environmentally friendly manufacturing methods. By using biodegradable chemicals, cosmetic formulations can naturally disintegrate, keeping waste from building up in ecosystems over time. The components, which are often derived from renewable resources, contribute in achieving sustainability objectives and lessen the environmental effect of a product (Rahimpour and Hamishehkar 2012).

Biodegradable ingredients in cosmetics can decompose naturally in the environment, lowering pollutants and their negative ecological effects. Natural sources such as plants and minerals tend to be the source of these components. Such as, bacteria create microbial biosurfactants, which are utilized in cosmetics as environmentally acceptable substitutes for synthetic surfactants. They are less harmful and biodegradable (Ahmadi-Ashtiani et al. 2020). Utilizing renewable resources, reducing waste, and decreasing carbon footprints are the main goals of sustainability in cosmetics. This includes choosing recyclable or biodegradable packaging and obtaining materials ethically. By using sustainable supplier selection processes, companies may collaborate with suppliers that adhere to high ethical and environmental standards, ensuring that sustainability is promoted throughout the supply chain (Acerbi et al. 2023).

However, because natural compounds cannot maintain the same shelf life or efficacy as synthetic preservatives, it might be difficult to ensure microbiological safety in biodegradable cosmetics. Natural preservatives like essential oils are being investigated as a solution to this problem. They preserve the microbiological safety of cosmetics while upholding sustainability standards. Without using artificial chemicals, these preservatives aid in halting the growth of dangerous microorganisms (Rahimpour and Hamishehkar 2012). Hence, it is crucial to balance sustainability, biodegradability, and microbiological safety when developing cosmetics that are safe for people and the environment. To achieve these two objectives, the cosmetics industry is increasingly looking to cutting-edge alternatives like bio-based polymers and natural preservatives (Bikiaris et al. 2024).

Preservation and formulation are key ingredients for the biodegradability and sustainability of cosmetics. Cosmetics need preservatives to prevent microbial contamination, which can lead to spoilage and health problems. However, many conventional preservatives are environmentally harmful and non-biodegradable. Researchers are exploring the antibacterial properties of natural preservatives, such as plant extracts and essential oils. While these alternatives are more environmentally friendly, they may need larger doses or specific combinations to match the effectiveness of synthetic preservatives. Sustainability, biodegradability, and microbiological safety must all be balanced. Comprehensive evaluation and creative formulation techniques are necessary since natural preservatives cannot always provide the same level of protection as synthetic substitutes. These ingredients' efficacy and safety are assessed using integrated approaches to testing and assessment (IATA), which makes sure they satisfy sustainability goals and microbiological safety regulations (Arshad 2024; Ashtiani et al. 2016).

16.2 Meta-Analysis of Studies Unraveling the Impact of Microbial Contamination in Cosmetics

Microbial vulnerability in cosmetic and pharmaceutical products is an important factor in assessing product performance and quality. It can result in alterations in the physical and chemical properties of the cosmetic product, affecting its shelf life and potentially posing a risk to consumers. Both sterile and non-sterile dosage forms are susceptible to microbial instability. To counter this issue, a broad range of preservatives and preservative combinations are used. However, there have been reports of microbial resistance to these preservatives in certain cases. Water activity control of the product has emerged as a novel and alternative approach to minimize microbial instability (Dao et al. 2018). A noteworthy concern is the absence of production and expiration dates on cosmetic products, which is surprising considering the effectiveness of cosmetic preservatives, which can diminish over time. Furthermore, cosmetics comprise inorganic and organic compounds, essential minerals, and growth factors, and they are frequently kept in moist environments such as bathrooms, creating favorable conditions for the proliferation of microorganisms (Jairoun et al. 2020).

In this study, a hundred products that were cosmetics and personal care products were observed. The presence of microbial contamination, including *Staphylococcus aureus*, *Candida albicans*, *Pseudomonas aeruginosa*, *Escherichia coli*, yeast and mold, and aerobic mesophilic bacteria, was assessed for each product. A prominent amount of yeast and mold was also found. The contamination levels were compared to the European guidelines. The results showed that 15% of the samples exceeded the maximum limit of 1000 CFU/g for aerobic mesophilic bacteria, while 13% of the samples exceeded the limit for yeast and mold. Nine samples were contaminated with both yeast and mold and aerobic mesophilic bacteria. However none of the samples that were tested was found to be contaminated with *Candida albicans, Staphylococcus aureus, Escherichia coli, or Pseudomonas aeruginosa* (Jairoun et al. 2020).

The dataset of the study shows a comprehensive list of preservatives commonly utilized in cosmetics. These preservatives are accompanied by their maximum allowable concentrations, as dictated by the European Union's Cosmetics Regulation. The preservatives featured include formaldehyde and formaldehyde-releasing agents, parabens, phenoxyethanol, and others. Formaldehyde and formaldehyde-releasing agents serve as preservatives in cosmetics, acting to prevent the growth of fungi and bacteria. However, these preservatives are also known for their allergenic potential and can lead to skin irritation and other health concerns.

The maximum allowable concentration for formaldehyde in cosmetics is 0.2%, while formaldehyde-releasing agents must not exceed 0.05% in product formulations. Parabens are another common group of preservatives utilized to inhibit microbial growth in cosmetics. These preservatives demonstrate efficacy even at low concentrations and are relatively cost-effective. Nonetheless, concerns have arisen regarding their potential impact on human health, including endocrine disruption and a possible association with breast cancer. Maximum allowable concentrations for parabens in cosmetics range from 0.19% to 0.8%, depending on the specific type of paraben employed. Phenoxyethanol is often employed as an alternative preservative to parabens in cosmetics. It exhibits broad spectrum antimicrobial activity and is considered safe when used at low concentrations. The maximum allowable concentration for phenoxyethanol in cosmetics is set at 1%. In addition to listing the preservatives, the dataset also provides insights into the types of products in which each preservative is commonly utilized. For instance, rinse-off products like shampoos and shower gels often incorporate methylisothiazolinone, while leave-on products like moisturizers and sunscreens may contain imidazolidinyl urea. Overall, the dataset serves as a valuable resource for cosmetic formulators and manufacturers, ensuring compliance with safe and effective preservative usage. It emphasizes the importance of proper preservation to prevent the growth of harmful microorganisms, thereby ensuring consumer safety. Regarding the antimicrobial effects of cosmetic products, the dataset presents a detailed classification of various product categories. The classification encompasses distinct product classes such as leave-on products, rinse-off products, and oral hygiene products. Each product class is associated with specific applications, target microorganisms, and active ingredients. Leave-on products, for example, primarily target Gram-positive and Gram-negative bacteria. Active ingredients like triclosan, triclocarban, and zinc ricinoleate are commonly employed to combat these microorganisms. Rinse-off products, on the other hand, aim to address microorganisms such as *Streptococcus* mutans and other oral bacteria. Active ingredients utilized in rinse-off products include fluoride, chlorhexidine, and essential oils. This comprehensive classification provides valuable information for cosmetic formulators and manufacturers, enabling them to select and incorporate antimicrobial ingredients at safe and effective levels. By preventing the growth of harmful microorganisms, the proper use of antimicrobial ingredients ensures the overall safety and quality of cosmetic products. The most frequent microorganisms found in cosmetics include *Staphylococcus aureus, Escherichia coli, Klebsiella oxytoca, Burkholderia cepacia, Pseudomonas aeruginosa, Candida albicans, Enterobacter gergoviae, and Serratia marcescens,* as well as other fungi, yeast, and bacteria. These microbes can pose a risk of microbial infection if they are present in cosmetics at high levels. It was stated that microbial contamination can occur during the manufacturing phase (primary contamination) or during the use of consumer (secondary contamination); therefore, it is necessary to monitor and identify all conceivable sources of contamination to prevent the growth of harmful microorganisms (Arshad et al. 2023; Halla et al. 2018).

In a different study conducted in Egypt, 54 samples were taken, out of which 36 talcum powders and 18 body lotions were thoroughly analyzed with the aim of determining the overall levels of aerobic bacteria, coliform, and fungi in them. Additionally, anaerobic bacteria counts were also performed specifically for talcum powder samples, and tests were conducted to identify potentially harmful bacteria across all samples that were tested. The results indicated that talcum powders exhibited higher levels of contamination with bacteria and fungi compared to body lotions. Over 40% of the tested body lotions showed either no viable bacteria or less than 100 colony-forming units per gram (c.f.u./g). In contrast, all tested talcum powders contained more than 100 c.f.u./g. Moreover, 30% of the talcum powders were contaminated with 104 c.f.u./g, while none of the body lotions exhibited contamination to that extent. None of the body lotions showed the presence of coliforms, whereas 17% of the talcum powder samples contained coliforms within the range of 230–500 c.f.u./g. *Staphylococcus* species were detected in 18 samples of both talcum powders and body lotions, with three of these identified as *Staphylococcus aureus*. Among the talcum powder samples, three contained *Escherichia coli*, two contained *Enterobacter agglomerans*, and one contained *Citrobacter freundii*. In terms of fungal counts, 70% of the body lotions showed no fungal growth, while 83% of the talcum powders exhibited fungal contamination. Most of the contaminated talcum powders contained more than 100 fungal cells per gram. Regarding anaerobic bacteria counts in talcum powders, half of the samples showed either no counts or less than 100 c.f.u./g. Four samples were contaminated with *Clostridium perfringens*, although *C. tetani* was not detected in any of the samples. These findings demonstrate that talcum powders, compared to body lotions, are more prone to bacterial and fungal contamination. The presence of coliforms in talcum powders raises concerns about potential fecal contamination. The identification of *Staphylococcus aureus* and certain enteric bacteria highlights the importance of monitoring and maintaining product hygiene. These results emphasize the need for stringent quality control measures and regular testing to ensure the safety and microbial integrity of cosmetic products, particularly talcum powders (Jairoun et al. 2020).

Another study was conducted for used cosmetic products from a group of 500 cosmetic users. From these samples, a total of 101 bacteria (20.2%) were isolated. Utilizing the Vitek 2 device for identification purposes, the isolated bacteria were detected, *S. epidermidis (47, 46.5%), E. coli (16, 15.8%), S. hominis (17, 16.8%), K. pneumoniae (11, 10.9%), S. aureus (6, 5.9%)*, and *P. aeruginosa (4, 4.1%)*. Further analysis of antibiotic resistance phenotypes revealed that one isolate was identified as methicillin-resistant *S. aureus*, and it tested positive for the mecA gene. Among the Gram-negative bacteria isolates, ten exhibited resistance to carbapenem and extended-spectrum beta-lactamase. Notably, the blaOXA-48 gene, associated with resistance to carbapenems, was detected in three *K. pneumoniae* strains among these isolates. According to another study, it was concluded that cosmetic products could potentially contribute to severe infections related to factors such as poor personal hygiene, prolonged product usage, and bacterial contamination in the environment. The presence of MRSA and the identification of carbapenem and extended-spectrum beta-lactamase-resistant Gram-negative bacteria highlight the importance of hygiene practices and the need for vigilant monitoring of cosmetic products to prevent potential infections (Akgül and Bakan 2021; Arshad et al. 2014) (Fig. 16.2).

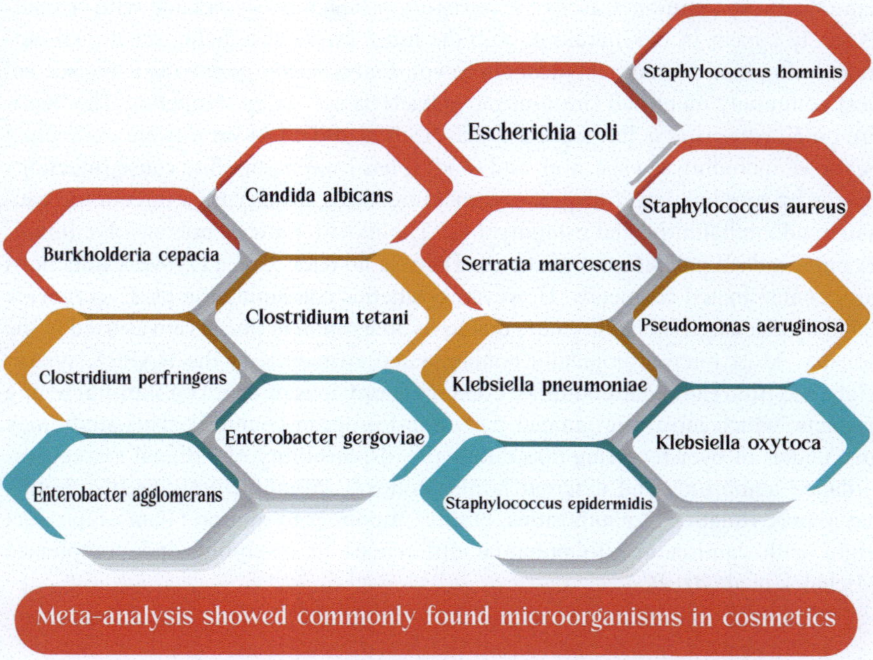

Fig. 16.2 An overview of commonly found microorganisms in cosmetics

From 2005 to 2018, the reports showed 104 cosmetics were at microbiological risk in the RAPEX database, with 103 intended for consumer use and 20 for children. Over half of cosmetics with microbiological risks were produced in countries like Germany, France, the Czech Republic, Greece, and India. Over 65% of the cosmetics were from one of the RAPEX group member states. Skin-cleaning products accounted for one-third of the recalled cosmetics, with shower and bath gels dominating. Skin care products were the most common, followed by hair and scalp products, makeup, and deodorants. The microbiological risk in 104 cosmetics was primarily related to Gram-negative bacteria, with over half (52.54%) being related to these bacteria. Gram-positive bacteria and fungi were observed in 5.93% and 1.69% of the products, respectively. Unspecified microbiological risks were found in 38.14% of the cases. In 67 out of 104 cases, products were withdrawn from the market, and 30 items were withdrawn from end-users. Fourteen cases ordered bans on sales, one reported a voluntary stop, eight consumers were informed, five products were destroyed, and import bans were placed on three products. The study found that most cosmetics contamination reports involve Gram-negative bacteria, specifically aerobic *Pseudomonas* and facultatively anaerobic *Enterobacter*. *Pseudomonas* and *Enterobacter* are found in various environments, including sewage, soil, water, and dairy products. In cosmetic products, *P. aeruginosa* is the most commonly found species, causing nosocomial infections and affecting the ear, burn wounds, respiratory tract, and urinary tract. Skin infections are less frequently described, with ecthyma granulosum, necrotizing fasciitis, and hot tub folliculitis being the most common. Cases of *P. aeruginosa* infections associated with cosmetics rarely appear in the literature, with the most severe case being a national outbreak in England and corneal ulcers in Egypt. *Enterobacter gergoviae* is the second most commonly identified Gram-negative bacteria species in cosmetics. The literature on *E. gergoviae* is limited, but it has been identified in various environmental locations, including sewage, soil, and food. It has been reported to cause infections such as bacteremia, lower respiratory tract infections, urinary tract infections, traumatic endophthalmitis, and osteomyelitis. The most reported cases involve immunocompromised patients and neonates. There is no data on *E. gergoviae* infections from contaminated cosmetics. However, cosmetics contamination by *E. gergoviae* is widely discussed in the literature due to its adaptation to preservatives used in the industry. Most microbiologically contaminated cosmetics in the RAPEX system originated from European countries, contrary to previous research on skin-lightening products, which mostly originated from non-European countries. This study may have under- or over-reporting bias due to the responsibility of national market surveillance authorities and external border controls. Manufacturers or distributors may report violations to authorities, but it's important to interpret data on product origin with caution due to potential differences in discipline and compliance (Michalek et al. 2019).

16.3 High-Quality Cosmetics Might Not Ensure Microbial Safety

Microbial contamination in cosmetics is a major worry and can result in large financial losses. Ingredients such as water can promote microbial growth. The results of the research revealed that both high-quality and low-quality cosmetics have bacterial and fungal development over the thresholds permitted by the FDA and other laws. High-quality cosmetics are not only pricey; they are created with high-quality raw ingredients, are stored and transported properly, and contain excellent preservatives to prevent microbiological development (Alshehrei 2023) (Figs. 16.3, 16.4 and 16.5).

16.4 Impact of Microorganisms in Sustainable Product Degradation

Microorganisms found in domestic surroundings are extremely adaptable to a wide range of situations and can degrade a wide range of product constituents. While mold development, color changes, foaming, and packaging anomalies such as bulging, leaking, or even bursting owing to gas generation are clear symptoms of severe contamination, more subtle modifications can also occur (Perry 2001).

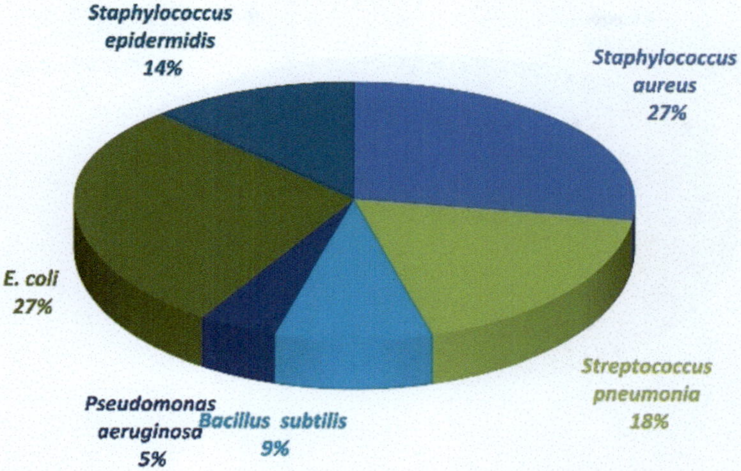

Fig. 16.3 Shows bacterial growth percentage in low-quality makeup products (Alshehrei 2023)

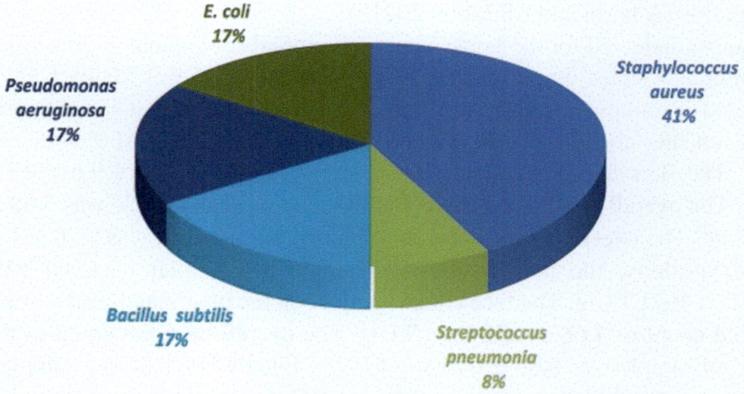

Fig. 16.4 Shows bacterial growth percentage in high-quality makeup products (Alshehrei 2023)

Microorganism contamination in products such as shampoo can result in unpleasant odors caused by the presence of yeasts and actinomycetes, as well as changes in the tactile qualities of the items. Cream formulations, for example, may experience visually unappealing changes in viscosity, which might reduce the overall performance of the product. Most cosmetic products contain a combination of organic and inorganic substances, moisturizers, vital minerals, and growth factors such as vitamins. These elements have the potential to foster the growth and multiplication of microbes. Unfortunately, the manufacture and expiration dates of beauty care items are frequently overlooked. As a result, the gradual loss of efficiency in cosmetic preservatives remains undetected. Furthermore, beauty establishments routinely interchange items without sanitary conditions. Moreover, microbial contamination is greatly increased while sharing tools and equipment among several people. Such

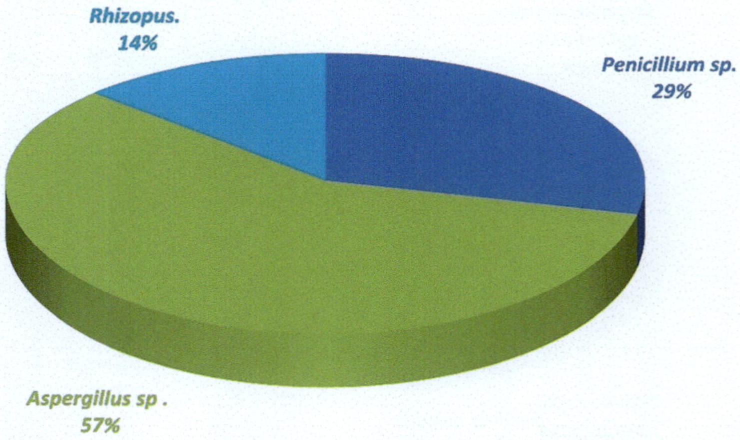

Fig. 16.5 Shows fungal growth percentage in low-quality makeup products (Alshehrei 2023)

as the usage of nail instruments, like clippers, might increase your risk of illnesses like hepatitis (Alharbi and Alhashim 2021b).

A study conducted for the assessment of microbial contamination in various cosmetic products states that a total bacterial count of 11.15 ± 1.51 log CFU/g was discovered in samples of body washing and cream. Hair cream samples and face washes, on the other hand, had a total mold and yeast count of 9.07 ± 1.04 log CFU/g. The skin smoothing cream had a total bacterial count of 9.6 ± 91.51 log CFU/g. The overall yeast and mold count in the hand soap sample was 3.68 ± 1.35 log CFU/g. The overall bacterial count in hair removal cream was 9.76 ± 1.48 log CFU/g. Similarly, the hair conditioner sample had a total bacterial count of 8.89 ± 1.51 log CFU/g. The face cream sample, on the other hand, had a total yeast and mold count of 7.68 ± 1.35 log CFU/g. The overall bacterial count in the hair styling gel sample was 6.89 ± 0.51 log CFU/g. Finally, the hair dye sample had a total yeast and mold count of 6.07 ± 1.04 log CFU/g. The foot cream sample had a total bacterial count of 7.15 ± 1.35 log CFU/g. And the lipstick sample had a total yeast and mold count of 4.68 ± 1.35 log CFU/g. These findings give useful insights into the levels of microbial contamination in the cosmetic goods under consideration, adding to a better knowledge of their safety and quality (Elzoughby 2022). It is essential to comprehend how microbes affect the sustainable breakdown of products. Their contribution to spoiling can have a major impact on the sustainability and quality of the product. We must develop methods for improving the life of products by addressing microbial effects. This will promote more sustainable activities and reduce environmental impact.

16.5 Risks Posed by Beauty Tools and Products

Numerous research studies in Nigeria, Iraq, and Iran have been conducted to explore the presence of microorganisms on salon beauty products, finding the prevalence of pathogenic bacteria and fungus on items like nail care tools, combs, cosmetics, hair stretchers, brushes, and hairpins. These investigations took samples from several beauty salons and cultivated them on selective microbial media, detecting *Staphylococcus, Bacillus, Microsporum, Trichophyton, Mucor, Aspergillus, Malassezia,* and *Streptococcus* species. The presence of such germs was highest on instruments such as clippers that had not been properly sterilized. Similar findings were obtained in other locations, including Nigeria, Iraq, and Iran, with *Staphylococcus aureus* being the prevalent bacterial species. Notably, the study emphasized the necessity of adequate sanitization in salons in order to minimize illness transmission. A number of sterilization methods, including the use of quartz sterilizer, UV sterilizer, and natural sterilization procedures, were tested. Despite efforts to maintain appropriate hygiene standards, the majority of salons used ineffective sterilization practices for their instruments. These findings highlight the need to adhere to the proper sanitation standards to reduce health concerns associated with salon equipment. Apart from pathogenic bacteria on cosmetic instruments, the literature extensively investigates microbial contamination in cosmetic items such as lipstick, mascara, eye shadow, foundation, blusher, eyeliner, face creams, and lotions. Because of their high concentration of chemicals, products foster microbial growth. Various university and beauty salon studies indicated the presence of both bacterial and fungal species in these cosmetics (Agarwal et al. 2020). Many studies have been conducted in the past to evaluate contamination levels/rates of contact lenses, lens covers, and other care accessories (Donzis et al. 1987).

In research conducted in Iran at the University of Raparin, samples of lipstick, eyeliner, mascara, and foundation were obtained from several salons, with bacterial growth outnumbering fungal development. *Staphylococcus, Candida*, and *Enterococcus* species were discovered, as well as *Candida albicans*. Similar discoveries were made at Benghazi University in Libya, where facial cosmetics such as kohl, eyeliners, and lipsticks were found to be contaminated with Staphylococcus and Candida. Another study in Tabriz, Iran, looked at eye and skin cosmetics such as lotions, mascara, powder, and eyeliner. *Streptococcus, Staphylococcus, Bacillus, Acinetobacter, Klebsiella, Citrobacter, Salmonella, Pseudomonas, E. coli, Candida, and Rhodotorula* were the most common bacteria, fungi, and yeast found in cosmetics. Notably, contamination levels in powder and mascara were greater, perhaps due to exposure to air and high moisture content, respectively. The number of users and whether or not they exchange cosmetic goods can impact the contamination rate. Research separated samples into categories used by a single individual, groups of persons, and expired items used by the general public. The risk of microbiological contamination is increased when there are more users of one product or when the same cosmetic product is shared. Research separated samples into categories used by a single individual, groups of persons, and expired items used by the general public. Personal-use items showed lower microbial content than public-use products, while expired products had the maximum contamination, with wide microflora and a small number of pathogenic organisms, such as *Candida albicans* and *P. aeruginosa,* were found (Agarwal et al. 2020).

Microorganism contamination in goods like shampoo can cause unpleasant odors, tastes, and texture changes, impacting product efficacy (Perry 2001). Most cosmetic products contain a wide range of ingredients, including inorganic and organic chemicals, vital minerals, moisturizers, and growth stimulants like vitamins. This concoction of substances generates a favorable environment for the development and survival of numerous bacteria. It is concerning, however, because the manufacture and expiration dates of beauty care items are frequently neglected or not carefully verified. As a result, the progressive loss of preservative efficacy in these products over time goes undetected. The common practice of exchanging beauty items within beauty salons, typically in an atmosphere without sufficient sterilization precautions, exacerbates the problem. The lack of strict hygiene and pollution control methods increases the likelihood of microbiological multiplication. When cosmetic equipment and tools are consistently used on several people without appropriate sanitization, they represent a major source of cross-contamination, allowing bacteria to spread between customers. Given the possible repercussions of such practices, it is critical that these concerns be addressed within the cosmetic business. Quality control procedures must be prioritized, including continuous monitoring of manufacturing and expiry dates, as well as ensuring that preservatives remain effective throughout the product's lifecycle. Furthermore, beauty salons should prioritize the implementation and enforcement of thorough sterilization processes in order to reduce the danger of microbiological contamination and transmission. By putting these steps first, the beauty care sector can make major gains toward providing a safer and more sanitary workplace for its clients (Edward et al. 2015).

A study's research emphasized the importance of container closure styles in minimizing microbiological contamination of cosmetics when used by customers. The study divided two types of cosmetics, shampoos, and skin lotions, into three groups depending on their preservation effectiveness: badly preserved, moderately maintained, and well preserved. Cosmetics that were badly kept had a significant degree of contamination during consumer use, while those that were moderately preserved had a substantially lower amount of contamination. Products in the well-preserved category, on the other hand, remained uncontaminated even after being used by customers. Furthermore, the study established a challenge test that effectively predicted the sensitivity of both shampoos and skin lotions to microbial contamination caused by customer usage. These findings highlight the critical importance of closure types in preserving the integrity and safety of cosmetic products during their intended usage (Brannan and Dille 1990).

There is a need to raise public knowledge about the risk of disease transmission through popular beauty salon instruments and supplies. As a result, according to the findings and suggestions of the research discussed here, the need to raise salon safety standards in sterilizing and preserving beauty products and tools. Furthermore,

Table 16.1 Potentially health hazards bacterial and fungal isolates found on various beauty salon equipment (Alharbi and Alhashim 2021b)

Microorganism	Type	Beauty Tool
Aspergillus niger	Fungus	Hair brushes, combs
Candida albicans	Fungus	Nail tools, makeup sponges
Escherichia coli	Bacteria	Contaminated water used in beauty treatments
Bacillus subtilis	Bacteria	Makeup applicators, sponges
Staphylococcus epidermidis	Bacteria	Makeup brushes, razors
Staphylococcus aureus	Bacteria	Makeup brushes, razors

we strongly advise the usage of unique cosmetic kits. If implemented, all of these guidelines should dramatically reduce the transmission of illness and disease through beauty salons. In terms of public awareness, the International Organization for Standardization issued criteria that should be followed in the sector of cosmetic production, from processing through client consumption and waste discharge (Alharbi and Alhashim 2021a) (Table 16.1).

16.6 Microbial Control Practices in Salons

Salons can play a role in the spread of viral, fungal, and bacterial diseases, as well as AIDS, herpes, hepatitis B and C, hair lice, skin and eye infections, and chronic fungal infections (Alharbi and Alhashim 2021a). Salons use a number of physical and chemical procedures to sterilize instruments and tools. Autoclaving, chemical reagents, boiling water, chemical reagents, autoclaving, quartz bead sterilization, UV radiation, and quartz bead sterilization are among the procedures used. Hot water sterilization is a common sterilization procedure that involves immersing instruments in boiling water for at least 3–5 min. However, research suggests that this approach may not completely eradicate germs. Autoclaving, on the other hand, emerges as the most reliable method for eliminating all microorganisms; nevertheless, it may not be suited for electrical equipment, and the sterilization process itself requires a relatively lengthy period. The Barbicide solution is now a commonly used disinfectant in salons, but it is important to remember that it is very hazardous to humans. Certain salons use sterilization procedures such as quartz sterilizers or UV sterilizers. The quartz sterilizer works by blowing air over quartz beads at a high temperature of 250 °C (482 °F) for a short period of 5–15 s. Although this approach can handle a large number of tools, it has the disadvantage of causing changes in the form and structure of the sterilized tools over time.

UV sterilization, on the other hand, is seen to be the safest solution for tools and brushes that cannot tolerate high-temperature sterilization. Its drawback is its capacity since it can only hold a certain number of tools at one time. To maintain proper hygiene, it is critical to store equipment in cool, dry places to reduce the possibility of bacterial and fungal development, which might potentially contribute to the spread of illnesses. Brushes should be kept in cosmetic bags, while metal and wooden instruments should be kept in closed plastic containers. In terms of predominance among service providers, around 35% choose UV sterilization, whereas approximately 20% prefer to sanitize their instruments using quartz beads. Surprisingly, just 1% of providers said they used an ultrasonic cleaning. It is important to note that none of these individual procedures have been definitely demonstrated to provide a reasonable degree of sterilization on their own, emphasizing the importance of combining treatments (Agarwal et al. 2020).

16.7 Enhancing Cosmetic Sustainability and Biodegradability with Natural Ingredients

Protein, carbohydrates, flavonoids, phenols, terpenoids, and steroids are phytochemical substances that can operate against human diseases (Sowmya et al. 2021). Cosmetic and pharmaceutical companies are increasingly interested in replacing synthetic antimicrobials in their topical treatments with natural alternatives. This tendency arises not just from rising consumer desire for natural agents but also from increased worry about bacteria's resistance to conventional antimicrobials. Antimicrobial properties of phenolic compounds, which are naturally synthesized by plants as part of their defense systems, have drawn attention. These chemicals have antimicrobial actions by interacting with the cell membrane or cell wall of bacteria, causing changes in membrane permeability and, eventually, cell death. Furthermore, phenolics can enter bacterial cells and stimulate the coagulation of their contents. Another benefit of phenolic compounds is their ability to extend the shelf life of a variety of items by suppressing the growth of harmful microbes (Ribeiro et al. 2015).

The prevention of microbial growth in cosmetics has become challenging due to a substantial rise in the resistance of hazardous bacteria to current preservatives. This increasing resistance emphasizes the need for creating preservation techniques that are efficient. As a result, finding safe, less toxic, biodegradable, and biocompatible active preservatives is seen as an essential priority. Natural antimicrobial preservatives are gaining popularity due to their outstanding qualities such as biodegradability, biocompatibility, and decreased toxicity. Cyanobacteria, like plants, algae, and eukaryotic microalgae, have been studied for the isolation of chemicals with potential use in the cosmetic and cosmeceutical fields. Spirulina platensis has been investigated as a valuable source of antibacterial, antiviral, and antioxidant chemicals; however, its efficacy varies depending on the extractive solvent. These chemicals have the potential to be used in the development of natural antibiotics against multidrug-resistant microorganisms. Spirulina has lately been employed in skincare products as a possible source of bioactive chemicals (Elzoughby 2022). The integration of bioactive extracts or phytochemicals

Table 16.2 Common natural substances commonly used as cosmetics are summarized in the following table; these compounds have the following advantages

Natural ingredient	Benefits	Reference
Rosehip oil	Anti-aging, improves skin texture and tone, rich in vitamins A and C	Hoang et al. (2021)
Jojoba oil	Moisturizing, balances oil production, resembles skin's natural sebum	Hoang et al. (2021)
Shea butter	Moisturizing, nourishes and protects the skin	Hoang et al. (2021)
Chamomile	Calming, anti-inflammatory, reduces redness	Hoang et al. (2021)
Green tea extract	Antioxidant, protects skin from free radicals, anti-inflammatory	Hoang et al. (2021)
Aloe Vera	Soothing, moisturizing, treats burns, acne, and dry skin	Liu (2022)

generated from various botanical sources serves a twofold role in the realm of cosmetics. First, these extracts contribute to general bodily well-being; second, they operate as active substances that can impact numerous biological processes in the skin, therefore imparting vital nutrients for skin health. Botanical goods are rich in a variety of useful elements, including vitamins, antioxidants, essential oils, hydrocolloids, proteins, terpenoids, and other bioactive chemicals. Depending on their precise makeup, these extracts can provide a variety of diverse qualities and advantages to the cosmetic compositions into which they are integrated. Plant extracts also show strong functioning against toxins and ROS and thus have antioxidant and antimicrobial properties (Ribeiro et al. 2015) (Table 16.2).

16.8 Role of Packaging in Microbiological Safety of Cosmetics

Inadequate hygiene during production can contribute considerably to bacterial contamination in cosmetic items. The final product may be significantly contaminated by microorganisms from the raw components. Furthermore, unsanitary production premises and employees might introduce germs into the product. This can raise the bacterial load of the product, posing a danger to human health if the product is utilized. To prevent bacterial contamination in cosmetic items, it is critical to maintain appropriate production practices and hygiene standards (March et al. 2018). According to one study, the type of container closure employed is an essential component in reducing microbiological contamination of cosmetics during consumer usage. The test classified two types of cosmetics (shampoos and skin lotions) as badly maintained, moderately preserved, or well preserved. Items that were badly maintained were substantially polluted during usage, whereas items that were moderately preserved had a low amount of contamination. After consumer usage, products designated as well preserved were not compromised. The study also showed that the proposed challenge test successfully predicted the vulnerability of shampoos and skin creams to microbial contamination from consumer usage. So the quality of product packaging also plays an important role in the prevention of microorganism contamination (Brannan and Dille 1990). Research shows how crucial hygiene and good manufacturing procedures (GMP) are to avoiding microbiological contamination. Cosmetics are very open to contamination from a variety of sources, such as handling by consumers and the environment (Choubey and Godbole 2017). The significance of packing as a barrier against microorganisms is shown by research on microbial penetration through packaging materials. Product shelf life is increased, and post-processing contamination can be avoided with efficient packaging (Choubey and Godbole 2017). Antimicrobial packaging is proven to be sustainable. It improves the safety and durability of cosmetics by extending the lag time and slowing the growth of microorganisms (Fadiji et al. 2023).

For instance, sustainable packaging materials might be a great choice since they offer protection with the least negative environmental impact. This includes utilizing plant-based polymers or recycled materials, which reduce the carbon footprint associated with package manufacture. By prioritizing eco-friendly solutions, cosmetic companies should promote a more sustainable and circular economy, reducing waste and protecting natural resources (Del Greco et al. 2024). Raw materials sourced sustainably are increasingly utilized in the cosmetics industry, serving both as packaging components and ingredients. This shift promotes environmentally sustainable production practices and reflects the industry's growing commitment to sustainability (Cinelli et al. 2019).

Microbial contamination of cosmetic items is a major source of worry. To address this issue, strong laws and regular and efficient microbiological monitoring are required. Furthermore, there should be increased surveillance of microbiological contamination, particularly in cosmetic goods used by those who are prone to diseases. Microbiological studies must follow known harmonized procedures, which include identifying and counting microorganisms cultivated in culture plates. The precision required determines whether genotypic or phenotypic approaches are used for microbial identification. Preservatives are commonly used to avoid bacterial contamination during product use. However, an ISO standard for defining the Minimum Inhibitory Concentration (MIC) or Minimum Bactericidal Concentration (MBC) of these preservatives is required. This guideline would allow for the most effective application of preservatives at quantities that would postpone the formation of preservative-resistant strains. The cosmetic industry may improve product safety, safeguard customers from possible dangers associated with microbiological contamination, and assure the continuous efficacy of preservatives in cosmetic formulations by employing these methods (March et al. 2018).

16.9 The Significant Impact of Green Packaging on Cosmetics and the Environment

The term "green packaging," often referred to as "sustainable packaging," refers to the use of materials and design techniques that reduce ecological footprint and environmental effects. Because of growing consumer awareness and regulatory pressure to reduce waste and promote sustainability, this packaging strategy is becoming more and more popular. Compostable, recyclable, or biodegradable materials are often utilized in green packaging. These materials aid in better waste management and lessen dependency on non-renewable resources (Branca et al. 2024). Reducing material consumption, increasing recyclability, and extending the packaging's entire lifecycle are the main goals of sustainable packaging design. This includes making sure the packaging can be readily dismantled for recycling and designed for reuse (Wandosell et al. 2021). Green packaging influences customer purchasing decisions by sending an environmentally friendly message. It helps companies demonstrate their environmental responsibilities and draws in clients who value sustainability (Wandosell et al. 2021).

16.9.1 Green Packaging as a Defense Against Microbial Intrusion

Additionally, green packaging can improve the microbiological safety of cosmetics. This is important because cosmetics must be free of dangerous microbes because they are frequently applied to delicate body parts. Certain eco-friendly green packaging materials are made to have antibacterial characteristics. Preventing the growth of bacteria, fungi, and other microbes can ensure the cosmetic products' lifespan and safety (Kaestner et al. 2023) sustainable packaging materials can be created with strong barrier qualities that will shield the contents from deterioration and contamination. This is especially crucial for organic and natural cosmetics, since they could be more vulnerable to microbial development (Wandosell et al. 2021). Chemical preservatives are frequently used in traditional packaging to ensure product safety. Green packaging can help customers with allergies or sensitive skin by lowering their need for these chemicals (Ibrahim et al. 2022) (Table 16.3).

16.10 Microbiological Quality Management Ensures Safe Cosmetics

MQM is an abbreviation for Microbiological Quality Management. It is a mechanism designed to assure the microbiological quality of cosmetic and detergent goods. The MQM consists of a collection of processes and practices intended to

Table 16.3 Microorganisms affecting cosmetics and green packaging prevention

Microorganism	Effect on cosmetics	Green packaging prevention	Reference
Staphylococcus aureus	• Product spoilage • Potential skin infections	Antimicrobial packaging materials can inhibit its growth	Dube and Dube (2023)
Pseudomonas aeruginosa	• Product spoilage • Cause infections, especially in damaged skin	Green packaging with antimicrobial properties can prevent its contamination	Branca et al. (2024)
Escherichia coli	• Cause product spoilage • Potential gastrointestinal infections if ingested	The use of biodegradable materials with antimicrobial agents can reduce the risk of contamination	Wandosell et al. (2021)
Candida albicans	• Causes spoilage • Can lead to fungal infections	Green packaging can include antifungal properties to inhibit growth	Kaestner et al. (2023)
Aspergillus niger	• Product spoilage • Can cause respiratory issues if inhaled	Incorporating antifungal agents in green packaging can prevent contamination	Dube and Dube (2023)
Bacillus subtilis	• Causes spoilage • Can produce harmful toxins	Green packaging materials with antimicrobial properties can inhibit its growth	Branca et al. (2024)

inhibit the development of microbes in goods. The MQM process begins with ensuring the microbiological purity of the raw materials utilized in product manufacture. This means that the raw materials must be devoid of pathogens with high virulence, such as *Staphylococcus aureus, Streptococcus pyogenes, Pseudomonas aeruginosa* and other *Pseudomonas* species, and *Klebsiella pneumoniae*.

The MQM comprises a validated product formulation with a preservative system that has been studied and proven to be effective in inhibiting microbial development. It also covers sanitary design and effective manufacturing hygiene. This implies that the manufacturing environment and equipment should be built to be easy to clean and sanitize and that all equipment is cleaned and sanitized appropriately. MQM contains a set of processes and practices aimed to ensure that items are manufactured in a way that inhibits microbial development. This contains comprehensive production activities for each stage, such as raw material addition, temperatures, speeds, mixing durations, sampling, equipment cleaning and sanitization, and bulk product transfer. Overall, the MQM is a complete method for ensuring the microbiological quality of cosmetic and detergent goods.

Manufacturers can take many procedures to assure the microbiological quality of their cosmetic and detergent goods. First, they can utilize raw materials that are microbiologically acceptable. This means that the raw materials utilized to make the items should be devoid of high-virulence microbial infections. Good plant hygiene and production practices must be implemented by firms. This involves keeping the production space clean and free of pollutants, as well as cleaning and sanitizing all equipment. Manufacturers can also employ sanitary design. This means that the manufacturing facility and equipment should be designed to be easy to clean and sanitize, with no regions where pollutants might collect. They can also utilize a preservative method that has been verified. This signifies that the product's preservative system has been evaluated and proven to be effective in preventing the growth of bacteria (Siegert 2012).

16.11 Adherence to Regulatory Measures Ensures the Purity and Quality in Cosmetics Manufacturing

Adherence to Good Manufacturing Practices (GMP) is required, as is a firm commitment to thoroughly inspecting raw materials, particularly those originating from natural sources, for any pollutants before their use. It is necessary to define the critical parameters for acceptance. In order to identify and manage contamination-prone locations, vigilance is essential. Given the inherent variation across units, the complexities of manufacturing facilities highlight the necessity for a personalized strategy. To prevent the spread of undesired levels of contamination, strict Quality Assurance (QA) processes must be used. The development and validation of complete cleaning and sanitization regimes are required. Finally, the human aspect contributes an element of uncertainty. Thorough training activities, as well as regular updates, are required to address this issue. Contamination in the course of active manufacturing is significantly reduced by strict adherence to GMP. Modern cosmetics production facilities attest to an unprecedented lack of germs in about 100% of the manufactured units. Manufacturers also work hard to create items that naturally inhibit microbial multiplication wherever possible. Following the conclusion of the manufacturing and packaging operations, the preservative system must survive standard microbiological challenges during the storage and utilization phases. Packaging design is important because it contributes to the reduction of contamination concerns (Perry 2001).

The European legislative framework includes an instrumental mechanism known as RAPEX that is meant to quickly notify stakeholders of non-food products that depart from European legal requirements and pose possible risks to consumer health and safety. This rapid exchange of information among member states and the European Commission about dangerous products has the potential to precipitate the removal of dangerous items from the market, consumer recall actions, or other interventions deemed appropriate by national competent authorities to mitigate consumer risk. Timely alerts outlining dangerous items are systematically cataloged in the publicly accessible European quick alert database, along with national-level consumer recommendations issued by EU member states. In Germany, for example, the Federal Institute for Occupational Safety and Health publishes a list of high-risk items on its official website, which is supplemented by findings from spot tests provided in yearly state reports. The European Cosmetics Regulation (EC, 2009) is the cornerstone for defining cosmetic product safety regulations. Historical practices included the use of compounds in personal care products (PCPs) that have subsequently been identified as problematic, resulting in their prohibition in modern PCPs. Notably, Annex II of the cosmetics rule contains over 1300 banned compounds and substance groups in PCPs, whereas Annex III lists over 250 substances that are restricted based on particular circumstances or quantities. The rule also includes extensive lists of permitted substances: Annex IV lists 153 colorants, Annex V lists over 50 preservatives, and Annex VI lists 26 acceptable UV-filters. Furthermore, some chemical combinations are forbidden. This regulatory framework covers approximately 1800 chemicals and requires PCP makers to comply. The law also imposes strict responsibilities on distributors (Art. 6) and establishes administrative processes for dealing with noncompliance and managing adverse consequences (Art. 26-28). When it comes to PCPs, the illegal inclusion of dangerous components can have an impact on health and the environment, adding to the cumulative body load via product consumption. The RAPEX database only includes goods with "high-risk" consequences for consumers. However, there is no generally applicable set of criteria for categorizing a non-compliant product risk as high or serious under the RAPEX framework. Cosmetics are often excluded from the classification and labeling standards established in the European Classification Labeling, and Packaging law (EC, 2008b). As a result, PCPs containing hazardous components that exceed categorization levels are not required to show hazard pictograms, as seen in other consumer goods such as detergents or paints.

The industry is responsible for assuring PCP safety, whereas national authorities are in charge of product conformity. Given the lack of pre-market external monitoring for PCPs, the importance of surveillance methods such as the RAPEX system inside this consumer product area is heightened. Personal care product oversight bodies have a significant problem because of the large number of regulated chemicals, a plethora of goods, brands, and manufacturers, the vast worldwide market, and intrinsic resource limits. This emphasizes the reliance on spot checks, consumer or toxicologist recommendations, and an understanding of the inherent constraints in ensuring comprehensive control (Klaschka 2017).

16.12 Conclusion

To conclude, cosmetics have developed from their historical roots as items aimed to improve appearance and encourage attractiveness. They have become essential components of modern daily skincare routines, and are available in a variety of formats for application to various body sections. Cosmetics, as defined by the FDA as things designed for beautifying, cleaning, and changing appearance, play an important part in personal care and self-expression. However, because cosmetics directly interact with the human body, their safety and quality are critical. While these items are typically safe, more attention is being paid to possible hazards such as microbial infection. Microorganisms can multiply in cosmetics, providing health risks to customers. This highlights the importance of thorough monitoring, regulation, and rigorous manufacturing practices in order to assure product stability, efficacy, and safety. Furthermore, questions regarding product quality and marketing practices have surfaced, leading to a request for tighter regulation. Data collection and analysis on adverse impacts are critical stages in protecting consumer well-being. Contamination with disease-causing microbes, especially antibiotic-resistant bacteria, must be addressed effectively to avoid serious health repercussions. Microorganisms have a significant impact on product quality. Cosmetics are exposed to different sources of contamination during manufacturing and everyday usage, necessitating continual monitoring to assure microbiological safety. In essence, the multidimensional character of cosmetics includes not only the quest for aesthetic enhancement but also the vital obligation of customer health and safety. The cosmetics business may continue to evolve while protecting the well-being of individuals who use these products if there is a concentrated commitment to best practices, constant research, and effective regulation. Furthermore, implementing environmentally friendly methods and innovative technology is the first step toward the advancement and betterment of the cosmetics field. We can significantly decrease the environmental effects of cosmetics by giving priority to natural preservatives, ecological packaging, and biodegradable components. In addition to helping the environment, this dedication to sustainability satisfies customer needs for safe and ethical beauty products.

16.13 Future Perspective

Bio-based ingredients are increasingly replacing traditional fossil-based components in the cosmetics industry. This change is driven by the need to reduce environmental impact and meet customer demand for natural products. Bio-based materials are sourced from plants, animals, microorganisms, enzymes, and organic crops that are free from artificial fertilizers and pesticides. Switching to natural bio-based ingredients has a significant impact on the environment. Plant, algae, and microorganism-based ingredients are not only renewable but also biodegradable. For example, the ecological impact can be decreased by substituting synthetic surfactants and emulsifiers with bio-based ones (Goyal and Jerold 2023).

For the development of more efficient, sustainable, and microbe-free cosmetics, advanced technology must be implemented in this field. Such as the use of digestate extracts high in surfactin in cleansing cosmetics has been researched recently. It has been demonstrated that these extracts, which come from biotechnological procedures, enhance the functioning and safety of cosmetic goods. They improve the items' inherent qualities and lessen their irritating effects, making them safer for consumers to use (Seweryn et al. 2023). Cosmetic products can be improved and more sustainable by utilizing biotechnological procedures and green chemistry concepts. This involves synthesizing substances using enzymatic procedures, which commonly require less energy and generate fewer byproducts than conventional chemical methods (Sasounian et al. 2024). Sustainable packaging solutions, such as biodegradable plastics, recycled materials, and refillable containers, can minimize waste. Innovations in packaging materials, like bioplastics made from corn starch or sugarcane, offer biodegradable alternatives to conventional plastics (Dube and Dube 2023).

Furthermore, natural preservatives can be used to ensure microbiological safety while preserving sustainability. Effective antibacterial qualities provided by ingredients like essential oils, plant extracts, and bio-fermented substances protect cosmetic goods against microbial contamination without the need for artificial preservatives (Seweryn et al. 2023). By using sustainable and biodegradable methods, the cosmetics industry has a significant opportunity to help protect the environment. Green chemistry, eco-friendly packaging, natural and bio-based materials, and natural preservatives all have the potential to make products that are not only safe for customers but also environmentally beneficial. A more environmentally friendly future for cosmetics will be made possible by ongoing innovation and commitment to sustainable development goals.

References

Acerbi F, Rocca R, Fumagalli L, Taisch M (2023) Enhancing the cosmetics industry sustainability through a renewed sustainable supplier selection model. Product Manufact Res 11(1):2161021

Agarwal SM, Caravaggio F, Costa-Dookhan KA, Castellani L, Kowalchuk C, Asgariroozbehani R, Graff-Guerrero A, Hahn M (2020) Brain insulin action in schizophrenia: something borrowed and something new. Neuropharmacology 163:107633

Ahmadi-Ashtiani H-R, Baldisserotto A, Cesa E, Manfredini S, Sedghi Zadeh H, Ghafori Gorab M, Khanahmadi M, Zakizadeh S, Buso P, Vertuani S (2020) Microbial biosurfactants as key multifunctional ingredients for sustainable cosmetics. Cosmetics 7(2):46

Akgül Ö, Bakan K (2021) The aerobic bacteria isolated from used cosmetic products and evaluation of antibiotic resistance. J Facul Pharm Ankara Univ 45(2):156–168

Alharbi NM, Alhashim HM (2021a) Beauty salons are key potential sources of disease spread. Infect Drug Resist 14:1247–1253

Alharbi N, Alhashim HM (2021b) Identification of pathogenic microbes in tools of beauty salon in Jeddah City. Biosci Biotechnol Res Asia 18(4):743

Almukainzi M, Alotaibi L, Abdulwahab A, Albukhary N, El Mahdy AM (2022) Quality and safety investigation of commonly used topical cosmetic preparations. Sci Rep 12(1):18299

Alshehrei FM (2023) Isolation and identification of microorganisms associated with high-quality and low-quality cosmetics from different brands in Mecca region-Saudi Arabia. Saudi J Biol Sci 30(12):103852

Arshad M (2018) Cogeneration policy to contest climate changes and the future of sustainable energy. In: IOP conference series: earth and environmental science

Arshad M (2024) Sustainability of renewable energy options as compared to coal-fired power plants in Pakistan. Green Low-Carbon Econ 2(3):162–173

Arshad M, Adil M, Sikandar A, Hussain T (2014) Exploitation of meat industry by-products for biodiesel production: Pakistan's perspective. Pakistan J Life Soc Sci 12:120–125

Arshad M, Ansari AR, Qadir R, Tahir MH, Nadeem A, Mehmood T, Alhumade H, Khan N (2022) Green electricity generation from biogas of cattle manure: an assessment of potential and feasibility in Pakistan. Front Energy Res 10:911485

Arshad M, Syed-Hassan SSA, Masood R, Ansari AR, Mumtaz A, Rahman A, Batool S, Atif FA, Muhammad A, Rehman AU (2023) Utilization of waste animal fat for sustainable biodiesel production. In: Climate changes mitigation and sustainable bioenergy harvest through animal waste: sustainable environmental implications of animal waste. Springer, pp 149–169

Ashtiani HRA, Bishe P, Lashgari N-A, Nilforoushzadeh MA, Zare S (2016) Liposomes in cosmetics. J Skin Stem Cell 3(3):e65815

Bashir A, Lambert P (2020) Microbiological study of used cosmetic products: highlighting possible impact on consumer health. J Appl Microbiol 128(2):598–605

Bikiaris N, Nikolaidis NF, Barmpalexis P (2024) Microplastics (MPs) in cosmetics: a review on their presence in personal-care, cosmetic, and cleaning products (PCCPs) and sustainable alternatives from biobased and biodegradable polymers. Cosmetics 11(5):145

Branca G, Resciniti R, Babin BJ (2024) Sustainable packaging design and the consumer perspective: a systematic literature review. Ital J Mark 2024(1):77–111

Brannan DK, Dille J (1990) Type of closure prevents microbial contamination of cosmetics during consumer use. Appl Environ Microbiol 56(5):1476–1479

Choubey S, Godbole S (2017) Methods for evaluation of microbiological safety, guidelines governing the quality and survey on microbial contamination of commercial cosmetic products. World J Pharm Med Res 3(8):85–94

Cinelli P, Coltelli MB, Signori F, Morganti P, Lazzeri A (2019) Cosmetic packaging to save the environment: future perspectives. Cosmetics 6(2):26

Dao H, Lakhani P, Police A, Kallakunta V, Ajjarapu SS, Wu K-W, Ponkshe P, Repka MA, Narasimha Murthy S (2018) Microbial stability of pharmaceutical and cosmetic products. AAPS PharmSciTech 19:60–78

Del Greco A, Bani M, Rampoldi G, Ardenghi S, Galli P, Strepparava MG, Russo S (2024) Re-evaluating beauty: attitudes and perceptions of eco-friendly packaging in beauty care products-a systematic review. Sustain Product Consumpt 52:458–468

Donzis PB, Mondino BJ, Weissman BA, Bruckner DA (1987) Microbial contamination of contact lens care systems. Am J Ophthalmol 104(4):325–333

Dube M, Dube S (2023) Towards sustainable color cosmetics packaging. Cosmetics 10(5):139

Edward SM, Megantara I, Dwiyana RF (2015) Detection of fungi in hair-brushes in beauty salons at Jatinangor. Althea Med J 2(4):516–520

Elzoughby MA-E (2022) In vitro evaluation of spirulina platensis extracts against pathogenic bacteria isolated from some cosmetic products. Ann Agric Sci Moshtohor 60(3):895–912

Fadiji T, Rashvand M, Daramola MO, Iwarere SA (2023) A review on antimicrobial packaging for extending the shelf life of food. PRO 11(2):590

Gilmore BF, Denyer SP (2023) Hugo and Russell's pharmaceutical microbiology. Wiley, New York

Goyal N, Jerold F (2023) Biocosmetics: technological advances and future outlook. Environ Sci Pollut Res 30(10):25148–25169

Halla N, Fernandes IP, Heleno SA, Costa P, Boucherit-Otmani Z, Boucherit K, Rodrigues AE, Ferreira IC, Barreiro MF (2018) Cosmetics preservation: a review on present strategies. Molecules 23(7):1571

Hoang HT, Moon J-Y, Lee Y-C (2021) Natural antioxidants from plant extracts in skincare cosmetics: recent applications, challenges and perspectives. Cosmetics 8(4):106

Ibrahim ID, Hamam Y, Sadiku ER, Ndambuki JM, Kupolati WK, Jamiru T, Eze AA, Snyman J (2022) Need for sustainable packaging: an overview. Polymers 14(20):4430

Jairoun AA, Al-Hemyari SS, Shahwan M, Zyoud S, e. H. (2020) An investigation into incidences of microbial contamination in cosmeceuticals in the UAE: imbalances between preservation and microbial contamination. Cosmetics 7(4):92

Kaestner L, Scope C, Neumann N, Woelfel C (2023) Sustainable circular packaging design: a systematic literature review on strategies and applications in the cosmetics industry. Proc Design Soc 3:3265–3274

Klaschka U (2017) Trust, but verify! Personal care products in the rapid alert system database RAPEX. Sustain Chem Pharm 5:30–41

Liu J-K (2022) Natural products in cosmetics. Nat Products Bioprospect 12(1):40

March GA, Garcia-Loygorri MC, Eiros JM, Bratos MA, de Lejarazu RO (2018) Microbiological quality in cosmetics. In: Analysis of cosmetic products. Elsevier, pp 585–597

Michalek I, John S, Caetano dos Santos F (2019) Microbiological contamination of cosmetic products–observations from Europe, 2005–2018. J Eur Acad Dermatol Venereol 33(11):2151–2157

Perry B (2001) Cosmetic microbiology. Microbiol Today 28:185–187

Rahimpour Y, Hamishehkar H (2012) Liposomes in cosmeceutics. Expert Opin Drug Deliv 9(4):443–455

Ribeiro AS, Estanqueiro M, Oliveira MB, Sousa Lobo JM (2015) Main benefits and applicability of plant extracts in skin care products. Cosmetics 2(2):48–65

Sasounian R, Martinez RM, Lopes AM, Giarolla J, Rosado C, Magalhães WV, Velasco MVR, Baby AR (2024) Innovative approaches to an eco-friendly cosmetic industry: a review of sustainable ingredients. Clean Technol 6(1):176–198

Seweryn A, Wasilewski T, Hordyjewicz-Baran Z, Bochynek M, Pannert D, Łukaszewicz M, Lewińska A (2023) Implementation of sustainable development goals in the cosmetics industry based on the example of cleansing cosmetics containing a surfactin-rich digestate extract. Clean Techn Environ Policy 25(9):3111–3125

Siegert W (2012) Microbiological quality management for the production of cosmetics and detergents. SOFW J Seifen Ole Fette Wachse 138(11):30

Smart R, Spooner D (1972) Microbiological spoilage in pharmaceuticals and cosmetics. J Soc Cosmet Chem 23:721–737

Sowmya S, Suba Sri M, Dineshkumar R (2021) In vitro therapeutic effect of spirulina extract. Asian J Biol Life Sci 10(3):583

Wandosell G, Parra-Meroño MC, Alcayde A, Baños R (2021) Green packaging from consumer and business perspectives. Sustain For 13(3):1356

Chapter 17
Cosmeceutical Legislation: Global and Pakistan View

Iram Liaqat, Husna Jurrat, Nazish Mazhar Ali, Muhammad Nauman Aftab, Sikander Ali, Sajida Naseem, Muhammad Arshad, and Guojing Yang

Contents

I. Liaqat (✉) · H. Jurrat · N. M. Ali
Microbiology Lab, Department of Zoology, Government College University, Lahore, Pakistan
e-mail: dr.iramliaqat@gcu.edu.pk

M. N. Aftab · S. Ali
Ikram-ul-Haq Institute of Industrial Biotechnology, Government College University, Lahore, Pakistan

S. Naseem
Division of Science and Technology, Department of Zoology, University of Education, Lahore, Pakistan

M. Arshad
Department of Basic Sciences, University of Veterinary and Animal Sciences, Jhang Campus, Lahore, Pakistan

G. Yang
Laboratory of Tropical Environment and Health, Hainan Medical University, Haikou, Hainan, China

S. Javed et al. (eds.), *Sustainable Cosmeceuticals*,
https://doi.org/10.1007/978-3-031-86087-4_17

Abstract Cosmeceuticals are hybrid products that blend cosmetic and pharmaceutical ingredients, claiming to improve attractiveness while also providing therapeutic or biologically active advantages. Cosmeceutical legislation is the legal and regulatory framework that governs the creation, marketing, and sale of cosmeceutical products. It focuses on analyzing the status of the regulation of cosmeceuticals in Pakistan against different countries. The study reveals particular primary and secondary regulatory systems in different areas of the world and exposes the legislator environment in Pakistan and recommendations aimed at bringing local regulation to the international level. The Drug Regulatory Authority of Pakistan (DRAP) oversees cosmeceutical regulation, ensuring the safety and efficacy of both cosmetics and medications. However, compared to developed countries, regulations surrounding cosmeceuticals are less explicit in Pakistan. This often leads to guidelines suited for regular cosmetics, with limited consideration for cosmeceuticals. Therefore, finding a balance between the legal frameworks to support safe products and innovation in the cosmeceuticals industry is crucial. Well-supported by Sustainable Development Goals, by leveraging natural resources and biodegradable materials, green extraction techniques, and recyclable packaging cosmetic industry can significantly reduce their environmental impact while meeting consumer demands for sustainable products. To protect consumers and foster corporate expansion, companies need to make sure that the laws and policies implemented locally are compliant with global standards. The regulators will have to adjust to new paradigms as a result of the emerging trends and technical instruments with initiatives. Regulators must adapt to new paradigms driven by emerging trends and technological advancements.

Keywords Primary and secondary regulatory systems · Drug Regulatory Authority of Pakistan (DRAP) · Legal frameworks · Cosmeceuticals industry

17.1 Introduction

Cosmetics are materials that are used to alter a person's appearance or preserve the integrity of specific body parts while also cleaning and protecting the exterior of the body (Russell 2011). "Kosm tikos" is a Greek word from which the term cosmetic is derived. The meaning of cosmetics is power, organization, and skill (Hasköylü and Öner 2023; Mishra and Kurmi 2023; Ullah et al. 2023).

Beauty consciousness cannot be attributed solely to modern times. In ancient times, beauty tips were also followed by personal grooming and enhancing overall appeal. The awareness of improving physical appearance, however, is gaining considerable attention in the current era of globalization, regardless of gender (Sartwell 2020; Kaur et al. 2021; Kuipers 2022).

Cosmetics are products that are used to clean and protect the external parts of the body, thus changing someone's physical appearance or maintaining certain aspects in good condition. Our external physical appearance plays an important role in our social interactions, as seen in many different cultures where fair skin complexion is treated as a symbol of beauty for both genders (Arshad et al. 2022; Celeiro et al. 2021; Mady et al. 2023). However, treating acne and boosting skin hydration for a younger glowing skin is all people looking for centuries, and cosmetics with such therapeutic properties are meant to do so.

The role of cosmetics with all necessary medical properties has gained importance in genders, males and females. To get certain benefits, people use cosmetics having natural and synthetic bioactive compounds promoting skin health, reducing inflammation, or enhancing tissue repair (Rybczyńska-Tkaczyk et al. 2023). Some bioactive compounds that help improve the skin texture and heal skin problems are listed in Table 17.1.

17.1.1 Fundamental Basis of Cosmetics and Cosmeceuticals

Cosmetics, throughout human history, have consistently formed a significant expression, evolving as civilization progresses. The use of various cosmetics for personal care is as ancient as human civilization (Haarmann 2020; Sullivan et al. 2023).

The demand for cosmetics has increased significantly worldwide over time. This is primarily due to the growing awareness about ways to enhance one's appearance from a bodily perspective. Today, cosmetics are commonly used for personal grooming and taking care of the body (Arshad et al. 2020).

These are the recommended temporary corrective pharmaceutical combinations for enhancing beauty through the use of these ingredients, which have essential properties for the quick care and maintenance of skin and hair. Human beings are very conscious about their appearance and fitness Lifestyle and diet play a significant role in physical well-being. Women have become increasingly conscious of supporting their physical appearance during their daily routine activities to appear more attractive and positive. Cosmetics have a rich history, dating back to ancient civilizations where natural materials were used for adornment and skincare. Over time, the evolution of cosmetics has been shaped by cultural practices and scientific

Table 17.1 Various bioactive compounds used in cosmeceutical products, their presence in these products, and their associated benefits

Bioactive compounds	Existence in cosmeceutical products	Benefits	References
Vitamin C	Face and eye serums, creams and masks	Anti-oxidant protection, more collagen synthesis and brightens skin	Boo (2022)
Green tea extract	Often found in creams and lotions	UV protection, anti-inflammatory properties, and anti-oxidant potential	Michalak et al. (2021)
Peptides	Present in anti-aging products	Enhance collagen production, improve skin texture	Ngoc et al. (2023)
Aloe vera	Used in skin and hair gels and cream	Soothing properties, hydrates skin and hair, and accelerates wound healing	Saleem et al. (2022)
Hyaluronic acid	Sold in the form of serum and moisturizers	Promotes skin elasticity, reduces fine lines, and hydrates dry skin	Juncan et al. (2021)
Omega 3 fatty acid	Exists in creams and oils	Enhances skin barrier function and reduces inflammation/swelling	Michalak et al. (2024)
Zinc	Common ingredient in ointments and sunscreen	Aids in acne treatment, anti-inflammatory properties, and UV protection	Kalasariya et al. (2021)
Probiotics	Present in serums and creams	Reduce skin acne and irritation, and enhance skin barrier function	Arora et al. (2024)
Retinol	Abundantly used in anti-aging formulations including creams and lotions	Prevents dryness, improves skin texture, and reduces wrinkles to ensure youthful skin	Mambwe et al. (2024)
Polyphenol	Majority of plant-based cosmetics and skincare products contain polyphenol	Reduces signs of aging and has well-reported anti-oxidant effect	Bharadvaja et al. (2023)

advancements. For example, the creation of cold cream by Galen (130–200 AD) marked a significant milestone in skincare. Lipstick is also one of the most preferred cosmetics for women. Abu al-Qasim Zahrawi (936–1013 AD) is credited with inventing the first lipstick for beauty (Fig. 17.1).

17.1.2 Economic Value of the Cosmetic Industry

The cosmetics industry does not demonstrate in any better way than the pharmaceutical industry how demand generates supply only during the past century, and the "beauty industry" has become one of the largest global industries (Kuipers 2022).

The cosmetics industry has recently focused on natural and herbal products. For example, aloe vera is known for its multifaceted benefits in skincare—constituting

Fig. 17.1 Different commonly used ancient cosmetics versus modern ones (Mishra and Kurmi 2023)

enzymes, polysaccharides, and vitamins—contributing to its effectiveness in various formulations. Aloe vera has known anti-aging properties, promoting the production of hyaluronic acid and collagen. All these help users curb the signs of aging including wrinkles and fine lines. Experimental evidence explains the consecutive use of aloe vera on the skin can improve its elasticity (Arshad et al. 2018; Witaszczyk and Klimowicz 2023).

As the market of cosmetics is growing, understanding the rules and regulations governing cosmetics and herbal cosmetic products is essential worldwide (Dubey et al. 2022).

The cosmetic industry is a high-value and evergreen multi-billion-dollar industry, which includes new and special products added every year. The global beauty products market has seen an average annual increase of 5%. It is interesting to note that the market for cosmetics and personal care products has shown consistent and stable growth since its inception, even in unstable economies (Arshad et al. 2020; Castro et al. 2023). These are driven by consumer demand for safer, eco-friendly, and sustainable alternatives. Ingredients like aloe vera have gained popularity for their anti-aging properties due to their ability to hydrate the skin, promote collagen production, and reduce the appearance of fine lines and wrinkles.

The issue is particularly important, given the significant economic impact and prominence of the cosmetics industry. According to Allied Industry Research, by

2027, the global cosmetics industry is projected to increase from $3802 billion in 2019 to $4635 billion. The annual expenditures on cosmetics in Pakistan amount to 101 billion rupees, which constitutes approximately 4% of the country's total domestic expenses. The cosmetic industry includes important products including haircare, skincare, perfumes, deodorants, toiletries, and makeup. Among these, skincare products rank highest in the list, accounting for approximately 36% of the global cosmetic market (Manjula 2022; Oduaran 2022; Bhadmus 2023).

The following list contains a large number of commonly used and recommended pharmaceuticals, especially in Pakistan.

- Hyaluronic acid
- Chamomile: oral
- Green tea, oral
- Vitamin C, oral and topical
- Sunscreen ingredients
- Grapeseed, oral
- Vitamin B-complex, oral
- Aloe vera, topical and oral
- Turmeric, topical and oral
- Retinol, topical
- Niacinamide
- Allium cepa
- Polyunsaturated fatty acids, oral
- Soya: oral and topical

17.1.2.1 Sustainable Development Goals (SDGs) and the Cosmetics Industry

The act of integration of Sustainable Development Goals (SDGs) in the world of beauty (cosmetics industry) is crucial for promoting sustainability and addressing ever-increasing environmental challenges. As per the SDGs, proposed and implemented in 2015, 17 goals are taken into account to ensure global prosperity, but not at the cost of the planet's serenity. Particularly in the case of cosmeceuticals or cosmetics sector, SDG 12: Responsible Consumption and Production and SDG 13: Climate Action provides a complete framework to encompass sustainability. These goals emphasize the need for sustainable practices throughout the supply chain management, starting from raw material sourcing to product disposal.

Green Chemistry in Cosmeceuticals

The milestone of sustainable cosmetics formulations is based on the adoption of green chemistry—a strategic environmental friendly process (Cannon and Warner 2009). Green chemistry primarily focuses on reducing hazardous substances, utilizing renewable resources, and advocating energy-efficient manufacturing processes (Franca and Ueno 2020). For instance, the use of biodegradable ingredients, especially organic based, in making cosmetics aligns well with the green chemistry fundamental concepts and sustainability goal 12.

The significant positive impact of green chemistry reducing environmental hazards and promoting sustainable practices fulfills customer's demands without neglecting the demands of the environment. In addition to the role of biodegradable plant-based compounds, innovative extraction techniques allow for the efficient use of plants and animals by-products, enhancing the socio-economical values of the overall industry. Supercritical Fluid Extraction (SFE), Ultrasound-Assisted Extraction (UAE), Microwave-Assisted Extraction (MAE), and Enzyme-Assisted Extraction (EEA) are enlisted as top techniques for promoting sustainability by further minimizing waste and promoting a circular economy (Chen et al. 2020).

Biodegradable Resources and Climate Impact

Biodegradable resources refer to the material that can be decomposed naturally, without adopting artificial decomposition and combustion processes, mainly through the action of bacteria and fungi. Mitigating pollution (land and air) by cost-effective waste management is well-adopted by the SDG 12 and SDG 13 acts. Stated in another way, eliminating microplastics from the formulation of synthetic cosmetics products including face, hair, and nail polishers can be a big step forward toward an ethical approach to reducing aquatic/water pollution or promoting a better aquatic ecosystem. By prioritizing biodegradable alternatives, the cosmetics industry can significantly reduce its carbon footprint and contribute to climate action. Using recyclable, non-toxic sustainable packaging material for making cosmetics retail and shipping ready also resonates with environmentally safe marketing standards (Table 17.2).

17.1.3 The Term Cosmeceutical and Associated Controversies

"Cosmeceuticals" are products that function as both cosmetics and medications. They are made by mixing materials for cosmetic and medicinal purposes (Srinivasulu et al. 2022). Cosmetic goods are used to treat a wide range of skin conditions, including scar healing, wrinkle reduction, and skin hydration maintenance. Cosmetics are not currently referred to be medicines or nutraceuticals. The concept of cosmeceuticals is a mixture of cosmetics and pharmaceutical products, due to the health aspects, and these products are designed to increase the condition and appearance of the skin that they are applied to, therefore falling in between cosmetics and pharmaceuticals. Cosmeceuticals, products that are between cosmetics and pharmaceuticals, can be identified as one of the major developments in the field of personal care and skin health. These products have the primary function of improving the

Table 17.2 The sustainable development goals (SDGs) (United Nations 2015)

Goals	Intent of suggested goals
Goal 1: No poverty	End poverty
Goal 2: Zero hunger	End hunger, achieve food security and improved nutrition Promote sustainable agriculture
Goal 3: Good health and well-being	Live healthy lives and Promote well-being for all at all ages.
Goal 4: Quality education	Ensure quality (inclusive and equitable) Promote education and lifelong learning
Goal 5: Gender equality	Empower all women and girls
Goal 6: Clean water and sanitation	Develop sustainable management of water and sanitation
Goal 7: Affordable and clean energy	Affordable, reliable, sustainable, and modern energy availability for all
Goal 8: Decent work and economic growth	Promote sustained, inclusive, and sustainable economic growth Offer employment and decent work
Goal 9: Industry, innovation, and infrastructure	Build resilient infrastructure Promote inclusive and sustainable industrialization Foster creativity and innovation
Goal 10: Reduced inequality	Reduce inequality within and among countries
Goal 11: Sustainable cities and communities	Make cities and human settlements
Goal 12: Responsible consumptions and production	Ensure sustainable consumption and production patterns
Goal 13: Climate action	Take urgent action to combat climate change Monitor and nullify the impacts
Goal 14: Life below water	Conserve and sustainably use the oceans, seas, and marine resources
Goal 15: Life on land	Protect, restore, and promote sustainable use of terrestrial ecosystems, sustainably manage forests, combat desertification, halt land degradation, and halt biodiversity loss
Goal 16: Peace, justice, strong institutions	Promote peaceful and inclusive societies for sustainable development Provide access to justice for all Build effective, accountable institutions
Goal 17: Partnerships for goals	Strengthen the means of implementation Revitalize the global partnership for sustainable development

look of the skin and providing other advantages as well. Therefore they can border cosmetics and drugs hence requiring strict regulation to determine their efficiency and safety (Agrawal and Chourasiya 2024). The legal requirements related to cosmeceuticals are distinct around the world due to the differences in attitude toward emerging opportunities, consumers, and competition.

The term cosmeceuticals has been coined to encompass an area that exists between the aforementioned two fields. Cosmeceuticals are hybrid products that are situated between cosmetics and pharmaceuticals, used to promote both physical health and beauty through their external or internal or both application. Natural cosmetics contain ingredients that can have medical and health effects on conditions such as wrinkles, pigmentation disorders, melasma (photo-protective), aging, skin inflammation, formation of wrinkles, and hair issues including thinning, as well as epilepsy (Lall et al. 2020).

Cosmetics are used in the industry to create and test effective and safe products, to provide benefits to the external appearance and to address and treat various skin conditions by offering different types of treatments (López-Hortas et al. 2021). Natural products, particularly active compounds from citrus fruits, are used for the treatment of various diseases, infections, and inflammations, and as protection against UV radiation and pollution. The hybrid of cosmetic and pharmaceutical compounds, known as cosmeceuticals, combines key ingredients for both treatment and beauty enhancement (Milam and Rieder 2021).

Cosmeceuticals can be defined as preparations that contain active ingredients used for treatment and have corrective effects on the skin surface when applied in conjunction with traditional cosmetics. These products have quick and effective effects on the skin and hair and are used for various treatments, such as damaged hair, wrinkles, photo-aging, dry skin, light spots, hyperpigmentation, etc. Working as a bridge between cosmetics and beauty care, they promise improvement in appearance. Currently, the cosmetics industry considers cosmeceutical as one of the fastest-growing segments in terms of personal care, and it is expanding on a large scale for individual exploration and research (Gupta et al. 2022). The aim of this paper includes outlining the current state of global cosmeceutical regulation and comparing it to Pakistan's practices, the main deficiencies, and recommendations.

Any substance intended to be applied to, poured upon, sprayed, sprinkled, or otherwise applied to the human body or any section of it to cleanse, beautify, increase attractiveness, or change appearance is considered a "cosmetic" under the terms of the Drug and Cosmetic Act 1940. The unique features and enormous market potential of nanomaterials make them widely used in contemporary beauty goods. Their growing use has, however, sparked several worries about environmental and public health safety (Rastogi et al. 2020). Few nations have, in retrospect, created their legal frameworks to govern the production and/or importation of cosmetics that use nanomaterials. The swift progress in science and technology necessitates updating our understanding of the safety of nanomaterials and addressing consumer concerns. This ultimately means that pertinent legislation needs to be reviewed and renewed on time.

17.1.4 Demand for Cosmetics According to Different Organizations

According to the Personal Care Association (PCA) of Cosmetic Europe, approximately 450 million individuals in Europe use various types of cosmetic products on a daily basis Innovation is a fundamental principle in this field (Khan et al. 2021; Pérez-Rivero and López-Gómez 2023).

Cosmetics according to FDA and European regulations are products that can be applied to numerous parts of the body including hair, face, eyes, lips, hands, nails, and feet for multiple purposes—typically for attraction and anti-aging (Rizzi et al. 2021). Metals gradually take their place as essential components of cosmetics (Sisti et al. 2021; Dubey et al. 2022).

The Greeks, Egyptians, and Romans used cosmetics based on lead (Pb) and white mercury (Hg). This ability is used to incorporate metals as components in cosmetics or to provide different colors for cosmetic products, or for UV absorbers and emollients (Wargala et al. 2021; Raza-Naqvi et al. 2022).

17.2 Cosmeceutical According to the Federal Food, Drug, and Cosmetic Act (FD&C Act)

According to the Federal Food, Drug, and Cosmetic Act (FD&C Act), there is no term like cosmeceutical. This term is only used for industrial purposes so that it can be associated with cosmetic products alongside treatment measures (Sarkar et al. 2023).

17.3 Cosmeceutical According to the European Union Cosmetics Directive (EUCD)

The European Union Cosmetics Directive (EUCD) defines cosmetics as any substance that aims to keep in contact with the different physical body parts (epidermis, hair system, nails, lips, and external genital organs) or with the teeth and mouth mucous membrane, specifically or primarily for their cleaning, perfuming them, alter their appearance, and keep them in good condition (Gupta et al. 2022).

Despite these definitions, the legal meanings of cosmetics vary widely in many countries. In some Western countries, cosmetics are generally understood as beautiful products, such as lipstick, mascara, eyeliner, highlighter, and other similar items. Cosmeceuticals can be defined as preparations that contain active ingredients used for treatment and have corrective effects on the skin's surface, particularly when used in conjunction with traditional cosmetics. These products have beneficial effects on the skin and hair and are used for various conditions such as damaged hair, wrinkles, photoaging, dry skin, light blemishes, hyperpigmentation, etc (Juncan et al. 2021).

While working between the fields of pharmaceuticals and beauty care, they promise improvement in physical appearance. Currently, cosmeceuticals are considered one of the fastest-growing segments in the personal care industry, expanding on a large scale for individual exploration and research. It is one of the projects advancing rapidly, demanding expansion in nanocosmetical research, exploration, and applications (Dal Zotto 2021; Jusup 2021; Gupta et al. 2022; Rathee et al. 2023).

17.4 Global Legislation on Cosmeceuticals

This book chapter compiles the literature relevant to the regulation of cosmeceuticals, providing information regarding various global standards.

17.5 Cosmetic and Cosmeceutical Rules and Legislation

Society considers a set of rules as a form of law. The purpose was to treat all equally and protect basic rights. Laws are necessary to ensure that within the scope of authority, a person gives a correct judgment and prevents wrongdoing. Public laws generally regulate matters affecting society, while private law governs relationships between individuals.

In the context of pharmacy, the rules and regulations of pharmacy provide guidelines and instructions for practicing pharmacy. They explain the best perspective for handling specific pharmacy situations. From a pharmaceutical viewpoint, the definition of "advertisement" under the 1956 Act encompasses all forms of communication intended to promote specific products or services. The advertisement and promotion of medicines are allowed only as long as they adhere to the guidelines provided by the regulatory authority, to ensure public safety and well-being.

Cosmetics products in Malaysia are regulated and controlled through the Cosmetics and Cosmetics Regulation 1984 Strict adherence to laws is necessary to ensure the quality of manufactured, sold, supplied, possessed, managed, or income-generating cosmetic products.

It is necessary to implement clear labeling with pharmaceutical products associated with the instructions of the Director of Pharmaceutical Sciences. To ensure the health status of employees or those in the vicinity, it is essential to adhere to Good Manufacturing Practices (GMP) for cosmetics manufacturing. Quality control and regular inspections should be conducted by the Quality Control Department to monitor the standards of registered products (Hassali and Shakeel 2020; Ganson et al. 2023; Mishra and Kurmi 2023).

Cosmeceuticals are currently approved as "over-the-counter" cosmetics in the United States but more attention is paid to them if they claim to have

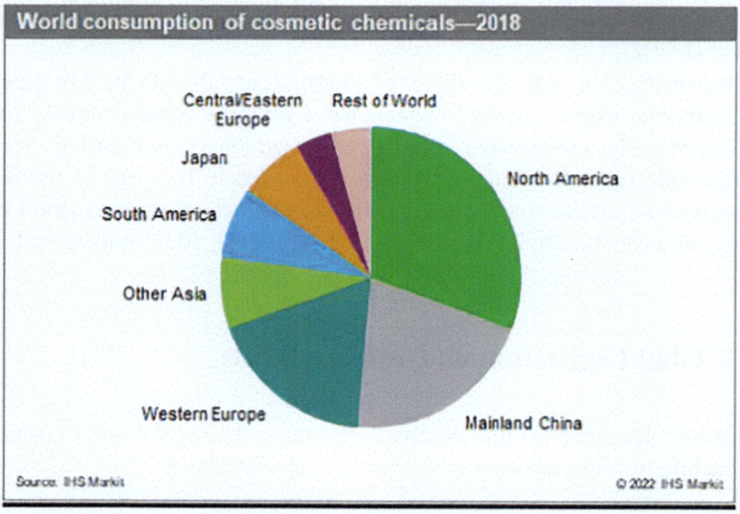

Fig. 17.2 Chemicals used in cosmetics and personal care products (Kaushik et al. 2023)

medication-like properties. The European Union's standards are more stringent as they require safety and efficacy reports before marketing the drug. The regulation of these markets varies from highly developed to relatively less developed markets in Asia. This chapter also includes an understanding of Pakistan's structure of regulation, analysis of the functioning of DRAP, and discovering gaps in legislation where the regulatory system may differ from the international one. The consumption of cosmetic chemicals by category and region, depending on value, is displayed in the following pie charts (Fig. 17.2).

North America currently consumes the most amount of substances used in cosmetics. However, it is anticipated that China will overtake all other global markets for cosmetic chemicals by 2023. North America, Western Europe, and Japan are expected to have growth rates ranging from 1% to 2%, whereas the Chinese markets are expected to rise at an average annual rate of nearly 8% between 2018 and 2023 (Gautam et al. 2022). Over 1000–1500 different chemical entities are recognized as cosmetic chemicals worldwide. These substances are grouped in this report into the primary categories mentioned above (Manful et al. 2024). Regardless of their chemical similarity, products within each functional category frequently compete with one another. The category of single-use additives comprises products such as antidandruff agents, silicone fluids and antiperspirants, sunscreens, hair conditioning polymers, and hair-setting resins (Balwierz et al. 2023). Large multinational makers of cosmetic chemicals now have a stronger competitive position due to the increased need for higher-performance, higher-purity products and data proving their safety and efficacy. These businesses have more resources available for technical support, research and development, and costly, time-consuming health and safety testing (Fig. 17.3).

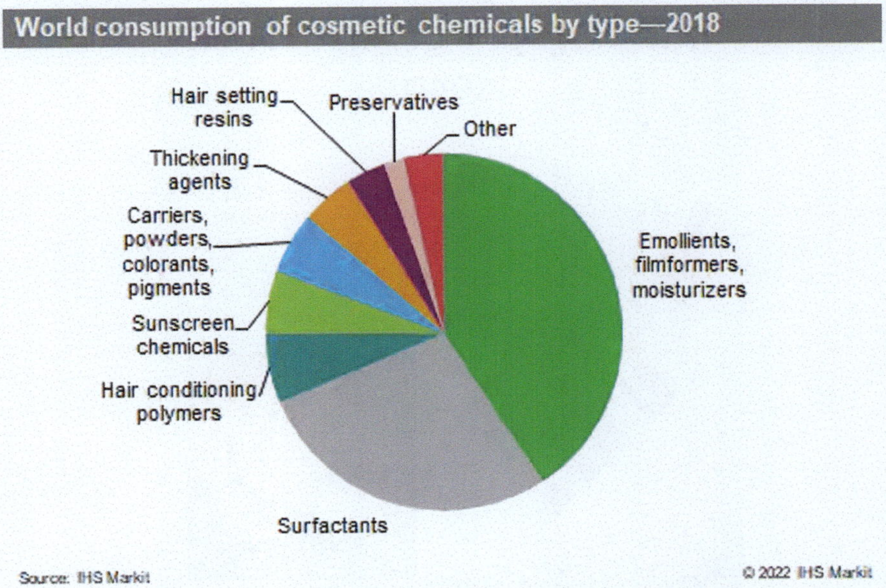

Fig. 17.3 Consumption of cosmetic chemicals

17.5.1 Global Legislation on Cosmeceuticals in the United States

In the United States, the FDA monitors cosmeceuticals as cosmetics provided the product does not present indications of treatment. The FDA does not have a special category for cosmeceuticals but regulates that those items that are marketed as cosmetics are safe to be used and labeled correctly. In essence, if the cosmeceutical has a therapeutic use, like treating skin conditions or changing body functions, it might be readjusted to a drug that undergoes extensive testing and approval. Knowledge of cosmetics and pharmaceuticals is required to define regulatory requirements. Depending on it, it is easy to distinguish cosmetics from medications. For instance, marketing a product to remove wrinkles can be categorized as cosmetic, but a product with a message of a severe reduction of wrinkles could fall under the drug category (Sukhnandan et al. 2023). The goal of the Food, Drug, and Cosmetic Act (21 USC 321-392 and associated regulations) is to stop items that are tampered with or mislabeled from being sold in the United States. On the other hand, the goal of the Fair Packaging and Labeling Act (15 USC 1451–1460) and its implementing regulations is to give consumers the ability to evaluate products in the marketplace and make decisions based on comparative value (Milam and Rieder 2021).

Fig. 17.4 Definition of cosmetics by The Drugs and Cosmetics Act, 1940

17.5.2 Definition of Cosmetics According to The Drugs and Cosmetics Act, 1940

Cosmetics and toiletries, according to the Drugs and Cosmetics Act of 1940, are defined as substances that are applied to the human body for the purpose of cleansing, beautifying, enhancing attractiveness, or altering its appearance, including rubbing, pouring, sprinkling, or otherwise applying to introduce them to the human frame, or for any other purpose (Pal et al. 2022; Sharma et al. 2024).

Cosmetics refers to anything that is used to cleanse, beautify, increase attractiveness, or protect the skin from germs and includes any subject related to the use of cosmetic ingredients (Barbar et al. 2024; Habib et al. 2024).

Despite this definition, in many countries, the legal implications of cosmetics are broader. In some Western countries, cosmetics are generally described only as beautiful products, such as lipstick, mascara, eyeliner, highlighter, and similar items (Sharmeen et al. 2021; Gupta et al. 2022) (Fig. 17.4).

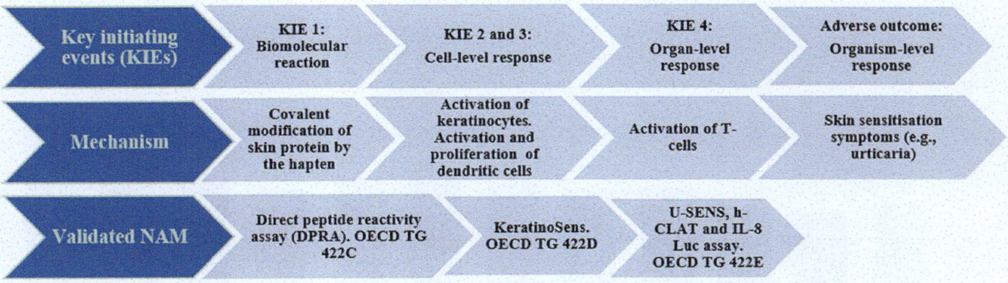

Fig. 17.5 Adverse outcome pathway of skin sensitization (Manful et al. 2024)

17.5.3 Global Legislation on Cosmeceuticals in the European Union

The European Union has more specific rules concerning cosmeceuticals through being under the European Medicines Agency (EMA) regulation and through country-specific offices (Lähteenmäki-Uutela et al. 2021). The laws governing cosmetics in the EU are capped under the general Cosmetics Regulation where all cosmetics that are sold in the EU including cosmeceuticals must meet high standards of safety as well as efficacy. Below is the diagram that indicates the additional feasible solutions for conducting skin sensitization tests including microphysiological test systems. To test the potential for skin sensitization of capsaicin, a common naturally occurring component in cosmetics, a microfluidic model was created. Notably, NAMs must be standardized and accepted for regulatory risk assessments to include them (Manful et al. 2024). Numerous NAMs pertinent to skin sensitization testing have been standardized as a result of extensive research, policy tools, and the work of the EU Reference Laboratory for Alternatives to Animal Testing (EURL ECVAM) (Fig. 17.5).

According to the EU Commission, the main piece of European law controlling completed cosmetics products in the EU is Regulation (EC) No. 1223/2009 (Pistollato et al. 2021). The framework of chemical and cosmetics laws set forth by the European Union is mandatory for all Member States and is implemented at the national level. To market it, manufacturers have to prepare a Product Information File (PIF) that has to include safety reports, a list of ingredients, and data on the product's effectiveness (Khan et al. 2023). This regulation makes sure that cosmeceuticals are safe for human use and that they are effective in their claimed purpose through pre-market testing and the continued monitoring of the products in the market (Fig. 17.6).

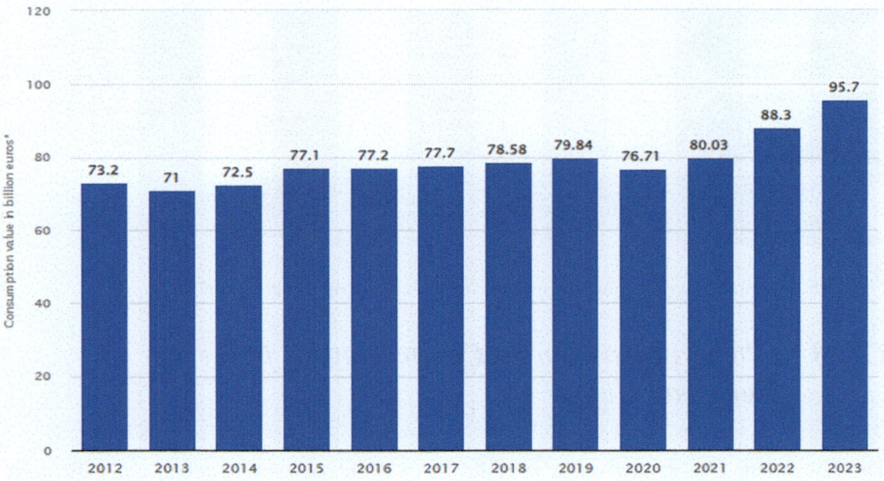

Fig. 17.6 Consumption value of cosmetics and personal care in Europe, Statista (2023)

17.5.4 Global Legislation on Cosmeceuticals in Asia

It can therefore be noted that Asia offers a complex legal environment for cosmeceuticals. Some regions such as Japan and South Korea already have structures that have already defined some of these products while other regions might not be so developed (Arnesdotter et al. 2021). Cosmeceuticals are regulated by the Pharmaceutical Affairs Law of Japan and before a product can be marketed some safety and efficacy must be proved. South Korea also has one namely the Ministry of Food and Drug Safety (MFDS), which is responsible for the regulation of cosmeceuticals and making sure they do not breach safety and quality standards (Mishra and Kurmi 2023). There is, however, the fact that in other Asian countries, the regulation can be completely different, which contributes to rather chaotic product quality and safety.

17.5.5 Global Legislation on Cosmeceuticals in China

The cosmetic regulations of China fall under the administration of the National Medical Products Administration (NMPA). Registration, safety assessment, labeling, and good manufacturing practices (GMP) requirements are included in the framework for cosmetics. Herbal cosmetics also adhere to these rules and regulations, which include additional instructions for the use of herbal ingredients (Anjum et al. 2024).

17.5.6 Global Legislation on Cosmeceuticals in India

The Drugs and Cosmetics Act of 1940 and the Drugs and Cosmetics Rules of 1945 regulate cosmetics in India. These regulations encompass various aspects of cosmetic products, including manufacturing, labeling, and ingredient restrictions. In India, herbal cosmetics also adhere to the same rules and regulations, but additional instructions are also provided by the Indian Pharmacopoeia Commission for Ayurvedic, Siddha, and Unani (ASU) systems. Cosmetics and herbal cosmetics in India are regulated under the Drugs and Cosmetics Act of 1940 and the Drugs and Cosmetics Rules of 1945. The Central Drugs Standard Control Organization (CDSCO) is the primary regulatory authority responsible for the protection, standards, and efficacy of cosmetics in India (Sahu et al. 2024).

17.5.7 Global Legislation on Cosmeceuticals in Pakistan

In Pakistan, the agency responsible for the regulation of cosmeceuticals is the Drug Regulatory Authority of Pakistan (DRAP). According to the present role of DRAP, not only medicines but cosmetics are also under its domain of safety and efficacy (Khan et al. 2023). Nonetheless, regulation guidelines are less concerning cosmeceuticals in comparison with the advanced nations. This often leads to the formation of rules that may be suitable for normal cosmetics without paying much attention to cosmeceuticals.

Cosmetics are substances used to clean and protect the body's exterior, altering a person's appearance or preserving the integrity of specific parts. Products labeled as "cosmeceuticals" are both drugs and cosmetics. They are made by combining medical and cosmetic substances. Cosmeceutical products are used to treat a variety of skin issues, such as wrinkle reduction, scar healing, and moisture retention. Cosmetics are not yet described with terms like pharmaceuticals or nutraceuticals. Cosmeceuticals contain retinoids, hydroxy acids, and antioxidants. Formaldehyde, quaternium-15, heavy metals, methylchloroisothiazolinone, methylisothiazolinone, parabens, lanolin, colophonium, and p-phenylenediamine (PPD) are common substances found in cosmetics. While many of these compounds are known to be harmless or have low toxicological risks, some include quantities of harmful chemicals that increase the likelihood of unfavorable reactions (Khan et al. 2023). According to a Karachi University research study, these items include highly volatile organic solvents (like camphor and terpineol), hazardous chemicals (including oxalic acid, sodium picramate, and titanium dioxide), textile dyes, and carcinogenic compounds (Sharma et al. 2024). Furthermore, heavy metals (such as PPD, lead, chrome, and nickel) included in these goods can puncture blood vessels and damage internal organs in addition to causing allergic reactions.

The General Cosmetics Act 2023 was introduced in Pakistan to govern the caliber and standards of cosmetics supplied there (McMullen and Dell'Acqua 2023). This covers the production, labeling, packing, distribution, storage, and sales of cosmetic products as well as related activities. The statute also penalizes anyone who engages in the counterfeiting of cosmetics. Cosmetics produced, processed, and packaged in the desecration of registered trademarks, labels, designs, and copyrights are referred to as counterfeit cosmetics (Arshad et al. 2023). The Pakistan General Cosmetics Regulatory Authority, a distinct agency within the Ministry of Science and Technology, is also established by the act (Bigambo et al. 2024). Its members have extensive backgrounds in general cosmetics, chemistry, dermatology, and the legal community. The act requires the implementation of GMP in the manufacture of cosmeceuticals. It helps in maintaining that products produced and controlled are of good quality all the time and reduces risks of contamination and defects (Malik et al. 2023). Stringent and specific guidelines are applied on the labels to ensure that the information that is provided to the customer is both informative and accurate when it comes to cosmeceuticals. These labels must contain data on the components used in the product, the way the product is to be used, the date of expiry, and if there is any caution that the buyer ought to be aware of regarding that particular product. It assists the clients in making wiser decisions and promotes the right application of the products.

The Pakistan General Cosmetics Bill, 2023, which establishes the Pakistan General Cosmetics Regulatory Authority, was approved by the Senate. This regulatory authority will set regulations for the creation, import, and export of general cosmetics in addition to supervising the standard, quality, labeling, and packaging of non-medicated cosmetic items (Tengli and Srinivasan 2022). The bill also offers additional protection to owners of registered trademarks and copyrights for cosmetic products by imposing criminal penalties on manufacturers of counterfeit goods, which include up to three years in prison and fines of up to PKR five million. The penal sanctions also include closing down the business and seizing the equipment used to produce the counterfeit goods (Arshad and Abbas 2018) (Fig. 17.7).

Ministry of Trade and Services, the government of Pakistan functioning through its constituent DRAP is the sole regulatory responsive authority for the safety,

Fig. 17.7 Advocates and Legal Consultants (2023)

efficacy, and quality of drugs and cosmeceuticals. "Our role at DRAP involves making sure that all the ones to be sold as drugs and the cosmeceuticals must meet the intended standards that are required of them before they are sold to the public." Cosmeceuticals with therapeutic claims describe a product that has cosmetic benefits but it also has therapeutic or medicinal purposes due to claims that it treats or prevents diseases or alters the structure of body functions. These are classified as drugs under the DRAP regulating mechanism (Glass 2020).

Cosmeceuticals with therapeutic effects require approval involving scientific evaluation to establish the effectiveness of the claims of the product. This process makes it possible for the products to be safe as well as efficient in their intended use, the respective pieces of legislation are the Cosmetic Act and The Drugs Act of 1976. Framework gives the legal authorities for medication management in Pakistan. It covers the regulation of drugs and cosmetics such as the importation, exportation, manufacture, distribution, and sale of the products. This legal act defines that the products which are determined to be in the category of cosmetics can be approved for their safety and quality but they will not be treated in the same way as drugs (Arshad 2018). Cosmeceuticals that are not intended to have therapeutic purpose for diseases, illnesses, or conditions are normally classified under the cosmetic sections of the Drugs Act. These products are safe for human use but do not have to prove that they are further beneficial for the body such as in the case of drugs.

Therapeutic cosmeceuticals will be implemented to ensure that products making therapeutic claims are officially registered with DRAP. In this context, there must be further details to enable the recognition of the product, its constituents, how it is made, and the health improvements that are promised by the product. Information about the product safety and effectiveness of the stated therapeutic qualities is required to be submitted. This may encompass but is not limited to, factors such as clinical trials, ingredients toxicity, and manufacturing details.

17.6 Drug and Cosmetic Rule 2020

As per previous studies, the cut-off date was September 2021, and there was no new "Drug and Cosmetic Rule 2020" in India or any other jurisdiction. It is necessary to stay updated with the current information on the official websites of related regulatory authorities so that you can be informed about any changes or updates in rules and regulations (Sharma et al. 2024).

The basic law regulating drugs and cosmetics for India is the Drugs and Cosmetics Act of 1940, and specific rules and requirements were outlined in the Drugs and Cosmetics Rules of 1945. To address current challenges, ensure product safety and quality, and align regulations with international standards, timely amendments can be made to these laws. In India, the Central Drugs Standard Control Organization (CDSCO) is the primary regulatory authority that oversees the safety, quality, and efficacy of medicines and cosmetics.

However, a concerning issue arises from the fact that the majority of cosmetic manufacturing units in Pakistan operate without proper authorization. The FDA (Food and Drug Administration) has identified permissible levels of ingredients in cosmetics, including As, Pb, Cd, and Hg, which are respectively set at 3 ppm, 20 ppm, 5 ppm, and 1 ppm. In 2018, it was discovered through quality inspection that out of 59 national and international cosmetic brands, only three had dangerously high levels of 074–44,292 ppm of lead. This finding has led authorities to take action against many Pakistani brands, especially those containing hazardous chemicals such as hydroquinone, mercury, or corticosteroids, which are present as key ingredients in beauty creams. This underscores the urgent need for tighter regulations and quality control to protect consumers from potential health risks (Raza-Naqvi et al. 2022).

According to Health Canada, in Nigeria and most sub-Saharan African countries, all cosmetic products were tested positive for lead (Pb) and beryllium (Be) in over 90% of cases, with 100% containing nickel (Ni). On average, at least four heavy metals were found in cadmium (Cd), Pb, mercury (Hg), nickle (Ni), beryllium (Be), arsenic (As), selenium (Se), and thallium (Tl). Environment and Climate Change Canada (ECCC) also tested approximately 49 different cosmetic products, with over 90% containing Pb and approximately 40% containing As.

The FDA (United States) has approved 10–20 mg/kg of Pb for cosmetics Health Canada has set a safe limit of less than 10 mg/kg for Pb, and 3 mg/kg for Hg. The German federal government has established an approved limit of 20 mg/kg for Pb and 5 mg/kg for Hg. In the United States, the FDA closely monitors the concentration of heavy metals in raw materials and finished cosmetic products. In developing countries, the regulatory system for monitoring is often flawed or non-existent.

For example, the registration protocol for cosmetic procedures should not be as complicated as that for medications. According to Good Clinical Practice (GCP) guidelines, it is necessary to demonstrate the desired activity of the cosmetic for the treatment of common skin defects or "conditions" with appropriate power and to ensure that safety requirements are optimal and there are no unexpected adverse effects. In the United States, this means that certain subcategories of drugs (cosmeceuticals) are registered as over-the-counter products. Due to a legendary legal dispute over regulatory labeling for topical minoxidil for male pattern baldness, it was claimed that the activity of formulating a product, rather than simply modifying it (commonly versus a drug), determines whether it is a cosmetic or a medicine. General Cosmetics Act, Pakistan General Cosmetic Authority, and Pakistan Standards and Quality Control Authority (PSQCA) play important roles in regulating cosmetic legislation in Pakistan (Table 17.3).

Table 17.3 Cosmeceutical legislations in Pakistan

Legislations	Year enacted	Short description	References
General Cosmetics Act	2023	Regulates standards and quality of cosmetic products. It includes: • Labeling • Packaging • Manufacturing • Distribution • Sales • Storage	Government of Pakistan Official Gazette
Pakistan General Cosmetic Authority	2023	Established under the general cosmetic act to • Oversee the regulation of non-medicated cosmetic products • Set standards for promoting the development of the cosmetics industry • Ensure compliance with safety and quality Standards	Government of Pakistan Official Gazette
Pakistan Standards and Quality Control Authority (PSQCA)	Ongoing	• Set regulatory rules for homemade cosmetics in Pakistan • Check product quality to ensure compliance with safety regulations	PSQCA official website https://www.psqca.com.pk/

17.7 Comparative Analysis

Several factors make it difficult for Pakistan to address the problems of cosmeceuticals regulation properly. They include inadequate resources to support the enforcement of regulations on cosmeceuticals, the absence of clear and complete policies in the cosmeceuticals sector, and the low levels of awareness among customers regarding the advantages and hazards related to the use of cosmeceuticals products (Khan et al. 2021). Such issues comprise variability in product quality and safety which can affect consumers' trust and properties of the market. From the perspective of international best practices, Pakistan's regulation can be considered relatively weaker in some aspects. For example, the United States and European Union have elaborate and strict legislation while Pakistan could be less precise and comprehensive in enforcement. From this comparison, one can conclude that Pakistan especially needs to improve the formulation of its regulation system to guarantee safe and effective cosmeceutical products.

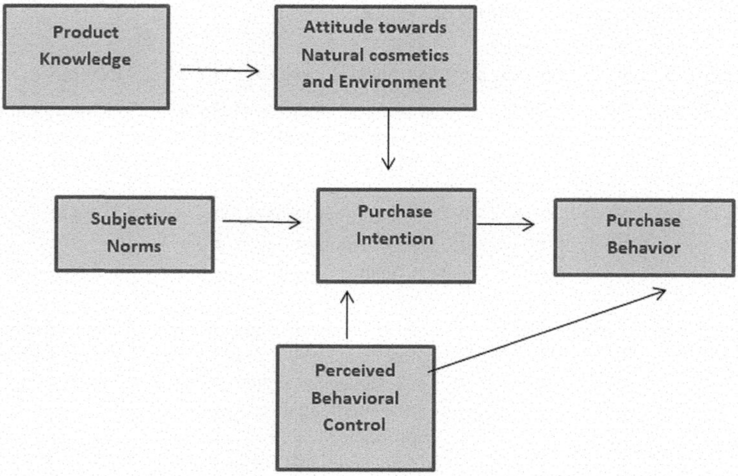

Fig. 17.8 Factors influencing purchased decision (Tengli and Srinivasan 2022)

17.8 Global Versus Local Standards

It is essential to compare the practices in the different countries, to highlight the main differences between the international requirements and the Pakistani ones (Pistollato et al. 2021). The United States and the European Union are the two major markets for cosmeceuticals, which have provided guidelines that address safety and efficacy, as well as processes for approving and regulating cosmeceutical products. However, the regulations set up in Pakistan could be relatively stringent; hence, there could be chances that the products differ as well as the protection of the consumers.

17.9 Impact on Market Practices

Regulations influence the nature of cosmeceuticals and the way they are promoted and received in various parts of the world. The downside of having fewer restrictive rules could be that the consumer trust in products could be low and the quality of the product might not be consistent, in comparison to areas that have an increase in the restrainment rules (Rastogi et al. 2020). Awareness of these effects will assist in analyzing the potential improvements of the regulation for consumer protection.

Customers are making more thoughtful decisions because they are more conscious of how their consumption habits affect the environment and their health (Tengli and Srinivasan 2022). As a result, it would make sense for cosmetics companies to go toward natural cosmetics. The majority of ingredients in natural cosmetics are derived from plants and naturally occurring elements. Fruit and plant extracts are often used to make natural cosmetics. Given that hazardous synthetic chemicals leak into lakes, oceans, or rivers, they are known to have an adverse influence on the health of people and other living things. These species are dependent on the water bodies for sustenance (Tengli and Srinivasan 2022). A research study indicated that perceived behavioral control also had a favorable impact on purchase intention. Both purchase behavior and attitude toward natural cosmetics and the environment were indirectly impacted by product knowledge through the influence of purchase intention. The findings indicate that there is no significant difference in the effects of variables on Purchase Intention and Purchase Behavior between male and female customers. This suggests that men and women have similar natural cosmetics purchase behaviors (Fig. 17.8).

17.10 Recommendations for Innovations in Cosmeceuticals

The cosmeceuticals industry is progressing very fast and new technologies have been developed including smart skincare products and enhanced formulations. Such innovations may also influence the future development of the regulations since the regulators will need to change their ways and practices to fit the new technologies while endeavoring to understand whether the technologies are safe and efficient in their use (Ali 2017). It utilizes biological processes to create active ingredients like hyaluronic acid and peptides more efficiently and sustainably. Moreover, nanotechnology, AI-powered solution, Micro capsulation, At-Home Devices, and the use of biodegradable materials for making cosmetics are highly recommended (Vaz 2021). There is a need for regulatory enhancements by increasing the stringency of its regulations, which could prove to be suitable for Pakistan, as the country could follow the global tendency and implement more precise regulations for cosmeceuticals (Pistollato et al. 2021). This entails developing stringent safety and efficacy standards as witnessed in the United States and Europe currently. Strengthening the regulation of cosmetics products by adopting appropriate rules and regulations for cosmeceuticals will also help increase people's trust in the products (Agrawal and Chourasiya 2024). Possible strategies to implement the policymakers' objectives may entail enhancing capacities to enforce the rules and regulations, possibly offering staff training involved in the regulation of cosmeceuticals, and possibly increasing consumer awareness of the products (Sharma et al. 2024). These can go a long way in closing the gap between national and international standards to offer products of high quality as well as meet the standards of safety that are expected of them.

17.11 Implications

The result of this study has relevant intervention points for politicians, manufacturers, and consumers. This work identifies the current and potential challenges in the regulation of Pakistan cosmeceuticals and recommends areas of improvement in line with existing legislation and other recommendations whereby the safety and efficacy of cosmeceuticals in the relevant market could be made more credible. It discusses the paramount need to balance the legal systems to promote safe products and innovation in the cosmeceuticals business. The industries must ensure that the local laws and policies adopted are standard with the international ones to safeguard the consumers and promote the growth of the business. These trends and technological tools with initiatives will emerge, and for this reason, the regulators will be forced to adapt to new paradigms. This will be beneficial for the regulator body and consumers since the quality of products will be intact and recommendations will increase the level of competition in the market.

17.12 Conclusions

In a nutshell, the disparities between the Pakistani and US rules and other Asian countries and their practicableness on the cosmeceuticals market have made clear. It further elaborates on the necessity for Pakistan to improve its capacity to regulate to international standards. The issues arising from cosmeceuticals' definition and their globalization are delicate and varied across the globe. The United States and the European Union have laid out very concrete safety measures to protect the consumers as well as provided policies in guaranteeing that such products are safe for people to use while Pakistan has the following regulatory loopholes that need to be closed. The comparative analysis also indicates that although Pakistan has made some efforts to establish and enforce regulations, much needs to be done in terms of improving the quality of such practices and making them closer to those that are adopted in more advanced markets. Thus, Pakistan can enhance the quality and safety of cosmeceuticals and restore the confidence of consumers and manufacturers in the market by enhancing the role of the regulatory authority and increasing the awareness of both sides. Adopting the recommendations, including strengthening the capacity of the regulatory body, training personnel, and revising the policies and procedures to reflect the best practices, will be significant in addressing the existing gaps.

References

Advocates and Legal Consultants (2023) Pakistan senate passes 2023 general cosmetics bill, strengthening regulation and protection in cosmetic industry

Agrawal A, Chourasiya SS (2024) Natural products as cosmeceuticals in different countries: a regulatory perspective. Royal Society of Chemistry, London

Ali S (2017) The antecedents of consumer intention to purchase halal cosmetic products in Pakistan: using behavioural reasoning theory. Universiti Utara Malaysia, Kedah

Anjum V, Sharma K, Jha P, Potoroko I, Kumar R (2024) Regulations in China. In: Global regulations of medicinal, pharmaceutical, and food products. CRC Press, Boca Raton, pp 115–132

Arnesdotter E, Rogiers V, Vanhaecke T, Vinken M (2021) An overview of current practices for regulatory risk assessment with lessons learnt from cosmetics in theEuropean Union. Crit Rev Toxicol 51(5):395–417

Arora R, Kaur R, Babbar R, Dhingra S, Dhingra AK, Grewal AS (2024) Evolving advances in the cosmetic use of probiotics and postbiotics: health, regulatory and marketing aspects. Curr Pharm Biotechnol 25(11):1349–1361

Arshad M (2018) Cogeneration policy to contest climate changes and the future of sustainable energy. IOP Conf Ser Earth Environ Sci 200:012050

Arshad M, Abbas M (2018) Water sustainability issues in biofuel production. In: Perspectives on water usage for biofuels production: aquatic contamination and climate change. Springer, Cham, pp 55–76

Arshad M, Bano I, Younus M, Khan A, Rahman A (2018) Health concerns associated with biofuel production. In: Perspectives on water usage for biofuels production: aquatic contamination and climate change. Springer, Cham, pp 97–105

Arshad H, Mehmood MZ, Shah MH, Abbasi AM (2020) Evaluation of heavy metals in cosmetic products and their health risk assessment. Saudi Pharm J 28(7):779–790

Arshad M, Ansari AR, Qadir R, Tahir MH, Nadeem A, Mehmood T, Alhumade H, Khan N (2022) Green electricity generation from biogas of cattle manure: an assessment of potential and feasibility in Pakistan. Front Energy Res 10:911485

Arshad M, Syed-Hassan SSA, Masood R, Ansari AR, Mumtaz A, Rahman A, Batool S, Atif FA, Muhammad A, Rehman AU (2023) Utilization of waste animal fat for sustainable biodiesel production. In: Climate changes mitigation and sustainable bioenergy harvest through animal waste: sustainable environmental implications of animal waste. Springer, Cham, pp 149–169

Balwierz R, Biernat P, Jasińska-Balwierz A, Siodłak D, Kusakiewicz-Dawid A, Kurek-Górecka A, Olczyk P, Ochędzan-Siodłak W (2023) Potential carcinogens in makeup cosmetics. Int J Environ Res Public Health 20(6):4780

Barbar AY, Kakayi M, Aksu B, Demir B (2024) Safety use of cosmetics. Aurum J Health Sci 6(1):49–70

Bhadmus I (2023) Beauty and personal care sector: investment opportunities in Nigeria

Bharadvaja N, Gautam S, Singh H (2023) Natural polyphenols: a promising bioactive compounds for skin care and cosmetics. Mol Biol Rep 50(2):1817–1828

Bigambo P, Peter V, Mrango M, Kimbokota S (2024) Tanzania's batik industry: status, challenges, and opportunities. Afr J Sci Technol Innov Dev 16:1–14

Boo YC (2022) Ascorbic acid (vitamin C) as a cosmeceutical to increase dermal collagen for skin antiaging purposes: emerging combination therapies. Antioxidants 11(9):1663

Cannon AS, Warner JC (2009) Green chemistry: foundations in cosmetic sciences. In: Global regulatory issues for the cosmetics industry. William Andrew Publishing, Norwich, pp 1–16

Castro ML, Ferreira JP, Pintado M, Ramos OL, Borges S, Baptista-Silva S (2023) Grape by-products in sustainable cosmetics: nanoencapsulation and market trends. Appl Sci 13(16):9168

Celeiro M, Garcia-Jares C, Llompart M, Lores M (2021) Recent advances in sample preparation for cosmetics and personal care products analysis. Molecules 26(16):4900

Chen TL, Kim H, Pan SY, Tseng PC, Lin YP, Chiang PC (2020) Implementation of green chemistry principles in circular economy system towards sustainable development goals: challenges and perspectives. Sci Total Environ 716:136998

Dal Zotto A (2021) The Chinese cosmetics industry: an attractive market for foreign companies. A scenario of its characteristics and regulations with Chinese-English Terminological Records

Dubey SK, Dey A, Singhvi G, Pandey MM, Singh V, Kesharwani P (2022) Emerging trends of nanotechnology in advanced cosmetics. Colloids Surf B: Biointerfaces 214:112440

Franca CCV, Ueno HM (2020) Green cosmetics: perspectives and challenges in the context of green chemistry. Desenvolvimento Meio Ambiente 53:133–150

Ganson KT, Sinicropi E, Nagata JM (2023) Assessing Canadian regulation of muscle-building supplements: identifying gaps and recommendations for improvement to protect the health and well-being of young people. Performance Enhance Health 11(3):100255

Gautam P, Ghanghas D, Mittal P, Gupta V, Kanika D (2022) Regulatory framework of cosmetic in the European Union. Toxicol Commun 4:13

Glass GE (2020) Cosmeceuticals: the principles and practice of skin rejuvenation by nonprescription topical therapy. Aesth Surg J Open Forum 2(4):ojaa038

Gupta V, Mohapatra S, Mishra H, Farooq U, Kumar K, Ansari MJ, Iqbal Z (2022) Nanotechnology in cosmetics and cosmeceuticals—a review of latest advancements. Gels 8(3):173

Haarmann H (2020) Advancement in ancient civilizations: life, culture, science and thought. McFarland, Jefferson

Habib U, Pervaiz A, Murtaza A, Hussain A, Sarwar H, Ullah M, Bashir M (2024) Assessing the cumulative effects of heavy metal exposure from cosmetic products on local consumers. J Mountain Area Res 9:149

Hasköylü ME, Öner ET (2023) Fructans in personal care products. In: The book of fructans. Academic, New York, pp 275–294

Hassali MAA, Shakeel S (2020) Pharmaceutical and cosmeceutical marketing and advertising in Malaysia: an overview of current governing laws and regulations. Biomed J Sci Tech Res 26(3):19997–20006

Juncan AM, Moisă DG, Santini A, Morgovan C, Rus LL, Vonica-Ţincu AL, Loghin F (2021) Advantages of hyaluronic acid and its combination with other bioactive ingredients in cosmeceuticals. Molecules 26(15):4429

Jusup R (2021) The halal cosmetics industry: a comparison between Qatar and Malaysia (Master's thesis, Hamad Bin Khalifa University (Qatar)

Kalasariya HS, Patel NB, Yadav A, Perveen K, Yadav VK, Munshi FM, Jeon BH (2021) Characterization of fatty acids, polysaccharides, amino acids, and minerals in marine macroalga chaetomorpha crassa and evaluation of their potentials in skin cosmetics. Molecules 26(24):7515

Kaur L, Singh AP, Singh AP, Kaur T (2021) A review on herbal cosmetics. Int J Pharm Drug Anal 9(3):196–201

Kaushik M, Farooq U, Ali MS, Ansari MJ, Iqbal Z, Mirza MA (2023) Safety concern and regulatory status of chemicals used in cosmetics and personal care products. Dermato 3(2):131–157

Khan NH, Ullah F, Khan TA, Zafar U, Khan FA, Mustaqeem M, Ji XY (2021) Personal-care cosmetic practices in Pakistan: current perspectives and management. Clin Cosmet Investig Dermatol 14:9–21

Khan N, Ahmed S, Sheraz MA, Anwar Z, Ahmad I (2023) Pharmaceutical-based cosmetic serums. In: Profiles of drug substances, excipients and related methodology, vol 48. Academic Press, New York, pp 167–210

Kuipers G (2022) The expanding beauty regime: or, why it has become so important to look good. Crit Stud Fashion Beauty 13(2):207–228

Lähteenmäki-Uutela A, Rahikainen M, Camarena-Gómez MT, Piiparinen J, Spilling K, Yang B (2021) European Union legislation on macroalgae products. Aquac Int 29:487–509

Lall N, Mahomoodally MF, Esposito D, Steenkamp V, Zengin G, Steyn A, Oosthuizen CB (2020) Cosmeceuticals from medicinal plants. Front Pharmacol 11:1149

López-Hortas L, Flórez-Fernández N, Torres MD, Ferreira-Anta T, Casas MP, Balboa EM, Domínguez H (2021) Applying seaweed compounds in cosmetics, cosmeceuticals and nutricosmetics. Mar Drugs 19(10):552

Mady S, Biswas D, Dadzie CA, Hill RP, Paul R (2023) "A whiter shade of pale": whiteness, female beauty standards, and ethical engagement across three cultures. J Int Mark 31(1):69–89

Malik M, Hanif Y, Hussain A, Hashmi A (2023) Exploring perceptions of regulators regarding factors affecting the quality of nutraceuticals and cosmeceuticals: a qualitative study from Pakistan. Int J Health Sci 7(1):1910–1929

Mambwe B, Mellody KT, Kiss O, O'Connor C, Bell M, Watson RE, Langton AK (2024) Cosmetic retinoid use in photoaged skin: a review of the compounds, their use and mechanisms of action. Int J Cosmet Sci 47(1):45–57

Manful ME, Ahmed L, Barry-Ryan C (2024) Cosmetic formulations from natural sources: safety considerations and legislative frameworks in the European Union. Cosmetics 11(3):72. https://doi.org/10.3390/cosmetics11030072

Manjula ML (2022) Consumer buying behavior of cosmetic products-a study. Asian J Multidimensional Res 11(10):57–60

McMullen RL, Dell'Acqua G (2023) History of natural ingredients in cosmetics. Cosmetics 10(3):71

Michalak M, Pierzak M, Kręcisz B, Suliga E (2021) Bioactive compounds for skin health: a review. Nutrients 13(1):203

Michalak M, Błońska-Sikora E, Dobros N, Spałek O, Zielińska A, Paradowska K (2024) Bioactive compounds, antioxidant properties, and cosmetic applications of selected cold-pressed plant oils from seeds. Cosmetics 11(5):153

Milam EC, Rieder EA (2021) An approach to cosmeceuticals. Essential psychiatry for the aesthetic practitioner. Wiley, Hoboken, pp 42–48

Mishra L, Kurmi BD (2023) Cosmetics regulations and standardization guidelines. Pharmaspire 15:137–150

Ngoc LTN, Moon JY, Lee YC (2023) Insights into bioactive peptides in cosmetics. Cosmetics 10(4):111

Oduaran E (2022) Understanding the role of corporate social responsibility in brand image, brand loyalty, and buyer purchase intention in the American Cosmetic Skincare Industry (Doctoral dissertation, Trident University International)

Pal M, Saxena R, Kanya R, Singh D, Diwakar D, Chauhan AS, Sharma A (2022) A detailed analysis of herbal cosmetics. NeuroQuantology 20(15):6367

Pérez-Rivero C, López-Gómez JP (2023) Unlocking the potential of fermentation in cosmetics: a review. Fermentation 9(5):463

Pistollato F, Madia F, Corvi R, Munn S, Grignard E, Paini A, Worth A, Bal-Price A, Prieto P, Casati S, Berggren E (2021) Current EU regulatory requirements for the assessment of chemicals and cosmetic products: challenges and opportunities for introducing new approach methodologies. Arch Toxicol 95:1867–1897

Rastogi S, Sharma G, Kandasubramanian B (2020) Nanomaterials and the environment. In: The ELSI handbook of nanotechnology: risk, safety, ELSI and commercialization. Wiley, Hoboken, pp 1–23

Rathee P, Sehrawat R, Rathee P, Khatkar A, Akkol EK, Khatkar S, Sobarzo-Sánchez E (2023) Polyphenols: natural preservatives with promising applications in food, cosmetics and pharma industries; problems and toxicity associated with synthetic preservatives; impact of misleading advertisements; recent trends in preservation and legislation. Materials 16(13):4793

Raza-Naqvi SA, Idrees F, Sherazi TA, Anjum-Shahzad S, Ul-Hassan S, Ashraf N (2022) Toxicology of heavy metals used in cosmetics. J Chil Chem Soc 67(3):5615–5622

Rizzi V, Gubitosa J, Fini P, Cosma P (2021) Neurocosmetics in skincare—the fascinating world of skin–brain connection: a review to explore ingredients, commercial products for skin aging, and cosmetic regulation. Cosmetics 8(3):66

Russell R (2011) Why cosmetics work. Sci Soc Vision 7:186–204

Rybczyńska-Tkaczyk K, Grenda A, Jakubczyk A, Kiersnowska K, Bik-Małodzińska M (2023) Natural compounds with antimicrobial properties in cosmetics. Pathogens 12(2):320

Sahu A, Srivastava N, Jena D, Singh A, Kumar S, Chauhan I (2024) Important regulatory guidelines on natural products. Curr Trad Med 10(6):152–172

Saleem A, Naureen I, Naeem M, Murad HS, Maqsood S, Tasleem G (2022) Aloe vera gel effect on skin and pharmacological properties. Sch Int J Anat Physiol 5(1):1–8

Sarkar S, Pandey A, Pant AB (2023) Regulatory requirements for safety/toxicity assessment of cosmetics/nanocosmetic products: challenges and opportunities. In: Skin 3-D models and cosmetics toxicity. Springer, Cham, pp 149–176

Sartwell C (2020) Six names of beauty. Routledge, London

Sharma DK, Suresh A, Kumari R, Kumar S, Tripathy DB (2024) Critical review on metal-based cosmetic products. Macromol Symp 413(1):2200205

Sharmeen JB, Mahomoodally FM, Zengin G, Maggi F (2021) Essential oils as natural sources of fragrance compounds for cosmetics and cosmeceuticals. Molecules 26(3):666

Sisti A, Aryan N, Sadeghi P (2021) What is beauty? Aesth Plast Surg 45(5):2163–2176

Srinivasulu M, Chandra MS, Gooty JM, Madhavi A (2022) Personal care products—fragrances, cosmetics, and sunscreens—in the environment. In: Environmental micropollutants: advances in pollution research. Elsevier, Amsterdam

Statista (2023) Europe: cosmetics consumption value 2012-2018. In: Micropollutants. Elsevier, Amsterdam, pp 131–149

Sukhnandan S et al (2023) A comparative study of the regulatory framework for cosmetics in South Africa, Europe, the United States of America and India (Doctoral dissertation)

Sullivan DA, da Costa AX, Del Duca E, Doll T, Grupcheva CN, Lazreg S, Markoulli M (2023) TFOS lifestyle: impact of cosmetics on the ocular surface. Ocul Surf 29:77–130

Tengli A, Srinivasan SH (2022) An exploratory study to identify the gender- based purchase behavior of consumers of natural cosmetics. Cosmetics 9(5):101. https://doi.org/10.3390/cosmetics9050101

Ullah H, Aslam S, Mustafa G, Waseem A, de Freitas Marques MB, Gul Z, Ibrahim M (2023) Potential toxicity of heavy metals in cosmetics: fake or fact: a review. Int J Environ Anal Chem 2023:2217406

United Nations (2015) Transforming our world: the 2030 Agenda for Sustainable Development. https://sdgs.un.org/goals

Vaz CR (2021) The paradigm of sustainability in the cosmetics industry (Doctoral dissertation)

Wargala E, Sławska M, Zalewska A, Toporowska M (2021) Health effects of dyes, minerals, and vitamins used in cosmetics. Women 1(4):223–237

Witaszczyk A, Klimowicz A (2023) Usefulness of aloe vera as a potential ingredient of cosmetic preparations. Pomeranian J Life Sci 69(3):76–87

Chapter 18
Life Cycle Assessment Approach in Eco-cosmeceuticals

Iram Liaqat, Nazish Mazhar Ali, Saiqa Andleeb, Sikander Ali, Nasib Zaman, Dawood Ahmed, Muhammad Nauman Aftab, and Guojing Yang

Contents

I. Liaqat (✉) · N. M. Ali
Microbiology Lab, Department of Zoology, Government College University, Lahore, Pakistan
e-mail: dr.iramliaqat@gcu.edu.pk

S. Andleeb
Department of Zoology, University of Azad Jammu and Kashmir, Muzaffarabad, Pakistan

S. Ali · M. N. Aftab
Ikram-ul-Haq Institute of Industrial Biotechnology, Government College University, Lahore, Pakistan

N. Zaman
Centre for Biotechnology and Microbiology, University of Swat, Charbagh, Pakistan

D. Ahmed
Department of Medical Laboratory Technology, The University of Haripur, Haripur, Pakistan

G. Yang
Laboratory of Tropical Environment and Health, Hainan Medical University, Haikou, Hainan, China

S. Javed et al. (eds.), *Sustainable Cosmeceuticals*,
https://doi.org/10.1007/978-3-031-86087-4_18

Abstract The skin is the largest and most sensitive part of the body and acts as an external defense against danger. There are one billion bacteria per square centimeter in the skin. In addition to helping improve the environment, it produces healthy microbiome metabolites that are necessary to maintain skin moisture, elasticity, and strength. Changing the habitat of bacteria on the skin, either orally or through the use of specific probiotics, can help restore the microbiome of healthy individuals. Cosmetics are products used to beautify the human body, enhance beauty, or change appearance. Cosmetics are products that combine cosmetics and medicine. Environmentally friendly methods have become important in many industries, including cosmetics. People's demand for environmental products continues to grow, and companies need to conduct life cycle assessments (LCAs) to assess the impact of their products on the environment throughout their lifespan. Sustainable development goals (SDGs) will balance environmental, economical, and social issues. Sustainable development will help in building inclusive, resilient earth for animals and humans. SDGs will make development for future generation without compromising the needs of the present generation. To ensure sustainability, cosmetics must be environmentally friendly and purposeful. The cosmetic industry is utilizing new ingredients such as gold nanoparticles to improve these effects. The aim of all research is to use life cycle analysis to estimate the ecological impact of beauty cream production during its life cycle. This includes everything from raw materials to production, packaging, transportation, and postuse disposal.

Keywords Microbiome · Probiotics · Cosmeceutical · Gold nanoparticles · Environmentally friendly · Life cycle assessment

18.1 Introduction

18.1.1 Integumentary System

The integumentary system is an extremely important system of the body's hair, skin, and nails (Visha et al. 2023; Caon 2024). Skin is the largest and most complex organ of the body, which serves against external hazards as an external defensive barrier. Maintaining internal homeostasis skin plays a key role in stability through sensations, regulation, and hydration, as well as protection against exposure to external aggression, such as ultraviolet radiation and temperature fluctuations (Agrawal et al. 2023; Jiao et al. 2024; Kumar 2024).

18.1.2 Skin Microbiota

Skin is a mobile organ that communicates closely with the complex environmental system of microorganisms, known as the skin's microbiome. Bacteria, fungi, viruses, and mites are about one billion microorganisms per square centimeter present in skin, and these microbes are important for the maintenance of healthy skin by providing protection against harmful agents, modifying immune responses, and maintaining the function of the skin barrier (Chen et al. 2023; Smythe and Wilkinson 2023; Criton and Joy 2024). In addition to its contribution to restoring a balanced environment, it produces healthy microbiome metabolites that are essential to avoid skin dehydration and provide elasticity and resilience. These include natural acids, polysaccharides, and various proteins that play a fundamental role in skin metabolism, such as the production of hyaluronic acid through hyaluronan synthase, ceramide synthesis via sphingomyelinase, and enhancement of melanocyte activity (Sharma et al. 2022; Woo and Kim 2024; Yang et al. 2024). *Staphylococcus epidermidis* and other coagulase-negative staphylococci have been regarded as the primary inhabitants of the skin. Other bacteria include coryneforms of Actinobacteria; including the genera *Corynebacterium*, *Propionibacterium*, and *Brevibacterium* and the genus *Micrococcus* sp. With the exception of some *Acinetobacter* spp., gram negative bacteria are generally not isolated from the skin but are thought to be contaminant from the gastrointestinal tract. Other microbes on skin include fungal species *Malassezia* spp., which are especially present in sebaceous areas. The *Demodex* mites (such as *Demodex folliculorum* and *Demodex brevis*), microscopic arthropods, are also part of skin flora. *Demodex* mites feed on sebum/oil and colonize sebaceous areas of the face more frequently following puberty. The role of commensal viruses has not been studied in depth, and most of the research is limited by the available molecular and microbiological techniques (Arshad et al. 2018; Grice and Segre 2011).

Staphylococcus epidermidis, a common bacterium, perhaps exemplifies the most frequent interaction with glycerol, which is a crucial component of the skin's outermost layer. Efforts to alter the habitats of resident bacteria in the human body, through various experimental and previous studies, have revealed that the use of probiotics whether orally or specifically targeted can contribute to the restoration of the microbiome in healthy individuals. The mechanisms of action of probiotics include competitive exclusion of pathogens for adhesion sites and also modulation improvement of the intestinal mucosal barrier and gut health and neurotransmitter synthesis. Probiotic strains prevent the adherence of the pathogenic bacterial strains to the host cells by strengthening the barrier effect of the intestinal mucosa

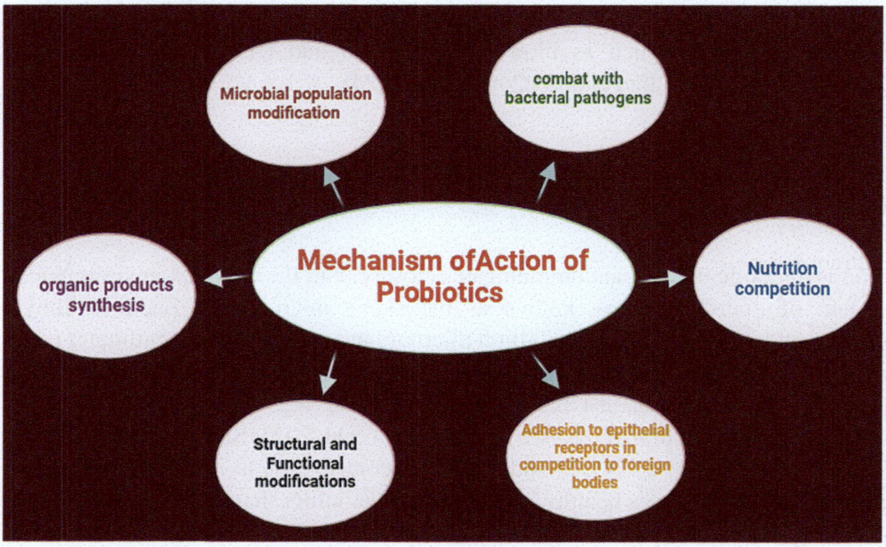

Fig. 18.1 Mechanism of action of probiotics (Theodorou et al. 2024)

Table 18.1 Postbiotics in cosmetics (Duarte et al. 2022)

Product name	Producer	Benefits
Lactolift	*Lactobacilli casei*	Antiwrinkle, skin moisturizer
Bifida	*Bifidobacterium longum*	Skin smoothing, skin conditioning
Sooth-BA	*Bifidobacterium*	Sensitivity and skin allergy reduction, skin brightening
Bebebiome	*Lactobacilli paracasei*	Skin regeneration, immunomodulation

(Fig. 18.1). This helps reduce inflammation and enhances immune modulation and epithelial barrier function through tight junction genes' upregulation (Baquero et al. 2021; Fang et al. 2021; Gowda et al. 2024).

18.1.3 Application of Postbiotics to the Skin

Postbiotics are products or metabolites of known bacteria or group of bacteria, which can be utilized for different purposes including overall health effect. Increase barrier function reduces transepidermal water causing increase in immunomodulatory properties, wound healing, and potential collagen stimulation as well as ceramide production. This has garnered significant attention from both scientific communities and the dermatocosmetic industry (Table 18.1) (Salminen et al. 2021; Araújo et al. 2024; Vinderola et al. 2024).

18.2 Cosmetics

18.2.1 American Federal Food, Drug, and Cosmetic Act

According to the American Federal Food, Drug, and Cosmetic Act, "Cosmetics are the articles intended to be rubbed, poured, sprinkled, or sprayed on, introduced into, or otherwise applied to the human body for cleansing, beautifying, promoting attractiveness, or altering the appearance" [FD&C Act, sec. 201(i)]. All cosmetics should be prepared and used according to the instructions labeled on cosmetics (Sisti et al. 2021; Rico et al. 2023; Serb et al. 2024).

18.2.2 European Regulation

European regulations define cosmetics as any product or mixture intended to be applied into the outside of the human body for the purpose of cleansing, changing the appearance, protecting, improving, or removing body odor (Hahladakis et al. 2023; Kumar et al. 2024). Emphasis should be on precautionary measures while using cosmetic products. Information relating to microbiological standards should be included in the cosmetic product safety report (Sarkar et al. 2023; Yossa et al. 2023; Abualhasan et al. 2024; Justo-Hanani 2024).

18.3 Cosmeceuticals

The word cosmeceutical is not recognized by the Food and Drug Administration (FDA) or the European Union (EU), but pharmacists, dermatologists, physicians, and dermatologists use it for the products that bridge the gap between cosmetics and pharmaceuticals. This term is commonly used in the field of skincare to refer to cosmetic products that can impact skin health relating it to medical benefits (Fig. 18.2) (Dhiman et al. 2024; Draelos 2024; Sathyaseelan et al. 2024; Shukla et al. 2024a, b). Cosmeceuticals contain bioactive compounds from natural sources (e.g., plants, animals, or bacteria), as well as synthetic substances (Gurumallu and Javaraiah 2021; Rohilla et al. 2023; Bharatia et al. 2024; Uğur et al. 2024). These cosmeceuticals are also used in teeth whitening, antiaging, and sun protection products. Cosmeceuticals are now a growing segment of the personal care industry (Awan et al. 2023; Mishra and Kurmi 2023; Hamdan 2024). Antiaging formulations for skin care also contain ingredients active between pharmaceuticals and cosmetics resulting in improvement in the appearance of the skin (Abdullahi et al. 2022; Altay Benetti et al. 2023; Darbre 2023; Gwanya et al. 2024).

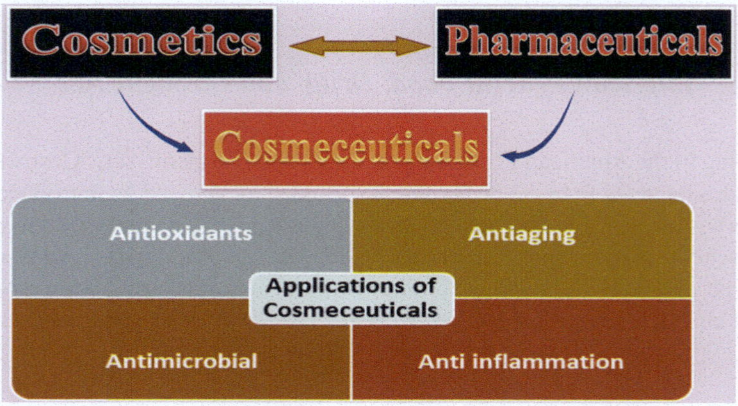

Fig. 18.2 Cosmeceuticals and their applications (Nirmal et al. 2024; Watts et al. 2024)

18.3.1 Important Personal Care Products: Cosmeceuticals

Countless tastes and scents have become integrated into everyday life, especially in food, drinks, and cosmetic products. Personal care products (such as soap, toothpaste, mouthwash, deodorant, lotion, and shampoo), perfumes, and other cosmetics are also included in pharmaceutical formulations (Sharmeen et al. 2021; Srinivasulu et al. 2022). In fact, taste and fragrance are incorporated to enhance or conceal the taste or odor of such products. In recent days, the revival of using natural and green consumerism has given a new trend for products based on plants, especially in the beauty (Saha et al. 2024; Wagener 2024; Zoninsein 2023).

18.3.2 Need of Cosmeceuticals

Cosmeceuticals create a defensive barrier in the body that protects skin from different chemicals, diseases, ultraviolet radiations, and mechanical shocks, together with regulating the body temperature (Kumar 2024; Valacchi and Ferrara 2024). To maintain normal physical activity of the skin, it is essential to have a well-balanced and hydrated diet, similar to other organs of the body. This nutrition, when combined with appropriate cosmetic products, will help sustain the skin's regular functions (Sahawneh 2024). Nevertheless, it must be remembered that these activities can be influenced by both intrinsic and extrinsic factors like medical conditions, vitamin deficiencies, and hormonal imbalances (Wu et al. 2023; Lingeswaran and Mohan 2024).

In this case, it is necessary to influence the use of active ingredients (AIs) and necessary medical products. Therefore, it is sometimes difficult to create a problem between cosmetics and medicine, as there are multiple boundaries between them.

Currently, customers specifically request cosmetic products with pharmaceutical properties, which makes categorizing the role of such specialized preparations in pharmaceutical or cosmetic products more challenging (Xu et al. 2023; Zeng et al. 2023; AlRuwaili et al. 2024).

The U.S. Food and Drug Administration (FDA) and the European Union (EU) do not have the authority to authorize cosmetic products before they are sold. However, these products must be effective and safe for consumers, and they should be properly labeled. According to this, the provision of accurate labeling and safety of products is the legal responsibility of companies (Iraldo et al. 2020; Morel et al. 2023; Nasrulloh and Qomaro 2024).

Cosmeceuticals not only include skincare and haircare products but also include specialized products for treating these conditions, such as photoaging, wrinkles, and hair loss, which are all signs of aging. Cosmetic products' international market is offering significant opportunities among many countries, particularly in the development and production of beautiful and effective products, along with the inclusion of new technologies (Muftiyanto and Hastuti 2023; Rohilla et al. 2023; Jiménez Bolaño et al. 2024).

18.4 Essential Oils

18.4.1 Sources of Essential Oils

Cosmetics include essential oils as essential or supplementary constituents. There are about 400,000 known species of ornamental and medicinal plants with about 60 families of essential oil plants (Kümmritz et al. 2023; Silalahi et al. 2023; Ebadollahi et al. 2020; Jayakodi et al. 2024).

18.4.2 Necessity of Essential Oils in Cosmetic Industries

Essential oil has been widely used around the world on a large scale in cosmetics products (Zaccardelli et al. 2021; Dilworth et al. 2024; Feng et al. 2024). In this way, a large quantity of essential oil is prepared at the global level to fuel the cosmetics industry as well as the pharmaceutical and aromatherapy industries (Bhattacharya et al. 2022; Assadpour et al. 2024; Mounira 2024; Sharma et al. 2024).

Mostly, demand comes from different markets: food and beverages (35%), fragrances, cosmetics, and aromatherapy (29%), household (16%), and pharmaceuticals (15%). As users have already been informed about the benefits of essential oils for health, a preference has been created toward foods and beverages containing these uplifting oils (Al-Harrasi et al. 2022; Gurkok and Sezen 2023; Khaliq and Mushtaq 2023).

18.4.3 Essential Oils Market

Due to increasing awareness among consumers about health issues, as well as the demand for natural and organic health products, the global essential oil market is also leading the way (Arshad 2024). It is possible to promote the demand for natural oils in fragrances, perfumes, cosmetics, thermal products, and relaxing applications. Essential oils and oleoresins are not only used in the food processing and sweetener industry but are also important in the flavor and aroma industry (Bolouri et al. 2022; Abere and Adetunji 2024; Aswandi et al. 2024). Flavor and fragrance manufacturers around the world are important buyers of the oil. Their sale has been introduced in the market and afterward it indicates a demand for essential oil (Kliszcz et al. 2021; Movahedi et al. 2024).

Especially in the cosmetics industry, essential oils like orange have generated significant revenue because they offer various health benefits, such as treating skin elasticity, strength, wounds, stretch marks, and other conditions (Chao 2021; Liu 2022). Lemon and orange essential oils also contain antibacterial properties, making them ideal ingredients for skin and hair care. In addition, the sterile properties of these oils create new opportunities for further economic development (Guzmán and Lucia 2021; Prakash et al. 2023; Grover et al. 2024; Verkerk 2024).

Some essential oils are produced very extensively (e.g., in 2008, orange oil production was ~51,000 tons, corn mint oil was ~32,000 tons, and lemon oil was ~9200 tons) (Schmidt 2020; Meng 2023). Essential oils derived from various fruits are the most popular. They produce the largest number of commercial natural flavors and fragrances (Bora et al. 2020). However, due to the rarity of some other produce, it is done on a very small scale. However, it is traded at very high prices. These prices can vary and may be related to shortages of raw materials, cutting issues, dependence on weather, or production constraints. The overall sales of essential oil industries represent several billion U.S. dollars in 2008 (Hughes et al. 2023; Botana and Cleaver 2024; Chauhan 2024; da Silva Ramos et al. 2024).

Europe holds the largest share in the global oil market, with Asia and North America following closely behind. Essential oils add a pleasant fragrance to candles and play a key role in the production of oils and other products (Afriani et al. 2024; O'Neil 2024; Putri and Putri 2024).

In 2017, the United States' petroleum industry saw an increase of approximately 10% in revenue, and it was expected to reach a market value of around 73 billion dollars by 2024. The essential oils that make up the majority of the U.S. market are orange essential oil, corn mint essential oil, and eucalyptus essential oil (Lucier and Parr 2020; Sharma et al. 2023; Sharmeen et al. 2021).

Furthermore, in the field of perfume manufacturing, France is not only rich in history but also serves as a major exporter of essential oils worldwide. During the past year, there has been an increase of approximately two billion U.S. dollars in the import price of essential oil and raisins from America. Some countries, unable to produce sufficient quantities of essential oils to meet local needs, must rely on income from other countries. For example, in 2017, Lebanon earned the highest oil revenue at the global level compared to any other country. Approximately 1.375 billion U.S. dollars, with Scandinavian countries such as Germany, the Netherlands, and the United Kingdom, account for approximately 80% of European orange revenue (Parvin et al. 2023; Abere and Adetunji 2024; MacCarthy et al. 2024).

18.4.4 Authenticity of Essential Oils

The production of essential oil on a large scale is one of the important issues associated with the production of crude oil. As a result, generally higher prices for oil are found in the international market (Hanieh 2024; Nagahama et al. 2024). The mixing of adulterated oil with essential oil is one of the clear reasons for the increasing demand for essential oil. Adulteration can potentially add harmful substances, especially in the area of immediate reactions (Shukla et al. 2024a, b). The presence of contamination and pollution is a cause for concern for both consumers and chemical companies. Therefore, verification is of utmost importance for both. Several techniques have been developed and implemented for the analysis of essential oils. While some have been simplified or modified for easier handling, others have maintained their significance and have consistently improved (Ni et al. 2021; Antunes Filho et al. 2023).

The long-term stability of their physical chemistry has been analyzed by measuring the essential oil's properties. These values indicate the standard of the oil and are important for determining its purity (Schmidt and Wanner 2020; Syafri et al. 2022). Solubility is indicated in terms of solubility in ethanol and its main properties (Fig. 18.3). On the other hand, the results of regular chemicals are used to estimate the main content of the oil, such as acid value, ester value, alcohol, peroxide, and ketone compounds, and are determined after distillation These parameters represent the properties of essential oil and are very helpful in assessing their quality (Akhter and Alam 2023; Peng and Shahidi 2023; Fotiadou et al. 2024).

18.5 Life Cycle Analysis (LCA) of Cosmeceuticals

Environmental-friendly approaches are becoming increasingly important in various industries, including cosmetics. Currently, there is a growing demand for sustainable alternatives concerning environmental issues related to cosmetic ingredients (Fig. 18.4). Sustainable development goals (SDGs) have been designed for implementation. The sustainable development goals (SDGs) are a set of new goals or targets or indicators for UN to frame agendas and policies for the next 15 years (Table 18.2). The implementation of sustainable development goals in the cosmetics industry is of primary importance. Starting with product design stage or phase to the formulation of fully functional cosmetics based on raw materials of natural origin is

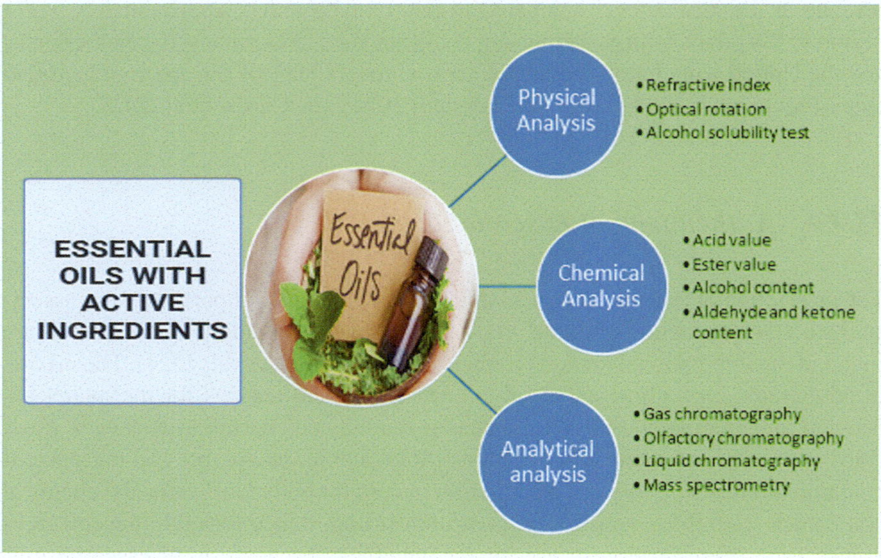

Fig. 18.3 Analysis of essential oils (Benchrifa et al. 2024; Repajić et al. 2024)

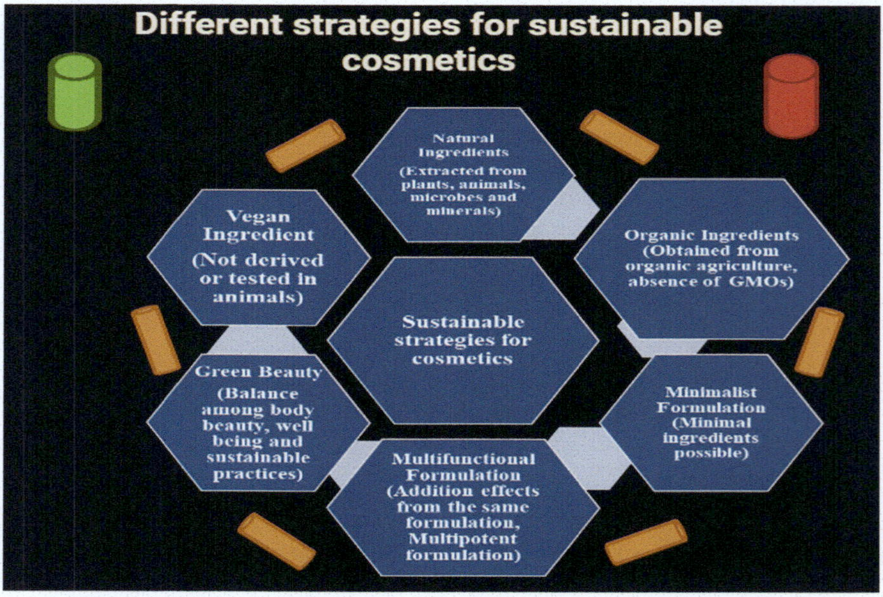

Fig. 18.4 Different strategies for sustainable cosmetics (Sasounian et al. 2024)

Table 18.2 17 sustainable development goals (SDGs) (Morsali and Yildirim 2024)

1.	No poverty
2.	Zero hunger
3.	Good health and well-being
4.	Quality education
5.	Gender equality
6.	Clean water and sanitation
7.	Affordable and clean energy
8.	Decent work and economic growth
9.	Industry, innovation, and infrastructure
10.	Reduced inequalities
11.	Sustainable cities and economies
12.	Responsible consumption and production
13.	Climate action
14.	Life below water
15.	Life on land
16.	Peace, justice, and strong institutions
17.	Partnership for the goals

preferred. The role of SDGs is crucial in environmental and consumer-friendly cosmetics products including cosmaceuticals.

Some of these components can have a negative impact on the environment, especially aquatic animals and plants, which has led to global demand for eco-friendly products. As market demand for environmentally friendly products continues to grow, companies need to conduct life cycle assessments (LCAs) to assess the environmental impacts of their products throughout their shelf life (Ajayi et al. 2024; Sasounian et al. 2024).

The cosmetics industry is a field that encompasses continuous advancement from chemical and mineral-based formulations to natural ingredients, along with bioactive properties. Like many other fields, terms such as natural and organic are prevalent. When choosing cosmetic products, consumers often seek clarity on terms such as paraben free, sulfate free, or silicone free. Many brands already offer a range of natural cosmetics, and many others are gaining expertise in this category (Pagels et al. 2022; Morganti et al. 2023).

Some plant extracts are also natural antioxidants. In cosmetic formulations, various types of plant extracts are commonly included, such as green tea, rosemary, or grape seed. The tea plant has been a tradition for centuries for medicinal use, and its extract plays an important role in the cosmetic market, primarily due to its antioxidant properties and phenolic compounds (Whitlock 2020; Hoang et al. 2021; Zohora and Arefin 2022; Masiala et al. 2024).

Another group that is recognized through natural ingredients is LG (both microwave and semiconductor), which has been used in formulations for medicinal and cosmetic purposes (Jahan and Ahmad 2024; Saini et al. 2024). Most of their cosmetic applications are related to their biological properties, such as their antioxidant, anti-inflammatory, antiaging abilities, along with cosmetic features like moisturizing and strengthening agents, and bioactive compounds in their composition (e.g., phycocyanin and flavonoids). Used in cosmetics, the three most common strains of algae are *Laminaria*, *Fucus* (brown algae), and *Chondrus* P (red algae) (Bharadvaja et al. 2023; Gomez-Molina et al. 2024; Leong et al. 2024).

It is necessary for cosmetics to be implemented in environmentally friendly ways and purposes in order to be recognized as sustainable. In this way, a comprehensive study on the environmental effects related to production has been suggested. Specifically, people who use antioxidants the most in cosmetics (such as vitamin C and green tea extracts) focus on *Fucus vesiculosus* (Limbu and Ahamed 2023; Suphasomboon and Vassanadumrongdee 2023; Ajayi et al. 2024). This seaweed extract is notable for its ingredients and its recognized antioxidant properties in phenolics and carotenoids (Carpena et al. 2023; Duan et al. 2023).

The environmental aspects and potential impacts of this action were assessed using the life cycle assessment (LCA) method, as it provides a systematic approach to analyze environmental effects related to materials and energy consumption during the life cycle of products or processes (Arshad et al. 2023). The concept of a lifecycle represents an essential criterion for analyzing the entire cycle of the social system's life, as sustainability aims to improve the overall quality of life without exceeding the sustainable level of resource utilization (Muazu et al. 2021; Sacchi et al. 2022; Pacana et al. 2023; Olanrewaju et al. 2024).

The cosmetics industry offers various products to enhance the appearance and feelings of consumers globally. Among these products, cosmetic creams are the most popular and widely used, including moisturizers, antiaging agents, and skin whitening and sun protection applications (Dube and Dube 2023; Shi 2023).

The cosmetic industry is using new ingredients such as gold nanoparticles (AuNPs) to increase their effectiveness. It is said that AuNPs possess antiaging and moisturizing properties for the skin (Pozos-Nonato et al. 2023; Rudolf et al. 2024). However, there are concerns about their potential toxicity, particularly particles as small as 20 nm have been found to have a greater ability to quickly enter compared to larger particles above 40 nm, as has been tested (Ghasemiyeh and Mohammadi-Samani 2020; Supe and Takudage 2021; Souto et al. 2022).

Such information exists that 28 nm AuNPs have a weak toxic reaction on the skin without hair, while no toxic reaction is apparent in the larger sizes of 49 and 73 nm AuNPs. This reaction is also obtained with cosmetic creams containing AuNPs, as the interaction between AMUPEs and cosmetic ingredients plays a significant role in toxicity (Saweres-Argüelles et al. 2023; Ko et al. 2024).

Studies have revealed the toxic and nontoxic nature of gold nanoparticles, which depends on their characteristics, preparations, and physicochemical properties. In this way, the overall effects of this have not been fully understood. Further studies are needed to focus on the changes in physical characteristics before and after treatment with vital media (Arif et al. 2024; Chakravarthy et al. 2024).

The production and harmful environmental effects of cosmetic creams should be assessed and understood. The aim of the investigation is to use life cycle analysis to estimate the environmental effects of cosmetic cream production throughout one's entire life. This includes everything from raw material acquisition, production, and packaging supply to transportation and handling of products after use (Ajayi et al. 2024; Manful et al. 2024).

The fundamental task of the LCA method is to provide essential information upon and can make environmentally friendly business decisions from it. For example, comparing the environmental impacts of different products, choosing different products, developing new products, choosing technologies that do not harm the environment, demonstrating layers of standard recycling, or choosing appropriate packaging (Ott et al. 2023; Subal et al. 2024). Pursued aim is to assess the environmental impacts of cosmetic creams on lives and to enhance the enthusiasm and excitement of marketing activities. Furthermore, increase awareness about environmental conservation will promote the use of eco-friendly products (Table 18.3).

18.5.1 Life Cycle Inventory Analysis and ISO 14044

Building on ISO 14040, ISO 14044 provides detailed requirements for conducting an LCA (ISO 14044). Specific criteria are required by ISO 14044 for choosing impactful categories, category indicators, and characterization models. This standard ensures that LCA studies are conducted with rigor and consistency (Arshad et al. 2023). ISO 14040 defines LCA as evaluation and compilation of inputs, outputs, and impacts of a product system through life cycle on environment.

The following steps are included in the inventory analysis:

1. Recognition of system boundaries: It includes determining which steps and activities are included in the inventory analysis, and which inputs and outputs are considered.
2. Collection of data: Information about the resources used within the limits of the system, for every operation and activity, input, and output. This can include raw materials used, energy consumption, water usage, disposal costs, and the impact of resources generated at every stage of life.
3. Converting numbers and figures into a standard unit: Converting written and measured data into standard measurements allows numbers to be compared and

Table 18.3 Phases of life cycle of product and LCA tool

Serial No.	Process	LCA tool
1.	Raw material extraction	From the extraction of raw materials needed to produced it until it is no longer used and thrown away or recycled
2.	Manufacturing and processing	
3.	Transportation	
4.	Usage and retail	
5.	Waste disposal	

added. For example, energy input can be converted to kilowatts per hour, while emissions can be converted to kilograms of carbon dioxide equivalent.
4. Quality check: This includes verifying numbers with industry experts or analyzing the impact of different assumptions and data sources.
5. Calculation of the inventory results: Software tools and libraries are used to calculate inputs and outputs at each stage of the lifecycle and to monitor the performance of the product throughout its lifecycle. It includes the use of special open-source software, in which pre-existing environmental data is included for various products and processes.
6. Documentation of the inventory: The methods and data used in the analysis will provide clear and consistent data to facilitate repeatability and comparison of results (Ncube et al. 2024; Subal et al. 2024).

18.6 Lipid Nanoparticles (LNPs) and their LCA

Solid lipid nanoparticles (SLNs), lipid nanoemulsions (LNEs), and nanostructured lipid carriers (NLCs) are different kinds of lipid nanoparticles (LN) used in cosmetic technology and pharmaceutical. All types of LPNs in these categories have a lipophilic (lipid) phase, but their identity can be determined by their core structure. Recently, a new generation was developed using LPHNs (lipid-polymer hybrid nanoparticles), and it is referred to as the core-shell nanostructure of the next generation (Agha et al. 2023; Mehta et al. 2023).

Mostly used sizes in the pharmaceutical and cosmetic fields represent the range of 150–300 nm. Mostly, in the production of LNPs, the majority of the components are approved by both the FDA and the EMA, which includes molecules of semisynthetic lipophilic components that are incorporated to enhance the encapsulation of active ingredients (AIs). According to this, these components are nontoxic and biodegradable (Mehta et al. 2023; Biscaia-Caleiras et al. 2024).

18.6.1 Advantages of Cosmeceuticals Nanoparticles

Cosmetic products can be used to obtain multiple benefits by using nano particles.

1. There is an increase in the stability of sensitive cosmetic active ingredients (CAIs) in it.
2. Effective protection against the harmful effects of ultraviolet (UV) rays.
3. To obtain essential and attractive cosmetic products.
4. Edit the output of the desired active components to the required extent.

Recently, the United Nations adopted and embraced the goals of sustainable development. Integration of specific principles for sustainable development is incorporated throughout the entire lifecycle of a cosmetic product, starting from product design and concluding with user usage. It is considered a crucial goal for many cosmetic industries (Prajapati and Kumar 2024; Shukla et al. 2024a, b).

18.7 Smart and Sustainable Nanomaterials

Lipid nanoparticles (LNPs) technology is a fertile ground that can contribute to sustainable development. Lipid nanoparticles are often reported in various technologies because they can fulfill the requirements of sustainable development, as they are mostly composed of renewable natural and biodegradable materials. LNPs technology could satisfy ambitious green growth. LNPs are often consisted of edible natural and biodegradable materials. Recent progress has been made via producing and using safe nanoingredients such as nanocellulose derivatives instead of synthetic nanopolymers or biosynthesis/green synthesis involving the use of less toxic and renewable reagents, solvents, and starting materials. Significant progress is evident in this field. For instance, instead of synthetic nano polymers, the production and utilization of sustainable nano materials such as nano cellulose derivatives or green chemistry processes have been implemented, incorporating safer and renewable reagents, solvents, and starting materials. Although nanotechnology has made significant advancements in the field of materials and toxic behavior understanding in recent decades, it has raised various concerns about safety and stability, especially regarding products prepared for the internal administration, such as pharmaceuticals (Arshad et al. 2014, 2023). Recently, regulatory agencies have expressed concerns about assessing risks and governing such types of nanomaterials that can respond to external stimuli, also known as "smart nanomaterials." The use of nanotechnology field in the last decade and quest to understand the environmental and toxicological behavior of nanomaterials has opened various concerns regarding the safety and sustainability applications, especially those products designed for internal administration including nano pharmaceutical products (Asghar et al. 2024; Saleh 2024).

Currently, the controlled delivery system has streamlined its process from pharmaceutical products to cosmetics. A new class of products was created by the manufacturers of LNPs who respond to stimuli, and it was commercialized so that it could be associated with cosmetic active ingredients (CAIs) that have the ability to treat various skin issues such as contact dermatitis and damage to the skin's image. In this way, it is essential to encourage the development of a safe, effective, and sustainable new smart nanocosmetic. It is necessary to establish a fast but flexible regulatory framework that can maintain innovation without creating undesirable barriers for the industry. In this regard, LNPs can be a suitable candidate because they meet the mentioned high standards (Shah et al. 2021; Parhi 2022; Draelos 2024; Giradkar et al. 2024).

18.8 Conclusion

The skin is a dynamic organ that interacts closely with the complex ecosystem of microorganisms known as the skin microbiome. Microorganisms are essential for healthy skin care by providing protection against harmful agents, modifying the immune response, and maintaining the barrier function of the skin. Natural acids, polysaccharides, and various proteins that play a fundamental role in skin metabolism, such as the production of hyaluronic acid through hyaluronan synthesis, the synthesis of ceramide via sphingomyelins, and postbiotics, are easily absorbed, rapidly proliferate, support, and restore our skin health. Research indicates that postbiotics function as signaling molecules, aiding in enhancing the skin's immune response, reducing inflammation, and supporting healthy restoration. Cosmetic skin care products are those that can only be applied to the outer surface of the skin. Since cosmetics cannot penetrate the epidermis into the dermis, they cannot directly treat any specific disease and the enhancement of melanocyte activity. On the other hand, cosmeceuticals are more potent and long-lasting skincare products. While cosmetics can be purchased at the counter, a prescription is required to obtain a cosmeceutical item. This is because cosmeceuticals contain higher levels of active ingredients that penetrate the epidermis and reach the depths of the dermis. When choosing between cosmetics and cosmeceuticals, there are many factors to consider, such as the following: Your skin care condition or concern; your desired results; price and self-treatment, or professional intervention. Life Cycle Assessment (LCA) is a method for evaluating the environmental impacts associated with all stages of the life of cosmetic products. LCA is an important tool to support ecodesign.

18.9 Future Perspective

The cosmetics industry and the research community are striving to adopt this concept and move toward sustainable production at every level. Life cycle assessment of cosmetic and cosmeceutical products is essential to assess sustainability, which appears as an industrial necessity, and the need to facilitate decision-making and evaluate the impacts of current strategies, products, and technology. Furthermore, this topic has also attracted the attention of consumers and various organizations. Concerns regarding the safety of cosmetic products, environmental impacts, and social effects, such as those arising from unfair trade, have emerged and made this subject a matter of great interest.

References

Abdullahi A, Tijjani A, Abubakar AI et al (2022) Plant biomolecule antimicrobials: an alternative control measures for food security and safety. In: Herbal biomolecules in healthcare applications. Elsevier, Amsterdam, pp 381–406

Abere BO, Adetunji CO (2024) Circular, bioeconomy, and gross domestic product perspective of essential oils in the food industry. In: Applications of essential oils in the food industry. Elsevier, Amsterdam, pp 275–284

Abualhasan M, Naffa L, Alarda RA et al (2024) Heavy metal and microbial testing of selected cosmetic products in the Palestinian market. J Environ Sci Health 42(1):1–15

Afriani L, Fajarianto O, Harimurti ER et al (2024) Aromatherapy candles as double protection: a modern solution to a classic problem. IJESS Int J Educ Soc Sci 5(1):155–160

Agha A, Waheed W, Stiharu I et al (2023) A review on microfluidic-assisted nanoparticle synthesis, and their applications using multiscale simulation methods. Discover Nano 18(1):18

Agrawal R, Hu A, Bollag WB (2023) The skin and inflamm-aging. Biology 12(11):1396

Ajayi SA, Olaniyi OO, Oladoyinbo TO et al (2024) Sustainable sourcing of organic skincare ingredients: a critical analysis of ethical concerns and environmental implications. Asian J Adv Res Rep 18(1):65–91

Akhter M, Alam MM (2023) Physical pharmacy and instrumental methods of analysis. Springer, Cham

Al-Harrasi A, Bhatia S, Chigurupati S et al (2022) Global herbal drug market and its regulations. In: Recent advances in natural products science. Routledge, London, pp 1–34

AlRuwaili R, Al-Kuraishy HM, Alruwaili M et al (2024) The potential therapeutic effect of phosphodiesterase 5 inhibitors in the acute ischemic stroke (AIS). Mol Cell Biochem 479(5):1267–1278

Altay Benetti A, Tarbox T, Benetti C (2023) Current insights into the formulation and delivery of therapeutic and cosmeceutical agents for aging skin. Cosmetics 10(2):54

Antunes Filho S, Dos Santos MS, Dos Santos OAL et al (2023) Biosynthesis of nanoparticles using plant extracts and essential oils. Molecules 28(7):3060

Araújo D, Silva AR, Fernandes R (2024) Emerging approaches for mitigating biofilm-formation-associated infections in farm, wild, and companion animals. Pathogens 13(4):320

Arif MD, Hoque ME, Rahman MZ et al (2024) Emerging directions in green nanomaterials: synthesis, physicochemical properties and applications. Mater Today Commun 40:109335

Arshad M (2024) Sustainability of renewable energy options as compared to coal-fired power plants in Pakistan. Green Low-Carbon Econ 2(3):162–173

Arshad M, Adil M, Sikandar A, Hussain T (2014) Exploitation of meat industry by-products for biodiesel production: Pakistan's perspective. Pak J Life Soc Sci 12:120–125

Arshad M, Bano I, Younus M, Khan A, Rahman A (2018) Health concerns associated with biofuel production. In: Perspectives on water usage for biofuels production: aquatic contamination and climate change. Springer, Cham, pp 97–105

Arshad M, Syed-Hassan SSA, Masood R, Ansari AR, Mumtaz A, Rahman A, Batool S, Atif FA, Muhammad A, Rehman AU (2023) Utilization of waste animal fat for sustainable biodiesel production. In: Climate changes mitigation and sustainable bioenergy harvest through animal waste: sustainable environmental implications of animal waste. Springer, Cham, pp 149–169

Asghar N, Hussain A, Nguyen DA et al (2024) Advancement in nanomaterials for environmental pollutants remediation: a systematic review on bibliometrics analysis, material types, synthesis pathways, and related mechanisms. J Nanobiotechnol 22(1):26

Assadpour E, Can Karaça A, Fasamanesh M et al (2024) Application of essential oils as natural biopesticides; recent advances. Crit Rev Food Sci Nutr 64(19):6477–6497

Aswandi A, Kholibrina CR, Kuspradini H (2024) Essential oils for cosmetics application. In: Biomass-based cosmetics: research trends and future outlook. Springer, Singapore, pp 151–173

Awan UA, Saeed RF, Ismail H (2023) Emerging trends of nanotechnology-based advanced cosmeceuticals. e-Journal Surf Sci Nanotechnol 22(2):86–97

Baquero F, Saralegui C, Marcos-Mencía D et al (2021) Epidermis as a platform for bacterial transmission. Front Immunol 12:774018
Benchrifa M, Mabrouki J, Ariss A (2024) Study of the physicochemical properties of lavender essential oil (Lavandula Stoechas). In: Advanced systems for environmental monitoring, IoT and the application of artificial intelligence. Springer, Cham, pp 329–341
Bharadvaja N, Gautam S, Singh H (2023) Natural polyphenols: a promising bioactive compounds for skin care and cosmetics. Mol Biol Rep 50(2):1817–1828
Bharatia R, Gupta N, Upadhyay P et al (2024) A review on cosmeceuticals. Int J Pharm Professional's Res 15(1):128–141
Bhattacharya S, Saha T, Das P et al (2022) Aromatic plants: role and uses in human prosperity and sustainability. Pharma Innov J 11(5):341–348
Biscaia-Caleiras M, Fonseca NA, Lourenço AS et al (2024) Rational formulation and industrial manufacturing of lipid-based complex injectables: landmarks and trends. J Control Release 373:617–639
Bolouri P, Salami R, Kouhi S et al (2022) Applications of essential oils and plant extracts in different industries. Molecules 27(24):8999
Bora H, Kamle M, Mahato DK et al (2020) Citrus essential oils (CEOs) and their applications in food: an overview. Plants 9(3):357
Botana F, Cleaver L (2024) The economics of the manuscript and rare book trade ca. 1890–1939. Arc Humanities Press, Ashland, p 134
Caon M (2024) Lecture 3: the integument. In: Lecture notes, worksheets, and exercises for basic anatomy and physiology. Springer, Cham, pp 25–30
Carpena M, García-Pérez P, García-Oliveira P et al (2023) Biological properties and potential of compounds extracted from red seaweeds. Phytochem Rev 22(6):1509–1540
Chakravarthy KK, Brahmanandam PS, Uma G et al (2024) Toxicity associated with gold nanoparticles: a review. Trends Biomater Artif Organs 38(2):121–127
Chao J (2021) Skincare for your soul: achieving outer beauty and inner peace with korean skincare. Mango Media Inc., Hyderabad
Chauhan K (2024) Commercial significance of medicinal and aromatic plants of India: importancia comercial de las plantas medicinales y aromáticas de la Índia. South Florida J Environ Anim Sci 4(1):2–23
Chen Y, Knight R, Gallo RL (2023) Evolving approaches to profiling the microbiome in skin disease. Front Immunol 14:1151527
Criton VS, Joy S (2024) Beyond skincare routines: follow your gut to healthy skin–a review of the interplay between gut microbiome and skin. J Skin Sex Trans Dis 6(1):5
da Silva Ramos A, Félix PH, Teixeira-Costa BE (2024) Use of essential oils in food industries—application as functional compounds in food matrices and packaging. In: Essential oils. Elsevier, Amsterdam, pp 61–78
Darbre PD (2023) Cosmeceuticals. In: Personal care products and human health. Elsevier, Amsterdam, pp 79–95
Dhiman S, Mannan A, Mehta D et al (2024) Nanodrug delivery systems for cosmetics. In: Nanotechnology and drug delivery. Jenny Stanford Publishing, Singapore, pp 469–505
Dilworth LL, Riley CK, Stennett DK (2024) Plant constituents: carbohydrates, oils, resins, balsams, and plant hormones. In: Pharmacognosy. Academic Press, New York, pp 49–74
Draelos ZD (2024) Cosmeceuticals: procedures in cosmetic dermatology series. Elsevier, Amsterdam
Duan X, Subbiah V, Xie C et al (2023) Evaluation of the antioxidant potential of brown seaweeds extracted by different solvents and characterization of their phenolic compounds by LC–ESI-QTOF–MS/MS. J Food Sci 88(9):3737–3757
Duarte M, Oliveira AL, Oliveira C et al (2022) Current postbiotics in the cosmetic market—an update and development opportunities. Appl Microbiol Biotechnol 106(18):5879–5891
Dube M, Dube S (2023) Towards sustainable color cosmetics packaging. Cosmetics 10(5):139

Ebadollahi A, Ziaee M, Palla F (2020) Essential oils extracted from different species of the Lamiaceae plant family as prospective bioagents against several detrimental pests. Molecules 25(7):1556

Fang Z, Li L, Zhang H et al (2021) Gut microbiota, probiotics, and their interactions in prevention and treatment of atopic dermatitis: a review. Front Immunol 12:720393

Feng A, Si T, Du B et al (2024) Antiviral potential of essential oils and their plant sources, delivery methods, and drug development strategies with special focus on their mechanisms: a comprehensive review. Phytochem Rev 2024:1–23

Fotiadou R, Vougiouklaki D, Houhoula D et al (2024) Improvement of the oxidative stability of edible oils through enzymatic esterification with hydroxytyrosol-rich extract. Eur J Lipid Sci Technol 126:2400014

Ghasemiyeh P, Mohammadi-Samani S (2020) Potential of nanoparticles as permeation enhancers and targeted delivery options for skin: advantages and disadvantages. Drug Des Devel Ther 14:3271–3289

Giradkar V, Mhaske A, Shukla R (2024) Nanocrystals: a multifaceted regimen for dermatological ailments. Part Part Syst Charact 41:2300147

Gomez-Molina M, Albaladejo-Marico L, Yepes-Molina L et al (2024) Exploring phenolic compounds in crop by-products for cosmetic efficacy. Int J Mol Sci 25(11):5884

Gowda V, Sarkar R, Verma D et al (2024) Probiotics in dermatology: an evidence-based approach. Indian Dermatol Online J 15(4):571–583

Grover S, Aggarwal P, Kumar A et al (2024) Utilizing citrus peel waste: a review of essential oil extraction, characterization, and food-industry potential. Biomass Convers Biorefinery 2024:1–22

Gurkok S, Sezen S (2023) Uses of essential oils in different sectors. In: Essential oils: extraction methods and applications. Wiley, Hoboken, pp 207–228

Gurumallu SC, Javaraiah R (2021) Bioactives in disease prevention and health promotion: exploiting combinatorial effects. Curr Bioactive Compd 17(4):299–317

Guzmán E, Lucia A (2021) Essential oils and their individual components in cosmetic products. Cosmetics 8(4):114

Gwanya HY, Mbuyiswa AAN, Phokwe OJ et al (2024) Harnessing the potential of helinus integrifolius in cosmeceutical research: toward sustainable natural cosmetics. Cosmetics 11(4):126

Hahladakis JN, Iacovidou E, Gerassimidou S (2023) An overview of the occurrence, fate, and human risks of the bisphenol-A present in plastic materials, components, and products. Integr Environ Assess Manag 19(1):45–62

Hamdan R (2024) The impact of marketing communication tools for skin care products on Jordanian adolescents (Master's thesis, İzmir Ekonomi Üniversitesi)

Hanieh A (2024) Crude capitalism: oil, corporate power, and the making of the world market. Verso Books, Brooklyn

Hoang HT, Moon JY, Lee YC (2021) Natural antioxidants from plant extracts in skincare cosmetics: recent applications, challenges and perspectives. Cosmetics 8(4):106

Hughes LJ, Morton O, Scheffers BR et al (2023) The ecological drivers and consequences of wildlife trade. Biol Rev 98(3):775–791

Iraldo F, Griesshammer R, Kahlenborn W (2020) The future of ecolabels. Int J Life Cycle Assess 25(5):833–839

Jahan A, Ahmad IZ (2024) Algae-based cosmeceutical products in the market–present scenario and future perspectives. In: Algal farming systems. Apple Academic Press, Palm Bay, pp 357–372

Jayakodi Y, Thiviya P, Gamage A et al (2024) Antioxidant activity of essential oils extracted from Apiaceae family plants. Agrochemicals 3(1):57–69

Jiao Q, Zhi L, You B et al (2024) Skin homeostasis: mechanism and influencing factors. J Cosmet Dermatol 23(5):1518–1526

Jiménez Bolaño DC, Insuasty D, Rodríguez Macías JD et al (2024) Potential use of tomato peel, a rich source of lycopene, for cancer treatment. Molecules 29(13):3079

Justo-Hanani R (2024) The spread of nano-specific risk regulation: the EUs international regulatory influence. In: Governing nanotechnology safety. Edward Elgar Publishing, Cheltenham, pp 57–83

Khaliq FA, Mushtaq A (2023) The five most traded compounds worldwide: importance, opportunities, and risks. Int J Chem Biochem Sci 23:185–194

Kliszcz A, Danel A, Puła J et al (2021) Fleeting beauty—the world of plant fragrances and their application. Molecules 26(9):2473

Ko J, Kim JU, Choi S et al (2024) Recent advances in nanotechnology-mediated noninvasive transdermal and topical delivery of proteins. Small Sci 4:2400175

Kumar MA (2024) The skin. Techniques in small animal wound management. Wiley, Hoboken, p 1

Kumar M, Mandal UK, Mahmood S (2024) Dermatological formulations. In: Dosage forms, formulation developments and regulations. Academic Press, New York, pp 613–642

Kümmritz S, Klocke B, Krähmer A (2023) The potential use of steam distillation residues from medicinal plant material as a natural agricultural agent. In: Medicinal agroecology. CRC Press, Boca Raton, pp 123–139

Leong EJ, Tan LF, Yap VL et al (2024) Cosmetological applications of Citrus limon: a mini-review. Indian J Nat Prod Resour 15(2):286–293

Limbu YB, Ahamed AJ (2023) What influences green cosmetics purchase intention and behavior? A systematic review and future research agenda. Sustainability 15(15):11881

Lingeswaran R, Mohan J (2024) Scientific interpretation of Hahnemann's theory on disease causation and homeopathy as a holistic medical system. Indian J Integr Med 4(3):71–73

Liu JK (2022) Natural products in cosmetics. Nat Prod Bioprospect 12(1):40

Lucier G, Parr B (2020) Vegetable and pulses outlook. United States Department of Agriculture

MacCarthy BL, Das S, Ahmed WA (2024) Smell the perfume: can blockchain guarantee the provenance of key product ingredients in the fragrance industry? Sustainability 16(14):6217

Manful ME, Ahmed L, Barry-Ryan C (2024) Cosmetic formulations from natural sources: safety considerations and legislative frameworks in the European Union. Cosmetics 11(3):72

Masiala A, Vingadassalon A, Aurore SG (2024) Polyphenols in edible plant leaves: an overview of their occurrence and health properties. Food Funct 14:1752

Mehta M, Bui TA, Yang X et al (2023) Lipid-based nanoparticles for drug/gene delivery: an overview of the production techniques and difficulties encountered in their industrial development. ACS Mater 3(6):600–619

Meng Y (2023) SC-CO2 processing: extraction, scale-up and impregnation of essential oil from mentha× piperita

Mishra L, Kurmi BD (2023) Cosmetics regulations and standardization guidelines. Pharmaspire 15:137–150

Morel S, Sapino S, Peira E et al (2023) Regulatory requirements for exporting cosmetic products to extra-EU countries. Cosmetics 10(2):62

Morganti P, Lohani A, Gagliardini A et al (2023) Active ingredients and carriers in nutritional eco-cosmetics. Compounds 3(1):122–141

Morsali S, Yildirim F (2024) Environmental impact assessment of red mud utilization in concrete production: a life cycle assessment study. Environ Dev Sustain 26(5):12219–12238

Mounira GM (2024) Essential oils and their bioactive molecules: recent advances and new applications. In: Essential oils-recent advances new perspectives and applications. IntechOpen, London

Movahedi F, Nirmal N, Wang P et al (2024) Recent advances in essential oils and their nanoformulations for poultry feed. J Anim Sci Biotechnol 15(1):110

Muazu RI, Rothman R, Maltby L (2021) Integrating life cycle assessment and environmental risk assessment: a critical review. J Clean Prod 293:126120

Muftiyanto RTN, Hastuti I (2023) Increasing sales with the roles of marketing mix (a study of the cosmetics business). In: Proceeding of international conference on science health and technology. Elsevier, Amsterdam, pp 501–508

Nagahama N, González SB, Grech MG et al (2024) Effect of pruning intensity on biomass and essential oil yield in wild paramela (Adesmia boronioides Hook f.). J Appl Res Med Aromat Plants 42:100562

Nasrulloh N, Qomaro GW (2024) Legal responsibility on halal labelling of packaged-processes food in Bangkalan. In: 4th international conference on halal development (4th ICHaD 2023). Atlantis Press, Amsterdam, pp 11–22

Ncube A, Fiorentino G, Panfilo C et al (2024) Circular economy paths in the olive oil industry: a life cycle assessment look into environmental performance and benefits. Int J Life Cycle Assess 29(8):1541–1561

Ni ZJ, Wang X, Shen Y et al (2021) Recent updates on the chemistry, bioactivities, mode of action, and industrial applications of plant essential oils. Trends Food Sci Technol 110:78–89

Nirmal N, Demir D, Ceylan S et al (2024) Polysaccharides from shell waste of shellfish and their applications in the cosmeceutical industry: a review. Int J Biol Macromol 265:131119

O'Neil SK (2024) The United State' missed opportunity in Latin America: economic security begins closer to home. Foreign Aff 103:134

Olanrewaju OI, Enegbuma WI, Donn M (2024) Challenges in life cycle assessment implementation for construction environmental product declaration development: a mixed approach and global perspective. Sustainable Prod Consumption 49:502–528

Ott D, Goyal S, Reuss R et al (2023) LCA as decision support tool in the food and feed sector: evidence from R&D case studies. Environ Syst Decis 43(1):129–141

Pacana A, Siwiec D, Bednárová L et al (2023) Improving the process of product design in a phase of life cycle assessment (LCA). Processes 11(9):2579

Pagels F, Arias A, Guerreiro A et al (2022) Seaweed cosmetics under the spotlight of sustainability. Phycology 2(4):374–383

Parhi R (2022) Recent advances in microneedle designs and their applications in drug and cosmeceutical delivery. J Drug Deliv Sci Technol 75:103639

Parvin S, Reza A, Das S et al (2023) Potential role and international trade of medicinal and aromatic plants in the world. Eur J Agric Food Sci 5(5):89–99

Peng H, Shahidi F (2023) Metabolic, toxicological, chemical, and commercial perspectives on esterification of dietary polyphenols: a review. Crit Rev Food Sci Nutr 64(21):7465–7504

Pozos-Nonato S, Domínguez-Delgado CL, Campos-Santander KA et al (2023) Novel nanotechnological strategies for skin anti-aging. Curr Pharm Biotechnol 24(11):1397–1419

Prajapati A, Kumar R (2024) A comprehensive review of the use of nanoparticles in cosmeceutical science. J Pharm Insights Res 2(3):30–37

Prakash B, Dubey NK, de São José JF (eds) (2023) Plant essential oils: from traditional to modern-day application. Springer, Cham

Putri A, Putri PG (2024) Sensory analysis of several aromatherapy scented candle formulations using cinnamon essential oil. Open Access Res J Sci Technol 11(1):94–99

Repajić M, Elez Garofulić I, Marčac Duraković N et al (2024) Physico-chemical characterization of encapsulated fennel essential oil under the influence of spray-drying conditions. Processes 12(3):577

Rico F, Mazabel A, Egurrola G et al (2023) Meta-analysis and analytical methods in cosmetics formulation: a review. Cosmetics 11(1):1

Rohilla S, Rohilla A, Narwal S et al (2023) Global trends of cosmeceutical in nanotechnology: a review. Pharm Nanotechnol 11(5):410–424

Rudolf R, Majerič P, Jelen Ž et al (2024) The environmental impact of using gold nanoparticles and 3HFWC in cosmetics, as determined with LCA methodology. Rev Inorg Chem 44:409–419

Sacchi R, Terlouw T, Siala K et al (2022) Prospective environmental impact assement (premise): a streamlined approach to producing databases for prospective life cycle assessment using integrated assessment models. Renew Sust Energ Rev 160:112311

Saha S, Hazari M, Chaudhuri S (2024) Application of exopolysaccharides in cosmetics. In: Microbial exopolysaccharides. CRC Press, Boca Raton, pp 215–249

Sahawneh P (2024) Factors influencing skin health from within. J Integr Health 3(1):156–163

Saini KC, Rani A, Gupta SK et al (2024) Algae a potential source in cosmetics: current status, challenges, and economic implications. Indian J Microbiol 64(4):1445–1460

Saleh TA (2024) Trends in nanomaterial types, synthesis methods, properties and uses: toxicity, environmental concerns and economic viability. Nano Struct Nano-Objects 37:101109

Salminen S, Collado MC, Endo A et al (2021) The International Scientific Association of Probiotics and Prebiotics (ISAPP) consensus statement on the definition and scope of postbiotics. Nat Rev Gastroenterol Hepatol 18(9):649–667

Sarkar S, Pandey A, Pant AB (2023) Regulatory requirements for safety/toxicity assessment of cosmetics/nanocosmetic products: challenges and opportunities. In: Skin 3-D models and cosmetics toxicity. Springer, Cham, pp 149–176

Sasounian R, Martinez RM, Lopes AM et al (2024) Innovative approaches to an eco-friendly cosmetic industry: a review of sustainable ingredients. Clean Technol 6(1):176–198

Sathyaseelan S, Rao BH, Anushmati S (2024) Cosmeceuticals: a transit state from synthetic to natural. Indian J Pharm 56(1):42–51

Saweres-Argüelles C, Ramirez-Novillo I, Vergara-Barberan M et al (2023) Skin absorption of inorganic nanoparticles and their toxicity: a review. Eur J Pharm Biopharm 182:128–140

Schmidt E (2020) Production of essential oils. In: Handbook of essential oils. CRC Press, Boca Raton, pp 125–160

Schmidt E, Wanner J (2020) Adulteration of essential oils. In: Handbook of essential oils. CRC Press, Boca Raton, pp 543–580

Serb AF, Georgescu M, Onulov R et al (2024) Mass-spectrometry-based research of cosmetic ingredients. Molecules 29(6):1336

Shah P, Gandhi J, Kansara Y et al (2021) Leveraging nanotechnology in cosmeceuticals: formulation, characterisation, regulatory status and toxicity. In: Nanocarriers: drug delivery system: an evidence based approach. Springer, Cham, pp 411–456

Sharma G, Khanna G, Sharma P et al (2022) Mechanistic role of probiotics in improving skin health. In: Probiotic research in therapeutics: volume 3: probiotics and gut skin axis–inside out and outside in. Springer, Cham, pp 27–47

Sharma A, Kumar V, Mittal C et al (2023) Role of essential oil used pharmaceutical cosmetic product. J Res Appl Sci Biotechnol 2(3):147–157

Sharma C, Anwar A, Yuksel AK et al (2024) Essential oils: an essential component of cosmetic industry. In: Specialized plant metabolites as cosmeceuticals. Elsevier, Amsterdam, pp 221–238

Sharmeen JB, Mahomoodally FM, Zengin G et al (2021) Essential oils as natural sources of fragrance compounds for cosmetics and cosmeceuticals. Molecules 26(3):666

Shi R (2023) The influence of marketing strategy on consumer behavior in the cosmetics industry. Highlights Business Econ Manage 23:623–629

Shukla A, Kushwaha A, Sanju S et al (2024a) Essential oils toxicity and conflicts. In: Aromatherapy: the science of essential oils. Bentham Science, Sharjah, p 124

Shukla S, Mhaske A, Shukla R (2024b) Nanocarriers for the delivery of cosmeceuticals. In: Nanotechnology based delivery of phytoconstituents and cosmeceuticals. Springer, Singapore, pp 305–328

Silalahi M, Purba EC, Sawitri IGAR et al (2023) International trade of medicinal and aromatic plants (MAPs). In: Medicinal plants: biodiversity, biotechnology and conservation. Springer, Singapore, pp 289–306

Sisti A, Aryan N, Sadeghi P (2021) What is beauty? Aesth Plast Surg 45(5):2163–2176

Smythe P, Wilkinson HN (2023) The skin microbiome: current landscape and future opportunities. Int J Mol Sci 24(4):3950

Souto EB, Fangueiro JF, Fernandes AR et al (2022) Physicochemical and biopharmaceutical aspects influencing skin permeation and role of SLN and NLC for skin drug delivery. Heliyon 8(2):e08938

Srinivasulu M, Chandra MS, Gooty JM et al (2022) Personal care products—fragrances, cosmetics, and sunscreens—in the environment. In: Environmental micropollutants. Elsevier, Amsterdam, pp 131–149

Subal L, Braunschweig A, Hellweg S (2024) The relevance of life cycle assessment to decision-making in companies and public authorities. J Clean Prod 435:140520

Supe S, Takudage P (2021) Methods for evaluating penetration of drug into the skin: a review. Skin Res Technol 27(3):299–308

Suphasomboon T, Vassanadumrongdee S (2023) Multi-stakeholder perspectives on sustainability transitions in the cosmetic industry. Sustainable Prod Consumption 38:225–240
Syafri S, Jaswir I, Yusof F et al (2022) The use of instrumental technique and chemometrics for essential oil authentication: a review. Results Chem 4:100622
Theodorou IM, Kapoukranidou D, Theodorou M et al (2024) Cosmeceuticals: a review of clinical studies claiming to contain specific, well-characterized strains of probiotics or postbiotics. Nutrients 16(15):2526
Uğur Ş, Zieniuk B, Fabiszewska A (2024) Nutritional and medicinal properties of microbial oil. Appl Sci 14(10):4232
Valacchi G, Ferrara F (2024) Adverse health effects of UVR on skin and other organs. In: Environmental stressors and oxinflammatory tissues responses. CRC Press, Boca Raton, pp 219–234
Verkerk E (2024) Natural healing with aromatherapy: from history to home practice. Breath Master Solutions LLC, Brisbane
Vinderola G, Sanders ME, Cunningham M et al (2024) Frequently asked questions about the ISAPP postbiotic definition. Front Microbiol 14:1324565
Visha P, Das PK, Mukherjee J et al (2023) Special senses. In: Textbook of veterinary physiology. Springer, Singapore, pp 295–311
Wagener H (2024) Beautifully green: beauty vloggers' assessments of sustainable cosmetic products: a qualitative content analysis (Master's thesis, University of Twente)
Watts SS, Pal K, Asthana N et al (2024) Green synthetic approaches of caffeine extraction for cosmeceutical application: a review. J Mol Struct 1305:137733
Whitlock C (2020) Botanicum medicinale: a modern herbal of medicinal plants. MIT Press, Cambridge
Woo YR, Kim HS (2024) Interaction between the microbiota and the skin barrier in aging skin: a comprehensive review. Front Physiol 15:1322205
Wu Z, Tian E, Chen Y et al (2023) Gut microbiota and its roles in the pathogenesis and therapy of endocrine system diseases. Microbiol Res 268:127291
Xu X, Chen M, Zhu D (2023) Reperfusion and cytoprotective agents are a mutually beneficial pair in ischaemic stroke therapy: an overview of pathophysiology, pharmacological targets and candidate drugs focusing on excitotoxicity and free radical. Stroke Vasc Neurol 9(4):351–359
Yang Y, Huang J, Zeng A et al (2024) The role of the skin microbiome in wound healing. Burns Trauma 12:59
Yossa N, Huang S, Canida T et al (2023) qPCR detection of viable Bacillus cereus group cells in cosmetic products. Sci Rep 13(1):4477
Zaccardelli M, Roscigno G, Pane C et al (2021) Essential oils and quality composts sourced by recycling vegetable residues from the aromatic plant supply chain. Ind Crop Prod 162:113255
Zeng J, Yang K, Nie H et al (2023) The mechanism of intestinal microbiota regulating immunity and inflammation in ischemic stroke and the role of natural botanical active ingredients in regulating intestinal microbiota: a review. Biomed Pharmacother 157:114026
Zohora KFT, Arefin MR (2022) Tea and tea product diversification: a review. Turk J Agric Food Sci Technol 10(12):2334–2353
Zoninsein L (2023) How a whale becomes a molecule: a geography of modern olfaction. University of California, Berkeley

Correction to: The Emerging Role of Nanotechnology in Skincare

Sadia Javed, Ayesha Khan, Andleeb Shahzadi, Muhammad Arshad, and Shahzad Ali Shahid Chatha

Correction to: Chapter 15 in S. Javed et al. (eds.), *Sustainable Cosmeceuticals*, https://doi.org/10.1007/978-3-031-86087-4_15

The original version of the chapter was inadvertently published with an incorrect chapter title. The correct chapter title is as follows:

Nanotechnology in Skincare: Unlocking the Future

The updated version of this chapter can be found at https://doi.org/10.1007/978-3-031-86087-4_15

S. Javed et al. (eds.), *Sustainable Cosmeceuticals*,
https://doi.org/10.1007/978-3-031-86087-4_19

The manufacturer's authorised representative in the EU is Springer Nature Customer Service Centre GmbH, Europaplatz 3, 69115 Heidelberg, Germany. If you have any concerns regarding our products, please contact ProductSafety@springernature.com

Printed and bound by CPI Group (UK) Ltd, Croydon, CR0 4YY
01/10/2026
02225958-0016